FUNDAMENTALS OF WILDLAND FIRE FIGHTING

Third Edition

Edited By
Carl Goodson
Barbara Adams

Validated by the International Fire Service Training Association
Published by Fire Protection Publications, Oklahoma State University

ISBN 0-87939-148-0
Library of Congress 97-77361

Third Edition
First Printing, January 1998

Printed in the United States of America

Dedication

This manual is dedicated to the members of that unselfish organization of men and women who hold devotion to duty above personal risk, who count on sincerity of service above personal comfort and convenience, who strive unceasingly to find better ways of protecting the lives, homes and property of their fellow citizens from the ravages of fire and other disasters . . . **The Firefighters of All Nations.**

Dear Firefighter:

The International Fire Service Training Association (IFSTA) is an organization that exists for the purpose of serving firefighters' training needs. Fire Protection Publications is the publisher of IFSTA materials. Fire Protection Publications staff members participate in the National Fire Protection Association and the International Association of Fire Chiefs.

If you need additional information concerning our organization or assistance with manual orders, contact:

Customer Services
Fire Protection Publications
Oklahoma State University
930 N. Willis
Stillwater, OK 74078-8045
1 (800) 654-4055

For assistance with training materials, recommended material for inclusion in a manual, or questions on manual content, contact:

Technical Services
Fire Protection Publications
Oklahoma State University
930 N. Willis
Stillwater, OK 74078-8045
(405) 744-5723

THE INTERNATIONALFIRE SERVICE TRAINING ASSOCIATION

The International Fire Service Training Association (IFSTA) was established as a "nonprofit educational association of fire fighting personnel who are dedicated to upgrading fire fighting techniques and safety through training." This training association was formed in November 1934, when the Western Actuarial Bureau sponsored a conference in Kansas City, Missouri. The meeting was held to determine how all the agencies interested in publishing fire service training material could coordinate their efforts. Four states were represented at this initial conference. Because the representatives from Oklahoma had done some pioneering in fire training manual development, it was decided that other interested states should join forces with them. This merger made it possible to develop training materials broader in scope than those published by individual agencies. This merger further made possible a reduction in publication costs, because it enabled each state or agency to benefit from the economy of relatively large printing orders. These savings would not be possible if each individual state or department developed and published its own training material.

To carry out the mission of IFSTA, Fire Protection Publications was established as an entity of Oklahoma State University. Fire Protection Publications' primary function is to publish and disseminate training texts as proposed and validated by IFSTA. As a secondary function, Fire Protection Publications researches, acquires, produces, and markets high-quality learning and teaching aids as consistent with IFSTA's mission. The IFSTA Executive Director is officed at Fire Protection Publications.

IFSTA's purpose is to validate training materials for publication, develop training materials for publication, check proposed rough drafts for errors, add new techniques and developments, and delete obsolete and outmoded methods. This work is carried out at the annual Validation Conference.

The IFSTA Validation Conference is held in July in the state of Oklahoma. Fire Protection Publications, the IFSTA publisher, establishes the revision schedule for manuals and introduces new manuscripts. Delegates are selected for technical input by the Delegate Selection Committee. The Delegate Selection Committee consists of three Board members and two conference delegates; the committee is chaired by the Vice-Chair of IFSTA. Applications are reviewed by the committee, and delegates are selected based upon technical expertise and demographics. Committees meet and work at the conference addressing the current standards of the National Fire Protection Association and other standard-making groups as applicable.

Most of the delegates are affiliated with other international fire protection organizations. The Validation Conference brings together individuals from several related and allied fields, such as:

- Key fire department executives and training officers
- Educators from colleges and universities
- Representatives from governmental agencies
- Delegates of firefighter associations and industrial organizations
- Engineers from the fire insurance industry

Delegates are not paid nor are they reimbursed for their expenses by IFSTA or Fire Protection Publications. They come because of commitment to the fire service and its future through training. Being on a committee is prestigious in the fire service community, and delegates are acknowledged leaders in their fields. This unique feature provides a close relationship between the International Fire Service Training Association and other fire protection agencies, which helps to correlate the efforts of all concerned.

IFSTA manuals are now the official teaching texts of most of the states and provinces of North America. Additionally, numerous U.S. and Canadian government agencies as well as other English-speaking countries have officially accepted the IFSTA manuals.

TABLE OF CONTENTS

TABLES

Preface

Even though the title has been changed, this is the third edition IFSTA manual dealing with wildland fire fighting. It is intended to serve as a primary text for wildland firefighter candidates or as a reference text for personnel already on the job. Addressing the objectives for Levels I and II of NFPA 1051, *Standard for Wildland Fire Fighter Professional Qualifications* (1995 edition), this updated manual contains hundreds of new photographs and illustrations.

Acknowledgment and special thanks are extended to the members of the material review committee who contributed their time, wisdom, and knowledge to the development of this manual.

Chair
Mary Chambers
Albuquerque Technical Vocational Institute
Albuquerque, NM

Secretary
John Trenner
Monterey County Fire Training Officers Association
Carmel, CA

Steve Brown
Chico Fire Department
Chico, CA

John Roberts
National Advanced Resource Technology Center
Marana, AZ

Kirk Hale
Tualatin Valley Fire & Rescue
Aloha, OR

Mark Watts
Baton Rouge Fire Department
Baton Rouge, LA

Charles Anaya
Stillwater Fire Department
Stillwater, OK

Also assisting the committee were Debra Amesqua, Tallahassee Fire Department, Tallahassee, FL, and Jim Sorenson, USDA Forest Service, Atlanta, GA.

Special recognition is given to the National Interagency Fire Center (NIFC) in Boise, ID, and the Monterey County (CA) Fire Training Officers Association for their assistance in obtaining many of the photographs used in the manual. Also assisting with the process of obtaining photographs were the members of the Stillwater (OK) Fire Department, the Santa Rosa and Monte Rio (CA) Fire Departments, and Sam Goldwater, Angus Fire, Angiers, NC.

In addition, the following individuals and organizations contributed information, photographs, and technical assistance that were instrumental in the development of this manual:

Marty Alexander, Natural Resources Canada
Jess Andrews, Oklahoma State University, Fire Service Training
Tony Bacon, Novato (CA) Fire Protection District
John Bartlett, Fire Protection (Barricade), Inc., Jupiter, FL
Steve Baxman, Monte Rio (CA) Fire Department
Central Mat-Su (AK) Fire Department
Merribeth Carlson, California Department of Forestry and Fire Protection

John Craney, California Department of Forestry and Fire Protection
Patrick Costales, Hawaii Division of Forestry and Wildlife
Maris Gabliks, New Jersey Forest Fire Service
Tiana Glenn, National Interagency Fire Center, Boise, ID
Good Will Fire Company #1, New Castle, DE
Hugh Graham, Yates and Associates, Inc., Broken Arrow, OK
John Hawkins, California Department of Forestry and Fire Protection
Neil Honeycutt, California Office of Emergency Services
Hotshield USA, Inc., Victorville, CA
Alan Jeffrey, Canadian Interagency Forest Fire Centre
Bill Lellis, Larkspur, CA
Kathy Lord, Santa Rosa (CA) Fire Department
Dan Madrzykowsi, National Institute of Standards and Technology
Montana Department of Natural Resources and Conservation
National Wildland Fire Coordinating Group (NWCG), Boise, ID
Russian River (CA) Fire Protection District
Matt Silva, California Department of Forestry and Fire Protection
Sonoma (CA) Fire Department
David Stinnett, Wildfire Halprin, Los Angeles, CA
Bob Stratton, Santa Rosa (CA) Fire Department
Nancy Trench, Oklahoma State University, Fire Service Training
Don Uboldi, California Department of Forestry and Fire Protection
Bill Vandevort, Monterey (CA) Fire Department
Mike Wieder, Oklahoma State University, Fire Protection Publications

Last, but certainly not least, gratitude is also extended to the following members of the Fire Protection Publications staff whose contributions made the final publication of this manual possible:

Marsha Sneed, Associate Editor
Don Davis, Production Coordinator
Ann Moffat, Graphic Design Analyst
Desa Porter, Senior Graphic Designer
Connie Cook, Senior Graphic Designer
Ben Brock, Graphics Technician
Eric Barnum, Research Technician
Michael Huskey, Research Technician
Jack Krill, Research Technician
Jared Moravec, Research Technician
Mark Slight, Research Technician
Mike Spini, Research Technician
Dustin Stokes, Research Technician

Lynne C. Murnane
Lynne C. Murnane
Managing Editor

Introduction

Wildland fire may be defined as any unwanted fire involving outdoor vegetation. While it is often thought of as occurring in forests, rangelands, or crop fields, it may also occur in areas such as vacant lots, highway medians, parks, and golf courses. In addition, with residential development spreading into once pristine areas, a relatively new phenomenon has been created — the *wildland/urban interface*. This phenomenon has changed the nature of the wildland fire problem in some very significant ways. Both the life hazard and the potential economic losses in wildland areas have increased greatly, and the increase in human activity has multiplied the number and variety of potential sources of ignition.

A prime example of the wildland/urban interface fire problem was provided in October of 1991 when a wildland fire swept through the hills in Oakland and Berkeley, California — an area considered to be urban. In the first hour, 790 homes caught fire! At the height of the fire, one home was lost every 11 seconds! This fire killed 25 people, injured 150 others, and consumed over 1,600 acres (648 hectares [ha]) of vegetation. More than 3,800 homes and apartments were destroyed, resulting in an estimated fire loss of more than $1.5 billion — the most costly wildland/urban interface fire in U.S. history.

Another aspect of the increasing population in the wildland/urban interface is that municipal firefighters, primarily trained and equipped to fight structure fires, are more likely than ever to have to fight wildland fires. The reverse is also true — wildland firefighters are now more likely to have to fight structure fires.

Firefighters across North America are killed each year because they lack the training necessary for them to know how wildland fires are affected by differences in fuels, weather, and topography; to know how to safely and effectively fight these fires; or to recognize when they are in a life-threatening situation until it is too late. Statistics from the early 1990s show that as many as 37 percent of all firefighter fatalities were associated with wildland fires. In July of 1994, 14 wildland firefighters were killed on a single incident in Colorado. Statistics such as these make this manual necessary.

PURPOSE

The purpose of this manual is to provide a text from which firefighters can learn to safely and effectively fight wildland fires and structure fires in the wildland/urban interface and to protect exposed property. However, this manual is only a text, and no one should expect to become a fully qualified firefighter in either wildland or structure fires simply by reading this manual — it *must* be supplemented with hands-on training delivered by qualified instructors.

SCOPE

The scope of this manual is limited to Levels I and II of NFPA 1051, *Standard for Wildland Fire Fighter Professional Qualifications* (1995). The text addresses all sections of the first two levels of the standard, but it does not address Levels III or IV except where it is necessary to make the Level I and II material easier to understand. Firefighters wanting or needing to be trained to Levels III and IV must obtain that training from the National Wildland Fire Coordinating Group (NWCG) or other agencies qualified to deliver such training.

According to NFPA 1051, Wildland Fire Fighters I and II are defined as follows:

- ***Wildland Fire Fighter I.*** The person, at the first level of progression as defined in Chapter 3, who has demonstrated the knowledge and skills necessary to function safely as a member of a wildland fire suppression crew. The Wildland Fire Fighter I works under direct supervision.
- ***Wildland Fire Fighter II.*** The person, at the second level of progression as defined in Chapter 4, who has demonstrated the skills and depth of knowledge necessary to function under general supervision. This person shall function safely and effectively as a member of a wildland fire suppression crew of equally or less experienced fire fighters to accomplish a series of tasks. The Wildland Fire Fighter II can be called upon to provide leadership and temporary supervision for a small crew. The Wildland Fire Fighter II maintains direct communications with a supervisor.

Anyone being considered for placement in a program to become qualified as a wildland firefighter, must meet the entrance requirements of Chapter 2 of NFPA 1051:

CHAPTER 2 ENTRANCE REQUIREMENTS

2-1 General. Prior to entering training to meet the requirements of Chapters 3, 4, 5, and 6 of this standard, the candidate shall:

a. Meet the minimum educational requirements established by the authority having jurisdiction.

b. Meet the age requirements established by the authority having jurisdiction.

2-2 Medical requirements for entry level personnel shall be developed and validated by the authority having jurisdiction and shall be in compliance with applicable legal requirements.

2-3 Job-related physical performance requirements for entry level personnel shall be developed and validated by the authority having jurisdiction.

Other standards are addressed in part in this manual, but it is beyond the scope of the manual to address these standards in their entirety. Anyone needing to review the full text of the referenced standards should contact the organizations that publish those standards.

WILDLAND FIRE BEHAVIOR: FUEL, WEATHER, TOPOGRAPHY

1

LEARNING OBJECTIVES

This chapter provides information that addresses the following objectives of NFPA 1051, *Standard for Wildland Fire Fighter Professional Qualifications* (1995 edition):

Wildland Fire Fighter I

3-1 **General.** The Wildland Fire Fighter shall meet the job performance requirements defined in Sections 3-1 to 3-5 of this standard.

3-1.1* *Prerequisite Knowledge:* Fireline safety, use and limitations of personal protective equipment, agency policy on fire shelter use, basic wildland fire behavior, basic wildland fire tactics, fire fighters role within the local incident management system, and first aid.

3-5.3.1* *Prerequisite Knowledge:* Basic fireline safety, fire behavior, and suppression methods.

Wildland Fire Fighter II

4-1.1.1* *Prerequisite Knowledge:* The Wildland Fire Fighter II role within the incident management system, basic map reading and compass use, radio procedures, and record keeping.

4-1.1.2 *Prerequisite Skills:* Orienteering and radio use.

Chapter 1
Wildland Fire Behavior: Fuel, Weather, Topography

INTRODUCTION

Wildland fires can grow very rapidly, especially if they go undetected in their early stages, so timely control and extinguishment can significantly reduce the loss of life and property. This is particularly true of fires in the wildland/urban interface. Therefore, an understanding of the combustion process, the factors that influence fire behavior, and how fire influences and is influenced by the environment is essential to firefighter safety and survival, as well as to effective fire control and extinguishment.

The laws of chemistry and physics that apply to structure fires also apply to wildland fires; however, because structure fires are more contained and localized, the way fire behaves in structures is different. For instance, a backdraft is not possible in the wildland because the fire must burn within a virtually airtight confined space for backdraft conditions to develop. On the other hand, fires in the wildland are far more affected by weather conditions than are those within structures.

This chapter reviews the combustion process, the fuels involved in wildland fires, the effects of weather and topography on these fires, and some other factors that influence the behavior of wildland fires.

COMBUSTION PROCESS

Fire is actually a by-product of a larger process called combustion. *Combustion* is the self-sustaining process of rapid oxidation of fuel (chemical union of an oxidizer and another material), which produces light and heat. Wildland fuels have an abundant supply of oxygen available in the air. Rapid oxidation occurs in two forms: *smoldering fires* (burning without flame) and *steady-state fires* (unchecked rapid burning). Steady-state fires are sometimes also called *free-burning fires.*

Fuel may exist in any of the three states of matter: solid, liquid, and gas. However, only gases burn. The initiation of combustion of a solid or a liquid fuel requires its conversion into a gaseous state by heating. Fuel gases evolve from wood and other solid fuels by *pyrolysis* — a chemical change by the action of heat. So, even though fuel and oxygen are present in the wildland, heat must be added to liberate the fuel gases and initiate the combustion process (Figure 1.1).

Once the chemical reaction (combustion) has begun, some of the energy is released as heat and light. The released heat energy causes further pyrolysis, which adds more fuel and which may produce a self-sustaining process. Heat activates and sustains the chemical reactions needed for continued combustion (Figure 1.2).

Heat generated by a fire evaporates the moisture in the fuel and heats the fuel to its ignition temperature. The amount of heat required to produce these effects depends upon the physical and chemical makeup of the fuel (see Fuel section) and the percentage of atmospheric moisture (relative humidity) (see Weather section).

If the supply of oxygen available to the combustion process increases, then combustion intensifies. While the percentage of oxygen in the atmosphere does not vary appreciably from place to place, wind can effectively increase the amount of oxygen available to a fire and thereby increase the rate of combustion (see Weather section).

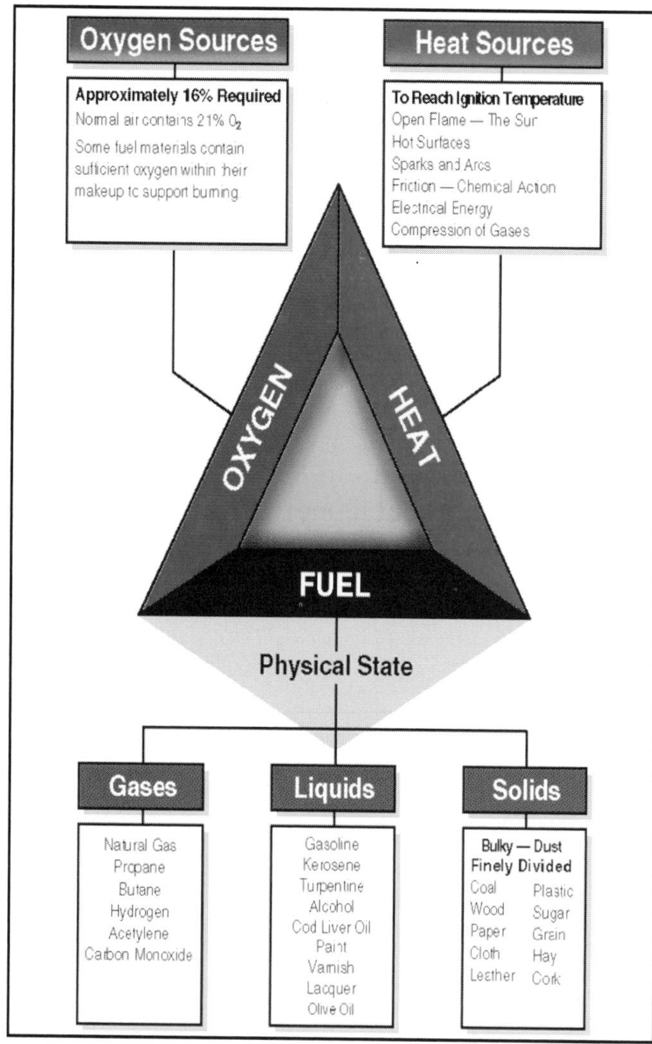

Figure 1.1 The fire triangle.

Heat Transfer

For sustained combustion, heat must be transferred from involved fuels to those not yet involved. Heat transfer takes place by one or more of three methods: *conduction, convection,* and *radiation.*

CONDUCTION

When two objects of different temperatures contact each other directly or through a heat-conducting medium, heat transfers from the warmer object to the cooler one until their temperatures are equal. Metals such as aluminum, copper, and iron conduct heat readily and can play a significant role in fire spread within structures. Fibrous materials such as plants and wood are poor conductors of heat; therefore, heat transfer by conduction has limited effect on the spread of wildland fires. However, if a vehicle equipped

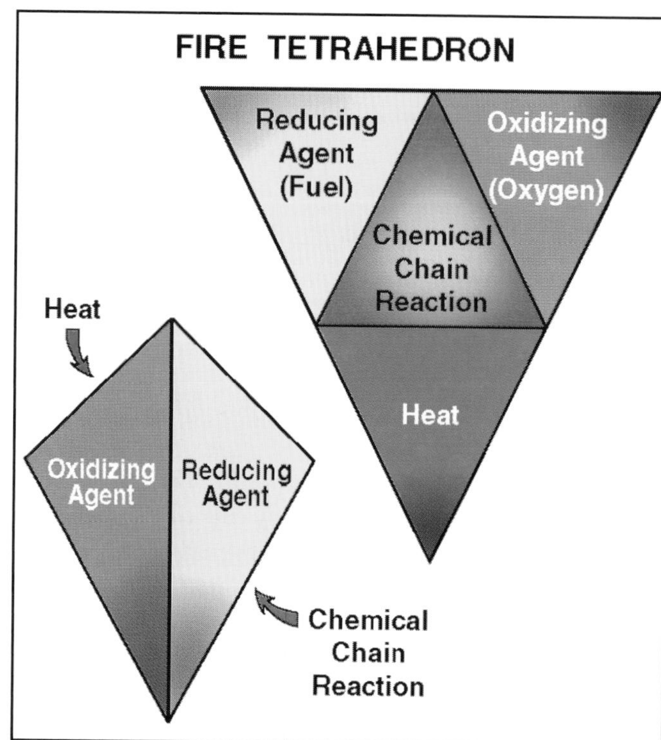

Figure 1.2 The fire tetrahedron.

with a catalytic converter stops in an area of tall dry grass, heat conducted to the grass in contact with the converter can ignite the grass (Figure 1.3).

Figure 1.3 Heat from catalytic converters can start fires. *Courtesy of NIFC.*

CONVECTION

Convection is the transfer of heat by the movement of any fluid — liquid or gas. Gases heated in a fire expand, become lighter, and rise. In a wildland fire, fire gases rise in a convective column, and cooler air flows in to replace them. As these gases draw up into the column, sparks, embers, and burning twigs are carried aloft. These burning materials fall back to earth as much as a mile

(1.6 kilometers [km]) downwind and can start spot fires ahead of the main fire (spotting) (Figure 1.4).

Direct flame contact is another form of heat transfer that involves convection, conduction, and radiation. When a substance is heated by flame contact, flammable vapors are released that may ignite and produce even more flame (Figure 1.5).

Figure 1.4 Convection may cause spot fires downwind of the main fire.

Figure 1.5 Direct flame contact preheats fuels upslope.

RADIATION

The best example of heat transfer by radiation is the heat from the sun. Although air is a poor conductor, it is obvious that heat energy can travel through air. Heat waves, sometimes called *infrared rays*, radiate in all directions from their source, and they travel through space until they are partially or totally absorbed by an opaque object. As the object is exposed to thermal radiation, it in turn radiates heat from its surface. Radiated heat is one of the major sources of fire spread in wildland fires. Burning fuels radiate heat to fuels immediately adjacent to the fire or across narrow draws or canyons. This radiated heat preheats and dehydrates those exposed fuels and initiates pyrolysis (Figure 1.6). Radiant heat is also responsible for many burn injuries to firefighters in wildland fires.

Figure 1.6 Radiant heating can increase fire spread.

Controlling Combustion

The combustion process is interrupted (and the fire extinguished) by one or more of the following means:

- Removing fuel
- Removing oxygen
- Removing heat energy that sustains the chemical reactions
- Inhibiting flame-producing chemical reactions

Clearing a space of all surface fuels down to mineral soil (dirt containing little or no organic material) is a common way of controlling and extinguishing wildland fires (removing fuel) (Figure 1.7). Wildland fires burn in the open air, so attempting to restrict the oxygen supply to a fire (removing oxygen) is usually limited to smothering relatively small fires with dirt (Figure 1.8). Cooling the fire with water (removing heat) is one of the most effective fire extinguishing methods (Figure 1.9). Aerial application of fire-retardant chemicals (inhibiting flame production) can also be effective when followed up by ground forces (Figure 1.10) (see Chapter 6, Fire Suppression Methods).

Figure 1.7 Firefighters learn to cut a fireline down to mineral soil. *Courtesy of Monterey County Training Officers.*

Figure 1.8 A firefighter throws dirt onto a fire to smother it. *Courtesy of NIFC.*

Figure 1.9 Cooling a fire with water is very effective. *Courtesy of NIFC.*

Figure 1.10 Flame production may be inhibited with retardant chemicals. *Courtesy of John Hawkins.*

FUEL

Fuel is any flammable or combustible substance available for a fire to consume. All forms of vegetation (grass, brush, trees, or field crops) — alive or dead — are fuel for a wildland fire. Wildland fuels are of many different types, and each type has different fire-behavior characteristics. Knowing the characteristics of different wildland fuels helps firefighters more accurately predict wildland fire behavior.

Several different systems for classifying wildland fuels are in use. One system is the National

Fire Danger Rating System (NFDRS), which lists 20 standard fuel models. The NFDRS models are aligned with 13 fire-behavior fuel models to predict large area daily fire danger ratings. In this system, fuels are classified into four broad groups: *grasses, brush, timber,* and *slash* (see Species Combinations section). The differences in fire behavior among these groups are principally related to the amount of fuel present and its distribution. For a more complete explanation of fire behavior models, see NFES 1574, *Aids to Determining Fuel Models For Fire Behavior*, available from the National Interagency Fire Center in Boise, Idaho.

Other systems in use classify wildland fuels by their moisture content and how various fuels react to changes in environmental moisture (see Fuel Moisture section). Others classify these fuels by their position in relation to the surface of the ground — under it, on it, or above it (see Fuel Position section). Still others classify fuels according to their size (see Fuel Size section).

Fuel Types

There is a variety of potential fuels in the wildland environment. Obviously, there are many varieties of plants, trees, and their litter, but there are also a number of different types of structures in wildland areas that are also fuel for wildland fires.

SPECIES COMBINATIONS

Among the many different types of live or dead vegetation that are available as potential wildland fuels, the general classes or groups are grasses, brush, trees, and slash. They may exist singularly or in combination.

Grasses. These consist of annuals such as rye grass and wild oats, perennials such as saw grass, and the various tundra species (Figure 1.11).

Brush. These are most often mature brush such as high pocosin, Alaska black spruce, buckeye, chamise, chaparral, coyote bush, manzanita, mesquite, sagebrush, and sugar bush (Figure 1.12).

Trees. Some are deciduous trees such as alder, ash, aspen, birch, cottonwood, dogwood, hickory, maple, and some oaks. Also typical are evergreens such as cedar, cypress, eucalyptus, fir, hemlock, live oak, pine, redwood, and spruce (Figure 1.13).

Figure 1.11 Typical prairie grasses.

Figure 1.12 A variety of brush species. *Courtesy of NIFC.*

Figure 1.13 A mature evergreen forest. *Courtesy of NIFC.*

Slash. *Slash* is the downed, dead residual material left on the forest floor after logging operations. Typical slash is composed of logs, treetops, limbs, and stumps (Figure 1.14). Depending upon its volume and distribution, slash is usually classified as light, medium, or heavy.

Figure 1.14 Slash can provide a significant volume of fuel to a fire. *Courtesy of NIFC.*

GEOGRAPHICAL DISTRIBUTION

Some species are found only in certain specific areas of the North American continent and Hawaii; others are found in more than one area. In general, the species that predominate in a given area are those that are native to that area. By general geographic area, the predominant species are as follows:

Eastern species. Predominant species combinations in the East are high pocosin composed of fetterbush, gallberry, bay, hickory, oak, maple, and pines (Figure 1.15).

Figure 1.15 A typical hardwood forest. *Courtesy of NIFC.*

Western species. Predominant species combinations in the West are grasses such as cheat grass, medusa's head, ryegrass, fescue, wild oats, and meadow foxtail; dwarf trees and woody shrubs especially adapted to dry summers and moist winters; and trees such as Douglas fir, eucalyptus, hemlock, live oak, and pines (Figure 1.16).

Northern species. Predominant species combinations in the northern U.S. and southern Canada are tall prairie grasses, sagebrush, cedar, Douglas fir, hemlock, jack pine and various hardwoods. In Alaska, open tundra and spruce forests with thick brushy understory predominate (Figure 1.17).

Figure 1.16 Oak savanna is common in the West. *Courtesy of NIFC.*

Figure 1.17 Species typical of Alaska and Canada. *Courtesy of NIFC.*

Southern species. Predominant species combinations in the South are varieties of pine, palmetto, bay, gallberry, and various hardwoods (Figure 1.18).

Southwest species. Predominant species combinations in the Southwest are live oak savanna with grassy understory, mesquite, sagebrush, tumbleweed, piñon, ponderosa pine, and mixed conifers (Figure 1.19).

Figure 1.18 Species typical of the southern U.S. *Courtesy of NIFC.*

Figure 1.19 Typical species in the desert Southwest. *Courtesy of NIFC.*

Southeast species. Predominant species combinations in the Southeast are saw grass "prairie" and "strands" in the Florida Everglades, tropical palms, moss-laden cypress, pine, and hardwoods (Figure 1.20).

Hawaiian species. Predominant species combinations in Hawaii are beard grass, broom sedge, fountain grass, guinea grass, molasses grass, and trees such as eucalyptus and pine (Figure 1.21).

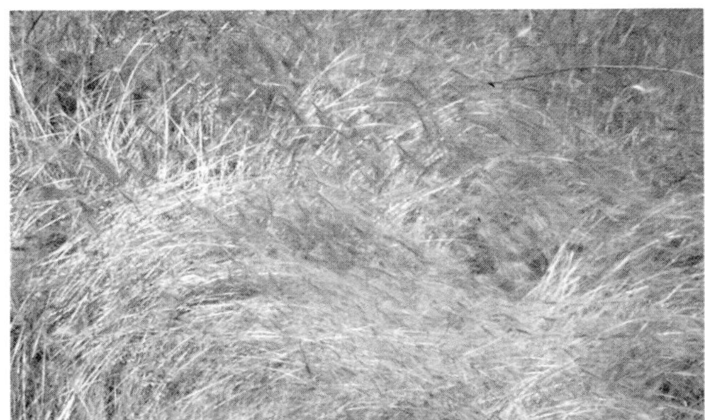

Figure 1.20 Saw grass typical of the southeastern U.S. *Courtesy of NIFC.*

Canadian Species. The species combinations in southern Canada are similar to those in the northern U.S. Prairie grasses, cypress, fir, hemlock, larch, pine, and spruce are most common. Northern Canada, including the maritime provinces, is characterized by aspen, birch, fir, maple, pine, poplar, and tamarack (Figure 1.22). Canada's west coast is a part of the Pacific Coast forest of North America. It consists primarily of cedar, hemlock, and spruce in the north and Douglas fir in the south (Figure 1.23).

Figure 1.21 Thick undergrowth typical of Hawaiian species. *Courtesy of Hawaii Division of Forestry and Wildlife.*

Figure 1.22 Typical species in northern Canada. *Courtesy of NIFC.*

Figure 1.23 Pacific Coast forests are predominantly Douglas fir.

COMMON STRUCTURES

Until recently, relatively few structures were located in undeveloped areas. Most were farmhouses and their associated outbuildings. Others were isolated commercial structures, such as sawmills, or recreational structures such as cabins or lodges. With the emergence of widespread residential and commercial development in the wildland/urban interface, the number and character of structures in the wildland have changed dramatically. See Chapter 7, Wildland/Urban Interface Fire Suppression. Although some of these structures have noncombustible roof coverings, the majority of the structures themselves are made of wood — another form of fuel for wildland fires (Figure 1.24).

Figure 1.24 A typical wooden structure in the wild. *Courtesy of NIFC.*

Fuel Characteristics

Regardless of the type of fuel, fire behavior is dependent on certain fuel characteristics. These characteristics include temperature, moisture, position, loading, continuity, compactness, and size.

FUEL TEMPERATURE

Fuel temperature is one factor in wildland fire behavior. Heat energy from the sun warms the earth's surface. The earth heats both the surrounding air and the wildland fuels, reduces their moisture, and brings fuels closer to their ignition temperatures.

Fuels exposed to heat from the sun can reach 150°F (66°C) and are more likely to burn than cool, shaded fuels (Figure 1.25). Because most wildland fuels must be heated to somewhere between 400°F and 700°F (204°C and 371°C) to ignite, solar heating alone will not cause ignition, but it makes ignition easier. Once a fire has started, radiant heat dehydrates and preheats adjacent fuels and makes them more susceptible to ignition.

Figure 1.25 Fuels in the shade are cooler than those in the sun. *Courtesy of Tony Bacon.*

FUEL MOISTURE

In the earth's environment, water is present in the form of precipitation, ground moisture, and atmospheric moisture (humidity). The moisture content of fuels changes constantly in response to the availability of water in the environment. Vegetation that is dead and dry ignites more readily and burns more intensely than vegetation that is green and laden with moisture. Initially, the heat absorbed by wet fuels drives off moisture through evaporation (Figure 1.26). As more heat is absorbed, evaporation continues, and the temperature of the fuel increases to the point of ignition.

Actively growing green vegetation absorbs water from the soil through its roots. The water circulates to the living parts of the plant. As a

Figure 1.26 A living plant showing signs of dehydration.

result, live fuels usually exhibit a relatively high moisture content (from 35 percent to over 250 percent of their dehydrated weight). If data from field measurements are unavailable for estimating live fuel moisture, Table 1.1 can be used for making rough estimates.

The percentage of fuel moisture varies according to plant species, age of plant, and weather conditions. For instance, extended periods of dry weather reduce the moisture content of vegetation, and old plants are usually drier than young plants. These conditions can reduce the moisture content of living fuel and make it much more susceptible to ignition.

Whether fuels are dead or live, their flammability is influenced by environmental moisture. Fuel moisture in live fuels depends primarily on the groundwater and the phase of the fuels' seasonal growth cycle. Fuel moisture in dead fuels depends on atmospheric temperature, humidity, and solar radiation. When plants die, most of the water in the plant tissues evaporates. Dead plants typically have a lower moisture content (1.5 to 30 percent) than live plants and are more susceptible to ignition. A prolonged dry spell reduces the moisture content of the fuels and increases their susceptibility to ignition.

TABLE 1.1
Live Fuel (Foliage) Moisture Content (Percent)

Moisture Content (%)	Stage of Vegetative Development
300	Fresh foliage, annuals developing early in the growing cycle
200	Maturing foliage, still developing, with full turgor
100	Mature foliage, new growth complete and comparable to older perennial foliage
50	Entering dormancy, coloration starting, some leaves may have dropped from stem; also indicative of drought conditions
<30	Completely cured (treat as dead fuels)

NOTE: Moisture content represents the weight of the fuel compared to its oven-dry weight.

Source: Fireline Handbook, Appendix B, Fire Behavior Supplement, NFES 2165 (PMS 410-1), 1989, page B-29.

Rain, snow, or hail can at least temporarily decrease the flammability of wildland fuels and alter their fire behavior. Continual or heavy rains soak these fuels and increase their moisture content. This makes both live and dead fuels less susceptible to ignition. By contrast, occasional rain showers are usually of such short duration that they have little effect on fuel moisture.

One system used to predict fire behavior is based on weather cycles and fuel types. It classifies dead fuels according to the amount of time it takes their moisture content to equalize with that of the surrounding air. This system classifies fuels as 1-hour, 10-hour, 100-hour, and 1,000-hour fuels (Figure 1.27). These time-lag classifications also reflect the average diameter of dead fuels: 1-hour = < ¼ inch (6.35 mm), 10-hour = ¼ to 1 inch (6.35 mm to 25 mm), 100-hour = 1 to 3 inches (25 mm to 76 mm), and 1,000-hour = 3 to 8 inches (76 mm to 203 mm).

Just as wood shavings are more readily ignited than is a large log because the shavings have a much higher surface area-to-mass ratio, the rate at which fuels absorb and liberate moisture is also a function of their size and shape (see Fuel Size section). For example, grass and pine needles are more readily affected by changes in humidity than are logs or slash. But regardless of size and shape, all wildland fuels are affected by daily and seasonal changes in humidity to a greater or lesser extent. Fuels may absorb moisture from dew forming as the atmospheric temperature drops during the night, but because solar heating during the day liberates moisture from the fuels through evaporation, fuels in the full sun may contain as much as 8 percent less moisture than those in the shade. However, fuels on north-facing slopes are less affected by solar heating, so their daytime moisture content is generally not as low as fuels on level or south-facing slopes (Figure 1.28).

Winds may also affect fuel moisture. Cool winds that reduce surface temperatures can sometimes

Figure 1.27 Typical dead fuels. *Courtesy of NIFC.*

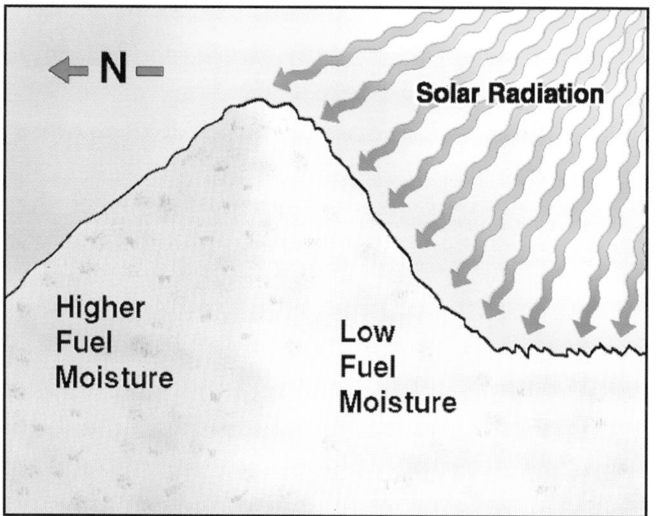

Figure 1.28 The direction a slope faces is a major factor in solar heating and evaporation.

reduce the rate at which wildland fuels liberate moisture. However, the warm to hot winds normally associated with wildland fires usually accelerate the loss of moisture by lowering relative humidity (percentage of atmospheric moisture) and increasing evaporation (see Relative Humidity section).

Knowing how quickly fuel moisture levels change in response to changes in weather conditions helps firefighters predict a fire's intensity, or heat release rate (HRR), and rate of spread (ROS). *Heat release rate* is the amount of heat generated in a given amount of time (see Intensity section). HRR is usually expressed in British thermal units or Btu (kilojoules [kJ]) per second per unit area. *Rate of spread* is the tendency of a fire to extend its horizontal dimensions in terms of the rate of increase in the total perimeter of the fire and the rate of forward progress of the fire front. ROS is usually expressed in feet (meters) per minute at the head (forward part) of a fire (see Rate of Spread section). In general, the drier the fuel, the more intensely it burns and the faster a fire spreads.

FUEL POSITION

How a fuel is positioned in relation to the ground is another way of differentiating among the types of wildland fuels. Based on position, the types of wildland fuels are (1) subsurface fuels, (2) surface fuels, and (3) aerial fuels. This classification system is useful because wildland fire behavior can be greatly affected by the position of the fuel.

Subsurface fuels. Subsurface fuels lie under the surface of the ground. Roots, peat, duff, and other partially decomposed organic matter are all examples of subsurface fuels. Subsurface fuels do not burn rapidly, but they are often difficult to extinguish completely. Buried roots or other underground fuels may burn their full length under the surface, and then the fire may resurface at some unexpected location (Figure 1.29)

Figure 1.29 Typical subsurface fuels.

Surface fuels. Surface fuels include needles, leaves, twigs, grass, field crops, brush (up to 6 feet [1.8 m] in height), downed limbs, logging slash, and small trees (Figure 1.30). A large percentage of the fuels on or immediately adjacent to the surface of the ground are small twigs, needles, or grass. These are sometimes called *flashy fuels* because they ignite easily and burn rapidly and almost completely when environmental conditions favor combustion.

Figure 1.30 Surface fuels are those on the ground.

Also included in this category are those fuels that are referred to as *ladder fuels* because they provide a way or "ladder" for a surface fire to climb into the aerial fuels above. Ladder fuels include hanging pine boughs, tree moss, tall grasses or brush in the understory beneath a canopy of aerial fuels, and downed dead limbs or logs in contact with surface fuels (Figure 1.31).

Figure 1.31 Ladder fuels bridge the gap between surface and aerial fuels.

Aerial fuels. Aerial fuels are physically separated from the ground's surface and sometimes from each other. Air circulates readily between these fuels and the ground, causing them to burn rapidly. Brush that is over 6 feet (1.8 m) tall, live or dead leaves and needles on tree limbs, branches, snags (standing dead trees), hanging moss, and lichen are examples of aerial fuels (Figure 1.32). The rate of fire spread in aerial fuels is generally inversely proportional to the horizontal distance between the fuels, but it can also be affected by the

Figure 1.32 Aerial fuels are those at the tops of the brush and trees.

spread of fire in the surface fuels below. An increase of distance between aerial fuels slows down the heat transfer process and decreases the rate of spread. Conversely, aerial fuels in close proximity to each other may have an increased rate of spread. Dry weather and winds may also affect the rate of spread and intensity of fire in these fuels.

FUEL LOADING

The amount of fuel (both live and dead) in a given area is referred to as the *fuel load* (also called *fuel volume*). It is commonly reported in terms of tons of fuel available per acre (tonnes per hectare [t/ha]), with ranges from less than 1 ton to more than 500 tons per acre (2.2 t/ha to 1 100 t/ha). For example, grass can range from ¼ ton to 1 ton per acre (0.56 t/ha to 2.2 t/ha); timber in some areas can average 600 tons per acre (1 320 t/ha). With all other factors that affect fire behavior being equal, areas of higher fuel loading will generate more heat than those with a lesser fuel load.

Fuel loading is a factor in the 13 fire-behavior fuel models. Each of the fuel types, ranging from grass to heavy logging slash, is classified according to fuel loading (volume) in tons per acre (tonnes per hectare), fuel bed depth, and moisture of extinction (see Table 1.2 on page 18).

This table can be used in combination with at-scene observations and any other available data to help firefighters make more informed estimates of expected fire behavior in a given situation. However, these are only estimates, and the factors that influence fire behavior can change very quickly. Firefighters must remain ready to deal with the unexpected. As noted earlier, other fire-behavior prediction systems also list 1,000-hour fuels — logs and other deadwood of 3 to 8 inches (76 mm to 203 mm) in diameter. These fuels add significantly to fire intensity, thus resistance to control. Since 1,000 hours translates into 42 days, this classification is of limited value for fire-behavior forecasts done in 12-hour increments.

FUEL CONTINUITY

Another factor that influences the spread of fire is continuity, which refers to the horizontal and vertical spacing of fuels. Fuels spread evenly over an area are called *continuous fuels*. A grain field is an example of a continuous fuel (Figure 1.33).

Patchy fuels, on the other hand, grow in clumps. Scattered desert vegetation or mowed and windrowed hay fields are examples of patchy fuels (Figure 1.34). They are separated by bare ground that has little or no flammable materials between the patches. Bare earth, green vegetation, plowed ground, roads, or marshy areas may break the continuity of fuels.

Fuel continuity influences the rate of spread in wildland fires. When fuels are close together, a fire spreads faster because of radiant heat transfer. A uniform continuity of fuel guarantees a relatively uniform and predictable rate of spread. When fuels are patchy, scattered, or separated by natural or

Figure 1.33 Grain fields can provide a continuous fuel.

Figure 1.34 Rate of fire spread will vary in patchy fuels.

man-made barriers, radiant heat may not be sufficient to preheat or ignite the surrounding fuels. The rate and direction of fire spread are generally less predictable in patchy fuels.

FUEL COMPACTION

Fuel compaction can be defined as the spacing between fuel particles. Fire spreads readily from piece to piece in highly compact fuels. However, it generally burns with both a low intensity and a low rate of spread because the air supply to the combustion process is limited. Wheat standing in a grain field burns more quickly and with a higher intensity than wheat tightly compacted into a bale of straw (Figure 1.35). Grass packed to the ground by snow does not burn as quickly or as intensely as grass that has not been compressed.

Figure 1.35 Highly compacted fuel does not burn well.

FUEL SIZE

Fuel size influences the rate of heat transfer and the change in moisture content. Heat and moisture transfers occur between the fuels and the environment at the surface of the fuel. The amounts of heat and moisture transferred are greater when the fuel surface is large. Depending upon weather conditions, this transfer of energy results in either an increase or a decrease in the fuel temperature and moisture. In turn, the change in temperature influences the combustion process.

Based on size, fuels may be described as light, medium, or heavy. Atmospheric conditions being equal, light fuels produce a relatively high ROS. As mentioned earlier, ROS is a measure of the forward spread of the involved area and is usually expressed in feet (meters) per minute at the head of the fire. In general, medium fuels tend to produce a moderate rate of fire spread, and heavy fuels produce a relatively low forward rate of spread. However, there may be exceptions to these general rules — pine needle litter may have a low ROS, while heavy brush may have a high ROS.

Light fuels. These surface fuels, also called *fine fuels, flashy fuels,* or *flash fuels,* are short grass and light brush (up to 2 feet [0.6 m]) that burn rapidly and with high intensity (Figure 1.36). Light fuels take on and give up moisture faster than heavier fuels. Fuel temperature, winds, and topography (physical shape of the land) greatly influence the rate of spread in light fuels (see Weather and Topography sections).

Figure 1.36 Fields of grass are usually considered *light* fuels.

WARNING

Historically, more firefighters are killed in light fuels than the other types.

TABLE 1.2
Description of Fuel Models Used in Fire Behavior

Fuel Model	Typical Fuel Complex	Fuel Loading (tons/acre) 1 hr	10 hrs	100 hrs	Live	Fuel Bed Depth (feet)	Moisture of Extinction Dead Fuels (percent)
	Grass and Grass-Dominated:						
1	Short grass (1 foot)	0.74	0.00	0.00	0.00	1.0	12
2	Timber (grass and understory)	2.00	1.00	0.50	0.50	1.0	15
3	Tall grass (2.5 feet)	3.01	0.00	0.00	0.00	2.5	25
	Chaparral and Shrub Fields:						
4	Chaparral (6 feet)	5.01	4.01	2.00	5.01	6.0	20
5	Brush (2 feet)	1.00	0.50	0.00	2.00	2.0	20
6	Dormant brush, hardwood slash	1.50	2.50	2.00	0.00	2.5	25
7	Southern rough	1.13	1.87	1.50	0.37	2.5	40
	Timber Litter:						
8	Closed timber litter	1.50	1.00	2.50	0.00	0.2	30
9	Hardwood litter	2.92	0.41	0.15	0.00	0.2	25
10	Timber (litter and understory)	3.01	2.00	5.01	2.00	1.0	25
	Slash:						
11	Light logging slash	1.50	4.51	5.51	0.00	1.0	15
12	Medium logging slash	4.01	14.03	16.53	0.00	2.3	20
13	Heavy logging slash	7.01	23.04	28.05	0.00	3.0	25

Source: Aids to Determining Fuel Models for Estimating Fire Behavior, NFES 1574, National Wildfire Coordinating Group, April 1982, p. 3.

Medium fuels. These fuels include brush up to 6 feet (1.8 m) in height and the grass understory (Figure 1.37). The tremendous amount of fuel contained within this classification can produce moderate- to very-high-intensity burning but with a somewhat slower rate of spread.

Heavy fuels. These are fuels such as heavy continuous brush, more than 6 feet (1.8 m) in height, and timber slash. Combustion characteristics of these fuels include high-intensity burning but generally a low-to-moderate rate of spread (Figure 1.38).

Figure 1.37 Scattered brush is considered *medium* fuel. *Courtesy of NIFC.*

TABLE 1.2
Metric Version
Description of Fuel Models Used in Fire Behavior

Fuel Model	Typical Fuel Complex	Fuel Loading (tonnes/hectare)				Fuel Bed Depth (meters)	Moisture of Extinction Dead Fuels (percent)
		1 hr	10 hrs	100 hrs	Live		
	Grass and Grass-Dominated:						
1	Short grass (0.3 meter)	1.65	0.00	0.00	0.00	0.30	12
2	Timber (grass and understory)	4.48	2.24	1.12	1.12	0.30	15
3	Tall grass (0.8 meter)	6.75	0.00	0.00	0.00	0.76	25
	Chaparral and Shrub Fields:						
4	Chaparral (1.8 meters)	11.23	8.99	4.48	11.23	1.83	20
5	Brush (0.6 meter)	2.24	1.12	0.00	4.48	0.61	20
6	Dormant brush, hardwood slash	3.36	5.60	4.48	0.00	0.76	25
7	Southern rough	2.53	4.19	3.36	0.83	0.76	40
	Timber Litter:						
8	Closed timber litter	3.36	2.24	5.60	0.00	0.6	30
9	Hardwood litter	6.55	0.92	0.34	0.00	0.6	25
10	Timber (litter and understory)	6.75	4.88	11.23	4.48	0.30	25
	Slash:						
11	Light logging slash	3.36	10.11	12.35	0.00	0.30	15
12	Medium logging slash	8.99	31.45	37.06	0.00	0.70	20
13	Heavy logging slash	15.71	51.65	62.88	0.00	0.91	25

Source: Aids to Determining Fuel Models for Estimating Fire Behavior, NFES 1574, National Wildfire Coordinating Group, April 1982, p. 3.

Figure 1.38 Thick continuous brush is considered *heavy* fuel. *Courtesy of NIFC.*

WEATHER

Because of its influence on wildland fire behavior, weather is usually a key factor in firefighter safety and survival on the fireline (area of fire fighting operations). A case in point is the 1994 South Canyon fire near Glenwood Springs, Colorado, in which 14 firefighters were killed. According to the investigative report, weather information was neither requested by nor supplied to the fire crews who were overrun by a sudden and unexpected blowup (sudden increase in fire intensity and spread) in one area of the fire.

The report concluded that weather "contributed significantly" to the accident and the resulting firefighter fatalities. Knowing the potential effects of weather on wildland fire behavior, continually monitoring the local weather during a fire, and getting frequent weather forecasts can be critical to keeping firefighters out of mortal danger (Figure 1.39).

Figure 1.39 A firefighter gathers weather data in the wildland. *Courtesy of NIFC.*

WARNING

Firefighter survival may depend on knowing the potential effects of weather on fire behavior. You must stay informed of current and expected weather in the fire area.

Weather is the state of the atmosphere over the surface of the earth. It is the result of the interaction of temperature, wind, relative humidity, and precipitation. Of the major factors that affect wildland fire behavior, weather is the most changeable. Local weather conditions may change within a matter of minutes, significantly affecting the rate of fire spread and the intensity of a wildland fire.

Temperature, relative humidity, wind, and precipitation act together to influence wildland fire behavior. An understanding of these elements begins with knowledge of the atmosphere.

Atmosphere

The *atmosphere* is a thick blanket of air, consisting of several layers, that surrounds the earth. The atmosphere extends from the earth's surface to several hundred miles (kilometers) into space. Beyond the earth's atmosphere is airless space.

Air is not uniformly distributed in the atmosphere. The outer limit of the atmosphere is thin, with very little air; therefore, the atmospheric pressure is low. In the lower atmosphere immediately above the earth, more air is compressed by gravity; therefore, the atmospheric pressure is higher.

The atmosphere is divided into several layers (Figure 1.40). The lowest layer of the atmosphere

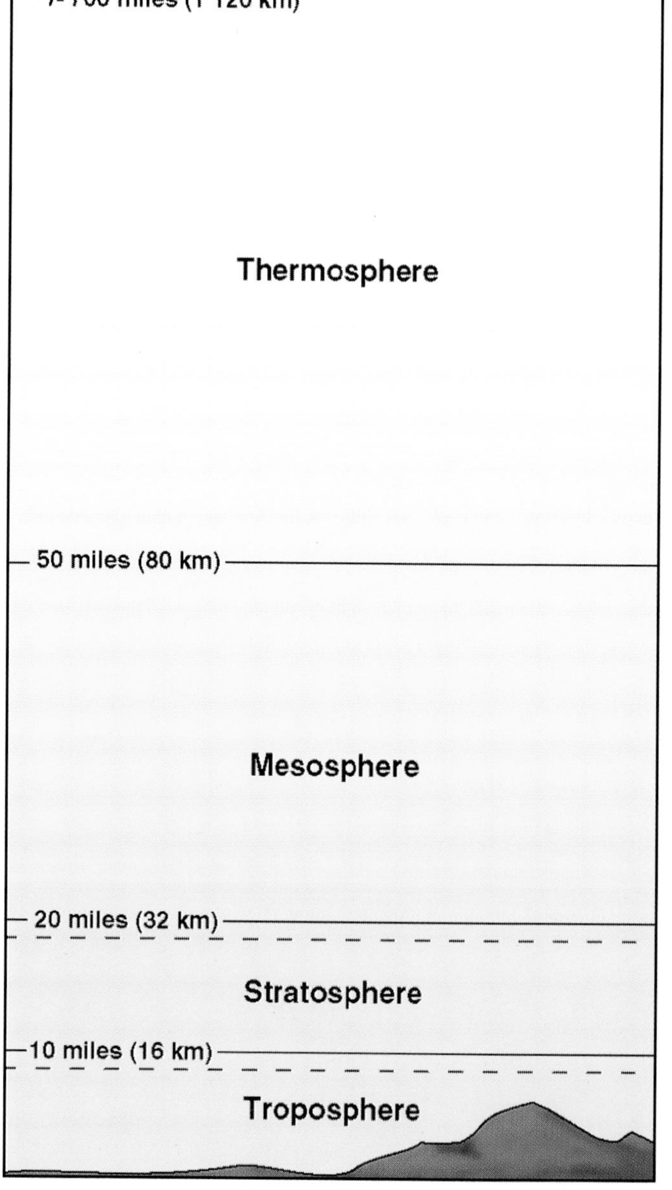

Figure 1.40 Earth's atmosphere is comprised of several layers.

immediately above the earth, up to about 10 miles (16 km), is the *troposphere*. It is the region of most changeable weather and most concern to firefighters. It is here that practically all clouds, storms, and other atmospheric changes that affect wildland fire behavior occur. Generally, temperatures in the troposphere decrease with elevation, but horizontal winds usually increase with elevation. However, there are exceptions. Warmer air at the surface rises through convection and mixes with the cooler air aloft. This rising and mixing motion makes the troposphere a sometimes turbulent layer, and turbulence can have a significant effect on wildland fire behavior.

Atmospheric Stability

Atmospheric stability can be defined as the atmosphere's resistance to the upward or downward movement of air. Unstable air encourages the vertical movement of air and tends to increase fire activity. Stable air discourages the vertical movement of air and tends to reduce fire activity.

In general, the temperature of an air mass decreases as the air rises and expands (decompresses), and the temperature of an air mass increases as the air descends and compresses. These changes in temperature are called *lapse rates*. The lapse rate for dry air masses is different from that of moist air masses. However, because actual conditions rarely represent either dry or moist conditions exactly, it is practical to assume an average lapse rate of about 3.5°F per 1,000 feet (-16°C per 305 m) of elevation.

Atmospheric stability can be determined by measuring the rate of temperature change with differences in elevation. The rate of change is called the *environmental lapse rate*. An environmental lapse rate of 5.5°F per 1,000 feet (-15°C per 305 m) or less in dry air is considered to be a *stable* atmosphere. An environmental lapse rate greater than 5.5°F (-15°C) in dry air is *unstable*.

WARNING

When the atmosphere is unstable, formerly calm fires may suddenly blow up and become very erratic.

Other indicators can also reveal important information about local atmospheric conditions. Steady winds indicate stable air; gusty winds are an indication of unstable air, except where mechanical turbulence (usually caused by terrain features) is the obvious cause. Dust devils (whirlwinds) are reliable indicators of instability near the surface (see Convective Winds section). Haze and smoke tend to rise in unstable air and to spread horizontally in stable air (Figure 1.41).

Different cloud formations also indicate atmospheric stability or instability. Cumulus clouds (rounded tops with a flat base) are characterized by vertical currents and therefore indicate unstable atmospheric conditions and the possibility of gusty or strong winds (Figure 1.42). The heights of cumulus clouds indicate the depth and intensity of the instability. Thunderheads (cumulonimbus clouds) moving into the area of a wildland fire may indicate strong, gusty winds that shift erratically as a storm

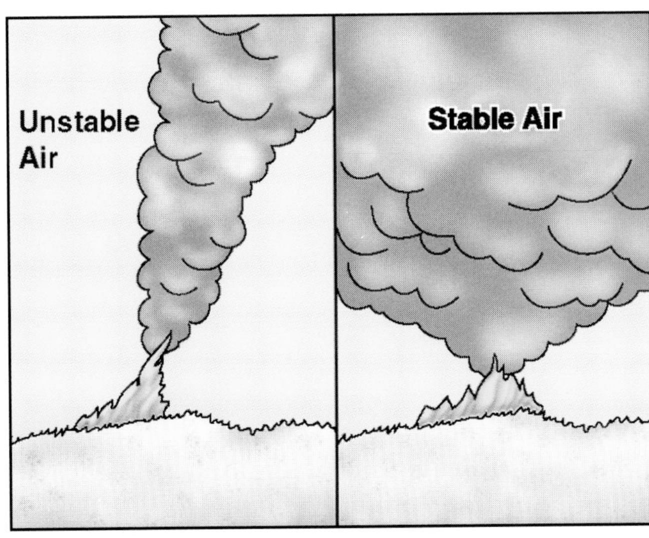

Figure 1.41 Atmospheric stability/instability affects how smoke behaves.

Figure 1.42 A typical cumulus cloud formation.

passes (see Thunderstorms section) (Figure 1.43). Stratus cloud sheets are indicative of stable layers in the atmosphere with little or no winds (Figure 1.44).

Figure 1.43 A typical cumulonimbus cloud formation.

Figure 1.44 A stratus cloud layer.

INVERSIONS

An *inversion* is a layer in the atmosphere where temperature *increases* with altitude and in which the air is extremely stable. The height at which rising smoke flattens out may indicate the elevation of the base of an inversion (Figure 1.45). The elevation of the cloud tops gives a good indication of the top of an inversion.

Air cooled primarily by contact with the earth's surface gradually deepens during the night and forms surface inversions, often referred to as *night inversions.* Night inversions are common during clear, calm, stable weather and are usually easy to identify. They trap smoke and fumes from vehicular traffic, industrial and residential chimneys, and wildland fires.

Figure 1.45 A clearly visible inversion layer. *Courtesy of NIFC.*

Inversions present a good opportunity for control of wildland fires because the relatively stable atmospheric conditions tend to reduce the level of fire activity. After sunrise, the night inversion begins to break up — winds increase, temperature rises, and relative humidity decreases. While this process is usually gradual, it can occur rather abruptly; when it does, fire activity can increase dramatically and threaten firefighters.

Topography plays a significant role in both the formation and intensity of night inversions. Surface layers are relatively shallow on mountain slopes and in open canyons or ravines where the cold air can drain away as it forms. This dense, cold air flows downward and gathers in valleys and small depressions (Figure 1.46). Patches of ground fog in surface depressions along highways are small-scale inversions.

Measurements of temperature and humidity can indicate the strength of a night inversion. Within an inversion layer, the temperature increases with elevation and may change as much as 25°F (-4°C) in 250 vertical feet (76 m). In mountain-

Figure 1.46 Ground fog forms in valleys and depressions. *Courtesy of Tony Bacon.*

ous areas, the elevation of the top of night inversions is usually below the main ridges, although the elevation varies from night to night. In the absence of a visible smoke or cloud layer, the top of the inversion can be found by measuring temperatures at various points up and down the slope. The warmest air temperature is at the top of the inversion, and temperatures decrease as one goes farther up or down the slope (Figure 1.47). Both the highest minimum temperatures and the least daily temperature variation of any level along the slope occur at the top of the inversion. The lowest nighttime relative humidity and the lowest nighttime fuel moisture are also found at this level.

Along the west coast of North America, the coastal or *marine inversion* is common in summer. Cool, moist air from the ocean spreads inland in a layer that may vary in depth from a few hundred to several thousand feet (meters) (Figure 1.48). It continues to spread inland until stopped by a much warmer, drier, and relatively unstable air mass. Although marine inversions may persist in some areas during the day, they are strongest and most noticeable at night. After sunset, fog and stratus clouds often form in the cool marine air and move inland into coastal basins and valleys. If the layer of cold air is relatively shallow, fog usually forms; if it is deep, stratus clouds are likely to form (Figure 1.49).

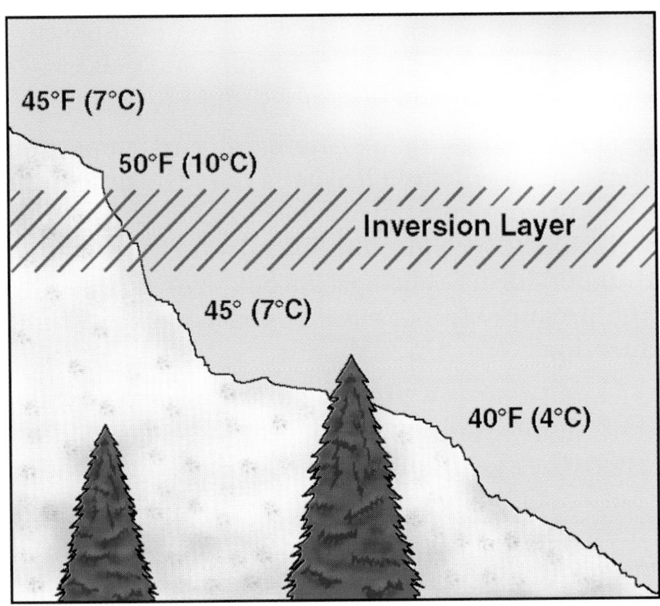

Figure 1.47 Temperatures decrease both above and below an inversion layer.

Figure 1.48 A typical marine inversion. *Courtesy of NIFC.*

Along the east coast of North America, the same weather phenomena operate. However, because the terrain along the eastern seaboard is not as mountainous as in the West, the interaction between weather and topography does not produce the same inversion patterns. The daytime sea breezes usually extend 5 to 6 miles (8 km to 9.7 km) inland but may extend as much as 25 miles (40 km) from the coast. Just as in the West, the wind may reverse at night and become a land breeze.

Figure 1.49 A stratus cloud layer along a coastline.

THERMAL BELT

The top of the inversion layer is known as the *thermal belt* because of the relatively higher temperatures found there (Figure 1.50). This area is characterized by the least variation in daily temperatures, the highest average temperature, and the lowest average humidity. Within the thermal belt, wildland fires can remain quite active during the night.

CAUTION: Firefighters must be especially careful at night because the dangers that accompany fighting an active fire are compounded by the hazard of darkness.

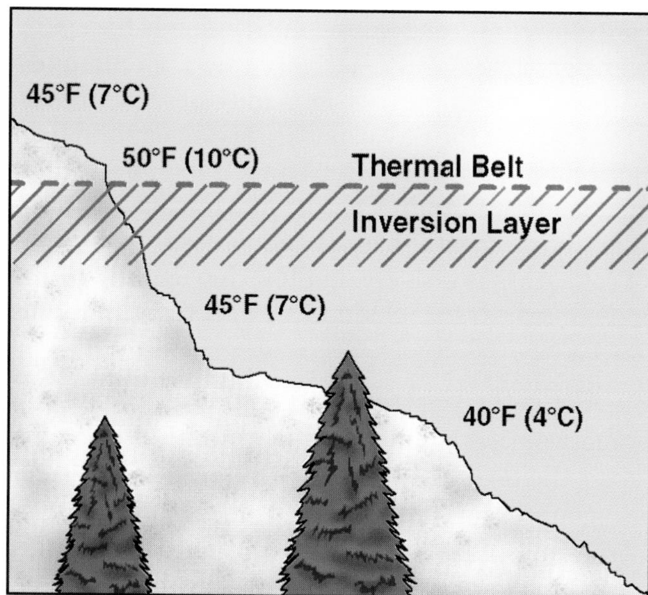

Figure 1.50 A thermal belt.

Below the thermal belt, fires are subdued by cool, humid, and stable air, often with downhill winds. Above the thermal belt, fires are somewhat subdued because temperatures decrease with elevation. However, the effects of these lower temperatures may be offset by stronger winds and less stable air as fires penetrate the region above the thermal belt.

THUNDERSTORMS

Thunderstorms are violent local storms spawned by some, but not all, cumulonimbus (high and anvil-shaped) clouds (Figure 1.51) They produce thunder and lightning, heavy rains, and sometimes hail. Thunderstorms also produce extreme convective activity in the atmosphere, which generates high-velocity updrafts and downdrafts on the ground (Figure 1.52).

Figure 1.51 An anvil-shaped cumulonimbus cloud. *Courtesy of NIFC.*

> **WARNING**
>
> **The sudden and often violent changes in wind speed and direction that accompany thunderstorms can be very dangerous to firefighters.**

Thunderstorms derive most of their energy from the release of latent heat in the condensation of water vapor. For each pound (0.4 kg) of water condensed from vapor, more than 1,000 Btu (1 055 kJ) of heat energy is released.

Thunderstorms are composed of one or more individual convective cells. A cell may range up to 10 miles (16 km) in diameter. Each cell in a cluster of cells may be in a different stage of development and may interconnect with others to form a cloud mass that can extend for 50 miles (80 km). Even though the bases of individual cumulus clouds may join to form a solid overcast, which tends to obscure the multicellular structure of the cloud mass, each cell has its individual identity and life cycle.

Two things make thunderstorms important elements in wildland fire behavior. The first is the fire-starting potential of cloud-to-ground lightning strikes. The second is the downdrafts generated by thunderstorms that spread out near the ground and produce strong, erratic, gusty winds of short duration.

Lightning can start wildland fires anywhere that thunderstorms occur, but it is most likely where the storms produce little or no precipitation that reaches the ground. These storms are sometimes called *dry thunderstorms*, and they occur mainly in the mountains of the West. Hundreds of lightning fires can be started during one day.

On the other hand, "wet" thunderstorms produce heavy precipitation that moistens fuels and may decrease the activity of fires. They may also reduce the risk of lightning-strike fires. However, if fires do start under these conditions, they may go undetected until the storm has passed. By then, these fires may have grown to major proportions.

Visual indicators of potential thunderstorm formation are as follows (Figure 1.53):

- Tall, building cumulus clouds
- Cauliflower shape of cloud tops
- Clouds with dark, flat bases

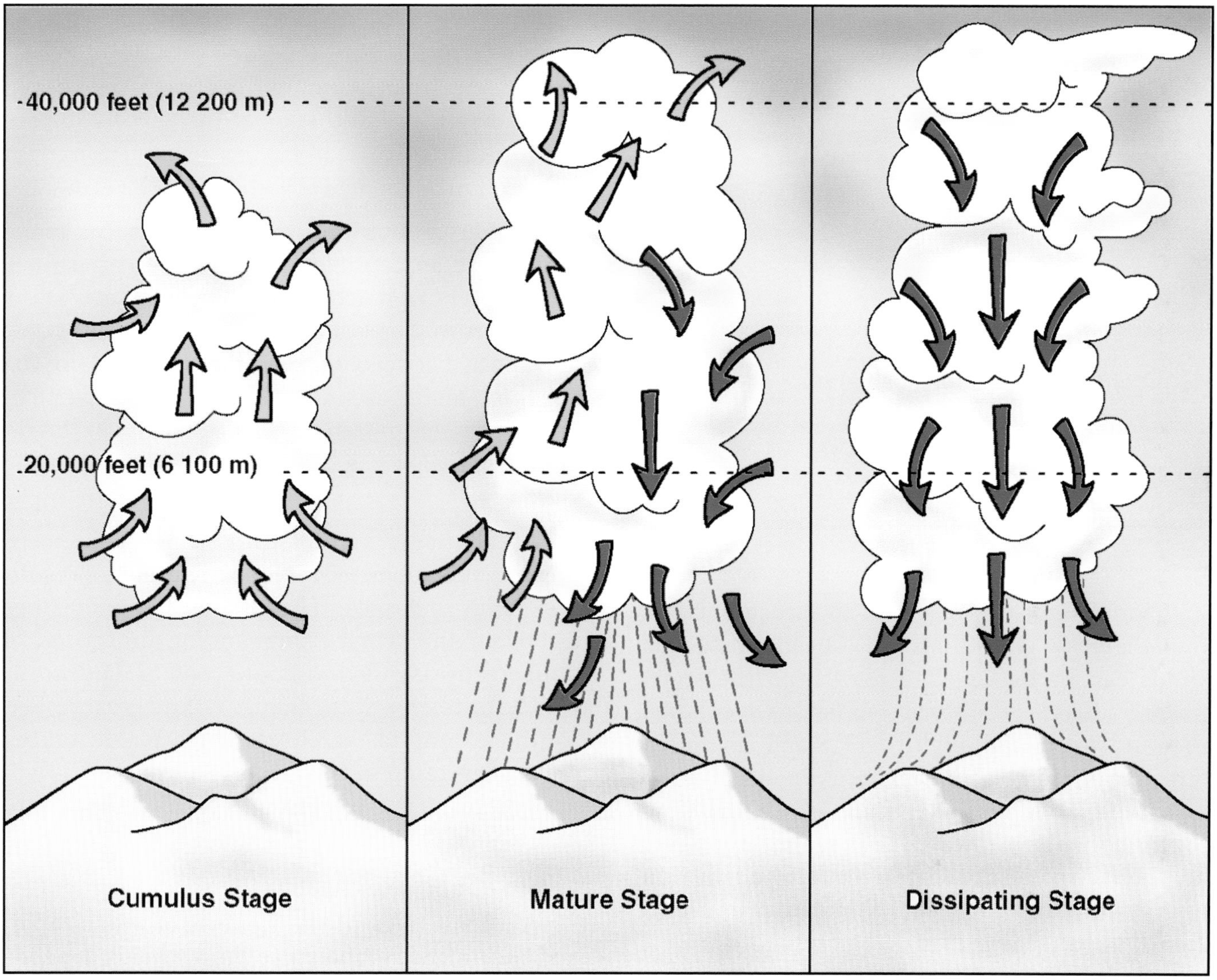

Figure 1.52 Thunderstorms produce both updrafts and downdrafts.

Figure 1.53 Indicators of thunderstorm formation.

- Rain that evaporates before reaching the ground (virga) or rain falling from the cloud bottom
- Anvil-shaped cloud with fuzzy appearance

The path of a thunderstorm is indicated by the direction in which the anvil-shaped cloud top is pointing — usually in the direction of the winds aloft. When downdraft winds produced by thunderstorms reach the ground, they usually spread horizontally in all directions (Figure 1.54). Wind velocities are often 25 to 35 miles per hour (mph) (40 kmph to 56 kmph) and may reach as high as 60 mph (97 kmph). Thunderstorm-produced surface winds tend to be strongest in the storm's direction of travel, but both wind speed and direction can be altered by topography and vegetation (see Wind section).

Figure 1.54 Thunderstorm downdrafts can spread in all directions.

Thunderstorm winds tend to spread outward from the center of the storm. As a storm approaches, winds can be expected to blow from the storm toward a fire. As the storm cell passes over a fire, the winds are usually highly erratic and can change direction unpredictably. Finally, as the storm moves away, the winds shift so that they are again blowing from the storm toward the fire. Thus, the winds may shift as much as 180 degrees between the time of a storm's approach and its departure (Figure 1.55).

Figure 1.55 Wind direction changes as thunderstorms pass.

Temperature

Temperature is a measure of the warmth or coldness of a substance — in this case, air. The main source of heat for the outside air is solar energy from the sun. Fuel and ground temperatures are primarily due to direct solar radiation. The temperature of the air at the earth's surface rises or falls because of contact (conduction) with wildland fuels, bodies of water, and the ground.

Higher ground and fuel temperatures make fuels more susceptible to ignition. Heated fuels ignite and burn much easier than those at a lower temperature. There may be more than 50°F (10°C) difference between temperatures of fuel in the sunshine and those in the shade. This is particularly true for flattened fuel beds and/or large-diameter fuels close to the ground and out of the winds.

As mentioned earlier, when a mass of air moves upward, it is subjected to less pressure. As a result, it expands and cools. If heat is neither gained nor lost by mixing with the surrounding air, the loss of temperature is called *adiabatic cooling*. The rate at which the temperature change occurs is controlled by the extent to which condensation is taking place. The temperature change in dry air is approximately 5.5°F per 1,000 feet (-15°C per 305 m) in altitude.

Relative Humidity

Air holds moisture in the form of water vapor, and the amount of moisture the air holds depends upon the temperature of the air. *Relative humidity* is the percentage of moisture in a volume of air relative to the total amount of moisture which that volume of air can hold at the given temperature and atmospheric pressure. For example, when the relative humidity is 63 percent, the air has only 63 percent of the moisture it could possibly hold at that temperature and atmospheric pressure. When air holds the maximum amount of moisture, the air is saturated, and its relative humidity is 100 percent (Figure 1.56).

Air can add moisture to fuels or remove it from them, depending on the relative humidity. If the relative humidity is high, air adds moisture to fuels, dampens them, and makes them less likely to burn. By contrast, when the relative humidity is low (<30 percent), the air absorbs moisture from fuels, dehydrates them, and makes them more susceptible to ignition. Naturally, these changes

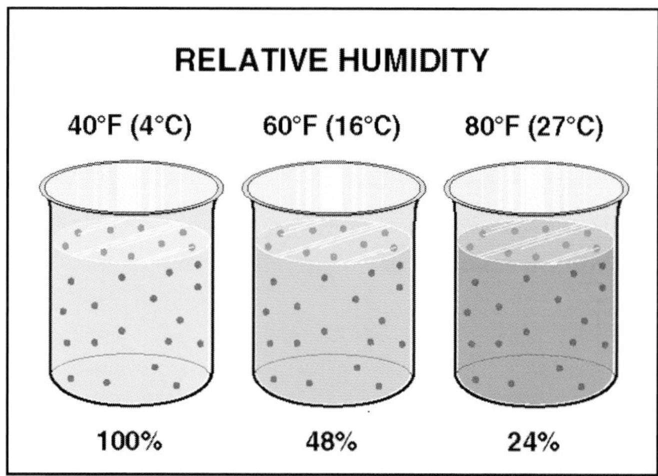

Figure 1.56 The relationship of humidity to atmospheric temperature.

occur faster in fine, 1-hour fuels than in heavier fuel types. In general, low humidity means an increase in fire activity — greater intensity, higher rate of spread, and more spot fires. When relative humidity drops below 30 percent, fires burn freely; when the relative humidity is below 10 percent, they burn vigorously, and the most extreme fire behavior is expected.

As a fire-behavior factor, relative humidity is most important nearest the ground where it has the greatest influence on fuel moisture. Near the ground, however, variations in relative humidity are influenced by factors such as season, time of day, slope, aspect (direction a slope faces), elevation, clouds, and type of vegetation as well as the moisture content of the air.

If they are available, special instruments can be used to measure the temperature and relative humidity. A sling psychrometer, part of a belt weather kit carried in the field, measures local temperature and humidity quite accurately (Figure 1.57). However, it may not show precisely the temperature and humidity except at the point of measurement.

Seasonal patterns of relative humidity are also seen. In the far West, fire season begins (usually about mid-April) after a normally wet winter and a relatively moist spring. It continues through the normally dry summer and early fall. Changes in relative humidity are quite noticeable. Early in the fire season when the sun is nearly overhead and nights are clear, the range of daily temperatures is greatest. Relative humidity levels are lowest in the afternoon because of intense surface heating during the day. Significant cooling after sunset and moisture in the soil and vegetation normally increase the relative humidity during the night.

Figure 1.57 A firefighter uses a sling psychrometer. *Courtesy of NIFC.*

In late summer and early fall in the West, both soil and vegetation have dehydrated even though solar heating is gradually diminishing. Both daytime and nighttime humidities become lower late in the season, so fire danger increases. Occasional summer rains may temporarily interrupt this progression, but they do not greatly change the overall seasonal pattern.

In other areas of North America, separate spring and fall fire seasons are separated by a relatively humid summer, and the range of daily temperatures is not as extreme as in the West. Nor is the cumulative dehydration of soil and vegetation as consistent, except during periods of drought. Because of periodic rains during the summer, the higher relative humidity tends to decrease fire danger. Evaporation from small lakes and ponds combine with peak transpiration from vegetation to keep daytime relative humidities higher in summer than in spring and fall, especially in the northern U.S. and southern Canada.

Regardless of the geographic location of a wildland fire, humidity follows certain basic patterns, and the effects on wildland fuels are the same. Cold air holds less moisture than hot air. When the air is cool, there is less transpiration of moisture from the fuels into the atmosphere. Because late-night and early-morning air temperatures are generally lower than those during the day and evening, wildland fuels (especially the light, flashy 1-hour fuels) retain more of their moisture. During these periods, fires are usually less active. While these effects are generally true, local microclimates may alter their onset or duration. Knowledge of local weather patterns helps firefighters anticipate these differences. In the absence of such local knowledge, firefighters may get useful data on humidity from a belt weather kit and/or from weather bulletins broadcast by local radio or television stations. Normally the reduced level of fire activity at night provides firefighters an opportunity to make significant headway toward control and extinguishment of a fire. If control and extinguishment are not achieved during the cooler nighttime hours, an increase in fire activity can be anticipated as temperatures again begin to rise the following day. With the increase in fire activity comes an increase in spotting and the rate of spread.

Wind

Wind is basically air in motion. Its principle characteristics are its direction, speed, and turbulence (gustiness). Surface winds are strongly affected by topography and by local heating and cooling. These factors are responsible for much of the wind's variability, and variability is why an adequate understanding of local wind behavior is very important. In North America, wind speed is measured in miles per hour (mph) or kilometers per hour (kmph), and the general wind direction is identified as the direction *from which* the wind is blowing. Thus, a north wind comes from the north and blows toward the south (Figure 1.58).

LOCAL WINDS

Local winds are a by-product of the diurnal cycle — daily heating and cooling patterns. Land masses heat more rapidly than bodies of water during the daytime and cool more rapidly at night. Darker soils absorb more solar heat than lighter soils. Bare soil absorbs more solar heat than grass-covered soil. In hilly or mountainous terrain, heating generally causes upslope winds; cooling causes downslope winds. In flat terrain, heating can produce whirlwinds or dust devils (Figure 1.59).

Figure 1.58 Wind direction nomenclature.

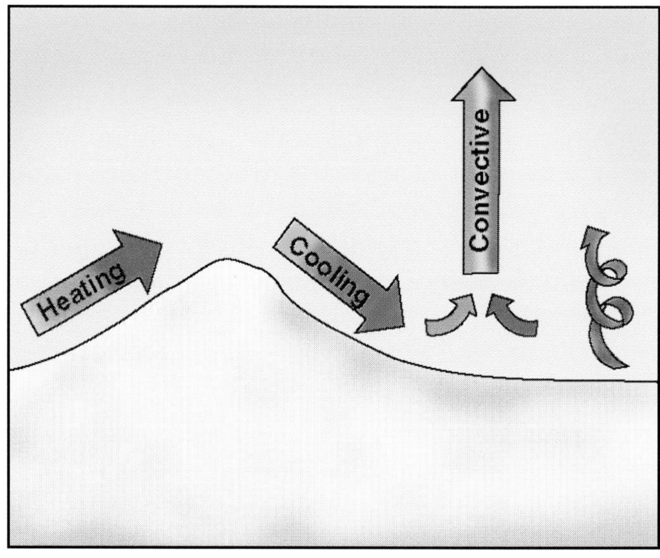

Figure 1.59 Winds caused by surface heating.

GENERAL WINDS

General winds are produced by large-scale high-pressure (area of high barometric pressure) and low-pressure (area of low barometric pressure) systems. Frontal, gradient, and gravity winds are examples of general winds that can significantly affect wildland fire behavior.

Frontal winds. Frontal winds develop in either of two types of air masses with different temperatures and moisture: cold and warm. The leading edges of the two air masses are called *fronts*. When a cold air mass replaces a warm air mass, the leading edge of the cold mass is called the *cold front*. Conversely, when a warm air mass replaces a cold air mass, the leading edge of the warm mass is called the *warm front*.

A cold front can have a dramatic effect on a wildland fire. When a cold front advances on a warm front, it produces strong, gusty winds. Wind direction can change sharply and distinctly (Figure 1.60). Surface winds cause longer horizontal flames that preheat fuels ahead of a fire. Strong winds with sudden changes in direction cause a wildland fire to behave erratically and increase the potential for spotting (fire caused by flying embers) downwind. These winds will also cause a fire to develop a very active head and flanks (sides) with high heat outputs. These factors make controlling such a fire extremely difficult.

Gradient winds. These winds occur at approximately 1,500 feet (457 m) above the earth's surface. They flow from an area of high barometric pressure toward areas of lower pressure. In a low-pressure system, the wind moves inward (counterclockwise) and rises (Figure 1.61). In a high-pressure system, the movement is outward (clockwise) and descending. However, in mountainous terrain the general direction of the wind can be altered by the topography. Therefore, knowledge of the topographical influences on wind is essential to predicting the effects of wind on wildland fire behavior (see Topography section).

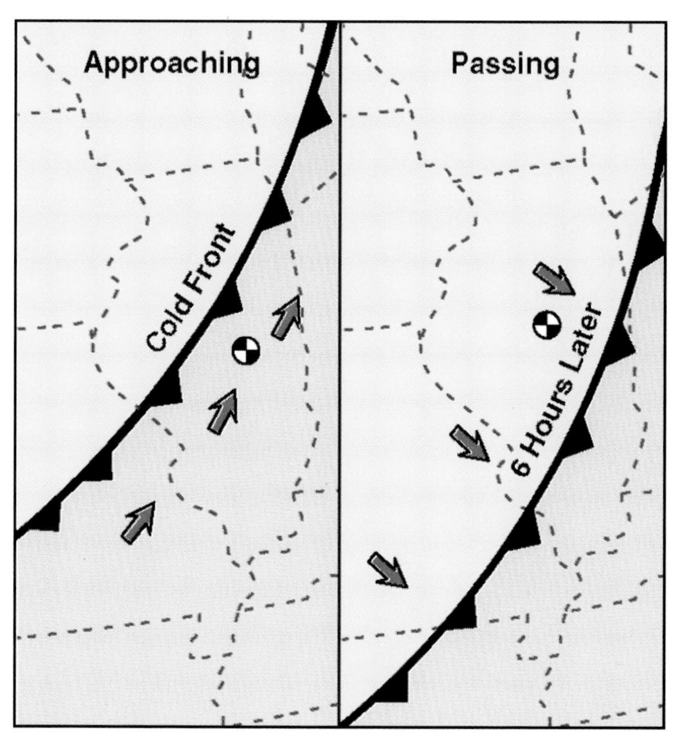

Figure 1.60 Wind speed and direction change sharply when a cold front passes.

Figure 1.61 Gradient wind circulation around low-pressure cells.

Like water in a stream, wind follows the path of least resistance across the earth's surface. Obstructions such as trees, large rocks, and ridges alter its flow. The obstructions can cause turbulence on the side of the obstruction that is sheltered from the wind (Figure 1.62).

Gravity winds. Gravity winds, also called *foehn winds*, are conducive to rapid wildland fire spread. These winds result from air being forced over mountain ridges by convection or high barometric pressure. The air then cascades downslope as gravity winds. As the air drops in elevation, the atmospheric pressure increases, which causes the air to compress and heat. The resulting winds are strong, hot, dry, persistent, and unfavorable for wildland fire control (Figure 1.63).

Gravity winds are often named according to their location. Perhaps the most well-known gravity winds are the Santa Anas of southern California. Other examples include the Chinook winds in the eastern Rocky Mountains and the North and Mono winds in northern California. These winds can turn wildland fires into fire storms (violent convection caused by a large continuous area of intense fire) (see Fire Storms section).

Figure 1.62 Effects of topography on winds.

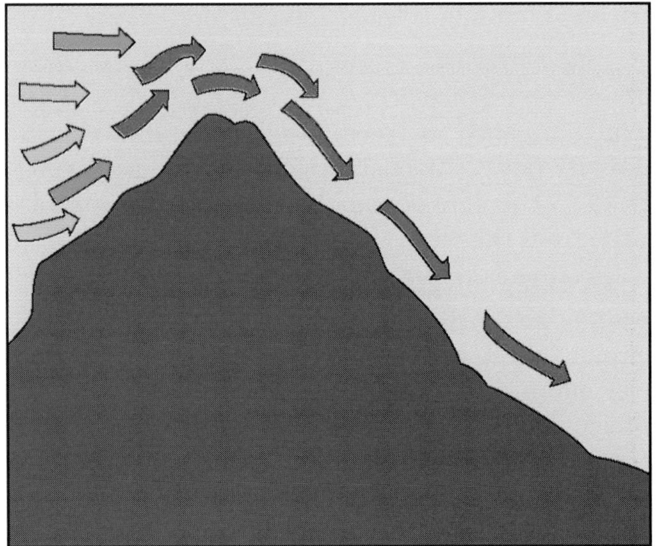

Figure 1.63 How gravity (foehn) winds are produced.

CONVECTIVE WINDS

Convective winds are caused by the localized heating of air that expands and rises while cooler, denser air descends to replace it. This concurrent ascending and descending air movement creates convective winds (Figure 1.64). Slope winds (both up and down), valley winds, land and sea breezes, and thunderstorm winds are all examples of convective winds.

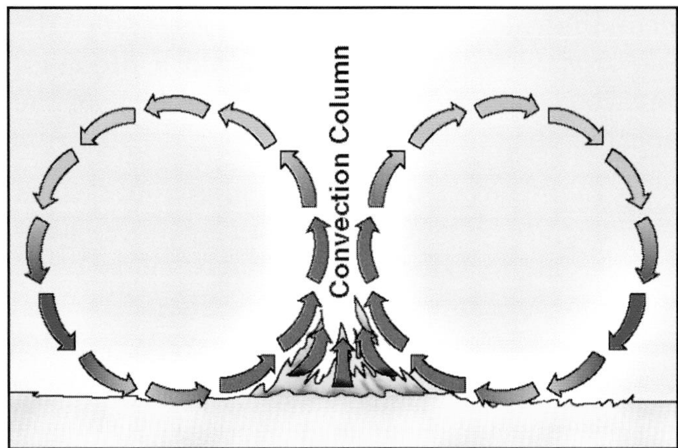

Figure 1.64 Wind circulation caused by a convection column.

Slope and valley winds. Wildland fires are greatly influenced by slope winds. Except for foehn conditions mentioned earlier, slope winds flow up during the day due to surface heating and flow down at night because of surface cooling. However, large wildland fires on a slope produce their own winds. The fires heat the air, and the hot air rises rapidly up the slope. To match the speed of the rising warm air, cold air moves down to the fires.

In the mountains, local winds begin as the morning sun preheats the eastern aspects (direction the slopes face). As the land heats, surface air rises, producing an upslope wind. The wind is quite localized early in the day and may average only 3 to 5 mph (5 kmph to 8 kmph). As solar heating continues, the upslope winds involve a greater area. By midday, with the sun directly overhead, the east, south, and west aspects are subjected to solar heating. By this time, upslope winds have developed sufficiently to be termed *up-canyon* or *valley winds* (Figure 1.65). Up-canyon or valley winds may eventually reach 7 to 10 mph (11 kmph to 16 kmph).

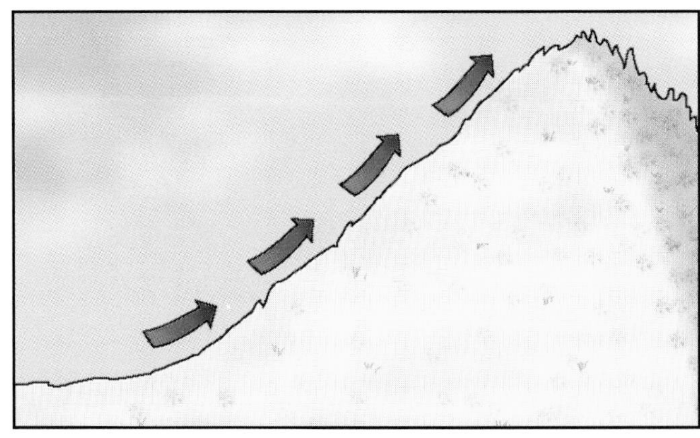

Figure 1.65 Typical diurnal slope winds.

In late afternoon the eastern and southern aspects begin a transition to downslope winds with increasing relative humidity because they are in shadow, despite continued heating on the western aspect. The ridge line cools first, then air cascades downslope. After sunset, the downslope winds intensify, flowing into canyons or valleys at an average of 5 to 7 mph (8 kmph to 11 kmph) (Figure 1.66).

During both the daily heating and the nightly cooling cycles, the thermal belt can remain quite windy. This, along with lower relative humidity and higher average temperature, allows fires to burn actively throughout the night in the thermal belt.

When wind flows through areas of least resistance, such as a steep V-shaped drainage, a chute (sloping channel), a saddle (ridge connecting two higher elevations), or a narrow canyon, wind speeds can increase significantly (see Chutes and Saddles section). These terrain features (sometimes referred to as *chimneys*) may also create turbulent updrafts in response to localized heating causing a chimney effect (Figure 1.67).

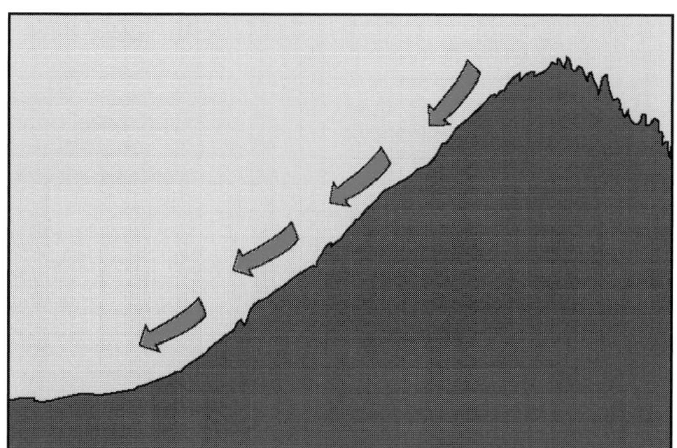

Figure 1.66 Typical diurnal valley winds.

WARNING

Fires in drainages or chutes may spread at an alarming rate, so these formations are always very dangerous locations during wildland fires. Wind-driven fires sweeping through these terrain features have been associated with many firefighter injuries and fatalities. These areas should *never* be used as safety zones.

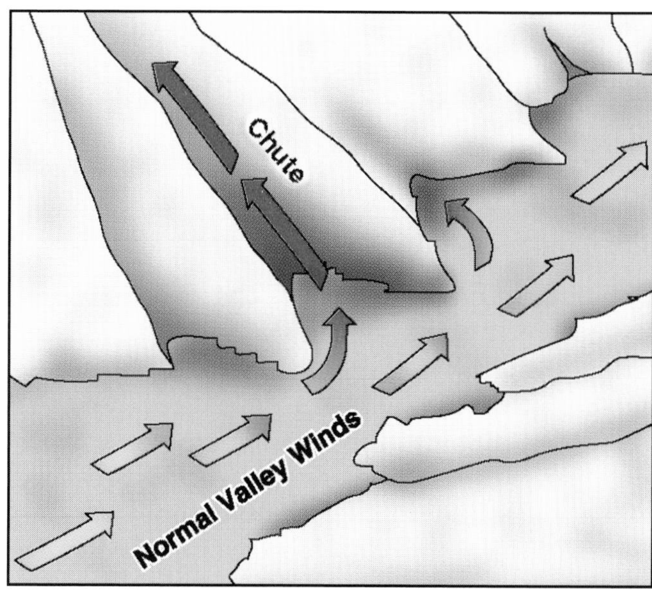

Figure 1.67 Wind flow in a chimney.

Land and sea breezes. When land surfaces along a coastline are heated by solar radiation during the day, the air over the land expands and rises. In the process, atmospheric pressure over the land becomes lower than that over the nearby ocean. As a result of this localized pressure difference, cooler, denser air from the ocean's surface is drawn inland to replace the air rising in a convective column. As the rising air expands and cools adiabatically, it flows seaward aloft to replace air that has moved toward shore, thus completing the cycle (Figure 1.68).

The landward sea breeze generally begins around midday, strengthens during the afternoon,

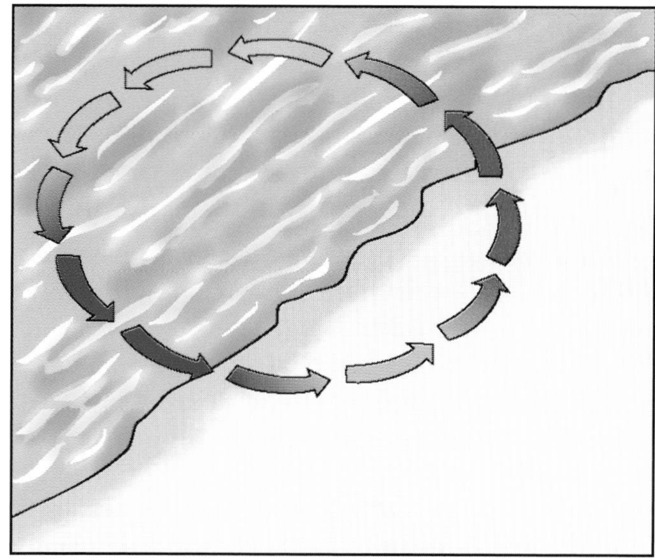

Figure 1.68 Wind flow related to land/sea heating.

and ends around sunset, although the timing can vary considerably because of local atmospheric conditions such as cloudiness and general winds. The breezes begin at the coast, gradually pushing farther and farther inland during the day, and reach maximum penetration about the time of maximum inland temperature.

Land breezes at night are the reverse of the daytime sea-breeze phenomenon. At night, land surfaces cool more quickly than water surfaces, and air in contact with the land then becomes cooler than that over the ocean. An increase in air density over the land causes atmospheric pressure to become relatively higher than that over the water. This pressure difference causes air to flow from the land to the water.

In the eastern and southeastern United States, thunderstorms frequently develop as sea breezes move inland from the coast. This movement results in weather very similar to cold fronts: strong shifting winds, cooler temperatures, higher relative humidities, and possibly heavy rains. Erratic winds associated with these "sea-breeze fronts" sometimes cause fire-control and safety problems on wildland fires in the East and Southeast.

Along the Pacific coast, sea breezes may attain speeds of 10 to 30 mph (16 kmph to 48 kmph), but fog or low clouds, very cool temperatures, and high humidity often move inland in the process. While these winds may initially increase fire activity, the cooler temperatures and higher humidities eventually cause it to diminish.

Thunderstorm winds. Thunderstorm winds are a good indication of an unstable air mass. They generally arise as a result of extreme differences in localized heating of the air near the ground. They are very strong, often unpredictable convective winds and can produce whirlwinds and firewhirls.

Whirlwinds. As mentioned earlier, whirlwinds or duct devils are good indicators of instability as a result of intense local heating. They usually occur on hot days over flat, dry terrain when skies are clear and general winds are light. Because burned-over areas are black or dark gray, they are more susceptible to solar heating than unburned areas; so dust devils often form in the black. Firefighters should wear adequate eye protection when they are in these areas. Dust devils can increase a fire's intensity if they move into the flames. They can cause spotting by picking up burning materials in the black and depositing them in the unburned (green) area (Figure 1.69).

Firewhirls. These rather spectacular phenomena are spinning, moving columns of rising air and fire gases that carry smoke, debris, and flames aloft. They range from a foot or two (0.3 m to 0.6 m) in diameter to a small tornado in size and intensity (Figure 1.70). Firewhirls may be caused by the same conditions that create dust devils; however, they may also be caused by thunderstorms, intense heating within a fire, or wind shears (radical shifts in wind speed and direction between slightly different altitudes). Usually formed on the leeward (protected from wind) side of elevated terrain features, firewhirls can cause spotting. They are generally considered far more dangerous than dust devils. Because of the intensity with which firewhirls burn, a direct attack (fighting the fire at its edge) is likely to be ineffective and unsafe, so it is not recommended.

Figure 1.69 A dust devil (whirlwind) in the distance. *Courtesy of NIFC.*

EFFECTS OF WIND ON FIRE BEHAVIOR

The two most important weather-related elements affecting wildland fire behavior are wind and fuel moisture. Of the two, wind is the most variable and the least predictable. The rate and

Figure 1.70 Firewhirls can be extremely dangerous. *Courtesy of NIFC.*

direction of fire spread are mostly functions of wind speed and direction, so the expected winds must be considered in the development of any wildland fire-control plan. Wind has both direct and indirect effects on wildland fire behavior.

As a direct effect, wind intensifies combustion by increasing the amount of oxygen available to the fire. Wind bends the flames close to uninvolved fuels, dries them, and preheats them. This makes them more susceptible to ignition. Also, the wind may carry embers and sparks more than a mile (1.6 km) ahead of the fire into unburned fuels (spotting), increasing the rate of fire spread (Figure 1.71).

As an indirect effect, the wind either increases or decreases fuel moisture. Strong, dry winds absorb the moisture from the fuels, but cool winds help wildland fuels retain their moisture.

Because wind speed and direction can change rapidly, wind-induced fire behavior may also change rapidly. An unexpected shift in wind direction, for example, can change the direction and the intensity of a fire. Anticipation of these changes increases firefighter safety and helps ensure success of fire-control efforts.

Figure 1.71 A typical wind-caused spot fire.

In addition to knowing the effects of wind on fire behavior, it is also important that firefighters know and follow the standard Fire Orders (see Chapter 8, Firefighter Safety and Survival, for more information).

Fire Orders

- **Fight fire aggressively but provide for safety first.**
- **Initiate all action based on current and expected fire behavior.**
- **Recognize current weather conditions and obtain forecasts.**
- **Ensure instructions are given and understood.**

- **Obtain current information on fire status.**
- **Remain in communication with crew members, your supervisor, and adjoining forces.**
- **Determine safety zones and escape routes.**
- **Establish lookouts in potentially hazardous situations.**
- **Retain control at all times.**
- **Stay alert, keep calm, think clearly, act decisively.**

SOURCES OF WIND INFORMATION

Firefighters can obtain reliable information about existing wind patterns from various sources.

Among them are public weather forecasts, specialized fire weather forecasts from local offices of the National Weather Service, personal experience, and the direct observation of existing weather conditions.

Familiarity with typical seasonal wind patterns helps firefighters apply general weather forecasts to local areas. Observation of current local weather conditions helps firefighters make appropriate modifications in the forecasts received from more distant sources. Direct observations may include using portable weather instruments at a site or making informal observations of wind speed, temperature, or cloud formations in the general vicinity.

Seasonal and Daily Weather Cycles

In addition to the effects of individual environmental conditions on fire behavior, some seasonal and daily weather cycles should also be considered. Local weather conditions conducive to wildland fires generally follow predictable patterns. The susceptibility of wildland fuels to ignition and combustion depends to a large extent on typical seasonal weather patterns. During prolonged rainy periods or consistently cold and damp weather, wildland fires are usually not a problem. As seasonal weather patterns change, the condition of wildland vegetation also changes.

Daily weather cycles also affect fire behavior, and they, too, tend to be predictable. For every 24-hour period, it is possible to make general predictions about burning conditions. The following characteristics are only generalizations; local terrain and weather conditions may modify the burning conditions (Figure 1.72).

- From mid-morning until late afternoon (10:00 a.m. to 6:00 p.m.) is the time when fire behavior is normally most erratic, and fire intensity is likely to be high. This "heat of the day" period is when the relative humidity is low, the temperature is high, the fuel is dry, and the wind is strong. All of these factors are unfavorable for fire control.
- In the evening and nighttime hours (6:00 p.m. to 4:00 a.m.), the wind usually moderates, the air cools, relative humidity usually rises, and fuels begin to absorb moisture. These factors are favorable for fire control.

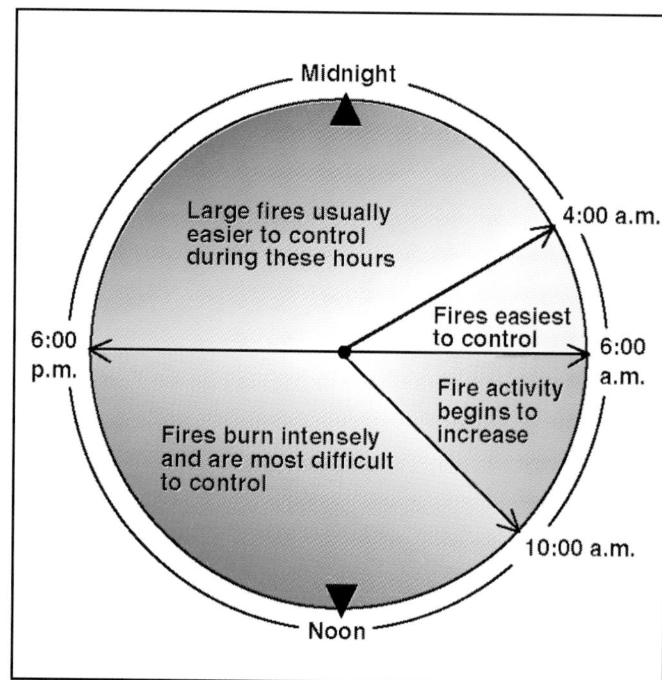

Figure 1.72 Burning characteristics at various times of the day.

- During the early morning hours (4:00 a.m. to 6:00 a.m.), wildland fire activity is usually at its lowest.
- From shortly after dawn until mid-morning (6:00 a.m. to 10:00 a.m.), the intensity of a wildland fire is likely to again increase. The wind usually increases, temperature rises, and controlling the fire becomes increasingly difficult.

Local winds may also vary according to the time of day. In mountainous terrain, daytime heating of the land produces an upward movement of air, creating up-canyon winds. At night, cooling of the land produces downslope winds. Similar winds can occur in depressions, valleys, and drainages. Wind direction and velocity in these areas can vary and may be erratic.

TOPOGRAPHY

Topography is composed of the features of the earth's surface, both natural and constructed, and the relationships among them. The topography in the immediate area of a fire affects both its intensity and its rate and direction of spread (Figure 1.73). It is much easier to predict the influence that topography has on a fire than the influences of fuel and weather. The topographical features of most significance in wildland fire fighting are *slope* and

Figure 1.73 Topography can greatly alter fire behavior.

aspect. It is also important for firefighters to be aware of the effects of wind channeling in various topographical features and to have skills in map and compass reading.

Slope

Slope is one topographical feature that has a pronounced influence on wildland fire behavior. Slopes can range from slight to steep depending upon their angle of elevation from horizontal, but slope is measured in "rise over run" expressed in percent. The percentage of slope is based on the ratio of vertical rise to horizontal distance. In other words, to find the percentage of slope, the amount of elevation change is divided by the horizontal distance and the result multiplied by 100. For example, a rise of 10 feet (3 m) divided by a horizontal distance of 100 feet (30 m) would represent a 10-percent slope. A 45-degree slope rises 100 feet (30 m) for every 100 feet (30 m) of horizontal distance, so it is a 100-percent slope (Figure 1.74). During training, firefighters should practice by calculating the percentage of slope of numerous nearby hills so they can develop a mental picture of what various slopes look like. This will allow them to more accurately estimate the percentage of slope of terrain features when they do not have access to a clinometer or a topographic map.

In the absence of winds, fires usually move faster uphill than downhill, therefore, the steeper the slope, the faster a fire moves. The increased rate of spread is due to several factors. On the uphill side of a fire, flames are closer to the fuel; consequently, the fuels dehydrate, preheat, and ignite sooner than they would if they were on level ground. Wind currents normally move uphill during the day and tend to push heat and flames toward new fuels. A fire burning upslope creates a draft that increases the rate of spread (Figure 1.75).

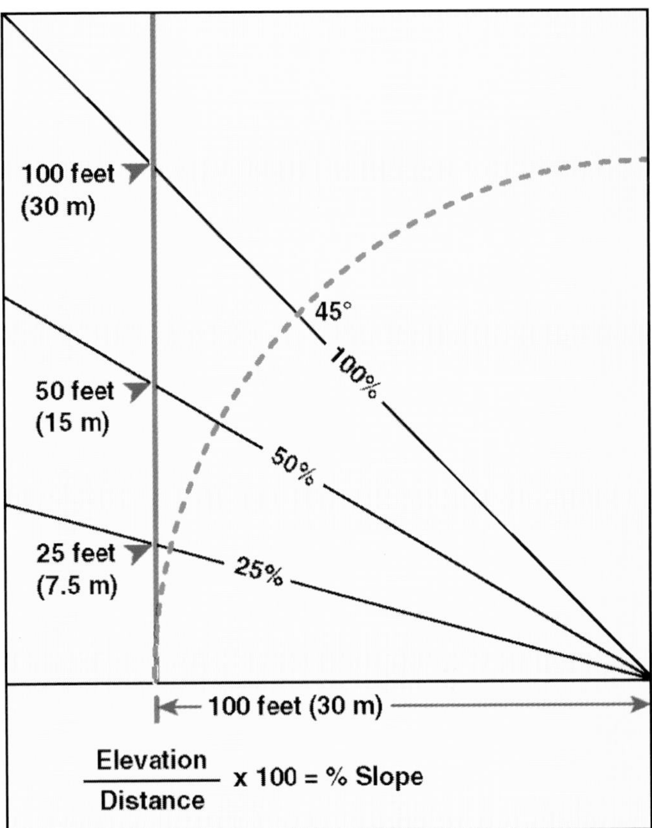

Figure 1.74 How slope percentage is calculated.

Because of these factors, wildland fires tend to burn much faster upslope than on level ground. These same factors work against a fire when it

Figure 1.75 Effect of slope on rate of fire spread.

burns from the top toward the bottom of a slope. Thus, firelines located just beyond a ridge from an advancing fire are often well situated to contain the fire (Figure 1.76). However, one concern about fires burning down steep slopes is the possibility of burning material rolling downhill, which can ignite fuel below the main fire (Figure 1.77). This increases the likelihood of firefighters being caught between fires above them and fires below them, so they need to pay close attention to the fire and have escape routes (pathways to safety) clearly identified (see Chapter 6, Fire Suppression Methods).

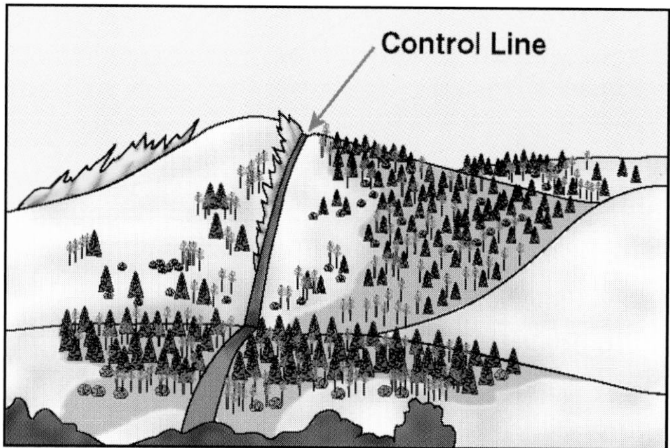

Figure 1.76 A fireline located just beyond a ridge can be very effective.

Figure 1.77 Firefighters must watch for burning material rolling downhill.

Aspect

A slope's *aspect* is the compass direction the slope faces: north, east, south, or west. The aspect of a slope determines the effect of solar heating on the slope's vegetation, air temperature, and moisture retention of the soil. In the Northern Hemisphere, the slopes facing south receive direct sun rays and become hotter than the slopes facing any other direction. Eastern, southeastern, southwestern, and western slope exposures have about equal solar heating as the sun moves across the sky from east to west. The higher temperature on the southern exposures results in lower humidity, rapid loss of fuel and soil moisture, and drier, lighter flashy fuels such as grass. Northern slopes tend to have lower temperature, higher humidity, and heavier fuel growth. The long daily exposure of southern slopes to the sun promotes the development of typical flashy fuels, which makes southern slopes more susceptible to fires than northern slopes (Figure 1.78).

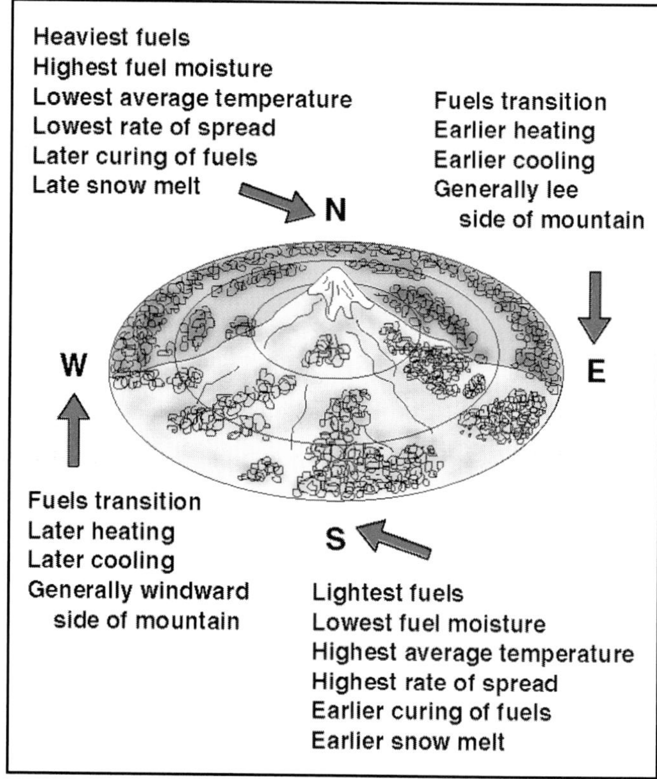

Figure 1.78 Aspect affects fire behavior in several ways.

Radiant heating also affects the moisture content of fuels on these slopes, and the steepness of the slopes affects the amount of solar radiation received. Surfaces perpendicular to solar radiation are subject to considerably more heating than slopes that are more parallel to the sun's rays. The angle at which solar radiation hits various surfaces changes with the time of day and with the time of year. There may be times of the year when steep northern aspects receive no direct solar heating at all.

Chutes and Saddles

A *chute* is a steep V-shaped drainage; a *saddle* is a common name for the depression between two adjacent hilltops (Figure 1.79). These terrain features, also called *gullies* and *arroyos*, can have a direct and dramatic effect on fire behavior. They can cause a fire to spread at drastically varying rates. For example, a fire burning slowly up a wide canyon may suddenly accelerate as it enters a chute or saddle. Chutes and saddles can also alter the flow of surface winds and thereby produce erratic fire behavior. But even in the absence of wind, the topographic effect of these formations can dramatically change a fire's rate and direction of spread by acting as chimneys.

Figure 1.79 A typical saddle between two peaks.

> **WARNING**
>
> **Even seemingly insignificant chutes and saddles, and those concealed by vegetation, have caused firefighter injuries and deaths.**

Wind Channeling

Convected air and superheated fire gases, like any other fluid, take the path of least resistance. This means that heat, smoke, and fire gases channel up through narrow canyons that act like natural chimneys. Deep, narrow canyons burn out rapidly because the radiant heat and fire embers generated by a fire on one side tend to ignite the other. The natural chimney effect increases the wind velocity. These effects combine to increase the rate of combustion.

> **WARNING**
>
> **When working in chutes, saddles, or narrow canyons, you must be especially watchful of the fire's behavior and be prepared to escape to a safety zone at any time.**

Map Reading

A topographic map is very useful in deciding where to locate control lines (barriers used to control a fire) on wildland fires. Topographic contour maps commonly show elevations, slope, and the contour of the land with its peaks, ridges, and drainages. These features are shown by concentric *contour lines* (Figure 1.80).

Each contour line represents a constant elevation on the ground surface so that all points on the same contour line are the same elevation. Contour lines that are close together mean that there is little horizontal distance between two adjacent elevations, so the slope is steep. Conversely, contour lines that are farther apart indicate that the slope is relatively flat. Peaks and hilltops are represented by a series of concentric circles, ovals, or irregular closed loops. *Ridges* are essentially elongated hilltops and may be indicated by long ovals. The end of a ridge is indicated by a series of U-shaped contour lines, and the bottom of the "U" points downslope. Canyons and drainages are indicated by a series of V-shaped lines, with the point of the "V" upslope. Other features shown on topographic maps include survey lines, roads, trails, streams, lakes, villages, towns, and cities.

Some maps are laid out in rectangular blocks with sides parallel to *latitude* and *longitude lines*. Locations, such as a fire's point of origin, are recorded by latitude and longitude. Describing locations by latitude and longitude also may be necessary when communicating with aircraft pilots. Degrees of latitude and longitude are reported in whole numbers and minutes to the nearest tenth.

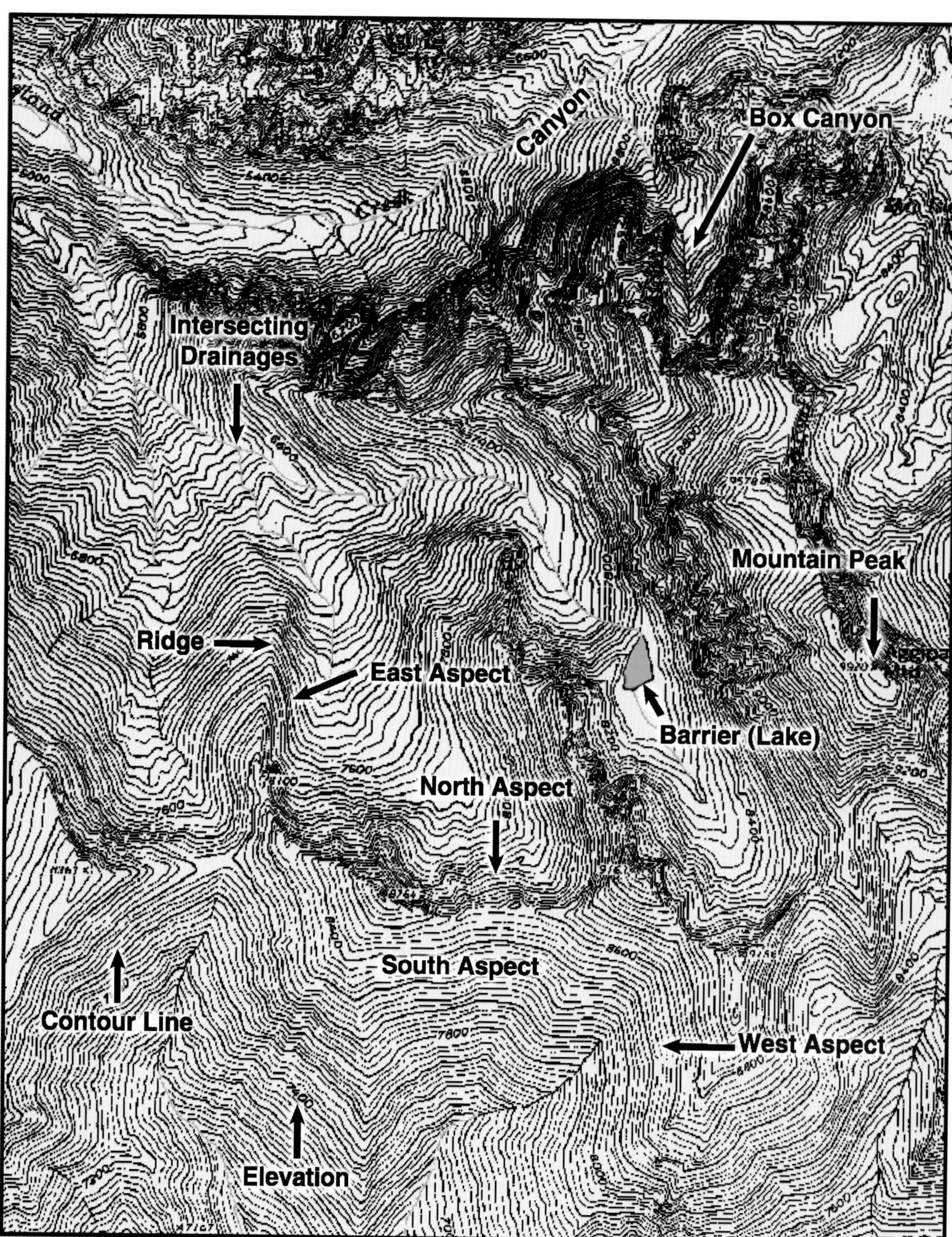

Figure 1.80 A typical topographical map.

Latitude is measured in degrees (0 through 90) north and south of the equator. Lines of latitude are parallel; therefore, the distance between two lines of latitude remains constant. Longitude is measured in degrees (0 through 180) east and west of the *prime meridian,* which runs between the North and South Poles through Greenwich, England. Lines of longitude are not parallel; the closer to the Poles, the smaller the distance between them (Figure 1.81). When specifying a position, latitude is normally given first. For instance, the location of Stillwater, Oklahoma, is identified as 36° 8' north latitude, 97° 4' west longitude.

General land survey maps used throughout most of the United States and Canada are laid out in ranges, townships, and sections. *Ranges* are north/south rows of townships, numbered east/west from the principal meridian of the survey on which the map was based. *Townships* are 6 miles square (93.6 square kilometers [km^2]), following due north/south and east/west lines. Each township is divided into 36 sections of 1 square mile (2.6 km^2) each (Figure 1.82). Each *section* is composed of 640 acres (259 ha) and is subdivided into quarter-sections of 160 acres (65 ha) each. One acre measures about 209 feet (64 m) square or about 70 by 70 yards (64 m by 64 m or about 0.4 ha). In more practical terms, each side of a square acre is about 2/3 the length of an American football field; each side of a square hectare is almost the full length of a Canadian football field (Figure 1.83).

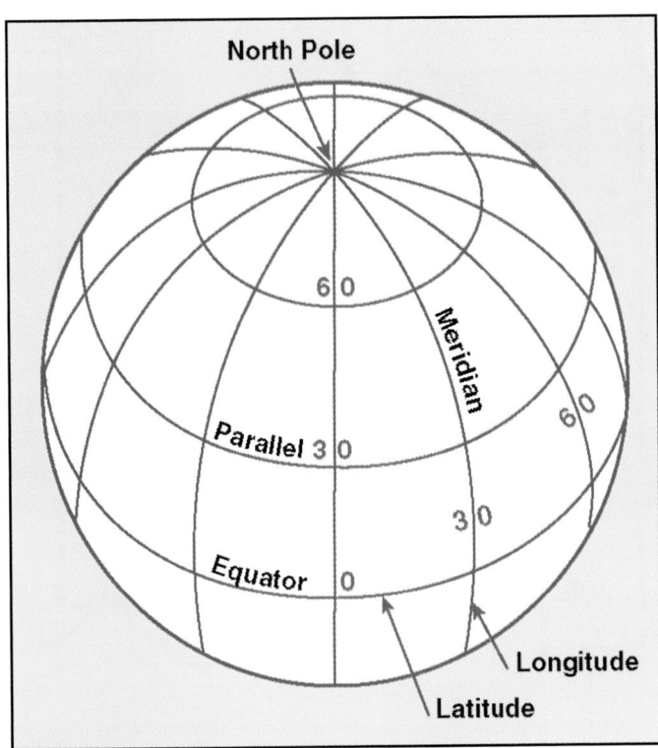

Figure 1.81 Lines of latitude and longitude.

Survey maps of most of the states east of the Mississippi River do not have ranges and townships. These states were mapped with the metes-and-bounds survey system. This system followed natural geographic features such as ridges, streams, and drainages. In actual use, locations may be described as being a certain number of miles (kilometers) in a given direction from some named feature or landmark. When locating fires in these areas, the feature or landmark most often used is a lookout tower.

Compass Reading

Many fires, especially those started by lightning strikes, occur in remote areas. When small, they may only be visible from a lookout tower or an aircraft flying overhead. If these fires are not visible from a road or trail, firefighters may have to use a map and compass to guide them to the fire's location.

Most hand-held compasses used in wildland fire fighting are of the liquid-filled type (Figure 1.84). Because all handheld compasses point to *magnetic north* and all fire-location information is based on *true north*, these compasses must be corrected for magnetic declination (deviation from true north). The *agonic line* (0-degree declination) runs north from a point in far western Florida through Alabama, Tennessee, Kentucky, Illinois, Wisconsin, Minnesota, Ontario, Manitoba, and the Northwest Territories (Figure 1.85). Areas east of the agonic line have a western declination because a compass needle in these areas will point west of true north. For the same reason, areas west of the agonic line have an eastern declination. For example, in California declination varies from about 14 degrees E in the southern part of the state to about 17 degrees E in the northern part. The declination in any given area changes so gradually that it is considered to be constant. Therefore, once the correction for declination has been made on a compass, it need not be done again as long as the compass is used in the same area. The declination

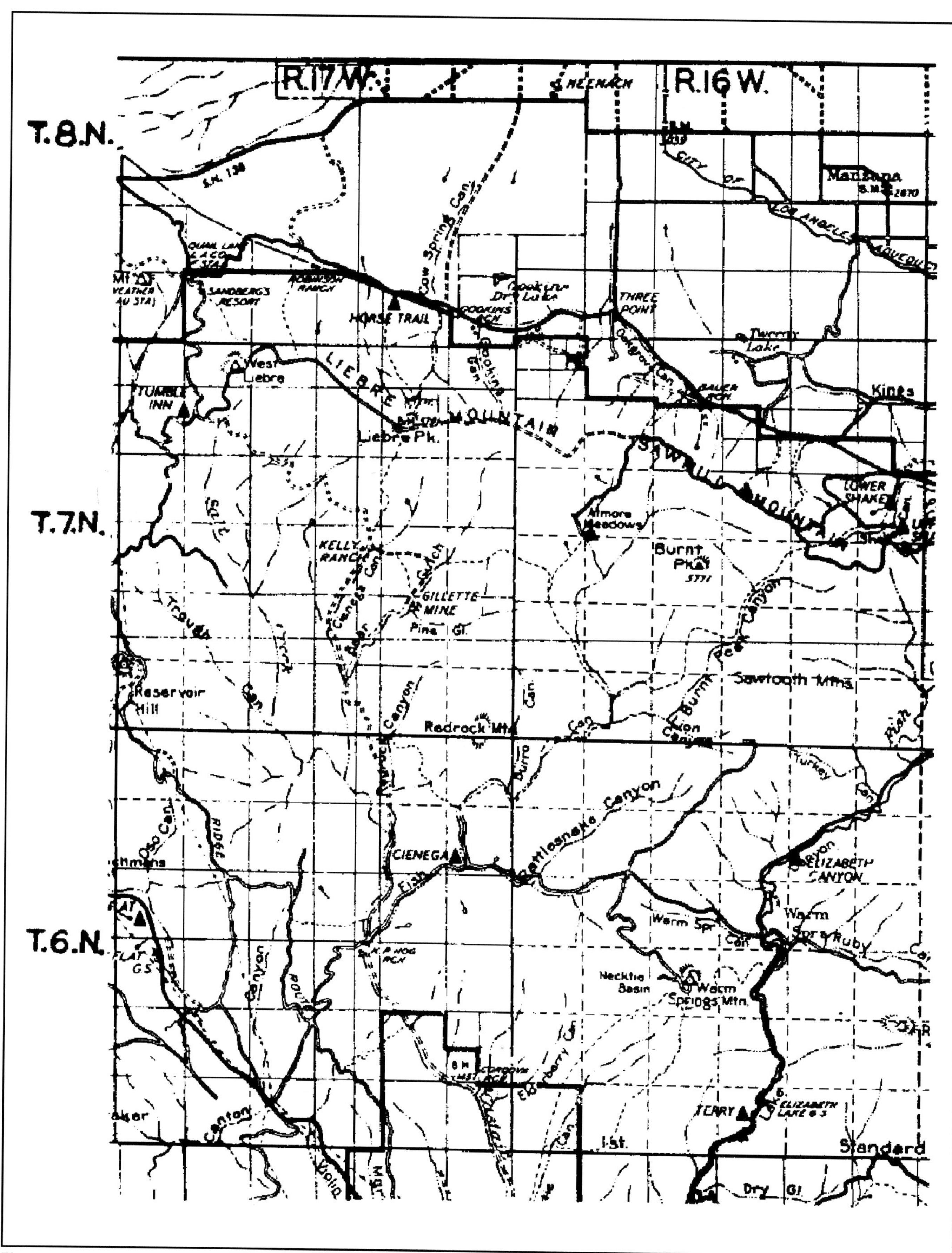

Figure 1.82 A typical survey map.

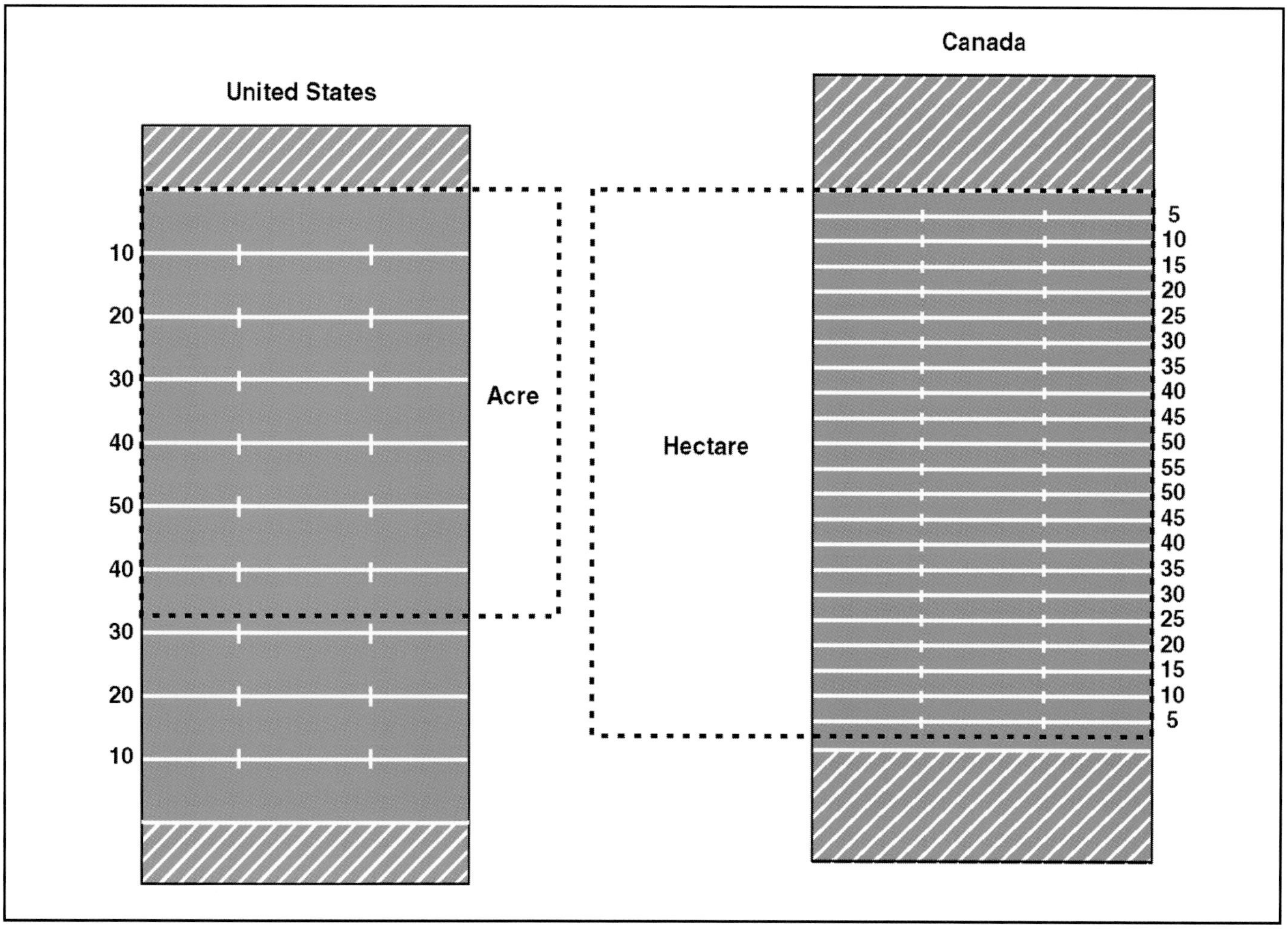

United States

Acre

Canada

Hectare

Figure 1.83 How U.S./Canadian football fields compare to acres/hectares.

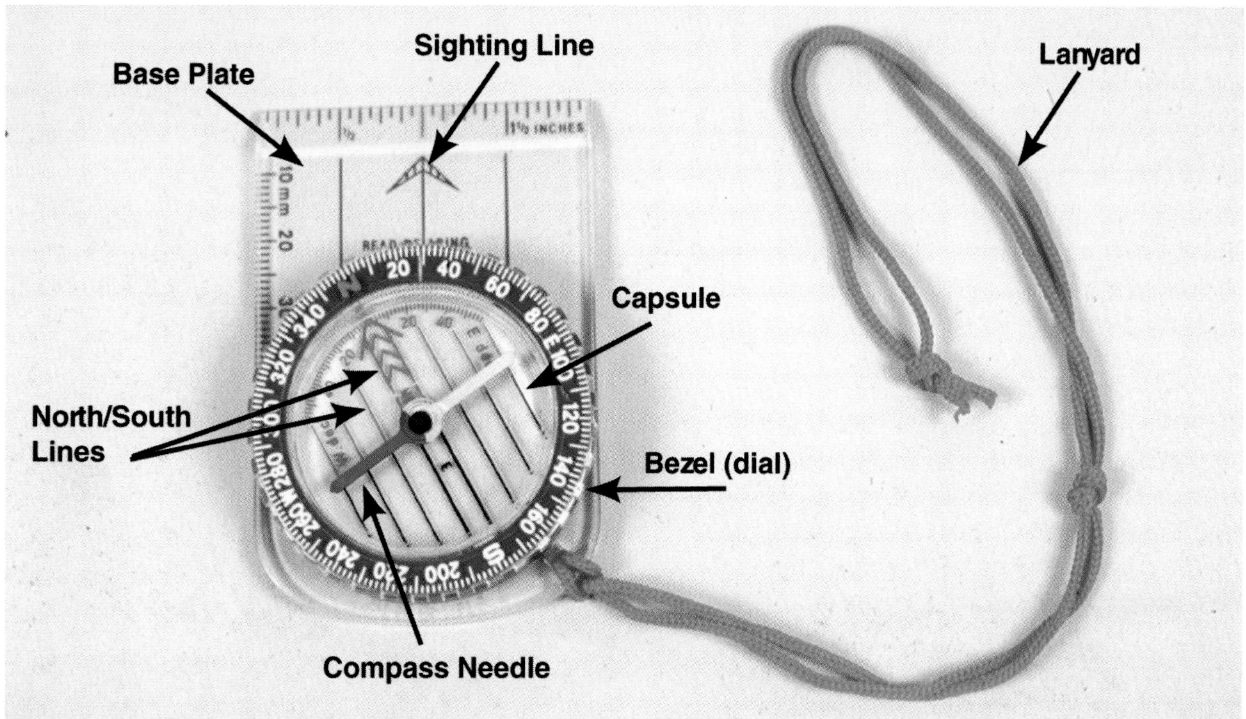

Figure 1.84 The parts of a liquid-filled compass.

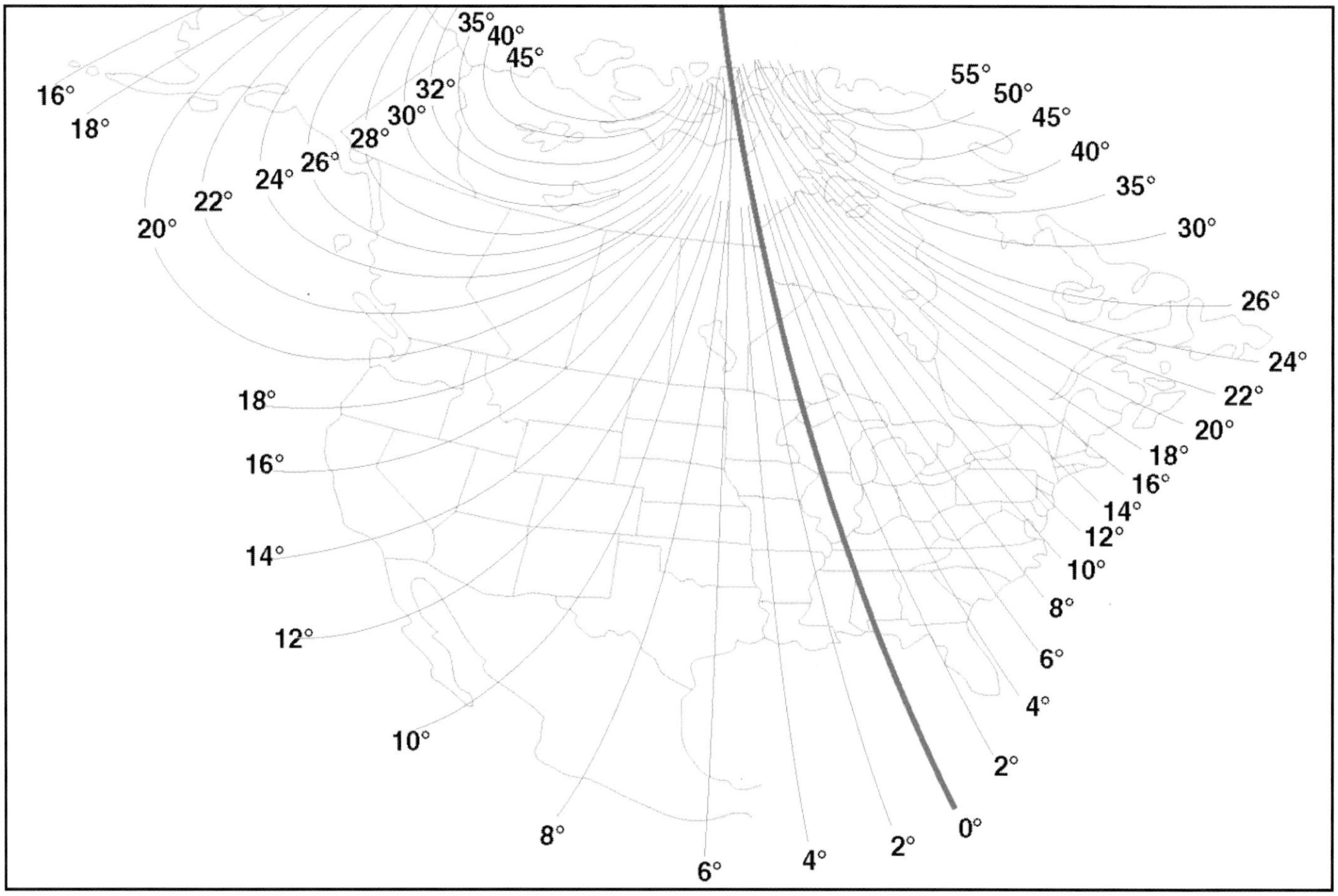

Figure 1.85 An isogonic map of North America.

for any area in North America can be obtained from an isogonic map available from the U.S. Geological Survey (USGS) at Box 25286, Denver, CO 80225.

To use a map and compass to find a fire that they cannot see from their current location, firefighters must be given the direction and distance to the fire. With this information, they can determine the direction they must travel to reach the fire. First, the compass must be placed on the map with the long edge of the base plate on a line drawn from their present location to the reported location of the fire (Figure 1.86). Next, the bezel (dial) must be turned until the north/south lines on the bottom of the capsule are parallel with the magnetic meridian lines on the map (Figure 1.87). Now, with the compass held level, it must be rotated until the north/south lines align with the compass needle (Figure 1.88). The sighting line on the compass base plate should now be pointing toward the fire. By extending their line of sight along this line, firefighters can identify a mountain top, rock formation, tall tree, or other prominent landmark toward which to walk.

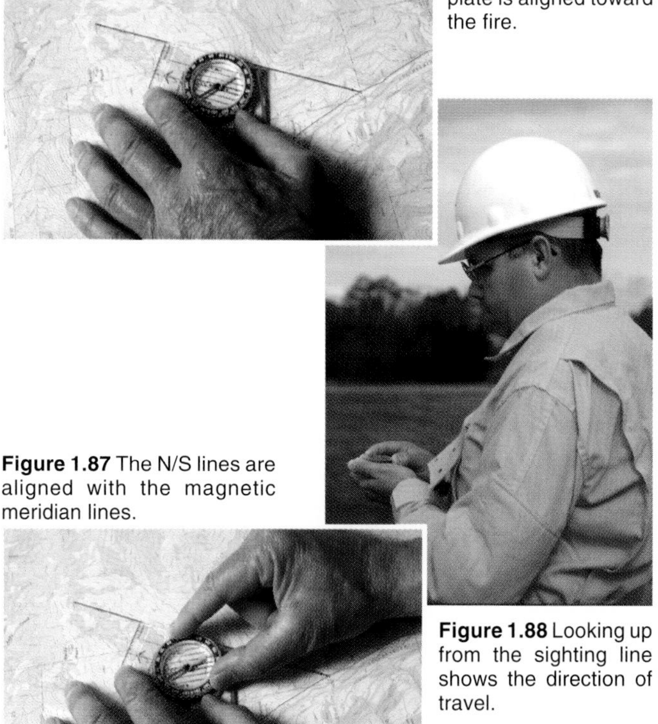

Figure 1.86 The base plate is aligned toward the fire.

Figure 1.87 The N/S lines are aligned with the magnetic meridian lines.

Figure 1.88 Looking up from the sighting line shows the direction of travel.

ADDITIONAL FIRE-BEHAVIOR FACTORS

Fuel, weather, and topography are the major factors affecting wildland fire behavior. Knowledge of these factors helps firefighters predict fire behavior more accurately and helps them control and extinguish fires safely and efficiently. In addition to the factors affecting wildland fire behavior already discussed, other factors should also be considered. Some of these factors are fire size, rate of spread, and intensity (flare-ups, blowups, fire storms, and area ignition). An additional effect on fire behavior can occur when the dead-fuel load has been increased by disease, insect infestation, or storm damage.

Fire Size

A large and intensely burning fire may not react to changing environmental conditions as quickly as a smaller fire might. To a certain extent, a large fire can create its own localized weather. While burning, it forms a convection column and develops winds in response to this convection column. The winds, thus created, preheat and dry fuels in advance of the flame front.

The strong convection column may increase the spread of a fire by starting spot fires ahead of the main fire by picking up sparks and burning embers and carrying them downwind. Spot fires can occur more than a mile (1.6 km) beyond the fire's perimeter, depending on conditions. The convective column may also cause the fire to *crown* (burn available aerial fuels).

However, if there is a sufficient *volume* of fuel available, regardless of the fuel's size, distribution, and compactness, a large and very intense fire can develop. If a forest has been subjected to disease or insect infestation or if it has suffered damage from snow, ice, and/or wind storms, the amount of dead fuel per unit area may be well above normal. Such a heavy fuel load supports a high-intensity fire that can consume vast areas of wildland very quickly.

Rate of Spread

Rate of spread (ROS) is the rate at which a fire is extending its horizontal dimensions. Depending on the intended use of the information, rate of spread is expressed in (1) rate of increase of the total perimeter of the fire or (2) rate of forward spread of the fire front. ROS is usually measured in feet (meters) per minute. When giving a report on conditions (see Chapter 6, Fire Suppression Methods), some of the more common terms used to describe ROS are as follows:

- ***Smoldering*** — Burning without flame and barely spreading
- ***Creeping*** — Burning with a low flame and spreading slowly
- ***Running*** — Spreading rapidly with a well-defined head
- ***Spotting*** — Starting new fires downwind of the main fire by sparks or embers being carried aloft by convection
- ***Crowning*** — Burning through the tops of trees or shrubs more or less independently of the surface fire

Fuel temperature also affects the rate of spread by increasing or decreasing the amount of additional heat required to produce pyrolysis and achieve ignition. Fuels that have been preheated require less additional heat energy to reach their ignition temperatures than fuels that have not. Unburned wildland fuels can collect significant heat energy through heat transfer from fires in close proximity, from solar radiation, or from the surrounding atmosphere. Combined with low relative humidity, fires generally spread faster when fuels are preheated.

As discussed earlier, the ROS can be greatly influenced by the degree of slope on which a fire is burning. When a fire is burning upslope, the steeper the slope, the faster the spread (Figure 1.89). Even when a fire is burning down a steep slope, the rate of spread may still be increased by burning materials rolling down into uninvolved fuels.

The rate of spread is also a function of the type of fuel in which a fire is burning. Some fuels burn faster than others. When a fire burns from one type of fuel bed to another, the ROS is directly affected by the change in fuel type. When a fast-moving fire burns from light, flashy fuels into an area characterized by heavier fuels, the ROS will most likely slow down. However, in the reverse situation, a fire burning from an area of heavy fuels into a clearing characterized by lighter fuels can increase ROS dramatically.

Figure 1.89 Slope can greatly increase the rate of fire spread. *Courtesy of Tony Bacon.*

Another major factor affecting the rate of spread is wind. A fire burning with the wind always spreads faster than one burning in calm conditions or against the wind. The typical upslope winds during the day can contribute to a fire spreading up a mountainside with devastating speed. Especially in V-shaped drainages and in saddles between hilltops, the wind can cause a blowup (sudden increase in fire intensity) that can put firefighters in serious jeopardy (Figure 1.90). Even in relatively flat terrain, the wind can cause the flames to parallel the ground instead of rising vertically (Figure 1.91). This preheats unburned fuels in the path of the fire, causing them to pyrolyze faster.

Although they are somewhat general, the standard guidelines regarding rate of spread are as follows:

- Forward rate of spread doubles as a fire front moves from heavier fuels to lighter fuels (for example, slash →×2 timber →×2 brush →×2 grass) *OR* conversely slows down as it moves to heavier fuels from lighter fuels.

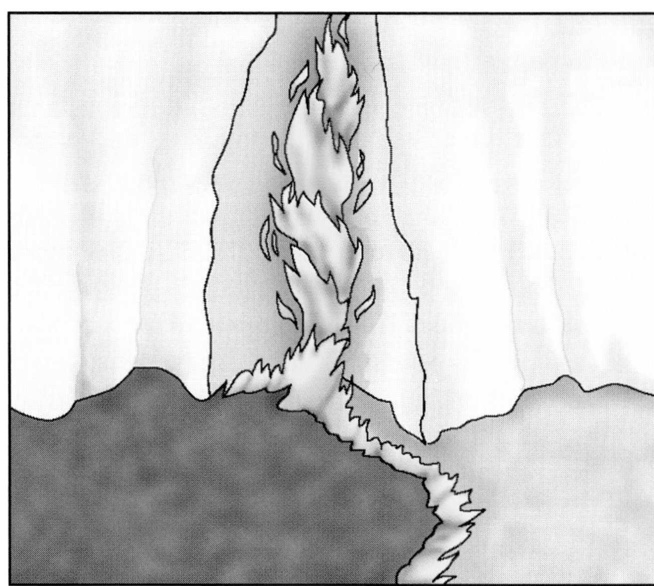

Figure 1.90 A fire may blow up when it reaches a saddle or V-shaped drainage.

Figure 1.91 On flat ground, the wind may cause flames to lay down.

- Rate of spread doubles for every 20 percent slope increase.
- Rate of spread doubles for every additional 10 mph (16 kmph) of wind above 10 mph (16 kmph).

Intensity

Intensity (also called *fireline intensity*) is the amount of heat a fire produces over time and distance. It is usually measured in Btu/second/foot (kilojoules/second/meter). The flame length at the head of a fire indicates the intensity. If the flame is long, the fire is intense. If the flame is short, intensity is low.

There is a direct relationship between fireline intensity, as indicated by flame length, and how difficult it may be to extinguish a fire. Controlling and extinguishing high-intensity fires is very difficult because they develop major convective columns that may cause spotting beyond control lines. Additionally, because these fires produce so much heat (Btu or kJ), significant numbers of resources are needed to extinguish them. The number of resources needed to deal effectively with expected fireline intensities is determined by the *burning index* (BI). The following table (Table 1.3) assigns a BI of from 0 to 110 based on the relationships between flame length, fireline intensity, and control strategies. The higher the BI, the more difficult suppression becomes; so the number of resources dispatched to any reported wildland fires should be increased accordingly.

Another useful way of looking at the relationship between flame length and control strategies is provided by Table 1.4. This table identifies the fire suppression limitations imposed by fires of various intensities.

FLARE-UPS

A *flare-up* is a sudden increase in fire intensity or acceleration in rate of spread. However, unlike a blowup, a flare-up is of relatively short duration and does not radically change existing control plans.

BLOWUPS

A *blowup* is a sudden increase in fire intensity or rate of spread sufficient to preclude a direct attack (attacking the fire at its edge). A blowup may force existing control plans to be changed. It may produce a violent convection column and other characteristics of a fire storm. Indicators of a possible blowup would be when a fire continues to rapidly grow in area or intensity despite fire-control efforts that might normally be expected to control it. A blowup is sometimes considered to be the wildland equivalent of a flashover in a structure fire.

TABLE 1.3
Burning Index/Flame Length/Intensity

Burning Index	Flame Length (ft or m)	Intensity (Btu/sec/ft or kJ/sec/m)	Comments
0–30	0–3 feet (0 to 1 m)	0–55 Btu/sec/ft (0 to 58 kJ/sec/m)	Most prescribed burns are conducted in this range.
30–40	3–4 feet (1 m to 1.2 m)	55–110 Btu/sec/ft (58 kJ/sec/m to 64 kJ/sec/m)	Intensity generally represents the limit of control for direct manual attack methods.
40–60	4–6 feet (1.2 m to 1.8 m)	110–280 Btu/sec/ft (64 kJ/sec/m to 308 kJ/sec/m)	Machine methods are usually necessary or indirect attack should be used.
60–80	6–8 feet (1.8 m to 2.4 m)	280–520 Btu/sec/ft (308 kJ/sec/m to 551 kJ/sec/m)	Prospects for direct control by any means are poor above this intensity.
80–90	8–9 feet (2.4 m to 2.7 m)	520–670 Btu/sec/ft (551 kJ/sec/m to 861 kJ/sec/m)	Heat load on people within 30 feet (9 m) of the fire is dangerous.
90–110	9+ feet (2.7+ m)	670–1,050 Btu/sec/ft (861 kJ/sec/m to 1 113 kJ/sec/m)	Above this intensity, spotting, firewhirls, and crowning should be expected.

Source: National Fire Danger Rating System (1978).

TABLE 1.4
Fire Suppression Limitations

Flame Length (ft or m)	Control Strategy
<4 feet (<1.2 m)	Can be attacked at the head or flanks by crews using hand tools. Handlines should hold the fire.
4–8 feet (1.2 m to 2.4 m)	Too intense for direct attack on the head by hand crews. Heavy equipment and/or aircraft needed. Fire is potentially dangerous to personnel.
8–11 feet (2.4 m to 3.4 m)	May present serious control problems (torching, crowning, spotting). Attack on the head will probably be ineffective.
>11 feet (>3.4 m)	Crowning, spotting, and major fire runs are probable. Attacking the head is likely to be ineffective and dangerous. Indirect attack may be the best alternative.

NOTE: *Torching* is a vertical phenomenon in which a surface fire ignites the foliage of a tree or bush. *Indirect attack* is a method of suppression in which a control line is located at natural or man-made barriers some distance from the fire's edge and the intervening fuel is burned out.

Source: NWCG S-290 Intermediate Wildland Fire Behavior.

FIRE STORMS

A fire storm is more than just a very large fire; it has certain characteristics that make it unique. It is characterized by a huge convection column with destructively violent indrafts near and beyond the perimeter caused by a large continuous area of intense fire. These indrafts often cause tornado-like firewhirls, uproot vegetation, and throw rocks and debris through the air. A sustained high-intensity fire with a growing convention column is an indicator of a possible fire storm.

AREA IGNITION

Area ignition may cause a blowup and/or contribute to the development of a fire storm. Area ignition occurs when a number of spot fires simultaneously or nearly simultaneously begin to influence each other to the point that they ignite the intervening fuels to become a single fire over a large area. An indication of possible area ignition would be several growing spot fires in close proximity to each other. Area ignition may also be thought of as the equivalent of a flashover on a massive scale.

FIRE-BEHAVIOR COMMON DENOMINATORS

The safety of all personnel, both firefighters and civilians, is always the primary consideration in any fireground decision. A thorough knowledge of wildland fire behavior can go a long way toward maximizing safety and minimizing risk. The Common Denominators of Fire Behavior on Tragedy Fires identified by the U.S. Forest Service (USFS) are critical for firefighters in the wildland to know. Also see Chapter 8, Firefighter Safety and Survival for more information.

Common Denominators of Fire Behavior on Tragedy Fires

- **Most incidents happen on the smaller fires or on isolated portions of larger fires.**
- **Most fires are innocent in appearance before the "flare-ups" or "blowups." In some cases, tragedies occur in the mop-up stage.**
- **Flare-ups generally occur in deceptively light fuels.**
- **Fires run uphill surprisingly fast in chimneys, gullies, and on steep slopes.**
- **Some suppression tools, such as helicopters or air tankers, can adversely affect fire behavior. The blasts of air from low-flying helicopters and air tankers have been known to cause flare-ups.**

2 FIRE APPARATUS AND COMMUNICATIONS EQUIPMENT FOR WILDLAND FIRES

LEARNING OBJECTIVES

This chapter provides information that addresses the following objectives of NFPA 1051, *Standard for Wildland Fire Fighter Professional Qualifications* (1995 edition):

Wildland Fire Fighter I

3-5.1 **Definition of Duty.** All activities to confine and extinguish a wildland fire, beginning with dispatch.

3-5.2 Assemble and prepare for response, given an assembly location, an assignment, incident location, mode of transportation, and the time requirements, so that arrival at the incident with the required personnel and equipment meets agency guidelines.

3-5.2.1* *Prerequisite Knowledge:* Equipment requirements, agency time standards and special transportation considerations (weight limitations), agency safety, and operational procedures for various transportation modes.

3-5.4.2* *Prerequisite Skills:* Proper use of hand tools, fire stream practices, and agent application.

Wildland Fire Fighter II

4-1.1.1 *Prerequisite Knowledge:* The Wildland Fire Fighter II role within the incident management system, basic map reading and compass use, radio procedures, and record keeping.

4-1.1.2 *Prerequisite Skills:* Orienteering and radio use.

4-3.1 **Definition of Duty.** Responsibilities in advance of fire occurrence to ensure that tools, equipment, and supplies are fire ready.

4-3.2* Maintain power tools and portable pumps, given agency maintenance specifications, supplies, and small tools, so that equipment is safely maintained, serviceable, and defects are recognized and repaired.

4.3.2.1 *Prerequisite Knowledge:* Maintenance procedures for power tools and portable pumps.

4-3.2.2 *Prerequisite Skills:* Power tool and portable pump preventative maintenance and repair.

4-3.3 Inspect tools and equipment, given agency specifications, so that availability of the tools and equipment for fire use is ensured.

4-3.3.1* *Prerequisite Knowledge:* Tool and equipment inspection guidelines.

4-5.2 Select fireline construction methods, given a wildland fire and line construction standards, so that the technique used is appropriate to the conditions and meets agency standards.

4-5.2.1 *Prerequisite Knowledge:* Resource capabilities and limitations, fireline construction methods, and agency standards.

4.5.5 Operate portable water delivery equipment, given an assignment at a wildland fire and operational standards, so that the proper portable pump and associated equipment is selected, desired nozzle pressure is attained, and flow is maintained.

4.5.5.1* *Prerequisite Knowledge:* Basic hydraulics, portable pump and system capabilities, operation of portable pumps, basic drafting, and associated equipment.

4.5.5.2 *Prerequisite Skills:* Placement, operation, and system set up.

Chapter 2
Fire Apparatus and Communications Equipment for Wildland Fires

INTRODUCTION

Fighting a major fire of any kind would be almost impossible without modern fire fighting apparatus and communications equipment. This is certainly true of wildland fires. Modern fire apparatus and communications equipment help firefighters control wildland fires quickly, effectively, and safely — thus reducing loss to life, property, and valuable natural resources.

In this manual, *fire apparatus* refers to those vehicles (including aircraft) used for fire suppression and for the transportation of firefighters and their equipment. Equipment includes the tools carried on the apparatus and the personal equipment of firefighters in the wildland (see Chapter 3, Wildland Fire Fighting Tools and Personal Protective Equipment, for more information on fire fighting equipment). Even though a piece of fire apparatus is able to carry and pump water and is equipped with hose and assorted tools, it is not necessarily effective on *all* types of fires. Sometimes a specially designed piece of apparatus is needed to most effectively perform specific functions. This is true of airport fire apparatus and is no less true of apparatus designed to fight wildland fires. This chapter focuses on apparatus designed especially for the suppression of wildland fires and on other resources used when fighting wildland fires.

Included in these categories are various types of mobile fire apparatus, aircraft, and a wide array of accompanying communications gear. Also included are the highly specialized pieces of mechanized equipment used for wildland fire suppression such as bulldozers and other heavy tractors. The last portion of the chapter deals with the array of communications equipment that may be used during the course of a wildland fire fighting operation.

STRUCTURAL FIRE ENGINES

The main purpose of structural fire engines is to provide adequate fire streams for attacking structure fires and other types of fires commonly encountered by municipal fire departments (Figure 2.1). Most municipal engines are built to comply with National Fire Protection Association (NFPA) Standard 1901, *Standard for Automotive Fire Apparatus.* NFPA 1901 contains an extensive list of minimum requirements for structural fire engines. Some of the more important requirements include the following:

- The minimum fire pump volume capacity is 750 gallons per minute or gpm (3 000 L/min). Pump sizes up to 2,000 gpm (8 000 L/min) are common in today's fire service.
- The minimum water tank size is 500 gallons (2 000 L). Tank sizes up to 1,000 gallons (4 000 L) are common. Apparatus with tanks

Figure 2.1 A typical structural fire engine. *Courtesy of John Hawkins.*

larger than 1,000 gallons (4 000 L) are considered mobile water supply apparatus and are covered later in this chapter.

- The minimum sizes and numbers of fire pump intake and discharge connections are specified in the standard.

Factors that limit the usefulness of most structural fire engines for wildland fire attacks include the following:

- Lack of pump-and-roll capabilities (ability to move while pumping water)
- Lack of four-wheel or all-wheel drive
- Inadequate approach and departure angles (front and rear overhangs too long)
- Limited tank capacity
- Limited ability to travel on rough terrain

While structural fire engines are generally designed for use in developed areas with paved or maintained gravel streets and readily available water supplies, they can also be effective on wildland fires, especially in the wildland/urban interface. Even though these units have the limitations mentioned earlier, they also have positive functional capabilities such as a greater pumping capacity than most dedicated wildland units. There are two primary situations that lend themselves to the use of structural apparatus in wildland fires: roadside fires and structure protection in the wildland/urban interface.

Structural Apparatus Activities on Wildland Fires

Structural fire engines can be very effective on those many fires that burn at the roadside or on a highway median strip (Figure 2.2). They can be used in a quick attack role on small nuisance fires by pumping from their tanks and using booster lines or smaller handlines.

From the relative safety of a paved street or road, structural engines can also be very effective on large roadside fires by being the anchor point for *progressive hose lays*. A progressive hose lay consists of extending hose from the apparatus to the fire's edge, extinguishing fire as the hose is extended, connecting another section, advancing, and extinguishing more fire (Figure 2.3). This procedure is limited only by the friction loss in the hose and the amount of water available to the engine. However, once a reliable water supply has been established (for example, water tender shuttle), an engine can supply a large volume of water for one or more progressive hose lays at considerable distances from a road or street. Given a relatively short response distance, an adequate water supply, and trained personnel, a structural engine used in this way can keep relatively small roadside fires from developing into major wildland fires.

One of the most effective and efficient uses of structural fire engines in wildland fires is for the protection of structures exposed to the wildfire. When structural triage (sorting) identifies a structure or group of structures as being defensible (tenable), an engine company can be assigned to protect the structure(s) from an approaching wildland fire. Procedures for accomplishing this de-

Figure 2.2 Structural fire engines are often used on roadside fires.

Figure 2.3 The crew of a structural engine begins a progressive hose lay.

fense are covered in detail in Chapter 7, Wildland/Urban Interface Fire Suppression.

Structural fire engines with open cabs are unsafe for use on wildland fires, including those in the wildland/urban interface. In addition to the lack of roll bars or cages, these units are vulnerable to being set afire from burning embers dropping into the open cab area.

Initial Attack Fire Apparatus

The initial attack fire apparatus are known by many different names throughout the fire service, including the following:

- Mini-pumper
- Midi-pumper
- Quick attack apparatus
- Booster apparatus
- Rapid intervention vehicle
- Attack pumper
- Quick response vehicle

An initial attack fire apparatus is basically a scaled-down version of a structural fire engine (Figure 2.4). The theory behind its design is that it is quicker and more maneuverable than the larger engines. Their maneuverability allows them to respond faster and attack fires at an earlier stage. NFPA 1901 contains requirements for the design of these vehicles.

An initial attack fire apparatus often consists of a water tank and pump mounted on a pickup truck chassis or on a small, medium-duty commercial truck chassis. These units often have greater off-road operating capability than conventional commercial vehicles. The fire pump must have a capacity of at least 250 gpm (1 000 L/min). Typically, it carries 100 to 300 gallons (400 L to 1 200 L) of water. Much of the same equipment carried on a larger engine is also carried on an initial attack fire apparatus, although in smaller numbers. Some initial attack fire apparatus are equipped with a turret gun that allows a master stream to operate from a location that would not accommodate a larger engine. Depending on local requirements, the initial attack fire apparatus may also be equipped with emergency medical equipment, extrication equipment, and wildland fire fighting equipment.

Figure 2.4 A traditional off-road engine used for initial attack. *Courtesy of NIFC.*

An initial attack apparatus operates with a crew of three to five people. Current standards require that new fire apparatus with a crew of more than three people must have a four-door cab to safely carry all crew members (Figure 2.5). However, older apparatus not meeting current standards may still be used.

Figure 2.5 An off-road engine with a four-door cab.

WILDLAND FIRE APPARATUS

All fire departments should have fire apparatus designed to fight the types of fires they confront most frequently, but few, if any, jurisdictions have no wildland fire potential. Even those fire departments with little or no undeveloped land within their boundaries may still fight wildland fires as a mutual aid resource. To accommodate these functions, dedicated wildland fire fighting apparatus may be needed.

Wildland fire fighting often requires a rugged, highly maneuverable vehicle that can go where large structural-type apparatus cannot (Figure 2.6). Because wildland fires often move rapidly, wildland fire apparatus must also have the ability to pump water while the vehicle is moving (commonly called *pump and roll*). This capability can be advantageous in mop-up (completing extinguishment) along the edge in grass fires. The exact design and size of wildland fire apparatus vary widely from region to region (Figures 2.7 a and b). These variations are necessary because of the differences in terrain and wildland fuels found in different regions. However, all wildland fire apparatus should meet the requirements in NFPA 1906, *Standard for Wildland Fire Apparatus*. The following sections discuss the design and construction of wildland fire apparatus.

Resource Typing

There are a number of different systems in use for classifying resources. All of these systems classify various kinds of resources into general categories (engines, bulldozers, water tenders, air tankers, etc.). However, the minimum standards that each system uses to define the various resource types within each category vary. These resource classification systems allow incident commanders (ICs) at wildland fires to know what type(s) of resources to request in order to achieve specific fireground goals and objectives. While there is more agreement than disagreement between the various systems, the differences are significant in some cases. Firefighters must know which classification system is used in their jurisdiction.

Regardless of which system is used, the various kinds of resources are classified by number according to the minimum standards established for each type. The lower the number assigned to a resource, the higher its capabilities. For instance, Type-1 engines are typical municipal fire engines, designed primarily for fighting structure fires. They must have a crew of four, a 1,000-gpm (4 000 L/min) pump, and 1,200 feet (366 m) of 2½-inch (65 mm) hose, etc. (Figure 2.8). Type-2 engines have a crew of three, a 500-gpm (2 000 L/min)

Figure 2.6 A modern off-road engine.

Figure 2.7a An off-road engine called a "brush breaker." *Courtesy of New Jersey Forest Fire Service.*

Figure 2.7b An unconventional off-road unit. *Courtesy of NIFC.*

Figure 2.8 A Type-1 engine with a crew of four.

pump, and 1,000 feet (305 m) of 2½-inch (65 mm) hose. All resource typing deals with *minimums*. Many engines of all types respond with more than the minimum requirements in personnel, hose, pump capacity, and tank size.

The same relationship between the number assigned to a resource and its capability holds true for all kinds of resources in all systems. Type-1 bulldozers are larger and more capable than Type-2s, which are larger and more capable than Type-3s, etc. The minimum staffing requirements (crew of four, etc.) refers to the number of personnel who must be maintained on a 24-hour basis.

There are no minimum requirements for designating apparatus equipped with Class A foam capabilities. The incident commander has to specifically request apparatus with foam capabilities if they are desired at the fire scene.

In general, Type-1 and Type-2 engines are used as structural engines because of their greater pump capacity and hose complement. Type-3 and Type-4 engines are most commonly referred to as brush or wildland engines because of their lightweight chassis, maneuverability, and smaller diameter hose. In the eastern U.S., these units are called *power wagons*. Even though Type-5, Type-6, and Type-7 engines have at least some fire fighting capability, they are rarely used in the western U.S. where they are normally referred to as *patrols*.

In some cases, Type-1 and Type-2 engines that were primarily designed for fighting structure fires may be adapted for fighting wildland fires. For example, an auxiliary pump and motor combination may be installed in the apparatus to give it pump-and-roll capability. In this modification, the intake of the auxiliary pump is connected to the water supply tank of the apparatus, and the discharge supplies the attack lines. These auxiliary units are also installed on other chassis, some with off-road capability.

Apparatus Fire Pumps

Apparatus fire pumps are necessary to provide the pressures needed to supply attack lines from the water tank mounted on the apparatus, from hydrants, or from auxiliary sources such as lakes, streams, ponds, and rivers. All fire pumps in use today fall into one of two categories: centrifugal or positive displacement.

There are many types of mounting arrangements and pump drive systems available on fire engines. When specifications are being drawn, the most important consideration is the use to which the engine will be put most often. Each arrangement has certain characteristics that make it more or less suitable for a particular fire department's needs.

AUXILIARY ENGINE-DRIVEN PUMPS

Auxiliary engine-driven pumps are powered by a separate auxiliary engine so they work independently of the vehicle's engine (Figure 2.9). This arrangement is very common on wildland fire apparatus because it offers the maximum amount of flexibility. With a separate engine, the pump can be mounted anywhere it fits on the apparatus. The pump speed is independent of the apparatus speed, which makes it ideal for pump-and-roll operations.

The auxiliary pump engine may be either gasoline- or diesel-fueled and may have its own fuel tank or be supplied from the apparatus fuel tank. Most of these pumps have capacities of less than 500 gpm (2 000 L/min) and are powered by engines that range from 15 to 50 horsepower (hp). Mobile water supply apparatus may have auxiliary engine-driven pumps with capacities up to 1,250 gpm (5 000 L/min).

Auxiliary-powered pumps have certain advantages and disadvantages. The advantages offered by these pumps are as follows:

Figure 2.9 An engine with an auxiliary engine-driven pump.

- Pumps are well-suited for pump-and-roll operations.
- Vehicle can be driven at road speed while still maintaining a charged line.
- Hoseline can be supplied if the vehicle's engine stalls.

However, auxiliary-powered pumps also have certain disadvantages:

- A separate fuel supply is required or the fuel available to the vehicle's engine is reduced.
- Communication may be difficult because of auxiliary engine noise.
- Additional engine maintenance is required.

POWER TAKE-OFF (PTO) PUMPS

In this arrangement, the pump is driven by a transmission-mounted power take-off unit (Figure 2.10). The pump gear case must not be mounted too far below the chassis so that it could be damaged when the unit is driven off the road. A skid plate should be installed to protect the pump if it extends below the chassis.

The shaft that delivers power to the PTO pump is independent of the gear in which the road transmission is operating, but it is engaged/disengaged by the clutch. When the driver/operator disengages the clutch to stop momentarily or to change gears, the pump also stops turning. The PTO pump permits pump-and-roll operation, but it is not as effective as the auxiliary engine type.

Because the pressure being developed is determined by the speed of the engine, the pressure developed by the PTO pump changes when the driver/operator changes the vehicle speed. If the unit is designed for pump-and-roll operation, a pressure gauge should be mounted inside the cab in view of the driver/operator. If the terrain over which the vehicle is being driven permits, vehicle speed can be varied according to the pressure gauge instead of the speedometer in a pump-and-roll operation.

Figure 2.10 A typical PTO unit protected by a skid plate.

Some advantages of PTO-driven pumps are as follows:

- A relatively high capacity is available; they are excellent for large-volume, stationary pumping.
- Pumps are easily engaged and operated.
- Little maintenance is required.

However, PTO-driven pumps generally have four distinct disadvantages:

- Apparatus and crew may be put in jeopardy because pumps will not function if the vehicle's engine stalls.
- Adequate volume and pressure may not be produced when the apparatus is driven slowly because pump speed is directly related to vehicle engine speed.
- Pumps mounted with insufficient ground clearance may hit rocks, stumps, and other objects.
- Extraordinary driver/operator skill may be required to control the vehicle over rough terrain while maintaining effective pump pressure.

Many PTO-driven pumps are located on the front bumper of the apparatus (commonly referred to as front-mount pumps) (Figure 2.11). Many rural departments prefer front-mount pumps because of their ease in positioning at water supply sites. The principles described earlier for other pump-and-roll operations also apply to front-mount pumps.

Front-mount pumps do have some disadvantages in addition to those listed for other PTO-driven pumps. These include the following:

- Because they are completely exposed to the atmosphere, the pump and gauges are particularly susceptible to freezing.

Figure 2.11 A front-mounted PTO pump. *Courtesy of NIFC.*

- Their vulnerable position on the front of the apparatus makes them susceptible to damage in frontal collisions.
- The pump operator works very close to the discharges and may be injured in the event of a hose failure.
- If the apparatus transmission accidentally slips into drive gear, the pump operator is in a vulnerable position in front of the unit.

MIDSHIP TRANSFER DRIVE

By far the majority of Type-1 and Type-2 engines and quints (apparatus equipped with fire pump, water tank, ground ladders, hose bed, and aerial device) have main pumps powered by a midship transfer drive system. In this arrangement, power is supplied to the pump through a transfer case between the transmission and the rear axle (Figure 2.12). By shifting into pump gear, power is diverted from the rear axle to the fire pump through a series of gears or a drive chain. The rated capacities of most midship-drive fire pumps on fire apparatus range from 1,000 to 2,000 gpm (4 000 L/min to 8 000 L/min). However, pumps with capacities up to 6,000 gpm (24 000 L/min) are available.

The advantages of a midship-drive pump are as follows:

- They are available in larger capacities than auxiliary engine-driven and PTO-driven pumps.
- The pump controls may be located on either the side or top of the apparatus.

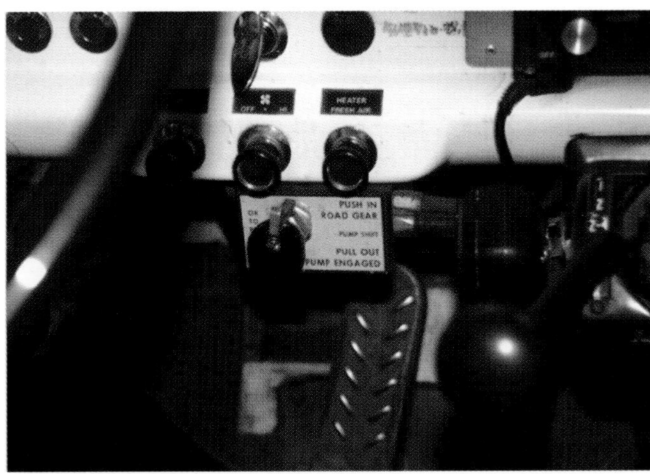

Figure 2.12 A typical transfer case control in the cab of an engine.

- There is only one engine on the apparatus to maintain.

The disadvantages of a midship-drive pump are as follows:

- They are usually not practical for pump-and-roll operations.
- They are located within the body of the engine and therefore are not as easy to service as front-mount or auxiliary engine-driven pumps.
- Plastic air actuation lines are susceptible to melting in high wildland fire temperatures, which may prevent shifting between pump and road gears.

Pump Intake and Discharge Capabilities

Apparatus that are used for wildland fire fighting operations have some specific requirements relative to their ability to intake and discharge water. The information contained in the following sections highlights these special needs.

DRAFTING CAPABILITIES

The ability to draw water from an unpressurized auxiliary (static) water supply source, such as a pond, lake, or swimming pool, into the fire pump is particularly important for wildland fire apparatus. Unlike apparatus used for fighting structure fires, wildland fire apparatus usually do not operate from pressurized water supply sources such as fire hydrants. Therefore, in order to refill the apparatus water tank or to operate fire streams for an extended duration, the apparatus often needs to draw water from an unpressurized source. This process is called *drafting*.

Drafting is done by connecting one or more sections of hard (noncollapsible) suction hose to the intake of the fire pump. Lightweight hard suction hose is most often used in wildland drafting operations (Figure 2.13). A strainer is attached to the end of the suction hose, and it is placed into the water. The strainer keeps debris from entering the pump. The strainer may be either the submersible or the floating type (Figure 2.14).

Most modern fire apparatus are equipped with centrifugal fire pumps. These pumps are incapable of drawing (drafting) water from an unpressurized source until they are primed (filled with water). However, once primed, they are capable of developing pressure and continuing the flow of water through the system. In order to draw the initial water into the centrifugal pump, a priming device is required. Nearly all modern fire apparatus use a small positive-displacement pump for this purpose.

Figure 2.13 Lightweight suction hose.

Figure 2.14 A typical suction strainer.

When the priming device is operated, it exhausts most of the air from inside the hard suction hose and the main pump, creating a partial vacuum. Because the pressure within the hose and pump is then lower than the atmospheric pressure, water from the static source is forced up through the suction hose into the pump. Once water fills the main pump, it is primed and capable of continuing the process without the aid of the primer. Most engines can pump their rated volume capacity when the pump intake is within 10 vertical feet (3 m) of the surface of the water. Most engines can only draft water from about 20 feet (6 m) below the level of the pump and only at about 60 percent of the rated capacity.

Apparatus with midship pumps are equipped with a suction intake on each side of the engine (Figure 2.15). Having an additional suction intake plumbed to the front and/or rear allows for maximum flexibility when positioning at a water supply source (Figure 2.16). Apparatus with front-mount pumps have one suction intake on the pump at the front of the engine (Figure 2.17).

PUMP-AND-ROLL CAPABILITIES

The ability to pump and roll is particularly useful for wildland fire apparatus. The speed with which wildland fires spread frequently requires firefighters to attack while on the move. Pump-and-roll operations are very effective when making a direct attack on low-intensity wildland fires in

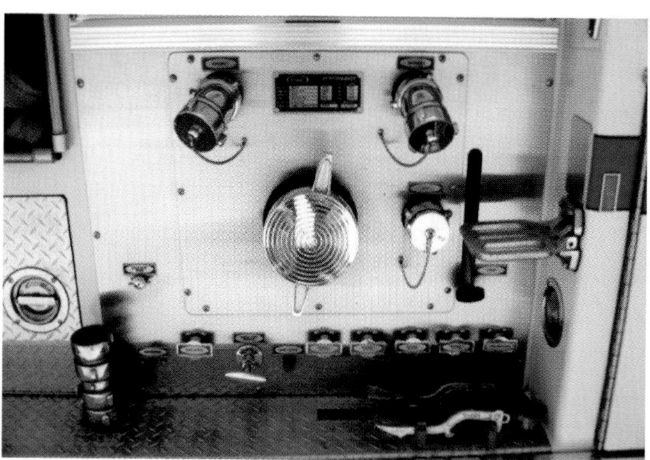

Figure 2.15 The main suction intake of a midship pump.

Figure 2.16 Some engines have front suction intakes.

Figure 2.17 Engines with front-mount pumps only have front suction intakes.

terrain that is suitable for the apparatus. Apparatus equipped for pump and roll with Class A foam are also well suited for laying a protective foam blanket over fuel and structures exposed to a wildland fire.

As described earlier, most pump-and-roll capable apparatus use either a power take-off or an auxiliary engine pump. Water is discharged during pump-and-roll operations by any one or more of the following means:

- Booster reel
- Short section of booster hose
- Short section of 1½-inch (38 mm) hose
- Ground sweep nozzles
- Remote control nozzles
- Master stream appliances

Booster reel lines are sometimes used for pump-and-roll operations because the length of the hose off the apparatus can be adjusted easily (Figure 2.18). Booster hose is also very durable — a necessity for wildland operations where the hose is subjected to much abuse. Booster hose should be at least 1 inch (25 mm) in diameter, and the hose may be either the low- or high-pressure type, depending on the preference of the individual agency.

Some departments prefer to equip the apparatus with short sections of booster hose connected directly to a small discharge on the apparatus (Figure 2.19). These discharges may be located on the front, rear, or sides of the apparatus. These lengths of hose range anywhere from 5 to 20 feet (1.5 m to 6 m). These short sections of hose allow firefighters to easily apply water or foam while walking alongside a slowly moving apparatus.

However, in flashy fuels with high heat output or long flame lengths, 1½-inch (38 mm) hose must

Figure 2.18 In light fuels, booster lines can be used for pump-and-roll operations. *Courtesy of NIFC.*

Figure 2.19 A short, preconnected booster line is handy for small nuisance fires. *Courtesy of NIFC.*

be used for safety. The relatively high friction loss in booster hose limits the quantity of water available from these lines. Booster lines must not be used when the thermal output of the fire exceeds the heat-absorbing capacity of the water being discharged (Figure 2.20). Booster lines should not be extended with unlined 1-inch (25 mm) forestry hose, except for doing mop-up in the black.

Ground sweep nozzles at the front or rear of a vehicle can be used when the fuel is relatively short and the fire intensity is low (Figure 2.21). Ground sweep nozzles may be mounted either at both corners of the bumper or as a single spray bar that spans the entire width of the bumper. Ground sweep nozzles are usually controlled from inside the cab.

Remote control nozzles, usually mounted above the front bumper, are also used for pump-and-roll operations (Figure 2.22). The nozzle is operated from within the cab of the apparatus. These nozzles are capable of flows of 10 to 300 gpm (40 L/min to 1 200 L/min). To avoid wasting water, they are usually operated in the low end of that flow range when attacking wildland fires.

Some larger pump-and-roll wildland apparatus may be equipped with master stream turrets on top of the apparatus (Figure 2.23). These may be used to knock down large bodies of fire close to a roadway. Apparatus equipped with both master stream turrets and Class A foam systems are very effective in laying protective foam blankets over structures exposed to a wildland fire. These mobile foam operations are commonly known as *drive-by* foaming operations.

Apparatus Water Tanks

NFPA 1906 requires wildland fire apparatus to carry a minimum of 125 gallons (473 L) of water on board. All water tanks must be partitioned with baffles to prevent the water from sloshing around

Figure 2.20 Booster lines are unsafe when heat output is high. *Courtesy of NIFC.*

Figure 2.21 Engines with ground sweep nozzles can be effective in light fuels.

Figure 2.22 An off-road engine uses a bumper-mounted remote control nozzle. *Courtesy of NIFC.*

Figure 2.23 An off-road engine uses a roof-mounted turret. *Courtesy of NIFC.*

inside the tank. Baffles provide stability when the vehicle is turning, stopping suddenly, or being driven on sloping terrain. Partially filled tanks are more dangerous than full tanks because the water can shift freely, exerting heavy forces if the tank is not properly baffled. The baffles must extend from wall to wall and cover at least 75 percent of the area spanned (Figure 2.24).

The water tank must be equipped with a fill opening/vent on top of the tank (Figure 2.25). This opening allows sufficient airflow for water to flow into the pump in a quantity that is at least equal to the rated capacity of the pump. The opening can also be used to fill the tank. The opening must be large enough to insert a 2½-inch (65 mm) hose and coupling into it. Screen should be provided inside the opening to catch debris that might be discharged into the tank. If there is a lid on the opening, it must be designed so that it blows off when pressure in the tank exceeds 2 pounds per square inch or psi (13.8 kilopascal or kPa). The tank fill line from the pump to the tank must be at least 1 inch (25 mm) in diameter.

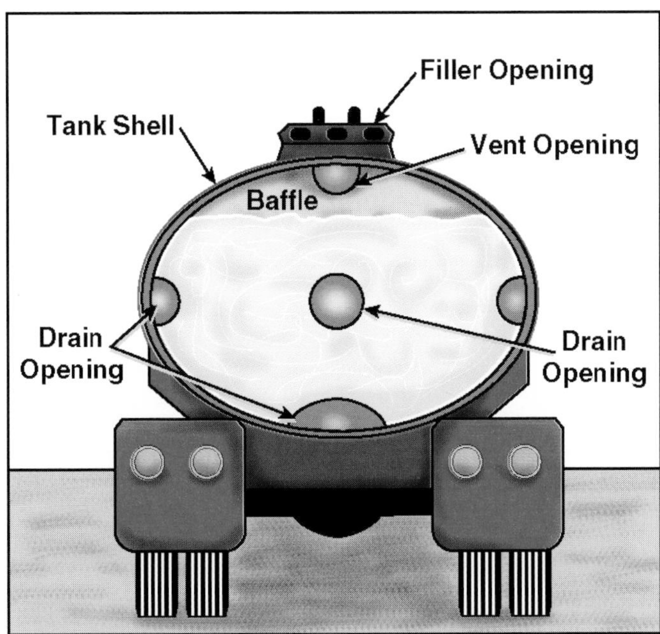

Figure 2.24 Baffles reduce the effects of water sloshing in the tank.

Figure 2.25 One form of tank-filler vent.

Historically, most water tanks were constructed of galvanized steel, but most modern tanks are constructed of stainless steel, aluminum, or polycarbonate materials. In particular, the polycarbonate tanks have become increasingly popular on all types of fire apparatus, including wildland apparatus. Their primary advantages are that they are very light in weight compared to metal tanks and do not rust or corrode.

Apparatus Chassis Requirements

Because of the rugged terrain and conditions in which wildland apparatus are expected to operate, they have very specific requirements for the apparatus chassis and body components. The following sections detail some of the more important considerations.

WEIGHT/HEIGHT/WHEELBASE

The combination of weight, height, and wheelbase must be considered in wildland apparatus design. These variables determine the vehicle's stability and maneuverability. Each of these design factors should be based on the terrain in which the vehicle is most likely to operate.

Weight limitations must be considered in apparatus design because the vehicle has to travel over roads and rough terrain and may have to cross marginal bridges to reach some areas. Excess weight can make the vehicle more likely to damage road surfaces, sink in mud or soft ground, break through the covering over septic tanks, and damage underground piping (Figure 2.26).

The total weight of water, tank, personnel, and equipment must be within the load limits of the vehicle chassis. This load limit is commonly known as the Gross Vehicle Weight Requirement (GVWR) for straight chassis vehicles. It is also known as the Gross Combination Weight Rating (GCWR) for tractor-drawn vehicles. Basic guidelines for calculating vehicle weight are as follows:

- Estimate 10 pounds (5 kg) per gallon (4 L) as the weight of the tank and water

Figure 2.26 Even "Hummers" can get stuck. *Courtesy of NIFC.*

Figure 2.27 Vehicle height and weight posted within the driver's view.

combined; for example, the design weight of a 500-gallon (2 000 L) tank and its water would be 5,000 pounds (2 268 kg).

- Allow for the weight of the crew (190 pounds [86 kg] per person) and all equipment such as pumps, hose, and tools that are either transported or permanently mounted on the apparatus.

It is recommended that the fully loaded vehicle be weighed on a truck scale to determine the actual weight of the vehicle. This figure can then be posted on a small placard on the dashboard where it is plainly visible to the driver/operator (Figure 2.27). This information allows the driver/operator to quickly determine whether the vehicle is safe to cross a bridge with a posted weight limit (Figure 2.28).

Figure 2.28 Vehicle weight must be compared to the bridge's weight limit.

Proper load balance should also be considered in the design phase. Improper load balance can make driving difficult and dangerous because it can cause erratic steering and difficult braking. It may also necessitate extra maintenance on brakes and suspension systems. A vehicle with improper load balance may be unstable laterally (side to side) as well as from front to rear. NFPA 1906 requires that the difference in weight on the end of each axle, from side to side, shall not exceed 7 percent when the vehicle is fully loaded.

Height and wheelbase considerations are, for the most part, interrelated. Vehicles that have a short wheelbase and sit high off the ground tend to be top heavy and unstable because the center of gravity is too high. Vehicles with a longer wheelbase that are lower to the ground tend to be more stable. However, the terrain in which the vehicle normally operates may dictate that it be high off the ground. In these cases, the vehicle can be made more stable by choosing a chassis with as long a wheelbase as possible. NFPA 1906 requires the vehicle to have a center of gravity that is no higher than 75 percent of the rear track width of the vehicle when resting on a flat, level surface.

ANGLES OF APPROACH/DEPARTURE

The *angle of approach* is defined as the smallest angle made between the road surface and a line drawn from the front point of ground contact of the front tire to any projection of the apparatus ahead of the front axle. Conversely, the *angle of departure* is defined as the smallest angle made between the road surface and a line drawn from the rear point of ground contact of the rear tire to any projection of the apparatus behind the rear axle (Figure 2.29). In other words, these angles give the amount of overhang in the front and rear of the vehicle.

The angles of approach and departure affect the road clearance of the vehicle when it must cross ditches or other obstacles. When any of the following obstacles are traversed, the apparatus may scrape if either of the overhangs is too great.

- Steep grades
- Large bumps
- High crowns on roads
- Ditches, stream beds, or arroyos

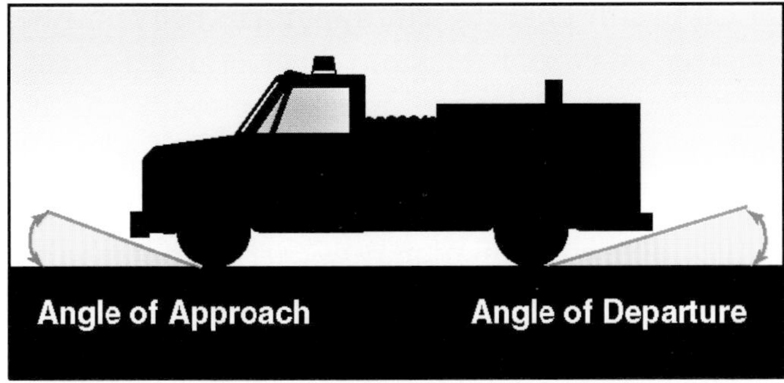

Figure 2.29 High approach/departure angles make a vehicle more maneuverable.

CREW CAB

Fully enclosed riding areas should be provided for all firefighters assigned to ride on the apparatus. To be specific, the firefighter riding area must be enclosed on all six sides. New apparatus are not designed without cab roofs or with open jump seats. The purchaser of a new apparatus must specify the maximum number of firefighters expected to ride on the apparatus when it is ordered from the manufacturer. The manufacturer is then required to provide an enclosed seat with a restraint system for each firefighter. Once the apparatus is in service, this number of firefighters should not be exceeded.

Older apparatus may have open riding areas for firefighters. These may be in the form of open cabs, unenclosed jump seats, or benches within the body or the rear of the apparatus (Figures 2.30 a and b). Firefighters riding in open cabs or

Figure 2.30a Open jump seats can be very dangerous.

Figure 2.30b Open crew benches can be very dangerous.

unenclosed jump seats should be wearing seat belts. Safety gates or bars may be added to the jump-seat entrance to provide an extra measure of security (Figures 2.31 a and b). The use of open riding areas on the rear of the apparatus should be discontinued as soon as economically feasible.

APPARATUS FUEL

Wildland fire apparatus may be powered by either diesel- or gasoline-fueled engines. There are few operational differences between the two types except that gasoline engines consume fuel somewhat faster than diesels, so they may have to be refueled more frequently during extended operations. Gasoline engines are also subject to stalling on slopes exceeding 35 percent. However, unused diesel fuel can deteriorate in the tank over a period of time.

Figure 2.31a A bar across the entry/exit opening adds to crew safety.

Figure 2.31b Jump-seat areas should have restraints.

There are other considerations regarding the vehicle fuel. If the apparatus is equipped with an auxiliary engine-driven pump, both the pump engine and the vehicle engine should operate on the same type of fuel. This eliminates any possible confusion when refueling either of the engines. Another consideration is that NFPA 1906 requires that the vehicle fuel tank be large enough to allow the apparatus to pump at full capacity for at least 2 hours without the need to refuel. However, the fuel supply may not last that long if pump-and-roll operations are used.

APPARATUS BODY

Wildland fire apparatus must have rugged bodies to withstand the harsh conditions under which they typically operate. The apparatus body may be constructed of rust-resistant steel, stainless steel, or aluminum. All nonaluminum metal surfaces that are not plated or made of stainless steel must be painted or coated to protect against rust.

Portable equipment may be carried in compartments or in mounting brackets on the outside of the apparatus body (Figure 2.32). Equipment carried within compartments is not required to be held in place by brackets, but it is advisable (Figure 2.33). Whether the equipment is carried in compartments or on the outside of the apparatus, it must be easily accessible to firefighters.

Figure 2.32 Tools and equipment are often mounted on the outside.

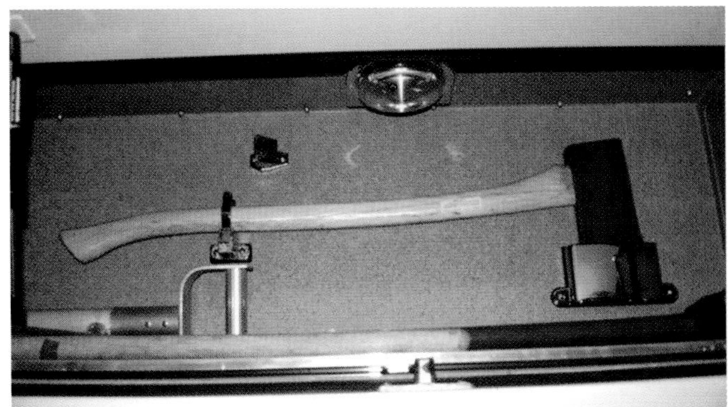

Figure 2.33 Most tools and equipment are mounted in compartments.

Any compartment larger than 2½ cubic feet (0.07 m^3) in volume must be weather resistant. The compartment should have adequate ventilation and provisions for drainage of moisture. Any electrical equipment or wiring within the compartment must be protected from damage.

ALL-WHEEL DRIVE

All-wheel drive capability allows the vehicle to maneuver under a wide variety of conditions including rough or uneven terrain, soft ground, and heavy ground cover. Should a front or rear tire lose traction or contact with the ground, the other wheels usually allow the vehicle to keep moving.

Most apparatus equipped with all-wheel-drive capability may be operated in either the two-wheel- or all-wheel-drive modes. The two-wheel mode is recommended when driving on dry pavement. Vehicles use less fuel when operated in the two-wheel-drive mode. When operating on wet, icy, or snowy pavement or off the road, the vehicle should be driven in the all-wheel-drive mode. On snowy or icy conditions, all-wheel drive provides greater traction to keep the vehicle moving, but it does *not* allow the vehicle to stop faster than a two-wheel-drive vehicle. The same travel distances between vehicles and speed precautions must be applied to both two-wheel- and all-wheel-drive vehicles.

Vehicles capable of all-wheel-drive operation should be equipped with appropriate mud and snow tires (Figure 2.34). These tires provide the maximum amount of traction on adverse terrain. All-wheel-drive vehicles are usually also equipped with skid plates that protect drive train components from possible damage caused by driving over rocks and other objects.

Figure 2.34 Tires designed for mud and snow are needed for off-road driving.

However, there are some disadvantages to all-wheel-drive vehicles compared to two-wheel-drive units. The higher center of gravity resulting from the need for higher ground clearance makes them more susceptible to rollover (Figure 2.35). Higher ground clearance also makes these vehicles taller, which makes it more difficult for them to be driven under a brush canopy or low-hanging tree limbs. In general, all-wheel-drive vehicles also have a wider turning radius that makes them less able to negotiate in some situations.

Figure 2.35 Off-road vehicles are prone to rollover. *Courtesy of NIFC.*

Foam Systems

Class A foam systems have become increasingly popular on wildland fire apparatus. While the tactics involved in using Class A foam to attack wildland fires are discussed in Chapter 6, Fire Suppression Methods, this chapter focuses on the apparatus-mounted systems used to generate foam for fire attacks. Portable foam proportioning systems using an in-line eductor and foam concentrate containers are not practical for wildland applications because they do not allow for pump-and-roll operations (Figure 2.36). Class A foam systems can be categorized into two general classifications: low-energy and high-energy systems.

Figure 2.36 Siphoning foam concentrate from containers limits mobility.

LOW-ENERGY FOAM SYSTEMS

Low-energy foam systems rely strictly on the pressure created by the fire pump to provide pressure to the foam solution flowing to the nozzle. Low-energy foam systems utilize the following elements to produce a foam fire stream:

- Water supply
- Fire pump
- Foam concentrate supply
- Foam proportioning system
- Fire hose and a nozzle (foam or fog)

The water supply for most Class A foam wildland fire fighting operations is from the apparatus water tank because of the need to remain mobile while attacking wildland fires. It is generally not recommended that an engine be connected to a hydrant or other external water supply source while attacking a wildland fire, although this may be done in some structure protection operations. Also, some types of foam proportioning systems, such as the around-the-pump proportioners, do not function properly when water is being supplied by a pressurized external source such as a hydrant.

Most wildland fire apparatus equipped with Class A foam systems use the main apparatus fire pump as the pressure source for the foam fire stream. In most cases this is a centrifugal fire pump. However, some smaller apparatus equipped with positive-displacement pumps may also have built-in foam systems.

The foam concentrate supply on wildland apparatus is held in permanently mounted tanks on the apparatus (Figure 2.37). The foam concentrate tanks may be integral cells within the apparatus water tank, or they may be completely independent of the water tank. Most wildland apparatus carry 10 to 100 gallons (40 L to 400 L) of Class A foam concentrate. This is enough concentrate to mix with numerous tank loads of water.

The foam proportioning system is responsible for injecting the proper amount of foam concentrate into the fire stream. Class A foam concen-

Figure 2.37 Foam can be stored in a separate foam tank on the apparatus.

trates are typically proportioned at a level of 0.1 percent to 1.0 percent of the total volume of the fire stream. Although a myriad of different types of foam proportioning systems are used on various types of apparatus, most wildland apparatus use some type of direct-injection system. Direct-injection proportioning systems monitor the volume of water flowing through the pump discharge piping and correspondingly inject an appropriate amount of foam concentrate into the water stream. The foam concentrate and water combine to make a foam solution as they flow through the hose towards the nozzle.

With low-energy foam systems, any standard fire hose used for plain water application may also be used for foam application. Standard fog nozzles or air-aspirating foam nozzles may be used to discharge the foam. These foam nozzles are designed to entrain air into the foam solution as it is discharged (Figure 2.38). The foam discharged from these nozzles is called *aspirated* foam (Figure 2.39). Foam solution discharged from a fog nozzle in a low-energy foam system does not have air entrained into it prior to discharge. This is called *nonaspirated* foam (Figure 2.40). Aspirated foam forms a thicker and longer-lasting blanket than nonaspirated foam. However, nonaspirated foam is extremely effective at penetrating deep into dense fuels that otherwise would shed plain water.

Figure 2.38 A typical air-aspirating foam nozzle.

Figure 2.39 Aspirating nozzles produce thick, frothy foam. *Courtesy of NIFC.*

Figure 2.40 Nonaspirating nozzles have greater stream reach than aspirating nozzles. *Courtesy of NIFC.*

HIGH-ENERGY FOAM SYSTEMS

High-energy foam systems differ from those previously discussed in that they introduce compressed air into the foam solution prior to discharge into the hoseline. The turbulence of the foam solution and compressed air going through the piping and/or hoseline creates a finished foam. In addition to forming the foam, the addition of compressed air also allows the foam stream to discharge a significantly greater distance than regular foam or water fire streams (Figure 2.41).

In the mid-1980s, the U.S. Bureau of Land Management (BLM) conducted research that led to the development of the high-energy Class A foam system now becoming common on structural and wildfire apparatus. This system uses a standard centrifugal fire pump to supply the water and a compressor to supply the air. A direct-injection foam-proportioning system is attached to the discharge side of the fire pump. Once the foam concentrate and water are mixed to form a foam solution, compressed air is added to the mixture before it is discharged from the apparatus and into the hoseline. This system is commonly called a *compressed-air foam system (CAFS)*.

There are several tactical advantages to using CAFS:

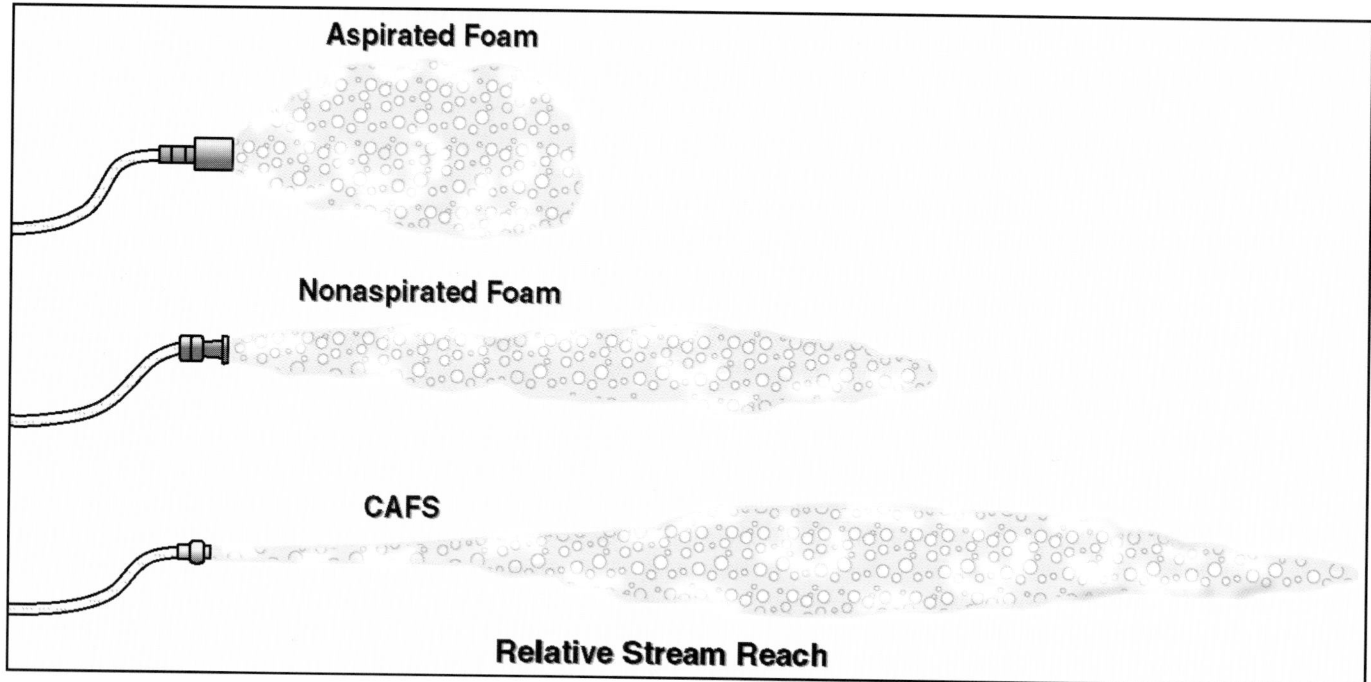

Figure 2.41 CAFS nozzles have greater stream reach than all other types.

- The reach of the fire stream is considerably longer than streams from low-energy systems.
- A CAFS produces uniformly sized, small air bubbles that are very durable.
- CAFS-produced foam adheres to the fuel surface and resists heat longer than low-energy foam.
- Hoselines filled with high-energy foam solution are lighter than hoselines full of low-energy foam solution or plain water.
- A CAFS allows safer fire-suppression action because an effective attack can be made from a greater distance than with standard fire streams.

A CAFS does have some inherent limitations:

- A CAFS adds expense to a vehicle and can increase the need for maintenance.
- Hose reaction can be erratic with a CAFS if foam solution is not supplied to the hoseline in sufficient quantities.
- The compressed air accentuates the hose reaction in the event the hose ruptures.
- Additional training is required for personnel who are expected to make a fire attack using a CAFS or who operate CAFS equipment.

Most apparatus equipped with a CAFS are also designed to flow plain water, should the choice be made to do that (Figure 2.42). In fact, most CAFS-equipped apparatus only flow foam through preselected discharges. Other discharges may be capable of flowing foam solution or plain water. The fire pump and proportioning system used on a CAFS-equipped vehicle are the same as described earlier for low-energy Class A foam systems.

In general, 2 cubic feet per minute (cfm) (0.06 m^3/min) of airflow per gallon per minute (4 L/min) of foam solution flow produces a very dry foam at flows up to 100 gpm (400 L/min) of foam solution. This produces a large amount of foam at a 10:1 expansion ratio. Most structural and wildland fire attacks using CAFSs are done with an airflow rate of 0.5 to 1.0 cfm (0.015 m^3/min to 0.03 m^3/min) per gallon (4 L) of foam solution. This rate allows for adequate drainage of solution from the blanket to wet the fuel and prevent reignition. It also prevents smoldering from occurring beneath the foam blanket.

For more information on all types of foam systems and foam fire fighting techniques, see the IFSTA **Principles of Foam Fire Fighting** manual.

Figure 2.42 A typical pump panel of a unit equipped with CAFS.

Apparatus Safety Features

Because wildland apparatus operate in extreme and varying terrains, they must be equipped with special safety features that are not necessarily found on other types of fire apparatus. These safety features are designed to protect both the apparatus and the firefighters who ride in them. The following sections highlight some of the more common safety features found on wildland fire apparatus.

BRUSH PROTECTION

Apparatus operated off-road often come in contact with varying types of brush and other objects. These apparatus must be protected from damage caused by this contact. Brush protection comes in two forms: undercarriage protection and outer body protection.

Undercarriage protection consists of skid plates to protect vital apparatus components that are exposed beneath the vehicle (Figure 2.43). Components that should be protected include drivetrain components, fuel tanks, and exposed portions of the engine and exhaust system. It is particularly important to shield the catalytic converter on vehicles so equipped. Catalytic converters become very hot and may start fires beneath the vehicle if it stops in dry grass.

Body protection varies depending on the terrain in which the vehicle is most likely to be driven. At a minimum, off-road vehicles should have heavy-duty, extended bumpers (Figure 2.44). Some departments provide protective cages over/around vulnerable portions of their apparatus such as headlights, warning lights, sirens, and grills (Figure 2.45).

Some wildland vehicles, often called *brush breakers,* are designed to knock down heavy brush and small trees as they go along. These vehicles are

Figure 2.43 A typical protective skid plate.

Figure 2.44 Off-road vehicles should have heavy-duty bumpers.

Figure 2.45 Headlights on off-road vehicles should be protected.

equipped with extensive rub rails and brush bars around and over the top of the entire vehicle (Figure 2.46). It is recommended that all special protective devices be bolted, not welded, in place. This facilitates easy repair or replacement should the devices become damaged. All of these devices should be attached directly to the vehicle chassis frame to provide the most stable level of protection.

Figure 2.46 A typical "brush breaker." *Courtesy of New Jersey Forest Fire Service.*

ROLL BARS

Vehicles that are operated off-road may be equipped with roll bars to provide protection to the vehicle and its occupants should the apparatus be involved in a rollover accident. At a minimum, one roll bar should be located over the passenger compartment (Figure 2.47). This prevents the cab from being crushed around the occupants. Additional roll bars, intended to limit damage to the rest of the apparatus, may be located over the apparatus body.

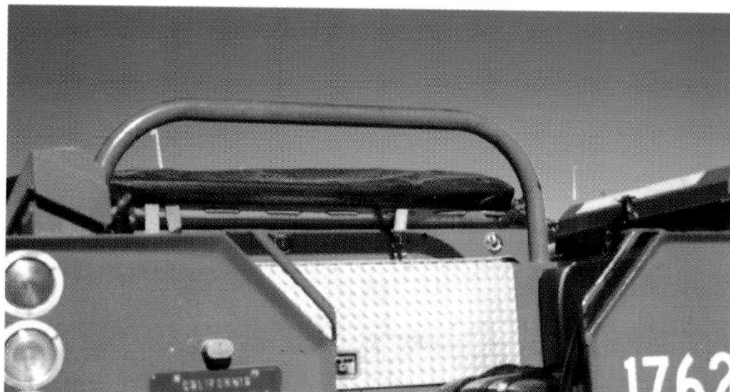

Figure 2.47 Crew riding areas should have substantial roll bars.

As with brush-protection devices, roll bars should be attached directly to the vehicle chassis frame. This provides the maximum level of protection should the vehicle overturn onto the roll bar. The roll bar should be of solid construction and capable of supporting the entire weight of the vehicle when fully loaded.

FIREFIGHTER RIDING AREAS

All of the NFPA standards related to apparatus safety, including NFPA 1500, *Standard on Fire Department Occupational Safety and Health Program*, and NFPA 1906, prohibit firefighters from riding outside of the vehicle cab while the vehicle is in motion.

> **WARNING**
>
> **Exterior riding positions for pump-and-roll operations are extremely dangerous, and this practice should be discontinued by those agencies still operating in this manner. Many firefighters have been seriously injured or killed when riding in these positions when the vehicle was involved in a collision or rollover accident.**

Firefighter safety must always be the top tactical priority. Any operational advantage gained by having firefighters ride on the outside of a moving apparatus ***does not*** outweigh the potential danger to which they are exposed (Figure 2.48). The only safe positions for firefighters during pump-and-roll operations are in or behind the cab while wearing a restraint system and walking beside the apparatus in view of the driver as the vehicle is driven slowly (Figure 2.49). As mentioned earlier, some departments use remote control nozzles mounted on the front bumper of the vehicle instead of firefighters riding in that position. With training and experience, firefighters can become just as effective controlling the nozzle from within the cab as they are using a handline nozzle from the exterior riding position.

Figure 2.48 Firefighters should *never* ride on the fenders or front bumpers of fire apparatus.

Figure 2.49 During pump-and-roll operations, firefighters should walk beside the apparatus. *Courtesy of NIFC.*

RESPIRATORY-PROTECTION SYSTEMS

Some wildland apparatus are equipped with respiratory-protection systems that allow firefighters working on the apparatus to breath clean air for extended periods of time. These respiratory-protection systems are similar to airline-breathing systems used in some hazardous materials incidents and confined space rescues. They allow the firefighter(s) to connect to one or more large-capacity (cascade-type) air cylinders. Depending on the size and number of cylinders attached to the system, several hours of breathing air may be available. These systems typically have some type of quick-connect coupling at the connection point on the apparatus (Figure 2.50). The firefighter wears a self-contained breathing apparatus (SCBA) facepiece and low-pressure regulator connected to the apparatus.

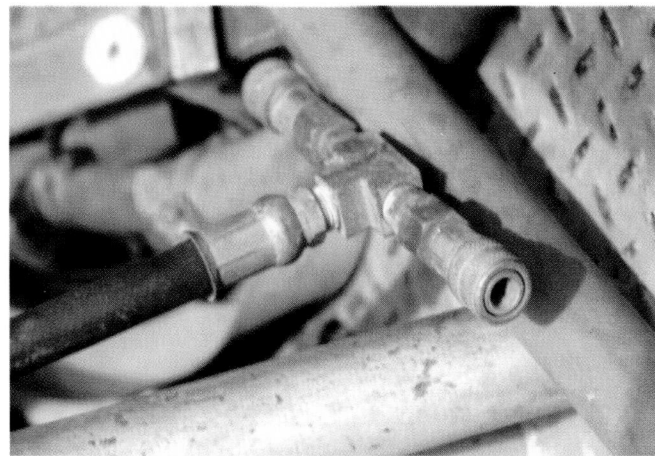

Figure 2.50 Some vehicles provide breathing air to firefighters during pump-and-roll operations.

These systems are most commonly found on apparatus that have exterior riding positions for firefighters to use during pump-and-roll operations. As stated earlier, this practice is not safe and is strictly forbidden by a number of NFPA standards. Some apparatus may have systems whose hose is long enough to allow firefighters to connect into the system while walking next to the apparatus during pump-and-roll operations. A few apparatus may be equipped with respiratory-protection systems for occupants of the vehicle cab. However, except for bulldozers and other mechanized equipment, it is generally not practical to drive the apparatus while wearing this equipment. Under some conditions, filter masks and other forms of respiratory protection may be of benefit to driver/operators in extended operations.

STEPPING SURFACES

Many wildland apparatus are configured in a manner that requires the firefighter to climb onto the body area during the course of normal operations. This may be necessary to pull hose, start a pump motor, fill the apparatus water tank, or access a tool. NFPA 1906 specifies that the maximum distance from the ground to the first step not exceed 24 inches (610 mm). Subsequent steps may be no farther than 18 inches (457 mm) apart. Steps should have a minimum surface of 35 square inches (22 582 mm^2) and have a skid-resistant surface (Figure 2.51). Steps should have a minimum depth of at least 8 inches (203 mm). If a ladder is used to access the top portion of the apparatus, the rungs must be at least 7 inches (178 mm) away from the body of the apparatus. All steps and ladders must be able to support at least 500 pounds (227 kg).

Figure 2.51 Vehicle steps should be designed to prevent slipping.

BACKUP ALARMS AND LIGHTING

Because wildland apparatus are frequently moved and repositioned on the fireline, it is important for them to have a properly working backup alarm system. A backup alarm system provides both audible and visual indications that the apparatus is being driven in reverse. The intention of this system is to warn people on the ground in the vicinity of the apparatus. This is a prime concern since statistics from NFPA and others show that most apparatus accidents occur during backing.

The backup alarm may be either electric or electronic in design. The alarm must emit a sound that is at least 87 decibels (dB). The Society of Automotive Engineers (SAE) standard J994, *Alarm-Backup-Electric-Performance, Test and Application,* contains the technical requirements for these alarms. Bright lighting should also come on when the vehicle's transmission is in reverse. This illumination gives the driver/operator more ability to see the area where the vehicle is being backed.

PROTECTION FROM EXHAUST SYSTEM

Mufflers and exhaust pipes, portable generators, and pump motors must be insulated or shielded (Figure 2.52). An unprotected muffler or exhaust pipe can cause serious burns to a firefighter who inadvertently touches or is thrown against either one. Hot exhaust pipes are also a potential ignition source and can start fires in vegetation.

On large vehicles, the exhaust pipe may be extended upward directly behind the cab to reduce the chances of starting a fire (Figure 2.53). It can also be insulated or shielded to prevent burns. However, care must be taken to ensure that the exhaust noise does not interfere with radio traffic or voice communications.

Figure 2.52 Engine exhausts should have protective shields.

Figure 2.53 Vertical exhaust pipes should also be shielded.

Figure 2.54 Pump operators should use hearing protection.

Figure 2.55 Catalytic converters may start fires in fine fuels. *Courtesy of NIFC.*

Figure 2.56 Having unit numbers clearly visible makes identification easier. *Courtesy of California Office of Emergency Services.*

A hearing-protection system in a pump-and-roll operation must be used because of the excessive noise from the vehicle's engine and exhaust. The type of hearing protection varies from one piece of apparatus to another. An industrial hygienist or a safety professional can conduct tests to determine what type of hearing protection is needed. Hearing protection is a recommended practice with any function in close proximity to a pumping operation (Figure 2.54).

Parking sites for vehicles with catalytic converters should be chosen carefully in wildland areas. Since the catalytic converters get extremely hot during operation, they can quickly start fires beneath the vehicles (Figure 2.55).

APPARATUS NUMBERING/MARKING

All wildland fire apparatus should have their unit numbers or radio designations plainly visible on all sides and tops of the apparatus (Figure 2.56). The purpose of these markings is to make apparatus easily identifiable from either the ground or air, which is particularly important on large incidents involving multiple resources. The markings should be in colors that are distinctly different from the rest of the apparatus. When possible, the markings should be reflectorized as well. Markings on the tops of apparatus should be at least 2 feet (0.6 m) in size, which makes them easily readable from aircraft (Figure 2.57). Markings on the sides of apparatus vary in size, depending on local preference.

Figure 2.57 Vehicles should have their numbers on their roofs as well. *Courtesy of NIFC.*

Portable Equipment Carried on Wildland Apparatus

The types and amount of portable fire fighting equipment carried on wildland fire apparatus depend on a variety of factors, including the following:

- Type of apparatus (structural engine, wildland apparatus, water tender, etc.)
- Amount of compartment space on the apparatus
- Types of equipment normally used by that department
- Number of firefighters normally assigned to that type of apparatus

Structural engines and water tenders should carry at least the equipment specified in NFPA 1901. NFPA 1906 contains a list of the minimum equipment that must be carried on wildland apparatus, which includes the following:

- One axe of any type
- One round-point shovel
- Two portable handlights
- One portable fire extinguisher with a minimum rating of 40 B:C
- Two spanner wrenches
- Two hundred feet (61 m) of fire hose (no size is specified)
- One nozzle suitable to the pump and hose on the apparatus
- One first-aid kit
- One hand-pump water extinguisher
- Two wheel chocks

All of this equipment must be mounted in compartments on the outside of the apparatus or in the cab using appropriate brackets. Tools carried in the cab must be securely mounted to prevent them from coming loose during acceleration, deceleration, or sudden direction changes.

In addition to the minimum equipment required by NFPA 1906, other types of portable equipment are commonly found on wildland apparatus. Depending on local requirements, these may include the following:

- Hard suction hose (for drafting)
- Hose packs (for progressive hose lays)
- Tees with shutoffs (water thieves)
- McLeods (scraping/raking tools)
- Fire rakes, swatters, or brooms
- Portable fire pumps
- Chain saws
- Pulaski tools
- Backpack water extinguishers
- Portable fire shelters
- Filled water canteens
- Spare gloves, goggles, breathing-air cylinders, etc.

More information on portable equipment is found in Chapter 3, Wildland Fire Fighting Tools and Personal Protective Equipment.

MOBILE WATER SUPPLY APPARATUS

Mobile water supply apparatus are used to provide water for fire fighting at locations where water supply from a piped system or auxiliary source is not available (Figure 2.58). In the Incident Command System (ICS), mobile water supply apparatus are called *water tenders,* and this is also their radio call designator. However, mobile water supply apparatus are still called *tankers* in many jurisdictions. In ICS, tankers are aircraft that drop water or fire retardant in aerial attacks.

Requirements for the design and construction of water tenders are in NFPA 1901. The basic definition of a water tender is a vehicle that carries at least 1,000 gallons (4 000 L) of water. Depending on local requirements, water tenders may be equipped to simply transport water for other apparatus, or they may be equipped to make fire attacks on their own. The following sections detail some of the more important features of water tenders.

Figure 2.58 A typical water tender. *Courtesy of John Hawkins.*

Types of Water Tenders

As with wildland engines and other resources, water tenders are classified into various types based on their *minimum* capabilities. In general, the three types of water tenders have pumps ranging from 50 to 300 gpm (200 L/min to 1 200 L/min) and water tank capacity ranging from 1,000 to 5,000 gallons (4 000 L to 20 000 L).

Water Tender Applications

Depending on local operating procedures and the capabilities of particular apparatus, there are three basic fireground applications for water tenders.

- Water shuttle operations
- Nurse tender operations
- Fire attack/exposure protection operations

Water shuttle operations involve the more-or-less constant movement of one or more water tenders between a water source (the fill site) and the location where the water is going to be used (the dump site). The fill site is usually a fire hydrant or a static water supply source. Typically, an engine is standing by at the water supply source ready to quickly fill the water tenders as they arrive (Figure 2.59). Once at the dump site, the water tender dumps its load into one or more portable water tanks and immediately returns to the fill site for another load (Figure 2.60). Attack fire apparatus draft from the portable tanks to supply fire fighting operations (Figure 2.61). For more information on conducting a water shuttle operation, see the IFSTA **Water Supplies for Fire Protection** manual or the appendix of NFPA 1231, *Standard on Water Supplies for Suburban and Rural Fire Fighting.*

Figure 2.59 An engine fills a water tender.

Figure 2.60 A water tender gravity dumps into a portable drop tank.

Figure 2.61 An attack engine drafts from a portable tank.

Nurse tender operations involve pairing the water tender with an engine to extend the engine's water supply. For example, a structural engine may be assigned to protect a structure from an approaching fire. If the amount of water in the engine's tank is not sufficient to allow it to complete its assignment, the water tender connects a supply line to the engine. When the situation requires that the water tender be positioned some distance from the engine, it can be connected by a long supply hose if the tender has an adequate fire pump. A water tender used as a refilling point for smaller wildland apparatus operating remotely from a water supply source also qualifies as a nurse tender operation. Depending on the size of the water tender and the wildland apparatus it is refilling, it may be possible for the water tender to refill as many as 20 wildland apparatus tanks of water before needing to be refilled itself.

Water tenders equipped with suitable fire pumps may be able to make fire attacks or perform exposure protection by themselves. In order to do these functions, the fire pump needs to be capable of providing adequate volume and pressure for attack hoselines. Under some fire conditions, water tenders are staged in an area of potential exposures to provide protection for that area while being available to refill wildland apparatus when required.

Pumping and Dumping Capabilities

There are three primary methods of getting water out of water tenders:

- Pumping the water through the fire pump
- Dumping the water through a quick-dump discharge
- Unloading the water by gravity through a gated dump valve

Pumping through the apparatus fire pump is generally done when water tenders are supplying attack lines or acting as a nurse tanker. The type of fire pump carried on a water tender is consistent with those described for wildland apparatus earlier in this chapter. The capacity of water tender fire pumps range from small 200-gpm (800 L/min) auxiliary engine-driven pumps to 5,000-gpm (20 000 L/min) midship transfer drive pumps. It should be noted that pumping water from the apparatus is generally not the most efficient method of filling portable water tanks during water shuttle operations.

The most efficient way to discharge water into a portable tank is through the use of a quick-dump discharge system (Figure 2.62). The quick-dump system connects directly from the apparatus water tank, through the apparatus body (if there is any), to the sides and/or rear of the apparatus. The quick-dump piping may be round or square and is generally at least 8 inches (200 mm) in diameter. The quick-dump system may rely on gravity to unload the water, or it may be jet-assisted by a small water line from the fire pump. The quick-dump discharge valves may be located on the dump itself or they may be remotely controlled from the apparatus cab.

Some water tenders are not equipped with any type of fire pump. These apparatus are generally

Figure 2.62 A water tender quick-dumps into a portable tank.

limited to gravity dumping their loads in water shuttle operations. If these apparatus have the appropriate connections, engines may be able to connect to them with hard suction hose and draft the water from the water tender's tank (Figure 2.63).

Figure 2.63 Engines may draft directly from water tenders.

Chassis Requirements

Apparatus specifications should match the right size vehicle chassis to the water- carrying capacity needed and vice versa. An overloaded vehicle may be subject to structural failure of the tank and/or the frame. It may also be unsafe to operate at otherwise normal road speeds. At no time should the total weight of a loaded water tender exceed the rated Gross Vehicle Weight (GVW) specified by the manufacturer of the chassis. Depending on the size of the water tank, the water tender may be a two- or three-axle straight-chassis apparatus or a tractor-trailer (Figures 2.64 a–c).

When water tenders are designed, a major consideration is keeping the vehicle's center of gravity as low as possible. The higher the center of gravity, the less stable the vehicle is. This is especially important with water tenders because of their heavy weight. As mentioned earlier, the water tank must be provided with adequate baffles (swash partitions) to prevent potentially dangerous load shifts during transport. Baffling is especially important when the tank is not completely full.

Figure 2.64a A typical two-axle water tender.

Figure 2.64b A typical three-axle water tender.

Figure 2.64c A tractor-drawn water tender works with an engine in the field. *Courtesy of NIFC.*

As described in NFPA 1231, converting tank vehicles designed for another purpose to mobile water supply apparatus is not recommended. The weight of a full tank of water will be different than the weight of a tank filled with the liquid that the vehicle was originally designed and built to carry. This weight difference can result in drastically different handling characteristics that could affect the safe operation of the vehicle. In addition, the material of which the tanks were constructed is often not compatible with water, so increased maintenance is likely to be required.

AERIAL APPARATUS

An aerial apparatus is a large vehicle with a powered aerial device that provides access to the upper levels of a structure and with the means to deploy elevated master streams (Figure 2.65). These apparatus have limited applications on wildland fires. Their primary uses are for exposure protection and for extinguishing structure fires that occur as a result of a wildland fire. All aerial apparatus fall under the requirements in NFPA 1901.

When aerial apparatus are used on wildland fires, they are most commonly used in the wildland/urban interface. At these fires, aerial apparatus may be positioned to provide exposure protection to a fairly large area. The elevated master stream on the aerial device may be used to wet down one or more structures. It may also be used to knock down an advancing fire before it reaches the exposures. If the aerial apparatus is not equipped with its own fire pump, it is necessary to assign a pumping apparatus to perform this function.

In wildland and wildland/urban interface fires, aerial apparatus can provide these additional capabilities:

Figure 2.65 Water tower operations may be needed in the wildland/urban interface.

- Additional personnel (to supplement or replace wildland crews)
- Elevated streams (to reach over barriers)
- Elevated vantage point (for better view of immediate area)

As with any other type of equipment, there are operational and safety issues to be considered when using aerial apparatus on wildland fires. The most common of these considerations are the following:

- Immobility of the equipment
- Firm ground required for jacks (Figure 2.66)
- Ladders not rated for work below horizontal
- Lack of off-road capability
- Relatively high center of gravity

Figure 2.66 Aerial jacks may sink in soft ground if plates are not used.

Aerial apparatus are very immobile once they are positioned and the aerial device is deployed. Should the need to reposition the apparatus become necessary, it takes several minutes to stow the aerial device, stow the stabilizers, and disconnect hoselines attached to the apparatus. In a fast-moving wildland fire situation, this amount of time could put the apparatus and the firefighters working on it in danger.

Some aerial apparatus are also designed to be operated as quints. In addition to having an aerial device, a quint is equipped with a fire pump, water tank, ground ladders, and fire hose (Figure 2.67). Many fire departments are experimenting with the

Figure 2.67 A typical "quint." *Courtesy of Sonoma (CA) Fire Department.*

quint as a replacement for traditional engine and ladder companies. However, because of the same operational and safety issues identified for aerial apparatus, quints also have very limited value in wildland fire situations.

Some departments also use a standard fire engine equipped with a 50- to 75-foot (15 m to 23 m) aerial ladder or elevated master stream device (Figure 2.68). These engines lack the space to carry a full complement of ground ladders or other truck company equipment; however, they can perform many of the same functions as aerial apparatus. In wildland situations where resources are spread thinly, quints may be used in place of wildland companies to provide exposure protection and to fight structure fires. These apparatus *should not* be operated in off-road conditions. Their high center of gravity makes them unsuitable for rough or uneven terrain.

Figure 2.68 An engine equipped with a telescoping master stream device.

HEAVY EQUIPMENT

In addition to standard fire fighting apparatus, a number of different types of heavy equipment may be used to support a major wildland fire fighting operation. This equipment is used primarily to construct firebreaks that may slow or halt the spread of an advancing fire. It is important that firefighters be familiar with the various types of equipment that are available to them and know their capabilities.

In many cases, heavy equipment used at wildland fires is not owned or operated by the fire department; it is contracted from other public agencies or private firms. During pre-incident planning, a survey should be conducted of all earthmoving equipment available within the jurisdiction to identify the capabilities and limitations of each type. Written agreements that specify the circumstances under which the equipment may be used and how the owner will be compensated should be signed with the equipment owners. These agreements should clearly identify the rights and obligations of both parties. Fire department personnel should inspect contract equipment on a regular basis (at least annually) and review the equipment operators' qualifications to operate at wildland fires.

All mechanized equipment used on wildland fires (including contract equipment) must be equipped with proper rollover protection, lighting, backup alarms, fire shelters, and personnel protective equipment before being placed in service (Figure 2.69). Operators should be trained both in the use of this safety equipment (and required to use it) and in wildland fire operations with their type of equipment. They should also have a basic understanding of wildland fire behavior.

There are three primary types of mechanized equipment used in wildland fire fighting covered in this chapter:

- Bulldozers
- Tractor-plows
- Road graders (maintainers)

Figure 2.69 A private dozer equipped with a roll cage. *Courtesy of NIFC.*

Depending on local conditions, other types of equipment may be used as well. Farm plows, large mowers, and similar types of equipment may be used to cut firebreaks if the fire is in terrain such as crop lands, pastures, and parks that is suitable for these implements. In many cases, their use is similar to the principles described for the other heavy equipment covered in this section.

Bulldozers

These crawler tractors with front-mounted blades have become known in the fire service simply as *dozers* (Figure 2.70). Three types of dozers are used in wildland fire fighting operations. Type-1 is also commonly known as *heavy*, Type-2 as *medium*, and Type-3 as *light* dozers. While each type has different minimum horsepower requirements, each has a crew of two.

The primary use for bulldozers is in the construction of firelines to halt or prevent the spread of a wildland fire. As mentioned earlier, bulldozers intended to be used in wildland fire fighting must be equipped with safety cages or cabs to protect the operator. Some are equipped with fully enclosed and air-conditioned environmental cabs. Because bulldozers are capable of operating in a wide variety of fuels, topography, and soil conditions, they are very well-suited to this application. The actual rate of fireline construction depends on several variables such as slope, fuel type, time of day, ground moisture, atmospheric temperature, fire behavior, age of dozer, size of dozer, condition of the dozer, and the skill and experience of the dozer operator. Depending upon a number of variables, a dozer can construct a single-pass fireline at approximately ½ mile (0.8 km) per hour over moderate terrain in medium fuel (Figure 2.71).

Figure 2.70 A typical fire service dozer. *Courtesy of New Jersey Forest Fire Service.*

Tractor-Plows

A tractor-plow is a vehicle used to cut a control line in somewhat the same manner as a bulldozer. However, most tractor-plows are smaller and more maneuverable than bulldozers. Tractor-plows may be tracked vehicles or rubber-tired vehicles (Figure 2.72). They pull a plow that is typically about 6 feet (1.8 m) wide. This cuts a narrower swath than the blade on most bulldozers.

Figure 2.71 A dozer builds a fireline in medium fuels. *Courtesy of NIFC.*

A Type-2 tractor-plow (TD-9, HD-6, D-3, Case 1150, JD 450) can cut a 6-foot-wide (1.8 m wide) fireline at a rate of up to 1 mile per hour (mph) (1.6 kmph). A Type-1 can double that rate. The rate, of course, depends on the same variables previously described in the section on bulldozers.

Figure 2.72 A typical tractor-plow. *Courtesy of NIFC.*

Road Graders (Maintainers)

Road graders, also called *maintainers*, are usually self-propelled and have the blade mounted just forward of the rear wheels (Figure 2.73). Road graders are not designed to be fire fighting equipment, but they can be effective in cutting a control line well ahead of a slow-moving fire in crop stubble for example. To be done safely, the line must be located far enough ahead of the fire front so that it can be completed well before the fire reaches the line. There are no ICS typing requirements for maintainers.

Figure 2.73 A typical road grader.

Heavy Equipment Transport

The various types of heavy equipment used to fight wildland fires are most often transported to the fireline on trucks or low-boy trailers (Figure 2.74). These transports may be required to have special permits to travel public roads and highways because they may exceed legal weight limits when fully loaded and may be heavy enough to damage the paved surface of highways. They may also exceed legal width limits.

Figure 2.74 Dozers are transported on low-boy trailers. *Courtesy of John Hawkins.*

AIRCRAFT

Aircraft of all types make it possible to transport personnel, equipment, and fire retardant and suppression agents over considerable distances and extremely difficult terrain very quickly. The primary use of aircraft is for quick initial attack on small fires to keep them from becoming large. The larger the fire, the less impact air tankers have. To be most effective for fire attack, air drops must be followed up by ground personnel.

While there are many different models of aircraft used in wildland fire fighting operations, each is in either of two major categories: *rotary-wing aircraft (helicopters)* or *fixed-wing aircraft.* Each of these is detailed in the following sections, as well as are the procedures for paracargo drops.

Rotary-Wing Aircraft (Helicopters)

Helicopters may be used for a variety of roles in wildland fire fighting operations. The different roles including the following:

- Transporting personnel, equipment, and supplies to the scene (including rappellers for initial attack)
- Dropping water, foam, or fire retardant chemicals on fires (Figure 2.75)
- Performing reconnaissance of the fire scene (including mapping)

Figure 2.75 A helicopter delivers its load onto a fire. *Courtesy of NIFC.*

- Transporting injured firefighters from the scene to a medical facility (Figure 2.76)
- Conducting aerial ignition of backfires and burnout operations (Figure 2.77)
- Filling remote portable water tanks to support relay pumping or hose lays in remote areas (Figure 2.78)

Some helicopters are equipped with aerial ignition devices that can drop burning liquid fuel onto vegetation to ignite it. These helicopters are used most often in the prescribed burning element of fuel-management programs. However, they may also be used for burning out or backfiring operations (see Chapter 6, Fire Suppression Methods).

Figure 2.76 An injured firefighter is loaded into a helicopter.

Figure 2.77 A helicopter in an aerial-ignition operation. *Courtesy of NIFC.*

Figure 2.78 A helicopter refills a portable tank. *Courtesy of NIFC.*

Fixed-Wing Aircraft

Fixed-wing aircraft are used for a variety of purposes in and around wildland fires. However, the two most common uses are for fire-scene reconnaissance and air drops of water, foam, or fire retardant. Less frequently, aircraft are also used to drop smoke jumpers or cargo. Light, single-engine aircraft are most often used for reconnaissance and for coordinating air drops (Figure 2.79). Larger aircraft, usually multiengine, are used for delivering water, foam, or fire retardants (Figure 2.80). Fire suppression agents (water and foam) are dropped directly onto the fire; fire retardants are usually dropped onto uninvolved fuels in the path of the fire.

In ICS terminology, large fixed-wing aircraft used for dropping water, foam, and fire retardants are called *tankers*. Some jurisdictions prefer to use the term *air tanker* to eliminate any confusion that may exist with ground-based mobile water supply apparatus (properly called *water tenders* in ICS but still called *tankers* in many jurisdictions). In the resource typing systems, requirements for various types of tankers are based solely on the *minimum* agent load capacity of the plane.

Figure 2.79 A typical light plane used for reconnaissance. *Courtesy of NIFC.*

Figure 2.80 A multiengine tanker makes a drop. *Courtesy of Tony Bacon.*

Extinguishing Agent/Fire Retardant Application

As stated earlier, aircraft may drop water, foam, or fire retardants. Water and foam are used primarily as extinguishing agents. Fire retardants are used to limit the spread of fire. Fire retardants are generally classed according to the characteristics they exhibit after being dropped on the fuel: short term or long term. For more information on the tactical use of aircraft for fire suppression, see the Air Operations section of Chapter 6, Fire Suppression Methods.

Paracargo Drops

Paracargo includes anything that is intentionally dropped or is intended to be dropped from an aircraft by parachute or other retarding device or in free fall. This is a good method for delivering equipment and supplies that are remote from, or inaccessible to, ground vehicles. In paracargo drops, the drop zone is a strip 200 feet (60 m) wide on each side of the flight path, 300 feet (90 m) short of the target in the direction of approach, and 1,300 feet (400 m) after the target in the direction of the aircraft's exit (Figure 2.81). The following procedure should be used when paracargo drops are needed:

- Mark target area with a large white or orange *T* in an open area, but not within 600 feet (180 m) of incident base (logistical center for the incident). The marking can be made of crepe paper or cloth and should be held down with stakes or rocks. The *T* should be at least 7 feet (2.1 m) long, with the top toward the wind. If there is more than one drop zone, mark each one with a number placed below and to the right of each *T* (Figure 2.82). Target markings must be removed when the operations are completed.
- Indicate wind direction with a paper streamer on top of a long pole.

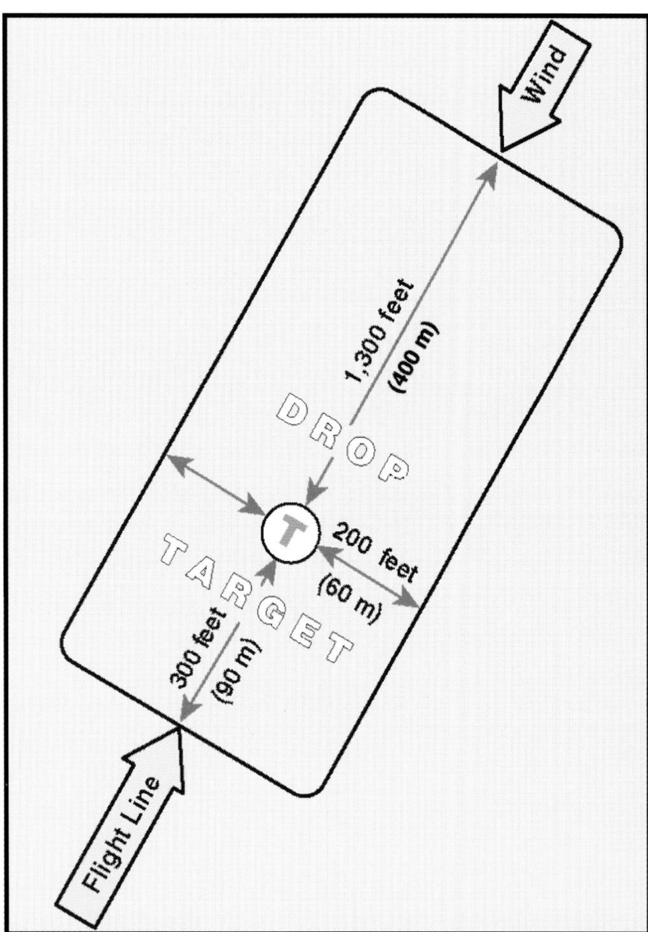

Figure 2.81 A typical paracargo drop zone.

Figure 2.82 Drop-zone markers are numbered if more than one zone is used.

- Clear all personnel, vehicles, and animals from the drop zone before cargo aircraft arrives.
- Keep drop zone clear until drop is complete.

Cargo is sometimes delivered to remote fireline locations by parachute (Figure 2.83). When this becomes necessary, the following procedures for drop-site selection, drop-site marking, and cargo retrieval should be used:

- ***Drop-site selection*** — Whenever possible, the drop site should meet the following criteria:
  - — The drop site should be a ridge top, meadow, or hillside. The bottoms of narrow canyons should not be used.
  - — For low-flying airplanes, the approach and exit should be clear of obstructions.
  - — Areas of excessive air turbulence, smoke, and clouds should be avoided.
  - — The drop site should be clear of snags, tall timber, and boulders.
- ***Drop-site marking*** — The site for a parachute drop should be marked in the same way as described earlier for paracargo drops.
- ***Retrieving cargo*** — When retrieving cargo dropped by parachute, the following procedures should be used:
  - — All cargo parachutes should be returned to base for repacking at the first opportunity.

Figure 2.83 Cargo is dropped to firefighters on the ground. *Courtesy of NIFC.*

— If parachutes are tangled in brush or trees, care must be exercised to avoid damaging them during recovery.

PREPARING APPARATUS FOR RESPONSE TO WILDLAND OR INTERFACE FIRES

In many cases, fire departments do not keep apparatus equipped and ready to handle major wildland and/or interface fires throughout the year. Before responding to these fires, it is necessary for firefighters to make some preparations to the apparatus and equipment. This is particularly true when responding to an incident that is very large or a great distance away. The following sections detail those preparations that should be made to structural apparatus responding to major wildland fires and to wildland apparatus responding to interface fires. While most of the items are common sense, they are certainly worthy of review as a reminder. A checklist of additional equipment should be prepared ahead of time so these items can be quickly and easily assembled when needed.

Preparing Structural Apparatus

Preparing structural apparatus to respond to a major wildland incident often requires a good deal of effort because these engines usually do not carry the necessary equipment year-round. Some departments keep these items ready to go in a "wildland kit" so they can be quickly loaded on apparatus; others add wildland fire fighting equipment at the beginning of the fire season (Figure 2.84). Once the engines are equipped for wildland response, they leave them equipped for the dual role for the remainder of the fire season. At a minimum, the following equipment should be added to structural engines before they respond to major wildland fires:

- Two lightweight hose packs, each containing 200 feet (60 m) of 1½-inch (38 mm) hose, shutoff nozzles, in-line tees with shutoffs, a hose clamp, and a spanner wrench
- One lightweight hose pack containing 200 feet (60 m) of 1-inch (25 mm) hose with nozzles to use for laterals off the in-line tee
- Wildland tools, including two McLeods, a Pulaski, a round-point shovel, and a file for sharpening tool blades

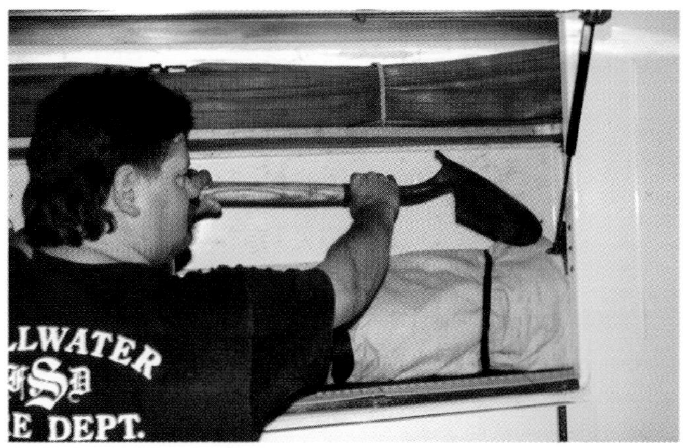

Figure 2.84 Wildland gear should be added to structural apparatus.

- Chain saw and chain sharpening tool
- At least one backpack water pump
- Water ejector for filling the booster tank from an auxiliary water source
- Fire shelters for all members of the company (plus at least one spare)
- Extra flashlight and portable radio batteries
- Nonperishable food and drinks
- Canteens and/or an insulated water cooler
- Road maps
- Agency fuel credit cards
- Spare drive belts, air filters, and engine oil

In addition to the minimum necessary equipment listed, the following items are also desirable to have on the apparatus:

- AM/FM radio
- Extra portable radios
- Cellular telephone
- Extra sets of personal protective equipment (PPE)
- Sleeping bags
- Portable awning

Preparing Wildland Apparatus

If there is a likelihood that wildland apparatus will be assigned to protect structures in the wildland/urban interface, the apparatus should be equipped for this assignment (Figure 2.85). If not normally carried on the apparatus, the following equipment should be added:

Figure 2.85 Structural gear may be needed on wildland apparatus.

Figure 2.86 Face-to-face communication is best. *Courtesy of NIFC.*

- Short section of soft suction hose for operating from a fire hydrant
- Ladder with sufficient reach to access the roof of a one-story building
- Pick-head axe or Halligan tool for forcible entry
- Portable pump
- Chain saw
- Water ejector for filling the apparatus water tank from an auxiliary water source
- Structural personal protective equipment (including SCBA) for all personnel

COMMUNICATIONS EQUIPMENT

Effective communication is critically important if wildland fire fighting activities are to be conducted safely and efficiently. Whenever the situation allows, direct, face-to-face voice communication is the preferred method of communication (Figure 2.86). However, the variety of environments in which firefighters in the wildland must attempt to communicate often makes the use of some form of electronic communications necessary. Because every form of fireground communication has certain advantages and disadvantages, no one method or system is optimally effective in every situation. This means that fireground communications may involve everything from face-to-face oral communication to the use of 800 megahertz (MHz) radios and very high technology satellite communications systems. Especially on large, complex incidents involving mutual aid units, all radio communications must be conducted according to the incident communications plan.

During pre-incident planning, the types and quantities of communications equipment needed by the anticipated resource requirements should be identified. Arrangements should also be made for obtaining the needed equipment (see Chapter 10, Fire Protection Planning). Agency officials should consult a communications specialist to become familiar with the inherent capabilities and limitations of the types of communications equipment available and to help determine the types and quantities that may be needed.

Radios capable of monitoring and transmitting on several frequencies selectively are the most flexible; however, adequate procedural controls must be established to prevent units from transmitting on any but their assigned frequency. As required under ICS, personnel should be trained to avoid using agency-specific codes or terminology and to use standard *clear text* (plain English) and general terminology to avoid misunderstandings between units and agencies (see Appendix A for clear-text terminology). Radio communications that are properly used and monitored provide the following advantages:

- A fire can be quickly surveyed and evaluated.

- Divisions/groups/sectors, air reconnaissance, lookouts, and dispatchers/ telecommunicators can be informed or consulted.
- Orders, plans, and information can be quickly given or received to meet changing conditions.

The remaining portion of this chapter is intended to familiarize the reader with the various means of communications and types of communications equipment that may be employed during the course of a wildland fire incident.

Fire Department Radios

Communications using some type of fire department radio equipment is the most common form of fireground communications at wildland fires (Figure 2.87). Radio provides instantaneous communication among fire fighting units, among fire units and the communications center, and among fire units and the rest of the fireground organization through the chain of command.

An initial attack fireground organization can usually operate effectively with all units on a single radio frequency. However, when the number of units from one agency assigned to a given incident increases or when units from many different agencies are assigned to one incident, the use of additional frequencies is usually necessary. Radio frequencies cannot be arbitrarily assigned and must not be unilaterally assumed by any operating unit during an incident. The assignment of frequencies reflects both pre-incident planning and the communications element of the incident action plan (plan stating overall objectives, strategies, and tactical actions) for that particular incident. Agencies that commonly work together should have written or working agreements for mutual frequency use and sharing. The number of frequencies needed on any given incident depends on the number of resources involved and the size of the management organization. Each of the following functions may require one or more radio frequencies on large incidents:

Figure 2.87 Portable radios make fireground communication more efficient.

- Command
- Tactical operations
- Support operations
- Air-to-ground communications
- Air-to-air communications

Depending on the magnitude of the incident, the following types of communications equipment may be used at a wildland fire incident:

- Base radios
- Mobile radios
- Portable radios
- Aircraft radios
- Radio repeaters

BASE RADIOS

Base radios are most often used at a fixed location such as the incident command post (CP) or the incident base (Figure 2.88). These are typically scaled-down versions of the radio equipment found in a fixed dispatch/telecommunications center. It is important for base radios to be capable of monitoring and transmitting on all the frequencies that will be used on a particular incident. Because of their critical role in the safety and effectiveness with which an incident is conducted and the chaos that would occur if they failed, base radios must be equipped with an uninterruptible power supply.

Depending on the geographical size of the incident and the required range for radio

Figure 2.88 The incident base may contain sophisticated communications equipment. *Courtesy of NIFC.*

communications, the base radio and other communications equipment (described later in this section) may be housed in a communications van equipped with a large dish antenna (Figure 2.89). Communications vans typically contain base radios, land-based and cellular telephones, computers, modems, and fax machines. If pagers or similar alerting devices are going to be used at the incident, the base radio should be able to activate these devices when notifications are required.

MOBILE RADIOS

Mobile radios are those mounted in vehicles. Typically, mobile radios are designed only to be used from the front seat of the vehicle cab. The operator speaks through a handheld microphone or through a headset that is part of a vehicle intercom system (Figure 2.90). Ideally, the radios in these vehicles should be able to communicate on any of the frequencies that may be used on an incident. At the very least, apparatus given specific functional or geographic assignments should be capable of communicating with each other and their command hierarchy. Most modern mobile radios are capable of transmitting and receiving on at least 32 frequencies, some on 200 or more. Most newer radio equipment is also capable of scanning all the frequencies programmed into the radio and

Figure 2.89 Communications vehicles make setup easier and faster. *Courtesy of NIFC.*

Figure 2.90 A firefighter uses a mobile radio.

setting one or more as *priority scan* channels. Having this capability is certainly an advantage when working on large incidents or incidents that involve multiple jurisdictions accustomed to operating on different radio frequencies.

Figure 2.91 Handheld portable radios are the most common fireground communications equipment. *Courtesy of NIFC.*

PORTABLE RADIOS

Portable radios, sometimes referred to as *walkie-talkies* or *Handie-Talkies,*™ are handheld radios that allow firefighters to remain in contact with each other, the apparatus, and the command staff when they are away from the mobile radio in the apparatus (Figure 2.91). Most portable radios have limited transmitting and receiving power, usually only 1 to 5 watts (compared to the 100 to 150 watts of a typical mobile radio). Therefore, portable radios also have limited range, perhaps less than 1 mile (1.6 km) when communicating from portable to portable. Communications distances between portable radios and mobiles or base stations vary depending on the topography and the capability of the mobile or base radio.

As with mobile radios, portable radios may be capable of operating on multiple channels. Newer portables are capable of handling up to 210 channels (Figure 2.92). Portables may also have *scan* and *priority* functions that allow the user to monitor multiple frequencies and select key frequencies for more frequent sampling.

Figure 2.92 Modern multichannel portable radios. *Courtesy of NIFC.*

RADIO REPEATERS

The range of portable radios may be extended using a repeater system. This system receives the signal from the portable, boosts its power, and then transmits the signal to the intended receiver. Two primary types of repeater systems are in common use. The first is a part of the mobile radio in the apparatus to which the crew using the portable radio is assigned. When a transmission is made on the portable radio, the repeater system in the mobile radio boosts the signal to the power the mobile radio is capable of and broadcasts the signal. The second type of repeater system is one in which fixed repeaters are located throughout a particular geographical jurisdiction (Figure 2.93). These repeaters pick up signals from base, mobile, and portable radios and boost their power so they are received by other radios.

Pagers

While they may have some fireground uses, pagers are most often used to notify volunteer, paid-call, and off-duty career firefighters to respond to the station or fire scene. Pagers are available in a wide variety of types and sizes and are

Figure 2.93 A fixed mountaintop radio repeater. *Courtesy of NIFC.*

capable of making contact with an individual or group of individuals selectively. Some pagers are activated by simply dialing a specific telephone number. Most pagers used in the fire service are activated by a transmitter tone from the dispatch/telecommunications center. Pagers provide information to the wearer in one of two manners: a voice message or a written display message.

It is conceivable that paging systems may be employed in large-scale, protracted wildland fire incidents. This is a particularly good way of contacting command staff or dispatching companies from staging to a particular location (Figure 2.94). If pagers are used on this type of incident, it is most desirable to have pager activation capabilities at the incident base.

Figure 2.94 A staging manager checks his pager.

Figure 2.95 Some fire apparatus are equipped with citizens band (CB) radios.

Alternative Communications Methods

Because of the long time frame and large-scale nature of many wildland fire incidents, alternative communication means may be employed during the course of operations. The use of each of these methods is highly dependent on the level of preparedness by the agency or agencies working the incident.

CITIZENS BAND (CB) RADIOS

Some small and rural jurisdictions rely on citizens band (CB) radios as their primary mode of mobile communication (Figure 2.95). This is primarily due to economics as many of these small departments lack the financial resources to purchase regular fire department mobile radios for their apparatus. In addition to their low cost, there are several positive attributes to the use of CB radios by fire departments:

- CB radios are better than having no radio communications at all.
- Most modern CB radios have 40 channels, which allows different parts of the organization to operate on different frequencies.
- Many personnel have CB radios in their personal vehicles, which can be a benefit when setting up a command structure.

In addition to these advantages, CB radios also have disadvantages:

- The public at large uses the same frequencies.
- The quality of the radio transmission may not be as good as with standard fire department mobile radios.
- The range of effective communication is generally less than that of standard fire department mobile radios.

HAM RADIOS

Ham radio clubs and individual ham operators have an extensive communications network. Through their base stations and mobile and portable radios, ham operators can access repeaters, satellites, and telephone systems. These operators and their equipment are often readily available on a volunteer basis, but they may take several hours to mobilize. Drills should be held with them before a major incident occurs in order to identify capabilities and limitations and to define roles.

In some jurisdictions a group called Radio Amateur Civil Emergency Services (RACES) may be available to assist emergency organizations. RACES is an organized group of ham operators who are ready and equipped to supplement or aid standard emergency radio systems when the need arises. Jurisdictions that have the availability of active RACES organizations are generally able to mobilize radio operators quicker than those areas where individual operators must be notified to mobilize. Operators in a RACES organization are also more likely to understand the needs of the emergency providers and are able to carry out these functions more quickly and easily.

LAND-BASED TELEPHONES

The public or field telephone has long been accepted as a means of communication between two points. If radio communication breaks down, telephones may have to be used. Even if good radio communications are in place, it may be advantageous to carry out certain functions (ordering resources, giving remote operation status reports, etc.) by telephone instead of tying up radio frequencies. Land-based telephones are most commonly used when the incident base or command post is located in a permanent structure that has telephone service. Land-based telephone service can also be installed in a command post or other location by telephone company representatives if it appears that an incident will be protracted.

CELLULAR TELEPHONES

Advances in cellular telephone technology have made telephone service much more available and useful to fire service personnel working on the emergency scene. Cellular telephone service is now available in the vast majority of geographical locations. Cellular phones may be handheld or mounted in a vehicle (Figures 2.96). They allow personnel to access the world telephone network without being hard-wired into a local telephone system. Telephone communications are transmitted as radio signals between the cellular phone and repeater/downlink equipment ("cell sites") that enter the call into the telephone system.

The use of cellular telephones is, however, not foolproof. If the emergency scene is in an area without a cell site, no service is available. Even within a cell-site area, there may be "dead spots" where

Figure 2.96 Cellular phones provide an additional means of fireground communication.

reception is not available. Even more troublesome are incidents in densely populated areas where routine cellular-call volume is normally heavy. Large-scale emergencies tend to generate a high volume of cellular telephone traffic from emergency providers and citizens alike. This increase in traffic can quickly overwhelm permanent cell-site equipment and block further calls from being made. When this happens, the cellular service provider should be notified so that additional temporary cell-site equipment can be activated to boost the system's capacity. Also, cellular phones are now available that detect overload in the land-based system and automatically transmit to a satellite link to complete coverage worldwide. Local jurisdictions should gather information on the availability of these services as part of their pre-incident planning.

FACSIMILE (FAX) MACHINES

On major wildland fires, the incident command post and/or incident base may have one or more facsimile (fax) machines. Fax machines can be very useful for transmitting and/or receiving written documents such as situation status reports and weather updates (Figure 2.97). They may also be used to obtain information on hazardous materials should these be involved in a wildland fire incident. Fax machines transmit their signals over telephone lines. They may be operated from either land-based or cellular telephone equipment.

COMPUTER MODEM

Computer modem equipment allows a computer at the incident command post, incident base, or other location to access databases and computer networks via a telephone line hookup (Figure 2.98). As with fax machines, computer modems may be attached to cellular or land-based telephone equipment. Computer equipment may have an endless variety of uses on an emergency scene. Virtually any function that would normally be performed in an office or dispatch/telecommunications center may be performed on location by using computer modem equipment.

Figure 2.97 Fax machines may be needed on large incidents. *Courtesy of NIFC.*

Figure 2.98 Computer modems facilitate data transfer. *Courtesy of NIFC.*

Advanced Technology Communications Systems

Ongoing research conducted by the communications industry, aerospace industry, and the military has allowed communications technology to continue to evolve. Unfortunately, the application of these technological advances in the fire service has not kept pace with the development. But, because these advancements are likely to become more available to the fire service as time goes on, they warrant a brief discussion in this manual. However, the pace at which technological advancements are being made can render very expensive equipment obsolete in a relatively short time. In addition, fire service agencies are highly interdependent under mutual aid agreements. Therefore, agencies should work together in joint purchase agreements to allow greater economies and to maintain equipment compatibility.

GEOGRAPHIC INFORMATION SYSTEM (GIS)

Geographic information systems (GIS) are designed to provide a computer-readable description of geographic features in a particular area. From a wildland fire fighting standpoint, they can provide valuable information on terrain, structure locations, fuels, water supply sources, streets, and previous fires. In this system a computer stores, maintains, and displays data on specific segments of the jurisdiction being covered (Figure 2.99). In urban/suburban areas, addresses and occupancy information on individual structures may be stored in the GIS. This information may be useful to dispatchers/telecommunicators, incident commanders, planning personnel, and technical specialists assigned to an incident.

Figure 2.99 A firefighter operates a GIS at an incident command post. *Courtesy of NIFC.*

MOBILE DATA TERMINAL (MDT)

Of the advanced communication technologies covered in this section, mobile data terminals (MDTs) are most commonly used by the fire service. MDTs are radio-operated computer terminals that link emergency vehicles with the dispatch/telecommunications center. An MDT looks like a small personal computer mounted on a pedestal near the vehicle dashboard (Figure 2.100). Many MDTs are equipped with status buttons that allow the dispatch/telecommunications center to stay apprised of the unit's status (en route, at scene, available, etc.) without the need for verbal radio transmissions. MDTs can be used to transmit dispatch information, incident/patient status information, special messages that are not appropriate for transmission over the airwaves, chemical information, and maps and charts. MDTs are capable of two-way communication.

Figure 2.100 Many fire apparatus now have MDT capability.

GLOBAL POSITIONING SYSTEM (GPS)

Global positioning systems (GPS) were originally developed by the U.S. military as a means of tracking troops in combat. This technology is now available to civilian emergency services. In a GPS, each vehicle is equipped with a radio transmitter. The signal that is transmitted bounces off a satellite and is received by an automatic vehicle locator (AVL) at the dispatch/telecommunications center. The position of the vehicle is then shown on a map of the jurisdiction using the system. These systems are generally capable of determining the location of a vehicle to within approximately 100 feet (30 m) of its actual position. Handheld GPS devices also allow units in the field to determine their exact positions (Figure 2.101).

Global positioning systems are typically used in conjunction with computerized dispatch/telecommunications systems, MDTs, and geographic information systems. There are several uses for the GPS. Two of the most common are tracking companies on the fire scene and dispatching the closest available companies to an emergency.

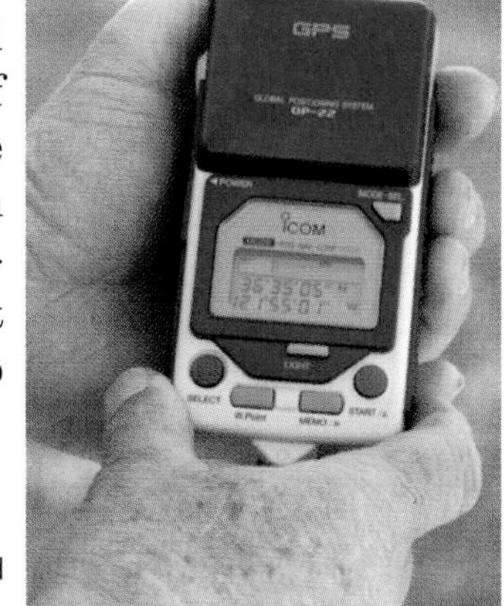

Figure 2.101 A typical handheld GPS unit.

WILDLAND FIRE FIGHTING TOOLS AND PERSONAL PROTECTIVE EQUIPMENT 3

LEARNING OBJECTIVES

This chapter provides information that addresses the following objectives of NFPA 1051, *Standard for Wildland Fire Fighter Professional Qualifications* (1995 edition):

Wildland Fire Fighter I

3-1 **General.** The Wildland Fire Fighter I shall meet the job performance requirements defined in Sections 3-1 to 3-5 of this standard.

3-1.1 *Prerequisite Knowledge:* Fireline safety, use and limitations of personal protective equipment, agency policy on fire shelter use, basic wildland fire behavior, basic wildland fire tactics, fire fighters role within the local incident management system, and first aid.

3-1.2* *Prerequisite Skills:* Basic verbal communications, the use of appropriate personal protective equipment.

3-3.1 **Definition of Duty.** Activities in advance of fire occurrence to ensure safe and effective suppression action.

3-3.2 Maintain assigned personal protective equipment, given the standard equipment issue, so that the equipment is serviceable and available for use on the fireline and defects are recognized and reported to the supervisor.

3-3.2.1 *Prerequisite Knowledge:* Maintenance of personal protective equipment including inspection, the recognition of unserviceable items, and proper cleaning procedures (including manufacturer's and authority having jurisdiction's recommendations).

3-3.3* Maintain assigned suppression hand tools and equipment, given tools and equipment, agency maintenance specifications, supplies and small tools, so that assigned equipment is safely maintained, serviceable, and defects are recognized and reported to the supervisor.

3-3.3.1 *Prerequisite Knowledge:* Inspection of tools and assigned suppression equipment, the recognition of unserviceable items, and safe maintenance techniques.

3-3-3.2* *Prerequisite Skills:* Sharpening and other maintenance techniques for assigned suppression equipment, and use of appropriate maintenance equipment.

3-5.4.2* *Prerequisite Skills:* Proper use of hand tools, fire stream practices, and agent application.

Wildland Fire Fighter II

4-3.1 **Definition of Duty.** Responsibilities in advance of fire occurrence to ensure that tools, equipment, and supplies are fire ready.

4-3.2* Maintain power tools and portable pumps, given agency maintenance specifications, supplies, and small tools, so that equipment is safely maintained, serviceable, and defects are recognized and repaired.

4.3.2.1 *Prerequisite Knowledge:* Maintenance procedures for power tools and portable pumps.

4-3.2.2 *Prerequisite Skills:* Power tool and portable pump preventative maintenance and repair.

4-3.3 Inspect tools and equipment, given agency specifications, so that availability of the tools and equipment for fire use is ensured.

4-3.3.1* *Prerequisite Knowledge:* Tool and equipment inspection guidelines.

4.5.4* Operate a chain saw, given an assignment at a wildland fire and operational standards, so that the proper tool is selected, and the assignment is safety completed.

4-5.4.1 *Prerequisite Knowledge:* Agency operational standards for tree size up, felling, limbing and bucking, chain saw safety, tool selection, and personal protective equipment used during saw use.

4-5.4.2* *Prerequisite Skills:* Proper tree size up, site preparation, handling and cutting techniques, use of wedges, and saw and equipment transportation.

4.5.5 Operate portable water delivery equipment, given an assignment at a wildland fire and operational standards, so that the proper portable pump and associated equipment is selected, desired nozzle pressure is attained, and flow is maintained.

4.5.5.1* *Prerequisite Knowledge:* Basic hydraulics, portable pump and system capabilities, operation of portable pumps, basic drafting, and associated equipment.

4.5.5.2 *Prerequisite Skills:* Placement, operation, and system set up.

Chapter 3
Wildland Fire Fighting Tools and Personal Protective Equipment

INTRODUCTION

As with any aspect of the emergency services, firefighters must be intimately familiar with the tools used and the personal protective equipment (PPE) worn. Knowledge of the proper selection, use, and care of the various tools used in wildland fire fighting aids firefighters in performing their job as efficiently and effectively as possible. Likewise, knowledge of the proper donning, care, capabilities, and limitations of personal protective equipment gives firefighters a better sense of which situations are tenable and which are not.

This chapter discusses the most common types of tools and portable equipment used by firefighters. Some jurisdictions may use other tools that are not covered in this manual. If that is the case, the jurisdiction should make sure that firefighters are given adequate information and hands-on training on each tool.

The personal protective equipment covered in this chapter is limited to those types specifically intended for use on wildland fire fighting operations. Many fire departments whose primary mission is to fight structure fires routinely have their firefighters attack wildland fires wearing either structural turnout clothing or little protective clothing at all. Neither of these practices is recommended. Also, personnel wearing wildland protective clothing should not attempt to make interior or close-up exterior attacks on structure fires. Fire departments that routinely respond to both types of fires should provide their firefighters with both types of protective equipment. For more information on structural protective equipment see the IFSTA **Essentials of Fire Fighting** or **Fire Department Occupational Safety** manuals.

HAND TOOLS

Although mechanized apparatus and equipment are widely used, hand tools also have an important place in fighting wildland fires. Some of the hand tools are conventional, and some are adaptations of conventional tools. However, some have been specially developed for fighting fires in wildland fuels. The selection and design of hand tools usually depend upon the situations where they are likely to be used and upon local preference. Generally, grass fires require more use of scraping or smothering tools, while brush fires require more cutting tools.

When carrying hand tools, firefighters typically walk and work 10 feet (3 m) apart for safety (Figure 3.1). Tools should be carried in the following manner:

- At the tool's balance point
- At the side close to the body, not on the shoulder

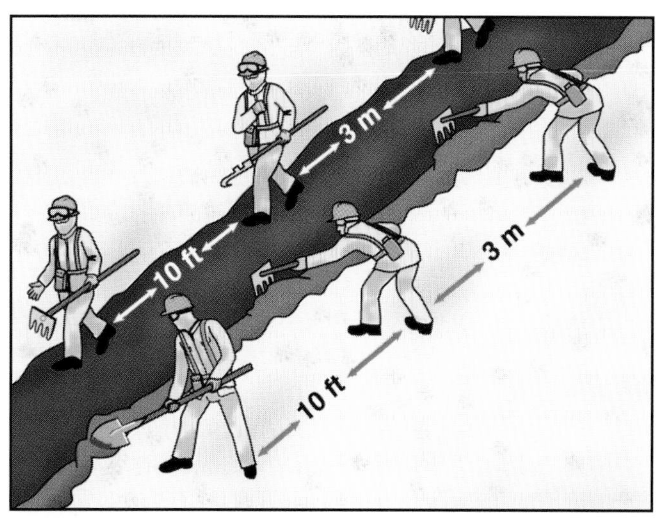

Figure 3.1 For safety, firefighters should walk and work 10 feet (3 m) apart.

- With cutting edges away from the body
- On the downhill side when walking across a slope

Cutting Tools

Cutting tools are primarily for fireline construction, including cutting brush and small trees (Figure 3.2). Normally, many hours of training and field experience are needed before the required degree of skill in using these tools is attained. If used improperly, these tools can be dangerous, so appropriate caution must be used at all times. The most common hand cutting tools for wildland fire fighting are the following:

- Axes
- Pulaski tools
- Brush hooks

Figure 3.2 A firefighter uses a Pulaski to construct a fireline.

AXES

There are two types of axes commonly used in wildland fire fighting: the single-bit and the double-bit. A single-bit axe has a cutting edge on one side of the head and a flat, striking surface on the opposite side (Figure 3.3). A double-bit axe has cutting edges on both sides of the head (Figure 3.4). Axes are effective in mop-up operations for felling snags, breaking stumps and logs, and driving wedges (single-bit axes only). Axe handles are made of wood or fiberglass. Cutting edges should be sharpened back approximately 2½ inches (65 mm) on each side with an even bevel on both sides (Figure 3.5).

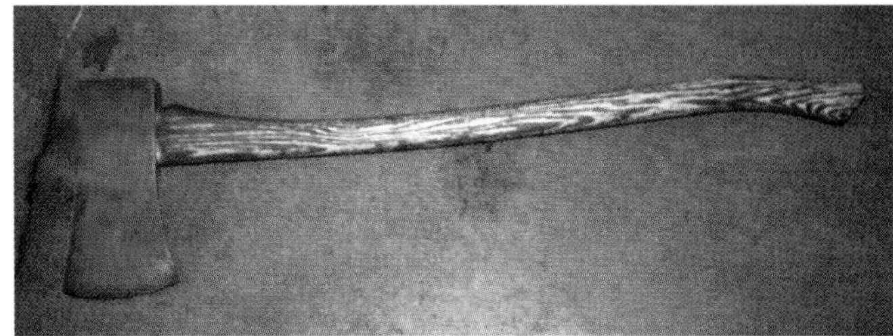

Figure 3.3 A typical single-bit axe.

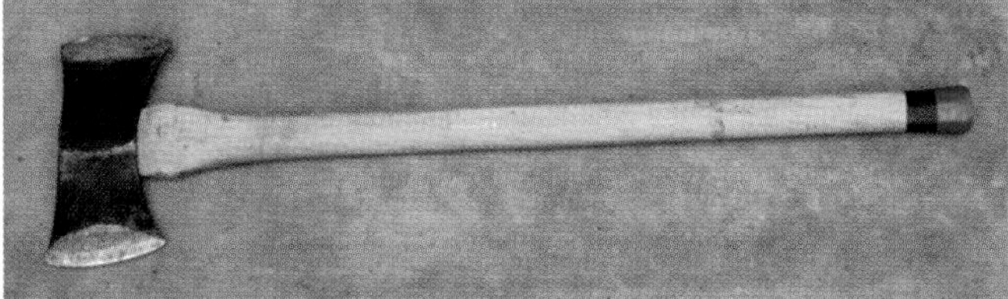

Figure 3.4 A typical double-bit axe.

PULASKI TOOLS

Pulaski tools are dual-purpose tools that have two different types of bits on the head (Figure 3.6). One side is a cutting edge similar to an axe blade. The other edge, called the grubbing edge, is used to dig roots and trenches. These are effective in cutting firelines and in mop-up operations. The cutting edge should be tapered 2 inches (50 mm) wide with an even bevel on each side. The grubbing edge should be beveled ⅜-inch (10 mm) wide on a 45-degree angle on one side of the head only (Figure 3.7).

BRUSH HOOKS

A brush hook, sometimes called a *brush axe*, is used to cut down small shrubs, trees, and tall grasses. The brush hook has a handle that is very similar to a standard axe. However, the head of the tool is in the shape of a *J*. A sharpened cutting edge is on the inside of the head (Figure 3.8). The brush hook is swung like an axe when cutting.

Scraping Tools

Scraping tools are used for fireline construction and mop-up operations. They can be used to clear

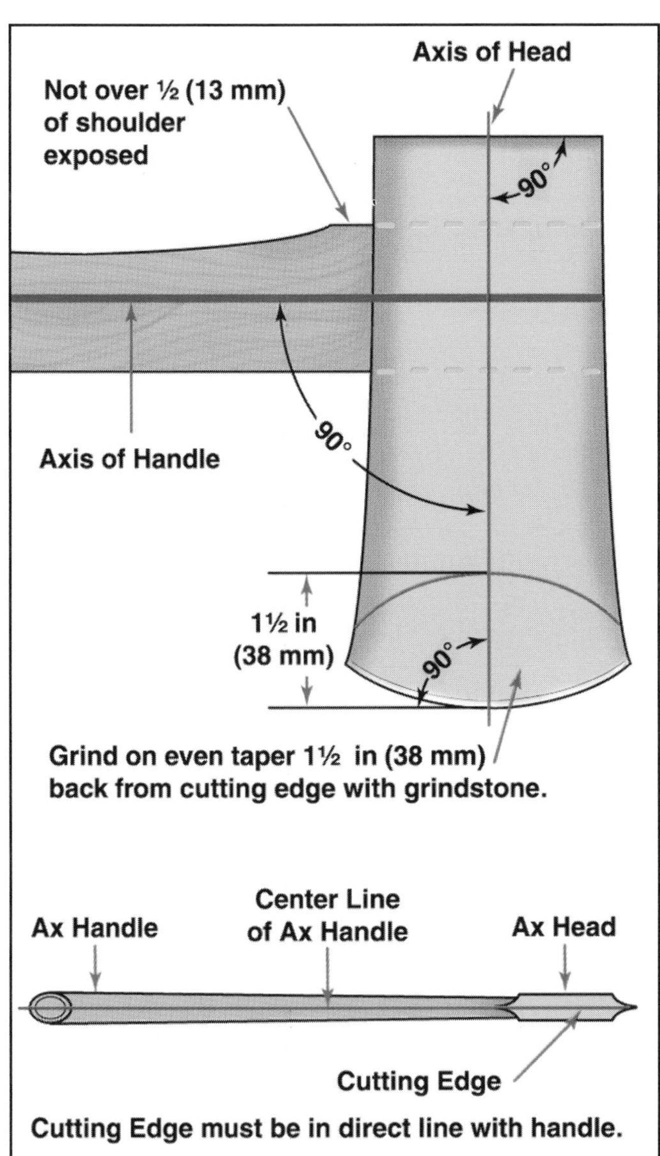

Figure 3.5 Axe blades must be sharpened to the correct angle.

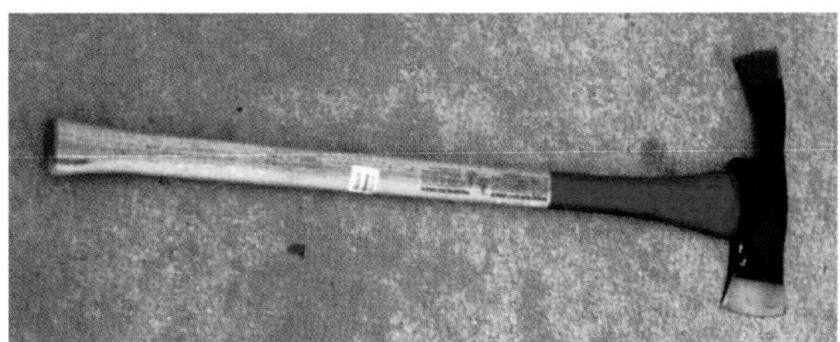

Figure 3.6 A typical Pulaski.

away small vegetation and debris to assist in making a fireline. They can also can be used to sift through and break up small vegetation and debris. The most common types of scraping tools used in wildland fire fighting are the following:

- Shovels

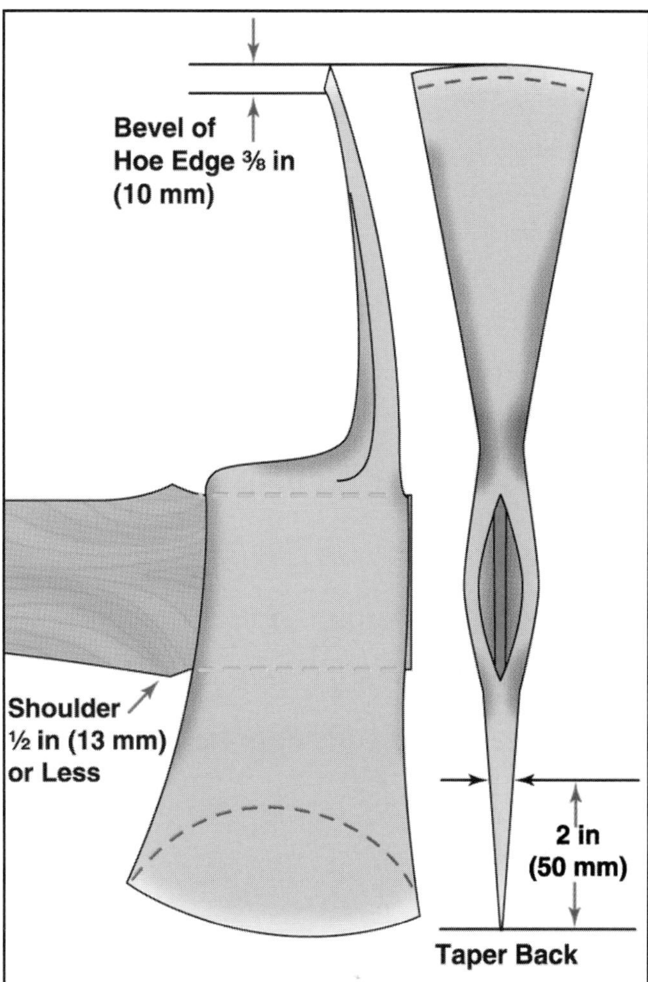

Figure 3.7 The proper bevel of a Pulaski tool.

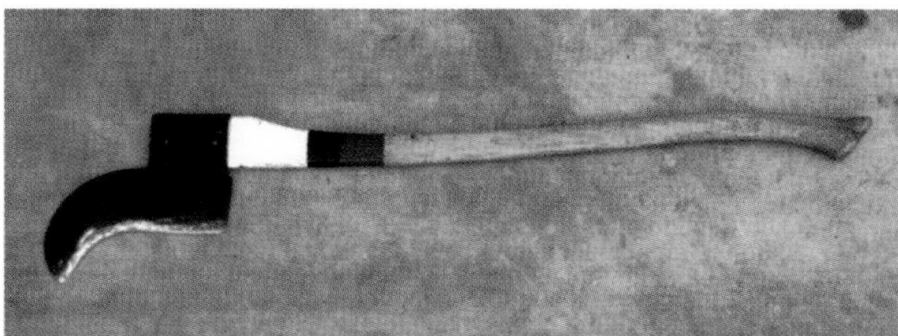

Figure 3.8 A typical brush hook.

- McLeod tools
- Fire rakes
- Hoes
- Combination tools
- Wire brooms

SHOVELS

Though a very common and relatively simple tool, shovels have a wide variety of applications to wildland fire fighting. The most common type of shovel used in wildland fire fighting is the short-handled, round-point shovel (Figure 3.9). Shovels may be used for the following:

- Digging
- Scraping
- Smothering
- Beating
- Cutting light fuels
- Throwing dirt

The blade of a round-point shovel should be sharpened starting 4 inches (100 mm) from the heel on each side of the blade. When properly sharpened, a subtle point is formed at the tip of the blade (Figure 3.10).

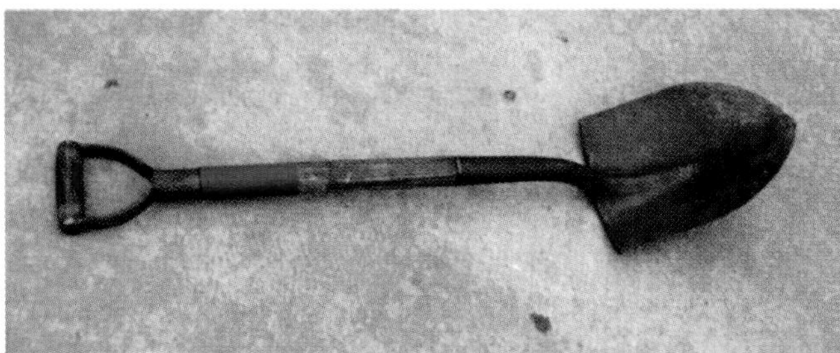

Figure 3.9 A short-handled, round-point shovel.

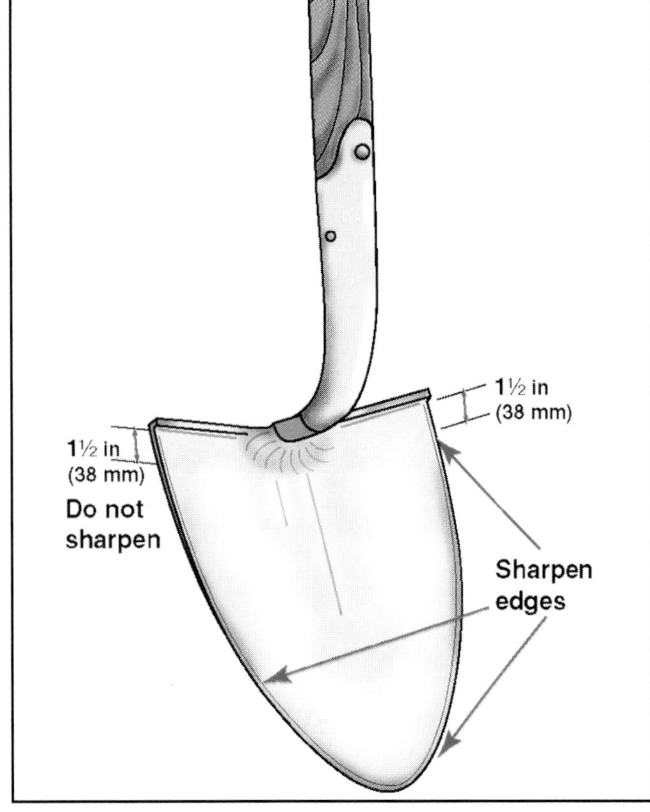

Figure 3.10 Shovel blades should be sharpened as shown.

MCLEOD TOOLS

The McLeod tool is used extensively in fireline construction and mop-up operations (Figure 3.11). One side of the head consists of a solid hoe blade for cutting grass, deep litter, and light brush. The hoe portion is also used for trenching and grubbing. The other side of the head consists of a rake with 5 to 7 tines. The rake tines are effective in raking pine needles, duff, and leaf mold. The hoe blade portion should be beveled to 45 degrees on the outside face of the blade (Figure 3.12).

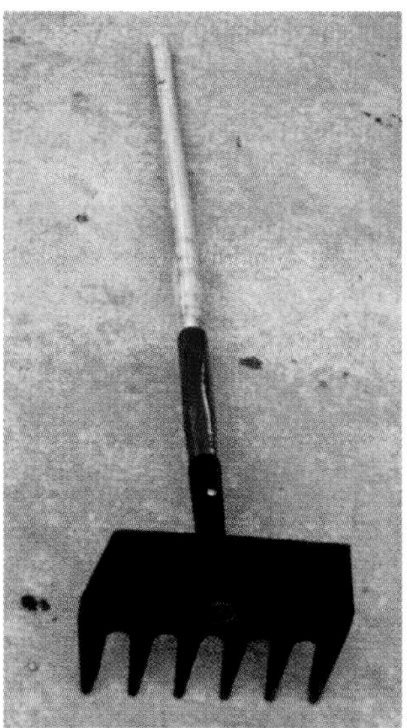

Figure 3.11 A typical McLeod tool.

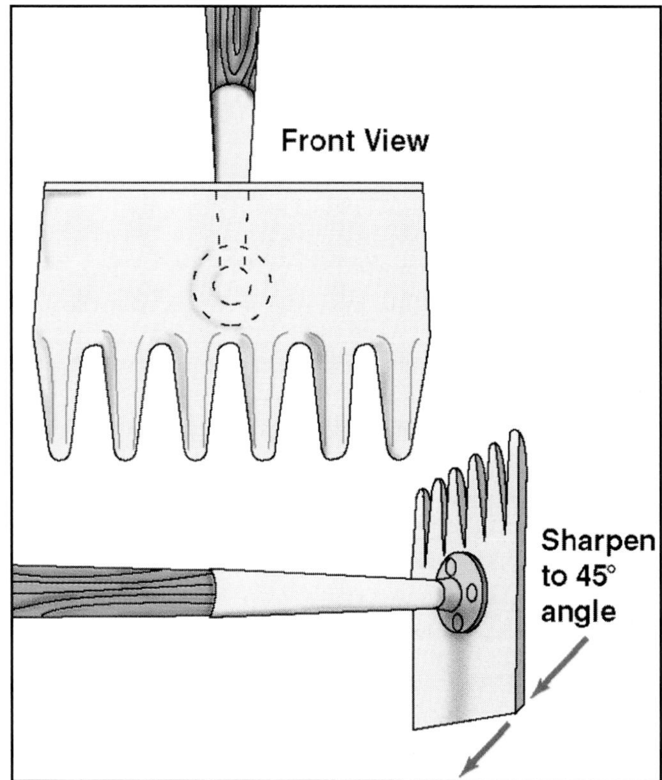

Figure 3.12 The proper bevel of a McLeod tool.

FIRE RAKES

The fire rake, sometimes referred to as the *council rake* or *council tool*, is used in fireline construction, mop-up, and burning-out operations in areas where the hoe function of the McLeod is not needed (Figure 3.13). It is well-suited to fireline construction in deciduous leaves. Fire rakes resemble standard garden rakes except that the tines are broader and triangular in shape.

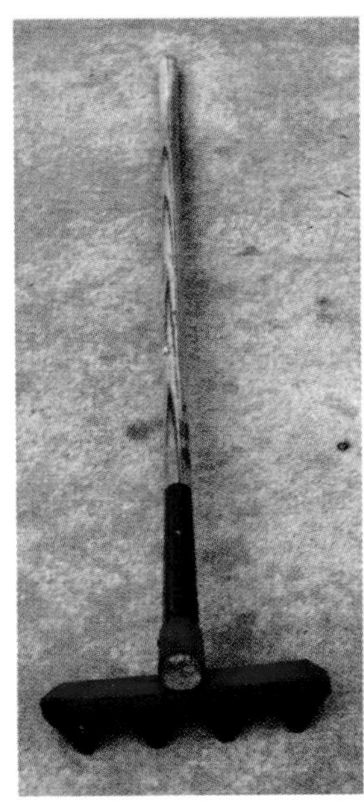

Figure 3.13 A typical council tool or fire rake.

HOES

In some jurisdictions, standard garden-variety hoes may be used in fireline construction and mop-up operations. One type of hoe that is often used has a flat blade on one side of the head and 2 or 3 tines on the opposite side (Figure 3.14). It is used and maintained in the same manner as the McLeod tool. The hazel hoe is a heavy-duty type of hoe used in wildland fire fighting. Other heavy-duty hoes are also used (Figure 3.15).

Figure 3.14 One type of hoe used in wildland fire fighting.

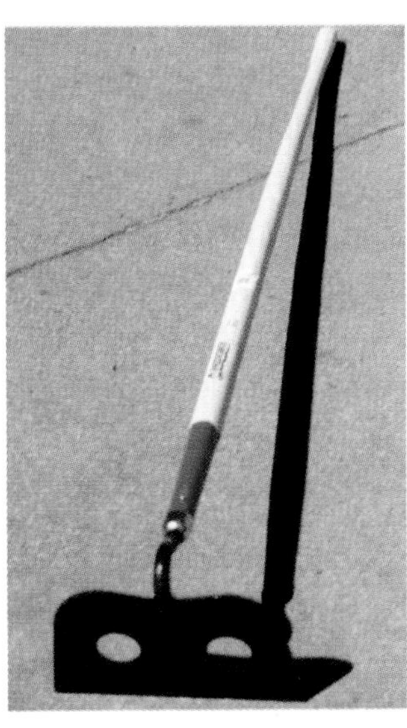

Figure 3.15 A typical heavy-duty hoe used in wildland fire fighting.

COMBINATION TOOLS

The *combination tool* is a versatile long-handled implement with a two-part, multiposition head (Figure 3.16). With the release of the locking collar, the configuration of the head can be changed so the tool can be used as a shovel, pick, hoe, or various combinations of these modes. However, the pick should not be used as a prying tool.

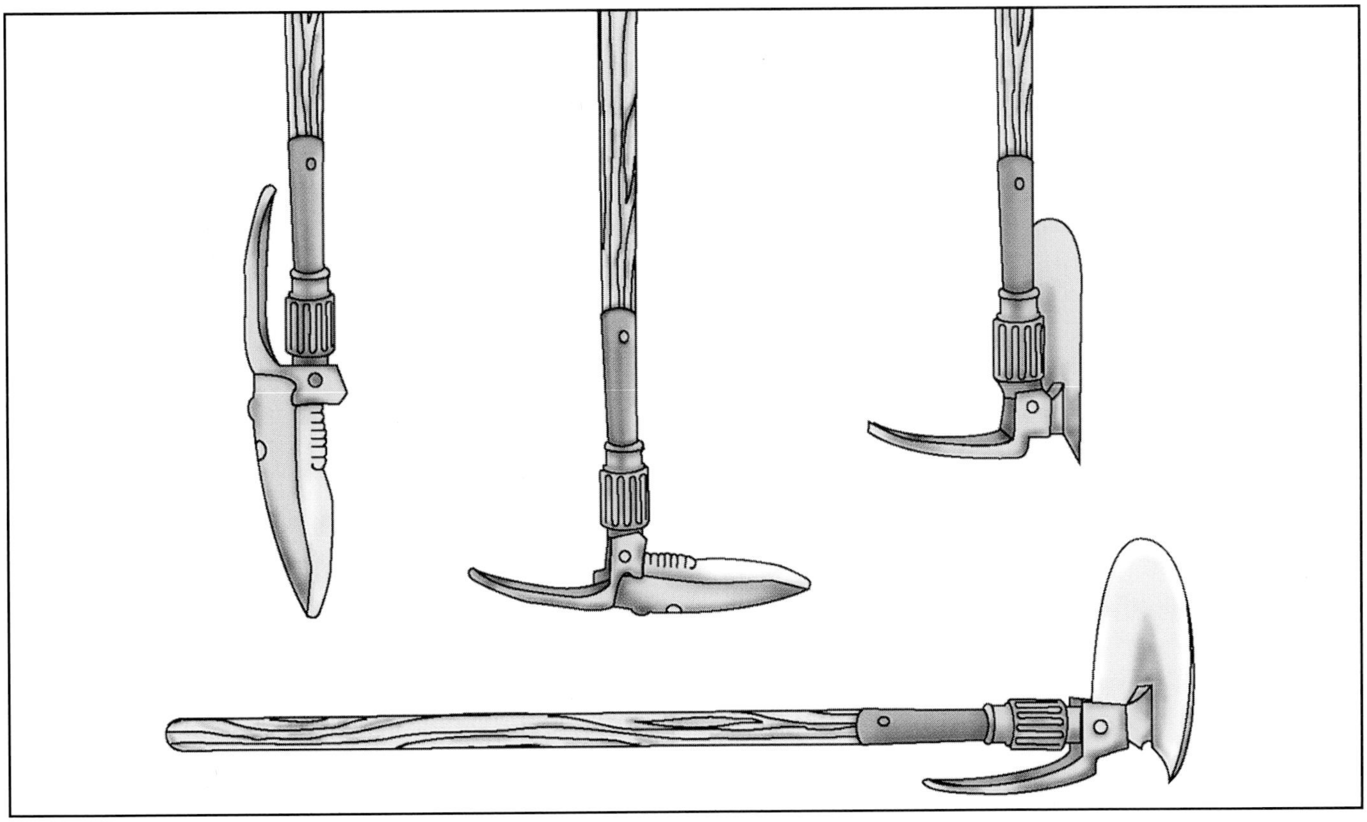

Figure 3.16 The combination tool in its various configurations.

WIRE BROOMS

Another tool designed for use in grass, grain, and moss fires is the wire broom. It resembles a push broom with wire bristles (Figure 3.17). The wire broom is especially effective in volcanic areas where light, sparse grasses protrude through a layer of small lava rocks. The grasses are literally swept away to create an effective fireline.

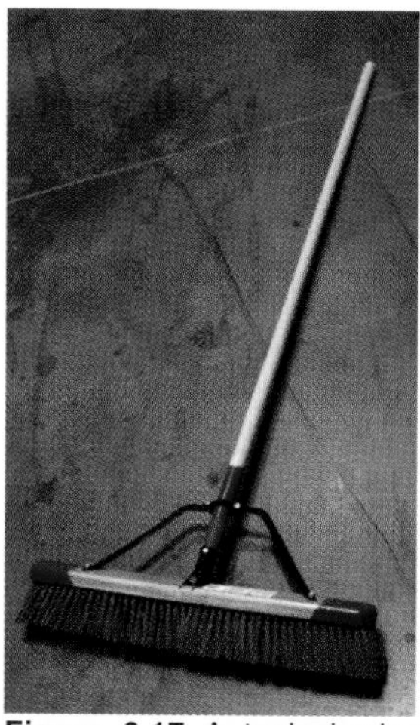

Figure 3.17 A typical wire broom used in wildland fire fighting.

Smothering Tools

Smothering tools are used to suppress fires in light fuels such as pastures, pine-needle litter, light hardwood litter, and light grasses. Smothering tools are very effective when used in conjunction with a backpack pump or fire rake. The smothering tool is used to knock down the flames, and then the fire is moped up with water from the backpack pump or by scraping with the fire rake. Although such things as wet spruce boughs can be used for this purpose, the tools used most often for smothering wildland fires are the fire swatter and the gunnysack.

FIRE SWATTERS

The *fire swatter,* sometimes called a *fire flail* or *flapper,* is a long-handled tool with a rubber or neoprene flap attached to one end (Figure 3.18). The flap is usually square in shape with each side being 16 to 24 inches (400 mm to 610 mm) in length. The flap may be replaced if it becomes damaged by heavy use. In use, the flap is swatted or dragged along the edge of a fire (Figure 3.19). If the fire is hit too hard, burning embers may be scattered into the unburned area and spread the fire.

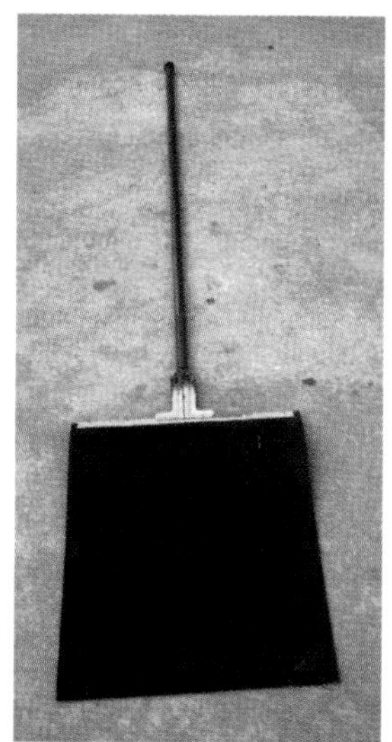

Figure 3.18 A typical fire swatter or flail.

Figure 3.19 A firefighter uses a fire swatter on a fire in short stubble. *Courtesy of NIFC.*

GUNNYSACKS

Gunnysacks are large, flat burlap bags that are thoroughly wet with water and then used to swat out a fire in much the same manner as a fire swatter. The burlap may need to be rewetted occasionally and must be kept moving at all times to prevent it from catching fire. Gunnysacks should be rinsed and hung to dry after use.

Hand Tool Care and Maintenance

The principles of care and maintenance are generally similar for all of the hand tools listed earlier. Handles should be checked to assure that they are smooth, free of cracks or splinters, properly aligned, and securely attached to the tool head (Figure 3.20). Wooden handles that are

Figure 3.20 A firefighter inspects the handle of a hand tool.

splintered should be sanded and refinished. Handles that are loose or cracked should be replaced.

Tool heads should be checked to ensure that they are not broken or cracked. Damaged or dull cutting edges should be sharpened using a hand file (Figure 3.21). Edges should be sharpened as described earlier for each type of tool. All cutting tool heads should be maintained free of paint or rust and given a light coating of oil. Tools with sharp cutting edges or pointed heads should have guards over the sharp edges when they are not in use (Figure 3.22).

Figure 3.21 A double-bit axe is sharpened in the field. *Courtesy of NIFC.*

Figure 3.22 The blades of cutting tools should be covered when not in use.

All tools should be stored in their mounting brackets on the apparatus. Tools stored in compartments should be arranged in an orderly manner so that they are easily accessible and not subject to damage.

CAUTION: Personnel must never ride with loose tools or equipment. Tools should be secured in brackets or carried in compartments.

BACKPACK PUMPS

A *backpack pump* is a type of portable fire extinguisher that carries plain water or a foam/water solution. It is used to attack small fires and hot spots and to overhaul areas that are not within reach of hoselines. The backpack pump is known by a variety of names, including the following:

- Pump can/tank
- Bladder packs
- Indian tank
- Trombone pump tank

Backpack pumps discharge water when the wearer operates the manual pump on the extinguisher. There are two basic operating mechanisms for these types of units. One type uses a sliding piston pump in the nozzle (Figure 3.23). A jet of water is discharged each time the wearer moves the pump handle out and back in. The second type uses a piston pump within the water tank itself. This piston pump is operated by pumping a lever arm on the side of the tank. As the lever is operated, it creates pressure within the tank. A hose with a squeeze-handle nozzle is held in the opposite hand. When the handle is squeezed, a continuous stream of water is discharged as long as pressure remains in the tank. As the pressure decreases, the lever is pumped to repressurize the tank.

Figure 3.23 A trombone-type backpack pump.

The actual tank of the backpack pump may be solid or collapsible. Most newer solid tanks are made of some sort of lightweight plastic or fiberglass material. Older models were commonly made of galvanized or stainless steel. Collapsible tanks, sometimes called *Fedco® tanks,* resemble a rubber or neoprene bladder (Figure 3.24). Both the solid and collapsible models carry 5 gallons (20 L) of water. One slightly different design is a collapsible, bladder-type worn as a vest rather than on the back (Figure 3.25). Because the weight is more evenly distributed with the vest design than with the backpack design, the vest type units may carry up to 8 gallons (32 L) of water.

Depending on local preference, noncollapsible backpack pumps may be stored either full of water or empty. Collapsible backpack pumps are usually stored empty. Regardless of the type, units that are stored empty should be checked periodically to make sure all seals are pliable and not dried out. All parts should be checked for dirt or rust that may affect their operation.

Figure 3.24 A firefighter wearing a bladder-type backpack.

Figure 3.25 A typical vest-type backpack. *Courtesy of Good Will Fire Co. #1, New Castle, DE.*

CHAIN SAWS

The chain saw is the most commonly used portable power tool in wildland fire fighting (Figure 3.26). Chain saws save many hours of labor, especially when working in fuels such as heavy brush, slash, and timber. In order to safely and efficiently use a chain saw, the firefighter must know the major parts of the chain saw, the basic principles of chain-saw operation and safety, and the proper care and maintenance of the chain saw.

Parts of a Chain Saw

Firefighters should be familiar with a number of major components of the chain saw. Each of these is outlined and briefly described as follows (Figure 3.27):

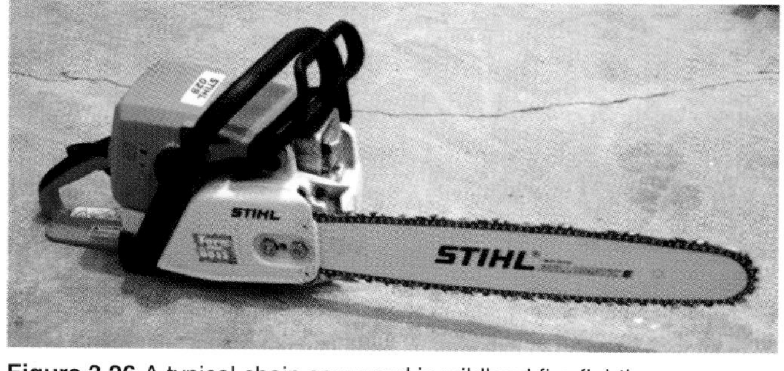

Figure 3.26 A typical chain saw used in wildland fire fighting.

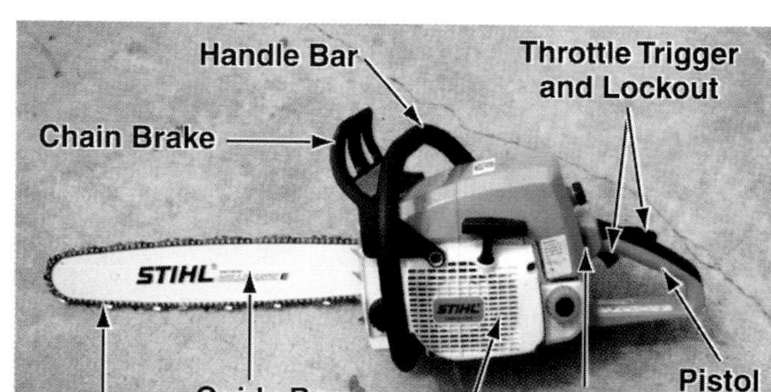

Figure 3.27 The major parts of a chain saw.

- ***Engine*** — Most chain saws used for wildland applications have a two-cycle, gasoline-powered engine. On most saws, the engines are controlled by an ON/OFF switch (Figure 3.28). Manufacturer's instructions should be followed for starting, operating, stopping,

Figure 3.28 The ON/OFF switch on a chain saw.

and maintaining the saw. As with any other two-cycle engine, these units operate on a gasoline/oil mixture. Only the proper mixture of fuel can be used in the saw without damaging the engine. Therefore, it is critically important that fuel cans be clearly labeled (Figure 3.29).

Figure 3.29 A clearly labeled safety can.

- ***Guide bar*** — The guide bar keeps the chain in line so that a straight cut may be made. It is grooved so the chain rides smoothly and securely around the track.
- ***Saw chain*** — The saw chain is composed of left and right cutters, drive links, and tie straps. Most modern saw chains use carbide-tip cutters for maximum strength, longevity, and cutting ability. Most saws also have an automatic oiling system to keep the saw chain and guide bar lubricated.
- ***Throttle trigger and lockout*** — The throttle trigger controls the amount of fuel entering the engine, which in turn controls the speed of the saw chain. The lockout prevents the throttle trigger from being depressed unintentionally.
- ***Choke lever*** — This lever activates the carburetor's choke to regulate the air-to-fuel mixture to the engine for easier starting.
- ***Pistol grip and handle bar*** — These devices are used to handle the saw. The throttle trigger and lockout are located on the pistol grip.
- ***Chain brake*** — This device is adjacent to the handle bar. The chain brake is activated by the chain binding in the material being cut, and it causes the saw chain to immediately stop moving.

Chain-Saw Operation and Safety

Because chain saws are potentially very dangerous, firefighters must follow basic operational and safety procedures whenever they use them. In addition to ensuring the firefighter's safety, following these procedures also results in getting the job done faster and more efficiently.

SAFETY PROCEDURES

One of the most basic safety procedures when operating a chain saw is to wear proper clothing and personal protective equipment. Firefighters typically wear their standard protective clothing when operating a chain saw (protective clothing for wildland fire fighting is covered in detail later in this chapter). All personal protective equipment should be worn when operating a chain saw. In addition to standard wildland turnout clothing, the following items should be worn when operating a chain saw:

- ***Shoulder pads*** — To reduce fatigue, the chain saw should be carried on the shoulder when it must be carried a long distance. Shoulder pads protect from bruises and cuts when the chain saw is carried in this manner (Figure 3.30).
- ***Leg protectors*** — Leg protectors, also called *chaps*, protect the operator's legs should the running chain come in contact with them (Figure 3.31).

Figure 3.30 Shoulder pads protect the firefighter's shoulders. *Courtesy of NIFC.*

Figure 3.31 Chaps protect the operator's legs. *Courtesy of NIFC.*

- ***Hearing protection*** — Hearing protection, in the form of earplugs or earmuffs, should be worn when operating a chain saw (Figure 3.32). Other firefighters in close proximity to an operating chain saw should also wear hearing protection.
- ***Eye protection*** — Safety glasses, goggles, or faceshields must be worn to protect the eyes against contact with wood chips or branches (Figure 3.33). The eye protection should be impact resistant and antifogging.

Figure 3.32 Hearing protection should be worn when operating noisy equipment.

Figure 3.33 To protect from flying chips, eye protection is a must. *Courtesy of NIFC.*

The following is a list of recommended safety measures when operating a chain saw or working near an operating saw:

- Keep the saw a safe distance from the legs.
- Never cut with the saw motor higher than chest level.
- Clear debris from the cutting location so that the tip of the guide bar does not hit it. While clearing debris, engage the chain brake or turn off the saw.
- Do not cut beneath a hazard overhead. Always look up before cutting.
- Initiate and complete all cuts with the chain at high speed.
- Watch for whipping branches and limbs when cutting small-diameter material.
- Cut limbs close to their crotch (point of attachment).
- Cut limbs so that a flat end is left. Do not leave pointed ends that may injure people if they lean or fall on them (Figure 3.34).

Figure 3.34 Limbs should be cut off square.

OPERATIONAL PROCEDURES

Another key factor in the proper and safe operation of a chain saw is to start the saw in the correct manner. The saw should be started at or very near the location where the cutting is to be done. There may be differences in starting procedures for different brands and models of saws, so the manufacturer's instructions must always be followed.

Once the saw is running, it is essential that the operator maintain control through proper stance and handling at all times (Figure 3.35). An operator must have firm footing before attempting to operate the saw. Loose materials, such as loose bark or rocks, should be removed from underfoot. The feet should be spread at least shoulder-width apart in a balanced stance. A firm grip on the saw should be maintained with both hands. Hands should be positioned so that they do not have to be crossed at any time. When possible, a second firefighter should work with the operator to pull debris away as it is cut (Figure 3.36).

Figure 3.35 Proper stance and handling are needed for chain-saw safety.

Figure 3.36 A second firefighter clears debris away as it is cut. *Courtesy of NIFC.*

Chain-Saw Care and Maintenance

Because chain saws tend to get heavy use under extreme conditions, it is important to use proper care and maintenance procedures to ensure the saw operates properly and is ready for its next use. Just as there are differences in starting procedures among different brands and models of saws, there are also different maintenance requirements. Manufacturer's recommendations for routine and periodic maintenance should be strictly followed.

In general, saws should be kept as clean as possible. Dirt, dust, and debris should be wiped from the saw or blown off if a compressed air source is available. Particular attention should be paid to dirt and debris around the openings for fuel or chain oil and the air filter (Figure 3.37).

Figure 3.37 Chain saws should be kept clean.

The three most common saw-care functions performed by firefighters in the field are as follows:

- Refueling the saw
- Purging the saw for travel and storage
- Lubricating the saw chain and guide bar

REFUELING THE SAW

As mentioned earlier, most chain saws are equipped with a two-cycle engine that runs on a gasoline/oil fuel mixture. The engine is lubricated through this mixture — using straight gasoline will damage the engine. Be sure to use the proper fuel can when refueling the chain saw. If the fuel has not been used for more than a few days, shake the can vigorously before pouring fuel into the saw. The saw manufacturer's instructions should be followed when fueling a chain saw in the field.

PURGING THE SAW

In order to store and transport chain saws safely, they must be purged of flammable liquids and vapors. The saw manufacturer's instructions for purging the chain saw should be followed.

LUBRICATING THE GUIDE BAR AND CHAIN

Most saws have automatic chain oilers, and some are equipped with a manual oiler to assist during heavy cutting operations. If the guide bar and chain are not properly lubricated, cutting efficiency is reduced and chain wear is increased. Only oil that is designed specifically for chain-saw lubrication or recommended by the saw's manufacturer should be used for this purpose.

The guide bar and saw chain lubrication should always be checked before and during chain-saw use. The chain oil reservoir should be refilled every time the chain saw is refueled. Occasionally, while the saw is running, the saw's engine should be revved up while holding the tip of the saw 1 to 2 feet (0.3 m to 0.6 m) from some object to check the amount of oil thrown off (Figure 3.38). If little or no oil is thrown off, supplement the automatic oiler with frequent use of the manual oiler until the automatic oiler can be repaired.

Figure 3.38 The operator checks chain oil operation.

PORTABLE FIRE PUMPS

Portable fire pumps are used to supply water from auxiliary water sources such as ponds, lakes, streams, or swimming pools. The water can be supplied directly to the fireground for fire suppression or exposure protection or to fill water tenders. Portable fire pumps are also used when the water-supply source is remote or inaccessible to fire apparatus.

Design requirements for portable fire pumps are outlined in NFPA 1921, *Standard for Fire Department Portable Pumping Units*. This standard identifies five classifications for portable fire pumps:

- ***Portable attack pumps*** — Used primarily for wildland fire fighting, exposure protection, and mop-up
- ***Portable combination pumps*** — Used for water supply operations, filling apparatus or portable water tanks, and wildland fire fighting; ***not*** intended for interior structural fire fighting
- ***Portable supply pumps*** — Used for filling apparatus water tanks or supplying water short distances to the fireground; ***not*** intended for direct fire fighting applications
- ***Portable transfer pumps*** — Used for filling apparatus water tanks; ***not*** intended for direct fire fighting applications
- ***Portable wildland pumps*** — Designed solely for wildland fire fighting applications

NFPA 1921 specifies different minimum pressure and volume requirements for each type of portable pump. However, the type most relevant to this manual is the portable wildland pump. The minimum specifications for these pumps are shown in Table 3.1.

The standard also specifies that the pump weigh no more than 80 pounds (36 kg) for each person intended to carry it (a one-person pump cannot exceed 80 pounds [36 kg], a two-person pump cannot exceed 160 pounds [72 kg], etc.). In reality, most portable wildland fire pumps weigh considerably less than the 80-pound (36 kg) maxi-

TABLE 3.1
Portable Wildland Pump Classifications

Nominal Suction Hose Size (in or mm)	Minimum Rated Performance Level: Capacity (gpm or L/min)	Minimum Rated Performance Level: Net Pressure (psi or kPa)	Minimum Close-Off Pressure (psig or kPag)
1½ inches (38 mm)	35 gpm (140 L/min)	150 psi (1 050 kPa)	195 psig (1 365 kPag)
2 inches (51 mm)	50 gpm (200 L/min)	150 psi (1 050 kPa)	245 psig (1 715 kPag)
2 inches (51 mm)	75 gpm (300 L/min)	150 psi (1 050 kPa)	245 psig (1 715 kPag)

Source: NFPA 1921, Table 3-1 (partial).

mum. They are designed this way because they may have to be carried long distances. Some wildland portable fire pumps are attached to a backpack frame to make them easier to carry; others are equipped with a handle (Figure 3.39).

Figure 3.39 Some portable pumps are designed to be carried by one firefighter.

Portable fire pumps may be powered by either a two-cycle or a four-cycle engine. Two-cycle portable fire pump engines use a gasoline/oil mixture and are similar to but more powerful than the engines described earlier for chain saws. Four-cycle portable fire pump engines run on gasoline and use a separate crankcase oil system for lubrication. Four-cycle engines are usually more powerful than two-cycle engines; however, they are also considerably heavier.

Most portable fire pumps used for wildland operations are centrifugal pumps. These are smaller versions of the centrifugal pumps described for fire apparatus in Chapter 2, Fire Apparatus and Communications Equipment for Wildland Fires. Positive-displacement pumps are not commonly used in portable wildland pumps.

Two basic styles of portable fire pumps are used in wildland fire fighting operations: land-based pumps and floating pumps. Land-based pumps are set near the water's edge, and a hard suction hose is run from the pump to the water (Figure 3.40). Land-based pumps are designed to be used fairly close to the apparatus they are carried on and are typically mounted in a cage or framework that allows one or more firefighters to carry them (Figure 3.41). Portable wildland fire pumps intended to be carried long distances are attached to backpack frames that allow firefighters to carry them on their backs. Some pumps are split into two pieces. One piece consists of the pump motor and the other is the pump itself. Each is carried on a separate backpack. Once the point of operation is reached, the two pieces must be coupled together according to the manufacturer's instructions before the pump is ready for use.

Figure 3.40 A land-based portable pump drafts from a stream. *Courtesy of NIFC.*

Figure 3.41 Firefighters carry a portable pump to the water source.

Floating pumps are designed to be operated from the water surface of the supply source (Figure 3.42). The pump and pump motor are mounted on a large flotation device. The fire pump intake is connected to a strainer on the underside of the flotation device (Figure 3.43). Some floating pumps are able to operate in water as shallow as 6 inches (150 mm) deep.

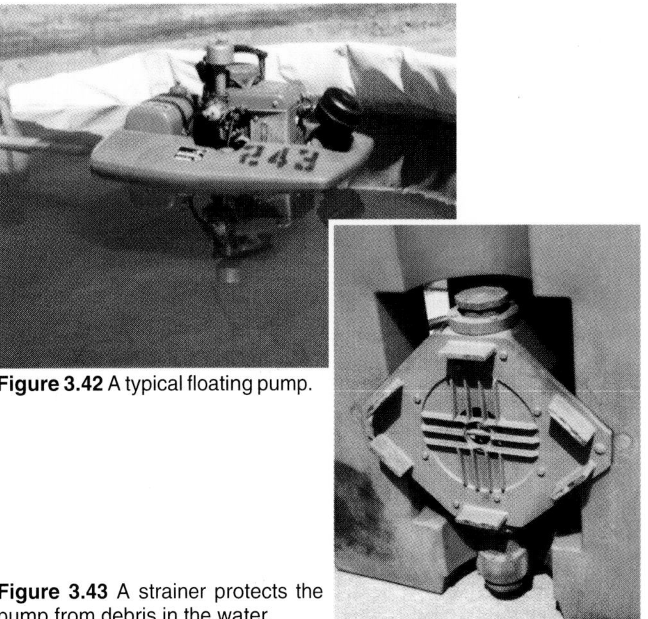

Figure 3.42 A typical floating pump.

Figure 3.43 A strainer protects the pump from debris in the water.

Pump Setup

Perhaps the most important step in setting up a portable fire pump is choosing a good operating location. Firefighters must choose a location that has a water source of sufficient depth and volume to supply the pump for the assigned task. The water source should be relatively free of dirt and debris so the pump will not become clogged during operation.

Environmental impact should also be considered when selecting a location for a land-based portable fire pump. This is particularly important when operating in ecologically sensitive areas. Leaky fuel systems or careless refueling operations may result in fuel spilling into the water or the soil beneath the pump. A plastic tarp or paper sleeping bag placed beneath the pump will catch/absorb any fuel that leaks or spills. When the operation is concluded, the contaminated tarp or sleeping bag must be disposed of properly.

Another common problem occurs when the pump is set up in areas with a high water table. Pump vibration can cause water to rise to the surface of the ground, and the pump can slowly sink into mud. This can cause mud to enter the flywheel housing or deposit on the engine cooling fins, causing the unit to overheat. The mud can also enter the starter cord housing and affect the recoil on future start-up attempts. To prevent these from happening, rest the pump on a small platform made out of logs or similar material.

Once the pump is in an appropriate location, suction and discharge hoses must be connected. Any other preparations, such as filling the fuel tank, are also done at this time. The pump engine is then started and operated according to the manufacturer's instructions. Most engines are manually started using a recoil-type, pull-cord starter. Whether a choke is required depends on the design of the engine.

Land-based portable fire pumps should be placed on a solid base as close to the water level as possible. While most portable pumps are capable of drafting from heights of 10 to 12 feet (3 m to 3.7 m), their pumping capacity is diminished proportionately with an increase in lift (Figure 3.44).

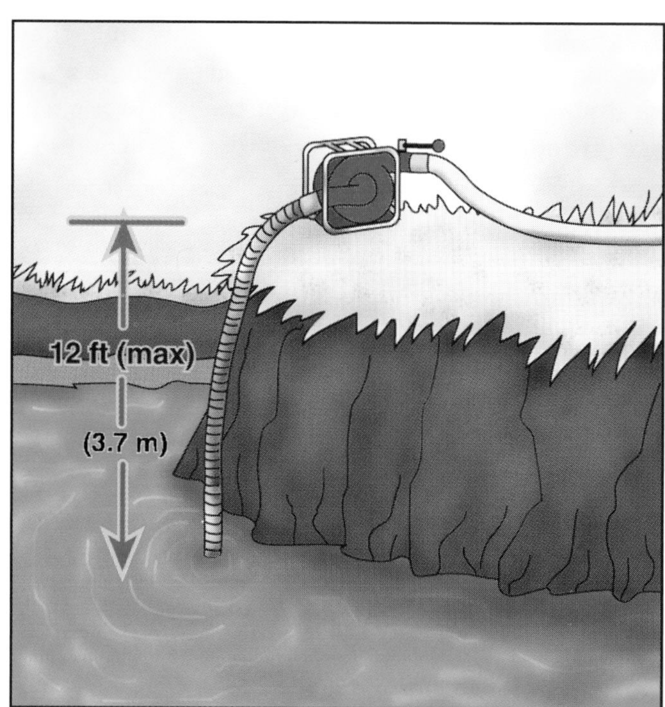

Figure 3.44 Portable pumps should be positioned as near the water surface as possible.

Associated Equipment

When deploying a portable fire pump, some additional pieces of equipment are needed to place the pump in operation. This equipment includes the following:

- Fuel can with the appropriate fuel or fuel/oil mixture
- Fuel supply line (with compatible hose and connections on can and pump) when the pump is capable of being supplied directly from the fuel can
- Tool roll containing pliers, wrenches, a rubber mallet, and other small tools that may be required to operate the pump
- Spare gaskets for any pump connections that may require them
- Suction hose and strainer

Supply and/or attack hose to connect to the portable fire pump and appropriate nozzles and fittings are also required. The amount and type of this equipment depend on local operating guidelines. Hose, nozzles, and adapters are discussed later in this chapter.

Series, Parallel, and Staged Pumping

There are situations where a single portable fire pump cannot supply water in sufficient volume and/or pressure to properly perform the assigned task. In these situations, series, parallel, or staged pumping using two or more portable fire pumps may be required.

SERIES PUMPING

Series pumping is required when a single pump is unable to provide adequate *pressure* due to excessive friction or elevation loss. It involves two or more pumps connected so that water is discharged from the first pump, through a hoseline, and into the intake of the second pump (Figure 3.45). Depending on the situation, the pumps may be located close to each other or be well apart. Series pumping results in a cumulative increase in pressure. For example, if two portable fire pumps rated at 200 psi (1 400 kPa) discharge pressure are connected in series, the discharge from the second pump will be approximately 400 psi (2 800 kPa), less any friction loss between the pumps. In this configuration, care must be taken to not exceed the rated pressure of the fire hose attached to the second pump because this may cause the hose to burst. In these cases, it is safer to place the pumps 100 feet (30 m) or more apart so that friction loss will reduce excessive pressures.

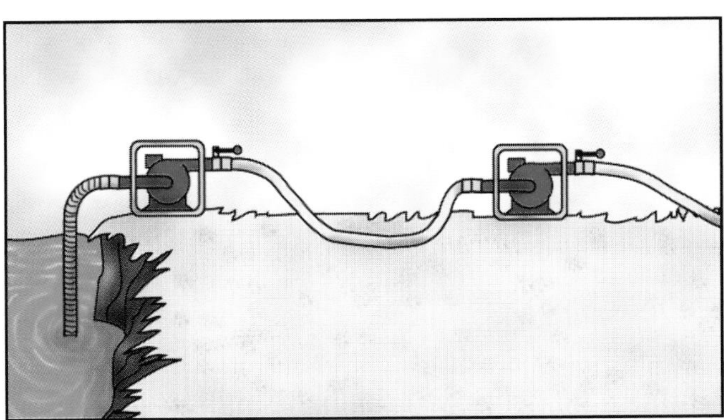

Figure 3.45 Portable pumps are often used in series.

PARALLEL PUMPING

Parallel pumping is used when a single pump is unable to supply an adequate *volume* of water. It is set up by placing two or more portable pumps side-by-side at a water-supply source. Water is discharged from each pump to a siamese appliance and into a single hoseline. Parallel-pumping operations are commonly used for filling water tenders quickly. It is most desirable to use pumps that have the same volume and pressure capabilities when doing parallel pumping. In addition, maximum efficiency is achieved if the size and amount of hose between each pump and the siamese are also the same.

STAGED PUMPING

Staged pumping is a form of series, or relay, pumping in which the two or more pumps involved are not actually connected to each other. The first pump takes the water from the source and pumps it through some distance of hose to a portable water tank. The second pump drafts from the portable water tank and sends the water further on down the line (Figure 3.46). The distance that water can be relayed in this manner is limited only by the number of pumps and portable water tanks available. The volume of water (in gpm or L/min) that can be moved in this manner is not more than the rated capacity of the smallest pump in the relay.

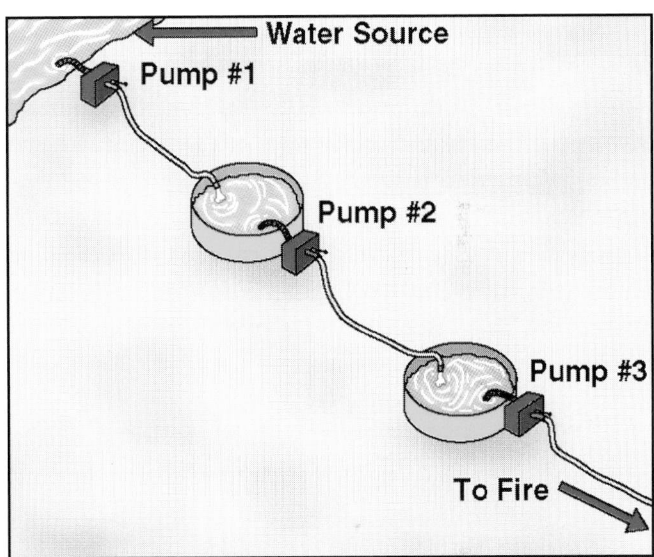

Figure 3.46 Portable pumps are sometimes staged.

Portable-Fire-Pump Care and Maintenance

The main concerns regarding portable-fire-pump care and maintenance are making any repairs that become necessary while operating in the field and doing routine maintenance in the station following every use of the pump. All such care and maintenance must be done according to the pump manufacturer's instructions. Any other significant repairs or servicing should be performed by a pump mechanic.

FIRE HOSE, NOZZLES, AND HOSE-RELATED EQUIPMENT

A variety of types of fire hose, nozzles, and hose-related equipment are used on wildland fires. Firefighters should be familiar with the types of hose and equipment their department uses.

Fire Hose

Numerous sizes and types of fire hose may be used during the course of a wildland fire. The size and type of fire hose used by any particular department depends on the types of fires experienced most often and the tactics used by that department. The four main categories of fire hose used for wildland fire fighting operations are as follows:

- Noncollapsible rubber hose
- Jacketed, lined fire hose
- Unlined fire hose
- Hard suction hose

NONCOLLAPSIBLE RUBBER HOSE

Noncollapsible rubber hose is more commonly referred to as *booster hose* or *hard line*. Booster hose is most often used in 100- to 200-foot (30 m to 60 m) lengths wound on a reel mounted somewhere on the apparatus (Figure 3.47). Some wildland fire apparatus have a 5- to 20-foot (1.5 m to 6 m) section attached directly to a discharge to more easily facilitate pump-and-roll operations and to use as a protection line. Booster hose are most commonly found in ¾- and 1-inch (19 mm and 25 mm) diameters. Because of its relatively small diameter and high friction loss, booster hose can only deliver a limited quantity of water. Generally, flows of less than 30 gpm (120 L/min) can be expected from booster lines. Rubber-covered hose is used for mobile attack on very low-intensity fires and for mop-up; fires of greater intensity require larger hoselines. Rubber-covered hose should never be extended with other hose for fire attack because there would not be enough water flowing to maintain firefighter safety in case of a blowup.

Figure 3.47 A typical booster reel on an engine.

Some departments carry garden hose and appropriate adapters on their wildland apparatus. Garden hose is most commonly either ½ inch or ⅝ inch (13 mm or 16 mm) in diameter, and flows of less than 10 gpm (40 L/min) can be expected. They are only useful for fires in very sparse fuels or for mop-up.

JACKETED, LINED FIRE HOSE

Jacketed, lined fire hose is intended for heavy fire fighting operations. This hose consists of an internal rubber waterway that is protected on the outside by a single or double layer of a woven fabric or an external rubber covering (Figure 3.48). The most common sizes used for wildland fire fighting applications are 1½-, 1¾-, 2-, and 2½-inch (38 mm, 45 mm, 50 mm, and 65 mm) diameter hose. The vast majority of wildland hose in the U.S. are lightweight versions of jacketed, rubber-lined fire hose designed specifically for wildland fire fighting. Lightweight jacketed and lined wildland fire hose is available in 1- and 1½-inch (25 mm and 38 mm) sizes.

Figure 3.48 Rubber-lined, double-jacketed fire hose.

UNLINED FIRE HOSE

Unlined fire hose, sometimes referred to as *forestry hose*, is also commonly used in wildland fire fighting. Unlined hose is made of a linen fabric woven to allow some of the water to weep through the hose. This provides protection against heat and flames while making a fire attack. Unlined hose is preferred by many departments because it is less expensive than lined hose and very lightweight and compact. One firefighter can carry 300 to 400 feet (90 m to 120 m) in a relatively small package (Figure 3.49). This is especially important when firefighters must hike long distances carrying hose and pumps to reach fires in remote locations. However, because this hose is unlined, it is less durable than lined hose and is more vulnerable to puncture. Unlined fire hose is best suited to fires in very light fuels. The most common diameters of unlined

Figure 3.49 This roll of unlined forestry hose is twice as long as the roll of lined hose.

hose are 1 inch and 1½ inches (25 mm and 38 mm), and it is most often purchased in 100-foot (30 m) lengths.

HARD SUCTION HOSE

Hard suction hose is used for drafting water from an auxiliary water source such as a pond, stream, lake, or swimming pool (Figure 3.50). Hard suction hose may be used with a portable fire pump or a fire department engine. This hose is designed to be rigid and noncollapsible because it must withstand a partial vacuum during drafting operations. Hard suction hose used on portable fire pumps is usually 1½, 2, or 2½ inches (38 mm, 50 mm, or 65 mm) in diameter.

Figure 3.50 Typical 2½-inch (65 mm) hard suction hose.

Nozzles

A variety of nozzles may be used for wildland fire fighting. These include standard structural fire fighting nozzles and those designed specifically for wildland fire suppression. Structural fire nozzles are most commonly used when structural fire apparatus are used to attack a wildland fire. However, some wildland fire apparatus are equipped with structural-type nozzles, and structural engines are often equipped with wildland-type nozzles. Any nozzle used for attacking an active fire should be able to produce a protective fog pattern for firefighter protection (Figure 3.51).

Figure 3.51 Nozzles should be capable of producing fog patterns for protecting the firefighters on the hoseline.

A variety of adjustable fog stream nozzles are also designed specifically for wildland fire fighting applications. Most of these are of the barrel design in which the nozzle is turned on or off and the discharge pattern is selected by twisting the barrel (Figure 3.52). However, for a variety of reasons, nozzles with bale shutoffs are more versatile than those with a twist-type shutoff. Most fog nozzles with bale shutoffs allow both the rate of flow and the discharge pattern to be adjusted as needed (Figure 3.53). Once a discharge pattern and flow rate have been selected, the nozzle will

Figure 3.52 A typical plastic forestry nozzle.

Figure 3.53 A typical adjustable fog nozzle.

automatically produce that flow rate and pattern each time the nozzle is opened. In other words, the nozzle does not have to be readjusted each time it is opened as is the case with twist-type nozzles.

Some bale-shutoff nozzles are manufactured with pistol grips (Figure 3.54). Pistol grips provide a good handle that allows for better control of the nozzle and more leverage when pulling hose. This is especially useful in developing a progressive hose lay. Pistol grips designed to be placed between the nozzle and the hose are also available (Figure 3.55).

One type of nozzle, the combination nozzle, is specifically designed for wildland fires and is recommended for mop-up only, not for attacking active wildland fires (Figure 3.56). Commonly called the *forester* nozzle, it has two separate discharge orifices. One is a solid-stream orifice, and the other is a spray-pattern orifice. Depending on the position of the shutoff control handle, the nozzle directs water through the solid-stream orifice or through the spray-stream orifice, or no water flows at all. The forester nozzle is sometimes called the "Death's Head" nozzle because of the number of firefighters who have died with a charged line and forester nozzle in their hands. It provides very low flow, and the spray orifice does not produce an effective water screen to protect the firefighter in case of a blowup.

Figure 3.54 Some nozzles have a built-in pistol grip.

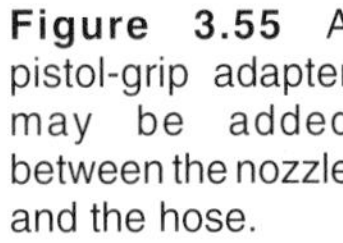

Figure 3.55 A pistol-grip adapter may be added between the nozzle and the hose.

Figure 3.56 The combination forester nozzle is recommended for mop-up only.

Compressed-air foam systems (CAFSs) use solid-stream nozzles in foam lines (Figure 3.57). Solid-stream nozzles provide the greatest amount of reach and accuracy with CAFSs.

Figure 3.57 CAFS is applied through a solid-stream nozzle.

Hose-Related Equipment

In addition to fire hose and nozzles, a variety of other types of equipment may be required to place fire streams in service at a wildland fire. Some of these are very similar to hose equipment found on structural fire engines, including the following:

- Double-male couplings to join two female ends (Figure 3.58)
- Double-female couplings to join two male ends (Figure 3.59)

Figure 3.58 A typical double-male adapter.

Figure 3.59 A typical double-female adapter.

Figure 3.60 A typical reducer.

- Reducers/increasers to join hoses of different sizes (Figure 3.60)
- Adapters for connecting hose with different threads (Figure 3.61)
- Hose clamps to stop the flow of water through a hose (Figure 3.62) (**NOTE:** Hose clamps for wildland applications are considerably more compact than those used in structure fire fighting.)
- Gated wyes to divide one hoseline into two hoselines of equal or smaller diameter (Figure 3.63)
- Siameses to merge two hoselines into one hoseline of equal or larger diameter (Figure 3.64)

In addition to these common types of hose equipment, there are several types peculiar to wildland fire fighting operations. The first of these is the hoseline tee, also called a *water thief* (Figure 3.65). This device is placed between sections of 1½-inch (38 mm) hose. The tee allows a 1-inch (25 mm) hoseline to be extended off the main 1½-inch (38 mm) hose. The tee may or may not be equipped with a shutoff valve to control the flow of water into the 1-inch (25 mm) lateral discharge.

Figure 3.61 Hose thread adapters are available in a variety of sizes.

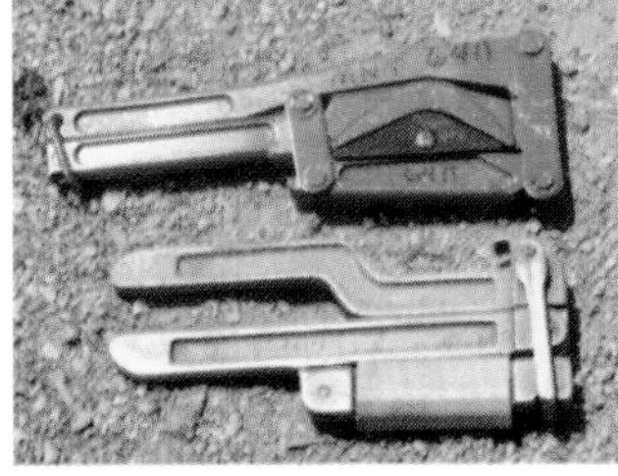
Figure 3.62 Typical forestry hose clamps.

Figure 3.63 A gated-wye allows a single line to be divided into two lines.

Figure 3.64 A siamese allows two small lines to supply one larger line.

Figure 3.65 A typical water thief.

Hoseline pressure-relief valves are used on 1½-inch (38 mm) hoselines from positive-displacement pumps, but they are also recommend for use with centrifugal pumps. This valve relieves line surges and excessive pressure on the pump due to kinked hose or to abruptly closing shutoff valves. The device bypasses water and relieves sudden line surges automatically to permit the use of shutoff nozzles. The pressure at which the relief valve will function can vary from 50 to 200 psi (350 kPa to 1 400 kPa) and is manually adjusted at the valve.

The check-and-bleeder valve is also used on wildland fires. This combination valve is used on 1½-inch (38 mm) hoselines being supplied by a portable fire pump. A spring check valve keeps water from flowing out of the pump before the pump motor is started. Once the motor is started, the bleeder valve is opened until the pump develops normal operating pressure. This valve also contains a 1-inch (25 mm) discharge similar to that used on a tee.

If a water source cannot be used from draft because it is too far below the level of the pump, such as on lifts of greater than 18 to 20 feet (5.5 m to 6 m), a water ejector may be used (Figure 3.66). An ejector is used where drafting is extremely difficult such as from deep cisterns or from comparatively high bridges. It can also be used when the engine cannot be located close enough to the water to be within reach of the hard suction. In these cases, the ejector can operate several lengths away from the engine (Figure 3.67). Water pumped from the engine to the ejector passes through a restriction (venturi) that increases the velocity of the stream. The partial vacuum thus created within the device draws some of the surrounding water into the ejector. The combined flow, which is greater than that being pumped by the engine, is discharged under low pressure. A water ejector's low-pressure stream allows the water tank to be filled at twice the rate of output from the engine.

Figure 3.66 A typical water ejector.

Foot valves are sometimes placed on the end of a hard suction hose being used to draft from a static water source. Most foot valves are equipped with a strainer that prevents large pieces of debris from being drawn into the pump. The foot valve itself is a spring-loaded clapper valve that prevents water from running out the end of the suction hose as the pump is being primed or when the pump is temporarily turned off.

Figure 3.67 A water ejector may be used to refill an engine's water tank.

FIRING DEVICES

As discussed later in Chapter 6, Fire Suppression Methods, wildland fires are often controlled by backfiring or burning out to consume fuel between a control line and the main body of a fire. However, these tactics are only used by qualified firefighters and only according to the incident action plan. These are not tactics for inexperienced firefighters and are never done without permission from command. When these tactics are needed and authorized, a variety of items may be used for ignition, including the following:

- Drip torches
- Fusees
- Items at hand

Drip Torches

The drip torch, also called a *burning-out torch* or an *orchard torch,* is perhaps the most commonly used ignition device for backfiring. The tank of the drip torch contains a fuel mixture that is ignited and spread onto the vegetation to be burned out (Figure 3.68). The fuel flows through a tube and

Figure 3.68 A firefighter uses a drip torch to start a backfire. *Courtesy of Monterey County Training Officers.*

saturates the wick, which burns continuously. The firefighter carrying the torch allows the burning fuel to drip onto the vegetation where the backfire is desired. The fuel in the drip-torch tank is generally one of three common mixtures:

- Two parts diesel fuel to one part solvent
- Two parts crankcase oil to one part gasoline
- Three parts diesel fuel to one part gasoline (**NOTE:** Fill only to the three-fourths level when using this mixture to allow for expansion.)

The following procedure is used when placing a drip torch into service:

Step 1: Remove the drip torch from its storage location, and shake vigorously to mix the fuel.

Step 2: Remove the lock ring in a vegetation-free area (Figure 3.69).

Step 3: Remove and secure the flow plug (Figure 3.70).

Step 4: Separate the tube from the fuel tank. Inspect the rubber gasket and the fuel level at this time.

Step 5: Set the tube upright on the torch in the opening provided, facing the opposite direction from the handle, and tighten the lock ring (Figure 3.71).

Step 6: Open the air vent three-fourths of the way.

Step 7: Wipe any spilled fuel from the exterior of the drip torch.

Step 8: Carry the drip torch in a completely upright, vertical position until the point of application is reached (Figure 3.72).

Step 9: Tilt the torch to spread a small amount of fuel on some ground litter or paper at the ignition point.

Step 10: Ignite the spilled fuel and light the torch from the ground fire (Figure 3.73).

Step 11: Tilt the torch downward to dispense fuel when each desired ignition point is reached (Figure 3.74). Experienced operators may carefully swing the torch side to side as they walk to spread the fire over a wider area.

When the drip torch is no longer needed, extinguish the wick or let it burn dry. Allow the entire unit to cool to ambient temperature. The unit can then be prepared for road travel by

Figure 3.69 The lock ring is removed. *Courtesy of Tony Bacon.*

Figure 3.70 The flow plug is removed. *Courtesy of Tony Bacon.*

Figure 3.71 The tube is secured with the lock ring. *Courtesy of Tony Bacon.*

Figure 3.72 The drip torch is carried upright to the point of application. *Courtesy of Tony Bacon.*

Figure 3.73 The torch is lighted with a fusee. *Courtesy of Tony Bacon.*

Figure 3.74 The torch is tilted down into the dry fuel. *Courtesy of Tony Bacon.*

removing the lock ring, placing the tube back inside the tank, and replacing the lock ring. The flow plug is also replaced and the tank vent closed.

Fusees

Fusees, also known as *road flares,* are also used for backfiring and burning out (Figure 3.75). Some fusees are designed specifically as backfire torches and are manufactured with a 6- to 8-inch (150 mm to 203 mm) handle attached (Figure 3.76). When using this type of fusee, or one to which a makeshift handle has been attached, operators do not have to stoop down. They can remain upright and are better able to see what is going on around them. Fusees burn phosphorous contained within the body of the device. Phosphorous burns very hot (1,400°F [760°C]) and easily ignites grass, twigs, leaves, and other light fuels. A fusee usually burns for 15 to 30 minutes. The following procedure should be used for lighting a fusee:

Figure 3.75 A firefighter holds a typical fusee or flare. *Courtesy of Tony Bacon.*

Figure 3.76 A fusee with a handle attached. *Courtesy of Tony Bacon.*

Step 1: Tape the fusee to a stick or tool handle if it does not have a handle so you do not have to bend over to light the fire (Figure 3.77).

Step 2: Grip the fusee in one hand, and remove the striker cap by the tapered end (Figure 3.78).

Step 3: Scrape the striker end sharply against the ignition end of the fusee in a downward motion, away from the body (Figure 3.79). Holding the fusee downwind, turn your head to the side when striking the fusee.

Step 4: Hold the fusee away from the body with the lighted end down so burning phosphorous does not drip onto the hand (Figure 3.80).

Step 5: Stay in the black once the backfire is started.

Step 6: Extinguish the fusee by burying the lighted end in the ground until the fire goes out. Dispose of fusees properly. Do not leave them where livestock or other animals can eat them.

Figure 3.77 A makeshift handle can be attached to a fusee. *Courtesy of NIFC.*

Figure 3.78 To use a fusee, the striker cap must first be removed. *Courtesy of Tony Bacon.*

Figure 3.79 The firefighter looks away as the fusee is lighted. *Courtesy of Tony Bacon.*

Figure 3.80 The lighted end of the fusee is kept away from the body and lower than the hand. *Courtesy of Tony Bacon.*

Items at Hand

It may sometimes be necessary for firefighters to light a backfire without a drip torch or fusee. In these cases, it is possible for firefighters to use items at hand to light a fire. Examples of these items include the following:

- Burning leaves or needles placed in unburned fuel with a shovel
- Burning oil-soaked rag wrapped around a stick (Figure 3.81)
- Matches or a cigarette lighter

Figure 3.81 A firefighter starts a backfire with a makeshift torch. *Courtesy of Tony Bacon.*

The basic principles of lighting a backfire with these materials are the same as those described for using drip torches and fusees. All of the same safety precautions should be observed as well. In addition to the usual safety procedures described with the other ignition methods, be careful with any materials that may roll or blow into fuels not intended for backfiring.

PERSONAL PROTECTIVE EQUIPMENT

Obviously, an important consideration in any emergency response is the type of protective clothing that the responders wear. Wildland fire fighting presents unique conditions that require specially designed protective equipment. The design of personal protective clothing for wildland fire fighting must balance the need for adequate protection from the hazards created by the fire with the need to wear the clothing for long periods in a very hostile environment.

The outer surfaces of personal protective clothing worn for wildland fire fighting must withstand significant levels of radiant heat, which is the primary threat to firefighter safety. However, because of the strenuous work done by firefighters, often for long periods of time, wildland protective clothing cannot be as heavy and bulky as that worn for structure fire fighting. In addition to radiant heat of the fire, firefighters in the wildland also operate in a wide variety of hostile ambient temperature conditions, ranging from below freezing to in excess of 100°F (38°C). While this wide range of ambient temperatures is reflected in the design of the protective clothing, these temperature variations are usually managed by varying what is worn beneath the protective clothing. A fully equipped wildland firefighter is protected by the following equipment:

- Protective outerwear (clothing)
- Gloves
- Protective footwear
- Protective helmet (with shroud)
- Eye protection
- Fire shelter
- Respiratory protection

In addition to the minimum equipment in the preceding list, some jurisdictions choose to provide firefighters with other types of equipment, including respiratory equipment. All of these, including their care and maintenance, are discussed in the remainder of this chapter.

Protective Outerwear (Clothing)

The protective outerwear worn for wildland fire fighting consists of a coat or shirt and trousers, or a one-piece garment (coveralls or jumpsuit). Depending on design and local preference, this outerwear may be worn over other clothing or directly over undergarments. The style of wildland fire fighting outerwear ranges from that which looks like scaled-down structural turnout clothing to that which resembles typical work clothing (Figures 3.82 a and b).

All wildland protective clothing should meet the requirements of NFPA 1977, *Standard on Protective Clothing and Equipment for Wildland Fire Fighting*. This standard contains detailed specifications for clothing design, performance, and testing criteria.

The most common material used in the construction of wildland protective clothing is Nomex® fire resistant material. Like most fabrics, Nomex® burns when exposed to direct flame impingement; however, it stops burning when the flame is removed. Instead of melting or burning to ash, it forms a char that continues to protect the wearer's skin. NFPA 1977 does not limit wildland protective

Figure 3.82a A typical two-piece wildland fire fighting ensemble.

Figure 3.82b A typical one-piece jumpsuit.

clothing to Nomex® construction; other fire-resistant materials such as Kevlar® aramid fibers, PBI® polybenzimidazole fiber, or fire-resistant cotton may be used. However, regardless of what material is used, it must meet the fire-resistive requirements of NFPA 1977.

All closures should be completely fastened when protective clothing is worn (Figure 3.83). This prevents hot embers from getting between the clothing and the firefighter's skin. Shirt sleeves or trouser bottoms must not be rolled up, nor trouser bottoms cuffed because hot embers may catch in them and damage the fabric and/or injure the wearer.

Figure 3.83 All closures on protective clothing should be fastened.

Whether the clothing is equipped with reflective trim is subject to local preference. Reflective trim is not required by NFPA 1977. Reflective trim is more commonly placed on helmets and other equipment worn over the protective clothing. If the department chooses to have reflective trim on the protective clothing, it is best placed on the sleeves near the wrist. Red-orange trim is not recommended because it may act as a heat sink. It also may not be distinguishable from the colors emitted by the fire. White or silver trim is recommended (Figure 3.84).

Careful consideration must be given to the clothing or undergarments worn beneath the protective garments. As stated earlier, what is needed beneath the protective garment depends on the design of the protective garment and the environmental conditions. Garments worn beneath protective clothing should have long sleeves and be made of cotton, wool, or flame/heat-resistant synthetic materials (Figure 3.85). Other synthetic fibers should not be worn because they may melt when exposed to heat. The amount of clothing worn beneath protective clothing must be adjusted to suit environmental conditions (more layers of clothing when it is cold, fewer when it is hot).

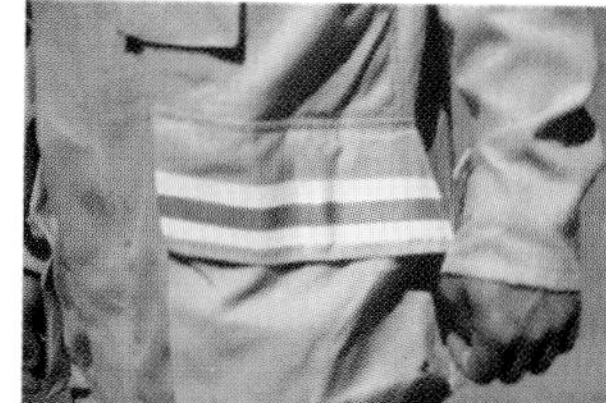

Figure 3.84 Typical reflective trim on a wildland jacket.

Protective clothing similar to a standard station uniform may only require that the firefighter wear conventional undergarments beneath it. These undergarments should basically meet the same requirements listed in the previous paragraph regarding clothing worn under protective clothing. These undergarments can absorb a large amount of perspiration and wick it away from the body to liberate heat.

All protective clothing should be cleaned with warm, soapy water as often as needed (Figure 3.86). Excessive dirt and foreign substances on or within their fibers causes Nomex® and other similar materials to lose some of their fire-retardant

Figure 3.85 A long-sleeve cotton undershirt should be worn under wildland clothing.

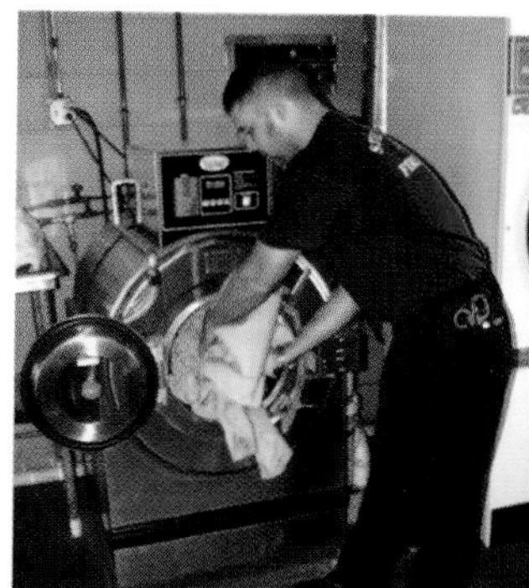

Figure 3.86 Wildland gear should be laundered according to the manufacturer's instructions.

capabilities. Personal protective equipment contaminated with petroleum products such as oil or liquid fuels should be thoroughly washed and dried before reuse. Clothing that is severely damaged should be replaced immediately.

Gloves

Specially designed gloves are essential in protecting the firefighter's hands against blisters, scratches, small cuts, and minor burns during routine fire fighting. However, they also play an important role in more severe conditions. Firefighters who need to deploy their fire shelters need the protection of good gloves so they can hold down hot shelter surfaces without being burned.

The design requirements for wildland fire fighting gloves are also contained in NFPA 1977. These gloves are designed to close tightly around the wrists to prevent hot embers or other foreign debris from entering the glove and injuring the wearer. Some wildland fire fighting gloves have an attached Nomex® wristlet. Gloves may be constructed of leather, Nomex®, other fire-resistive materials, or any combination of these (Figure 3.87).

The care of gloves is basically the same as that described earlier for protective outerwear. It is very important that firefighters replace any gloves that are damaged or worn through (Figure 3.88). Because they are one of the least costly pieces of protective clothing, gloves should be replaced as soon as their protective capability is in doubt.

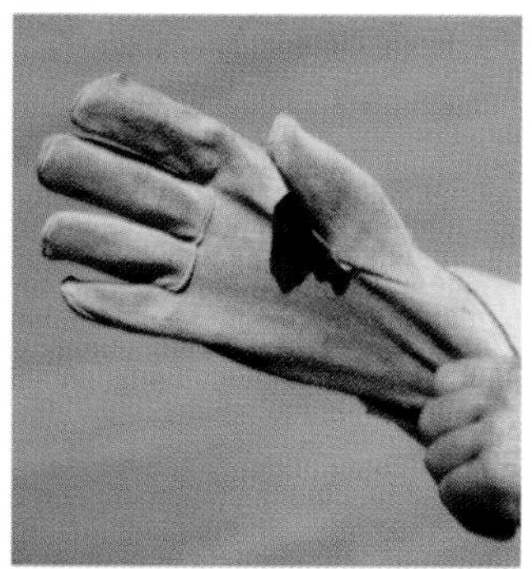

Figure 3.87 Wildland firefighters need sturdy gloves.

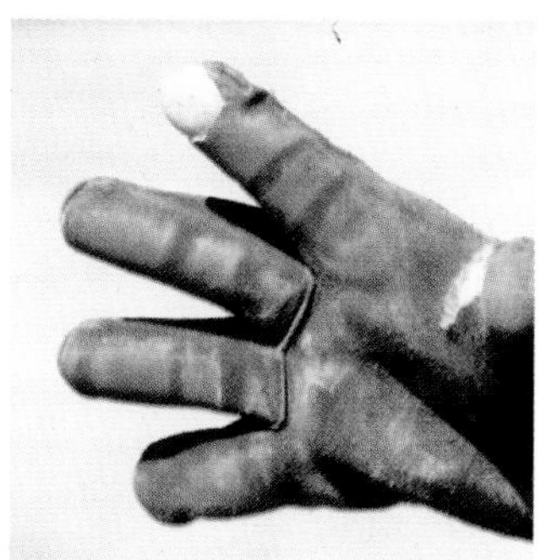

Figure 3.88 Worn or damaged gloves should be replaced.

Protective Footwear

As with the other types of protective equipment, the boots worn for wildland fire fighting are substantially different from those worn for structure fire fighting. Wildland fire fighting boots must be designed to allow the firefighter to operate in a wide variety of terrains. Because the firefighter may be required to wear these boots for days or even weeks at a time, wildland boots must fit very well. Minimum requirements for wildland fire fighting boots are also contained in NFPA 1977.

Nearly all wildland fire fighting boots are completely leather, with the exception of the soles. Synthetic materials that may melt when exposed to heat or that are easily frayed or damaged may not be used. Wildland fire fighting boots should be of the lace-up design, and even the laces should be made of leather to assure that they can withstand the same conditions as the rest of the boot (Figure 3.89). The boot should measure at least 8 inches (203 mm) from the bottom of the heel to the top of the boot.

Figure 3.89 A typical wildland fire fighting boot.

The design of the soles varies depending on the terrain most typical of the jurisdiction. Different sole designs are available for different terrains. Sole designs include flat bottoms, ripple soles, and lug soles (Figure 3.90). All of these styles have a walking heel with a heel breast of not less than ½ inch (13 mm). NFPA

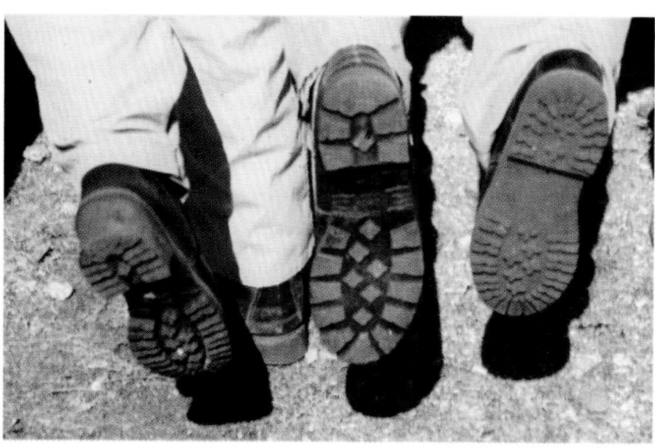

Figure 3.90 A variety of tread patterns are used on wildland boots.

1977 does not require wildland fire fighting boots to have steel safety toes. Steel-toed shoes and boots tend to be heavy, and they can be uncomfortable when worn for long periods of time. For these reasons most departments and firefighters prefer that their boots not have steel toes. However, if steel-toed boots are issued, they should meet the requirements of American National Standards Institute (ANSI) standard Z41, *Personal Protection - Protective Footwear*.

Boots should be maintained in good repair. Loose stitching or other deterioration should be repaired as soon as possible. Boots should be cleaned daily, if possible, and boot grease applied as necessary to keep the leather supple.

In addition to the boot itself, an important part of footwear is the sock worn inside the boot. Socks should be made of natural fibers, not synthetic. All medium-weight wool, or mostly wool, socks offer added thermal protection (Figure 3.91). Wool also wicks moisture away from the skin. This helps keep feet cooler and drier than other fabrics, which reduces the likelihood of developing blisters, a common fire fighting injury. Some firefighters prefer wearing a sock liner or lighter sock under the heavier sock when hiking long distances. Firefighters should take extra socks when dispatched to large-scale incidents. If they do not have extra socks, they should wash and dry the ones they are wearing daily.

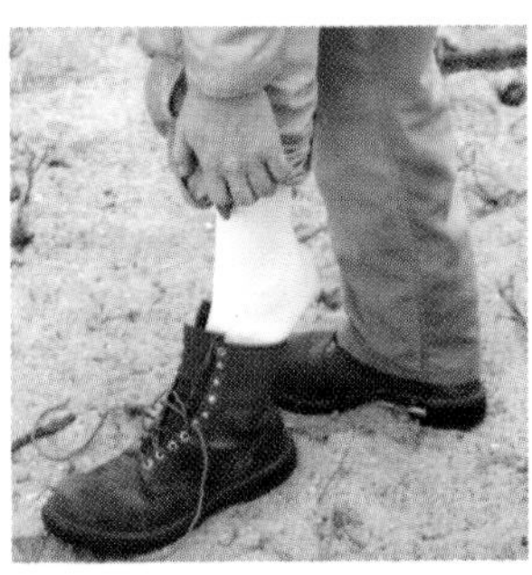

Figure 3.91 Plain cotton or wool socks should be worn in wildland boots.

Protective Helmet

Because of their size, shape, and weight, most structural fire helmets are impractical for wildland fire fighting operations. However, there are some lightweight structural helmets that are suitable for both types of fires (Figure 3.92). Using these combination helmets saves the cost of buying separate types and the space needed for storage on the apparatus. Most wildland protective helmets closely resemble a typical construction-type hard hat (Figure 3.93). The main purpose of the wildland protective helmet is to protect the wearer from bumps or falling debris. As with other types of wildland protective equipment, requirements for wildland protective helmets are contained in NFPA 1977.

Figure 3.92 A typical combination helmet.

Figure 3.93 A typical wildland helmet with goggles and shroud.

The protective helmet should have an adjustable suspension system to ensure a proper fit (Figure 3.94). The suspension system is designed to protect the wearer from a blow to the head by absorbing and dissipating the energy of the blow. It also provides ventilation space between the head and the helmet shell. The helmets must have a chin strap attached directly to the helmet shell. NFPA 1977 requires that at least 4 square inches (2 580 mm^2) of retroreflective material be on each side and rear of the helmet shell.

Figure 3.94 Helmet suspension systems must be adjustable.

Depending on local preference or requirement, the following items may also be a part of the wildland fire fighting protective helmet:

- ***Neck shroud*** — Protects the wearer from radiant heat and hot embers; similar to but slightly larger than the earflaps found on structural fire fighting helmets (Figure 3.95)

- ***Hard hat shoulder cape*** — Provides more extensive protection against heat and embers than the shroud (Figure 3.96)
- ***Headlamp*** — Used during nighttime operations; similar to lamps worn by search and rescue (SAR) teams (Figure 3.97)
- ***Goggle brackets*** — Hold goggles (discussed later in this section) on the helmet

Figure 3.95 A typical neck shroud.

Figure 3.96 A fully enclosed shroud protects the firefighter's face.

Figure 3.97 A battery-powered helmet lamp.

Other accessories may be added according to local preference or requirement. However, accessories added or modifications made to the helmet must not compromise the integrity of the helmet or void the manufacturer's warranty.

Helmets should be cleaned with warm, soapy water. Never use cleaning solvents on helmets because they may cause chemical degradation of the helmet shell. Clean headbands, suspension components, and chin straps daily during periods of heavy use.

Eye Protection

A myriad of eye hazards are present in most wildland fires. Airborne ash, embers, smoke, and other particles have the potential to injure firefighters' eyes. A wide variety of eye-protection devices are available to prevent such injuries. These include goggles, safety glasses, and helmet-mounted faceshields (Figure 3.98). It is generally accepted that eye goggles and safety glasses protect the wearer from about 85 percent of all eye hazards.

Requirements for the proper design of eye protection devices are contained in ANSI Z87.1, *Practice for Occupational and Educational Eye and Face Protection*. One of the major requirements of this standard is that the eye-protection device must afford the wearer a 105-degree field of view (Figure 3.99). The standard also contains requirements for things such as ventilation, impact/scratch resistance, and temperature resistance.

Figure 3.98 A variety of eye-protection devices may be used.

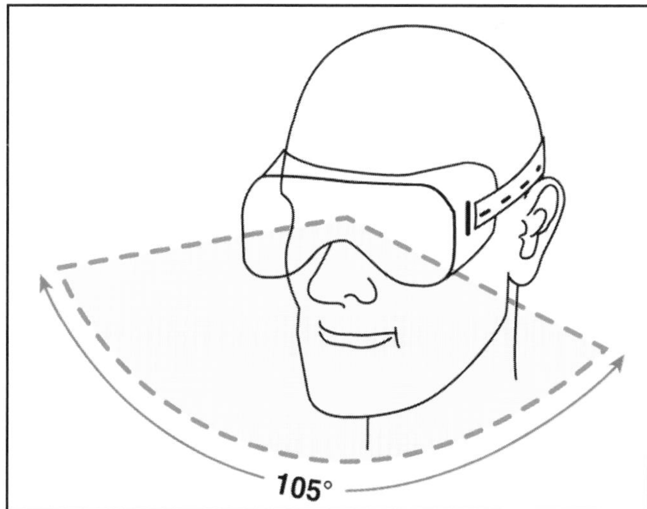

Figure 3.99 Goggles should provide a wide field of vision.

Goggles are the most commonly preferred form of eye protection for wildland fire fighting. They provide the greatest amount of protection against foreign objects contacting or entering directly into the eyes. Goggles are held in place by an elastic strap that is either worn around the back of the wearer's head or attached directly to the helmet. Most goggles are easily cleaned using soap and water or a cleaning solution suggested by the manufacturer. Goggles that are damaged or have excessively scratched lenses should be replaced.

Fire Shelters

Perhaps the most crucial piece of safety equipment that must be carried for all wildland fire fighting is the fire shelter. The fire shelter is an effective, life-saving device that allows firefighters to protect themselves in place should they be overrun by a fire with no option for escape.

> **WARNING**
>
> **Fire shelters are intended to be deployed only as a last resort to survive a fire entrapment. All other reasonable means of escaping the fire should be exhausted before deploying the shelter.**

However, for the fire shelter to be as effective as it is designed to be, it must be protected from damage and worn where it is easily and quickly accessible when needed. This means that firefighters should wear their fire shelters in a position where they can ***see*** their shelter at all times (Figure 3.100). Wearing the shelter in this way does not allow firefighters to sit on their shelters, perhaps damaging them. It also allows firefighters who are running to get the shelters out of their pouches quickly.

Figure 3.100 Firefighters should be able to see their fire shelters at all times.

A fire shelter consists of an aluminum foil outer skin that is bonded to a fiberglass cloth material using a nontoxic, high-temperature adhesive. The fire shelter has a pup-tent shape that lets the firefighter lie flat against the ground (Figure 3.101). This exposes less of the firefighter's body to radiant heat and more to ground cooling. A firefighter whose face is pressed to the ground is in the best position to breathe cool, clean air. The fire shelter's low profile reduces the amount of turbulence and flame contact to which it is exposed. Hold-down straps and a turned-in skirt around the edge of the shelter aid the firefighter in holding the fire shelter tightly to the ground (Figure 3.102).

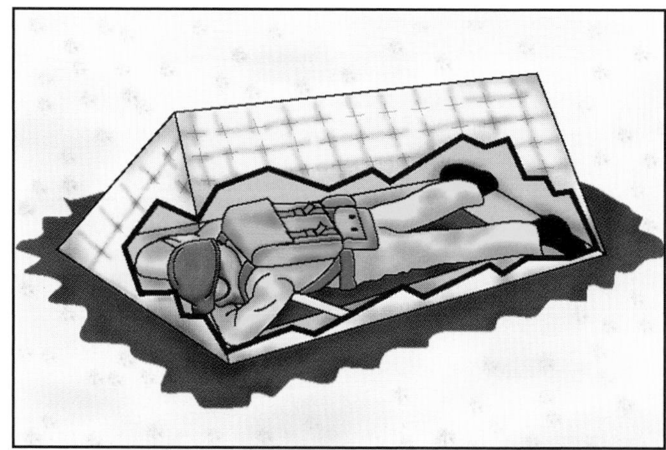

Figure 3.101 A firefighter inside a fully deployed shelter.

Figure 3.102 The shelter straps and skirt are clearly visible as the firefighter begins to deploy the shelter.

The fire shelter protects primarily by reflecting radiant heat. The foil outer skin reflects approximately 95 percent of the flame's radiant heat. The remaining 5 percent is absorbed in part by the heat warming the fiberglass cloth liner, part is radiated into the shelter space, and the remaining part is radiated back to the external environment. It is this 5 percent that gradually heats the inside area of the shelter. With prolonged exposure to a fire, the temperature inside the tent may exceed 150°F

(66°C). However, temperatures of this level can be survived for a prolonged period of time. For example, temperatures in ordinary dry saunas are often around 190°F (88°C).

The heat adsorbed by the fiberglass cloth is not a problem until temperatures reach about 500°F (260°C). If the adhesive that bonds the foil outer cover to the fiberglass cloth begins to delaminate, the fiberglass cloth may separate and drop down onto the shelter occupant. This may initially burn the occupant slightly, but the cloth cools quickly. The foil remains intact and continues to shelter the occupant. Should the shelter delaminate, movement inside the shelter should be kept to a minimum because the foil outer skin is more susceptible to damage.

The limiting factor on the shelter's durability is its melting temperature of 1,200°F (649°C). Because flame temperatures in a typical forest fire are around 1,100°F (593°C), with peak temperatures around 1,800°F (982°C), shelters can tolerate some flame contact, but they cannot endure for a prolonged period.

When the adhesive that bonds the foil to the fiberglass begins to thermally degrade, it releases carbon dioxide and carbon monoxide. However, the decomposing adhesive is only one-half as toxic as an equivalent amount of wood smoke. Even when the adhesive totally degrades, the main toxicant in the air is the smoke from the fire. When the smoke becomes excessive inside the shelter, it can be vented by carefully raising the edge of the shelter away from the flames. The freshest air is at the ground surface.

Firefighters should carefully inspect their fire shelters and carrying pouches at the beginning of each fire season and then at least every two weeks during the season. They should pay particular attention to abrasion damage on the shelter. The shelter should be removed from service if any of the following conditions are found:

- The storage bag has turned gray, gray stains are visible, or the shelter is not visible inside the bag.
- Aluminum particles are visible inside the bag.
- Tears exceeding 1 inch (25 mm) in length are detected along folded edges.
- Dents or punctures in the foil are over 1 inch (25 mm) wide, or ½ inch (13 mm) or more of foil is missing.
- The shelter has been deployed for any reason.

WARNING

According to the National Fire Equipment System (NFES), some fire shelters are approved for operational use and others are not. Fire shelters that are not compliant with NFPA 1977 should not be purchased or used. Shelters that are not approved should be removed from service immediately and used for training purposes only. Approved shelters are listed in NFES publication PMS 409-1 (June 1996), which may be ordered from the following location:

National Interagency Fire Center
Attn: Supply
3833 S. Development Avenue
Boise, ID 83705

For information on the techniques of fire shelter deployment, see Chapter 8, Firefighter Safety and Survival.

Respiratory Protection

Unlike structure fire fighting, respiratory protection is not a major part of the personal protective equipment package for wildland fire fighting. In most situations, the long duration and heavy nature of work faced in wildland fire fighting makes the use of self-contained breathing apparatus (SCBA) impractical. Also, it would be virtually impossible to refill and service SCBA units in the remote locations where wildland fire fighting operations often occur.

Respiratory protection for wildland fire fighting is usually limited to particulate respirators that filter out the airborne dust and ash created during fire fighting activities. Unlike full facepiece respirators, most particulate respirators used for wildland fire fighting only cover the nose and mouth. Filter masks include surgical-type masks, helmet shrouds that also cover the lower part of the face, and those that cover the lower face and neck but are not attached to the helmet (Figure 3.103).

Depending on user preference, the filters in the respirator can be for particulate matter, carbon monoxide, or both. However, scientific studies done in Australia during the late 1980s showed that firefighters in the wildland are unlikely to experience hazardous levels of CO exposure, so filter masks generally provide an adequate level of protection. Dust masks and respirator filters should be replaced regularly according to manufacturer's instructions.

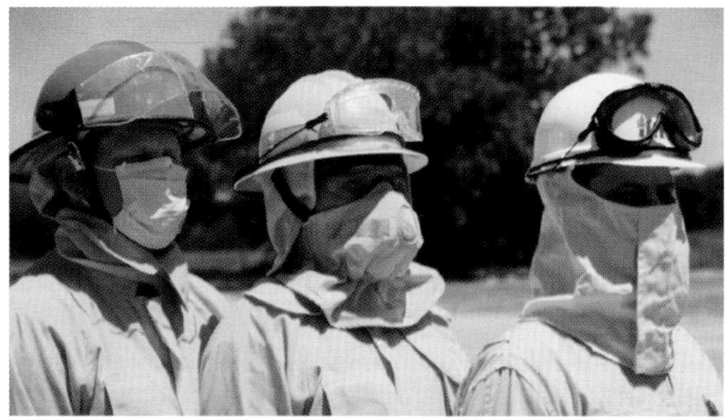

Figure 3.103 A variety of respiratory protection devices may be used.

WATER SUPPLY 4

LEARNING OBJECTIVES

This chapter provides information that addresses the following objectives of NFPA 1051, *Standard for Wildland Fire Fighter Professional Qualifications* (1995 edition):

Wildland Fire Fighter I

3-5.4.2* *Prerequisite Skills:* Proper use of hand tools, fire stream practices, and agent application.

Wildland Fire Fighter II

4.5.5 Operate portable water delivery equipment, given an assignment at a wildland fire and operational standards, so that the proper portable pump and associated equipment is selected, desired nozzle pressure is attained, and flow is maintained.

4.5.5.1* *Prerequisite Knowledge:* Basic hydraulics, portable pump and system capabilities, operation of portable pumps, basic drafting, and associated equipment.

4.5.5.2 *Prerequisite Skills:* Placement, operation, and system set up.

Chapter 4
Water Supply

INTRODUCTION

One of the most important resources for firefighters is an ample water supply. However, wildland fires often occur in areas where a public water supply system is not available, so firefighters must take advantage of auxiliary water sources near the fire or depend on water (and other agents) being transported to the fire. Having access to water sources and the efficient use of that water can significantly affect the outcome of efforts to control wildland fires. Firefighters must know the location of water sources within their response districts and be prepared to take full advantage of them. See NFPA 1231, *Standard on Water Supplies for Suburban and Rural Fire Fighting*, and the IFSTA **Water Supplies for Fire Protection** manual for more details.

This chapter discusses the various ways in which water is stored, accessed, and transported to fires. Included are discussions of piped systems, streams and impounded water, and engine water tanks. Also discussed are mobile water supply apparatus, drafting, and systems/equipment compatibility. The chapter also discusses water additives — their applications and their capabilities and limitations. Included is a discussion of the applications of Class A foam and gelling agents in the wildland.

SOURCES OF WATER SUPPLY

Water for fighting wildland fires may be available from piped systems, large permanent tanks, ponds, lakes, swimming pools, streams and impounded sources. A dry- hydrant system can make drafting from any of these sources easier. In the absence of these fixed sources, mobile water tenders and portable drop tanks may have to be used. During pre-incident planning, firefighters should identify all available water storage facilities to determine their usefulness. Formal agreements between the department and the owners of these water sources may be needed for those sources deemed potentially useful. Pre-incident planning visits allow firefighters to gather the following information on water-storage facilities:

- Location — address, property owner's name, etc.
- Total capacity — refill rate, if applicable
- Access — condition of roads, location of locked gates, etc.
- Tank fittings and valves — types, location, etc.
- Adapters needed — type, size, etc.

Piped Systems

Water may be available to wildland areas in or near cities and towns through a piped water supply system. These public or private systems consist of water mains (pipes) that carry the water from the source to the point of use — individual homes, businesses, or fire hydrants and standpipes (Figure 4.1). Most water systems now use at least 8-inch-diameter

Figure 4.1 A typical municipal fire hydrant.

(203 mm diameter) pipe for mains. In older systems, however, smaller mains may still be in use. It is important that firefighters become familiar with the capabilities and limitations of any public or private water systems in their response districts *before* these systems are needed for fire fighting. Of most importance are the locations of hydrants, the rate of flow to be expected, and the total volume of water available.

Another aspect of piped water systems with which firefighters should become familiar is the color-coding system used by the local water utility to mark hydrants and standpipes. While local entities are free to use whatever colors they choose, the most common practice is to paint standpipes and privately owned hydrants all red (Figure 4.2). The barrels of all public hydrants are usually painted a highly visible color (white, yellow, etc.). However, the caps and bonnets of individual public hydrants are usually painted a contrasting color to indicate the flow in gpm (L/min) that can be expected from that hydrant. For example, hydrants with red caps and bonnets may flow less than 500 gpm (2 000 L/min); those with yellow caps and bonnets, 500 to 1,000 gpm (2 000 L/min to 4 000 L/min); and those with green caps and bonnets, greater than 1,000 gpm (4 000 L/min) (Figure 4.3). Some entities also use blue on the caps and bonnets of hydrants that have very high static pressure such as those downslope from a storage tank (Figure 4.4). Again, it is extremely important that firefighters be familiar with all local water sources.

Figure 4.4 A blue top usually indicates excessive static pressure.

Even though these systems can supply large quantities of water for fire fighting purposes, the water should never be wasted. A small water system may be supplied by a well or surface reservoir with sufficient capacity for normal domestic consumption but be incapable of supplying the additional quantities needed for fire fighting. Excessive water use from piped water systems can reduce the volume and pressure needed for fire fighting operations. In addition, if a number of engines are connected to hydrants and all are pumping at or near capacity, the residual pressure in the water system may drop significantly. This can lead to cavitation in the fire pumps and to possible water main collapse.

Figure 4.2 Private fire hydrants are usually painted all red.

Figure 4.3 Green-top hydrants usually flow in excess of 1,000 gpm (4 000 L/min).

Streams and Impounded Water

Streams, aqueducts, irrigation canals, whether dammed or not, are all potential auxiliary water sources (Figure 4.5). Lakes, farm ponds, cisterns, and swimming pools are examples of impounded water sources available for fire fighting (Figure 4.6). However, it is sometimes difficult to position fire apparatus close enough to these sources to use them. During pre-incident planning, likely sites where streams can be dammed during a fire should be identified. Residents who have one or more

Figure 4.5 Rivers, streams, and irrigation canals can be valuable water sources.

auxiliary water sources on their property should be encouraged to construct suitable approaches for fire apparatus. A suitable approach is one that will support the weight of a fully loaded fire engine. The value of these auxiliary water sources is significantly enhanced with the addition of a dry hydrant connected to a pipe that extends under the water (Figure 4.7). See Dry-Hydrant Systems section later in this chapter.

Figure 4.6 Farm ponds are typical impounded water sources.

Engine Water Tanks

Water tanks on fire engines have a limited capacity, usually less than 1,000 gallons (4 000 L), so the tanks may have to be refilled several times during a large fire. In addition to refilling from public or private hydrants, engine water tanks can refill from streams and impounded water sources, especially if the sources have a dry hydrant with hose connections compatible with the fire engine.

Mobile Water Supply Apparatus

Mobile water supply apparatus (water tenders) are widely used to transport water to areas without piped systems or auxiliary water sources. In an emergency, almost any vehicle with a large-capacity tank can be used for this purpose. Some of the auxiliary water-carrying vehicles that may be used are street flushers, sprinkler trucks, ready-mix concrete trucks, milk trucks, units from construction contractors, and highway maintenance equipment (Figure 4.8). Auxiliary water-carrying vehicles may be rented or borrowed during emer-

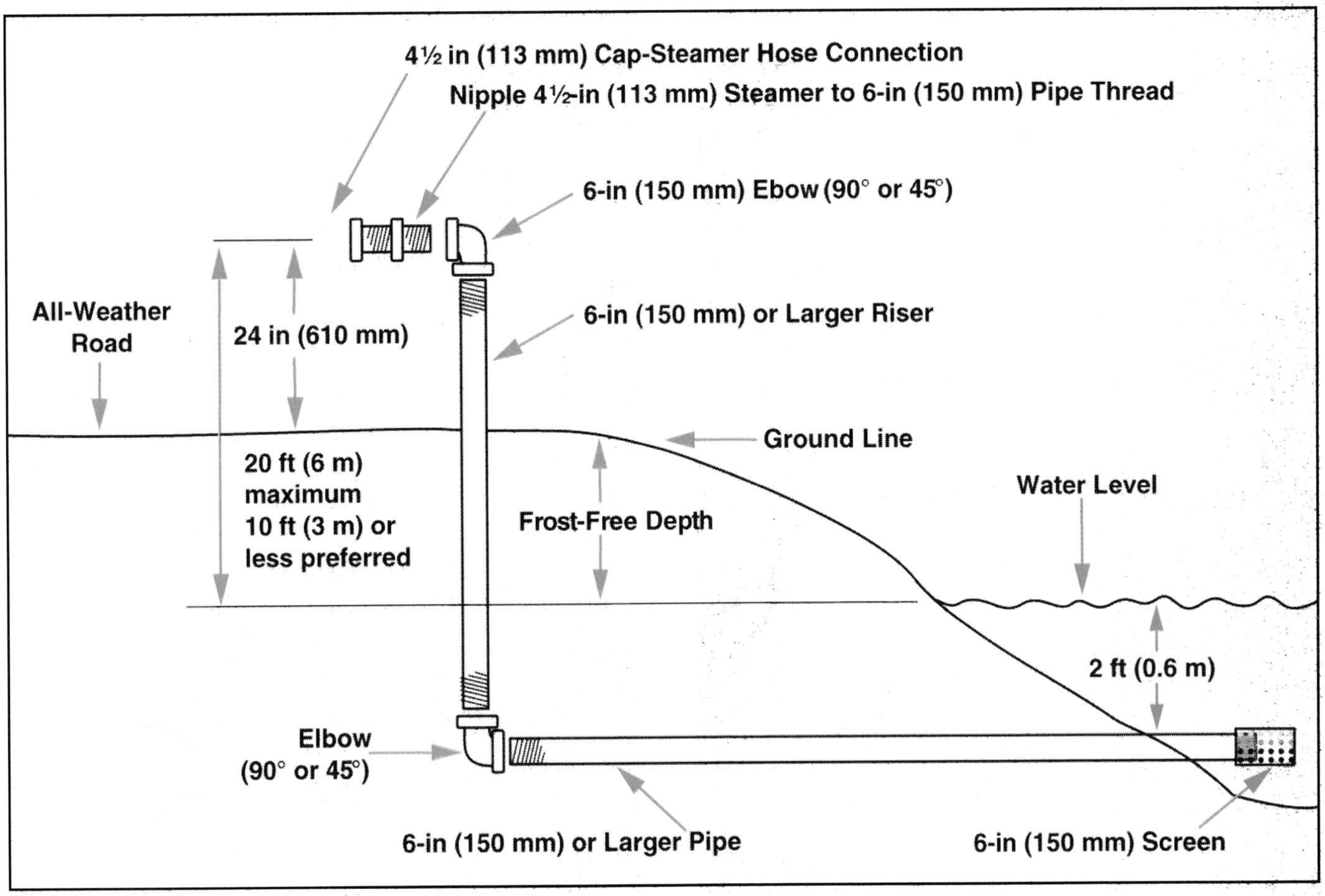

4½ in (113 mm) Cap-Steamer Hose Connection
Nipple 4½-in (113 mm) Steamer to 6-in (150 mm) Pipe Thread
6-in (150 mm) Ebow (90° or 45°)
All-Weather Road
24 in (610 mm)
6-in (150 mm) or Larger Riser
Ground Line
20 ft (6 m) maximum 10 ft (3 m) or less preferred
Frost-Free Depth
Water Level
2 ft (0.6 m)
Elbow (90° or 45°)
6-in (150 mm) or Larger Pipe
6-in (150 mm) Screen

Figure 4.7 A typical dry-hydrant installation.

Figure 4.8 Construction water trucks may be used for water shuttle operations.

gencies under prior agreements made with their owners during the pre-incident planning process.

As discussed in Chapter 2, Fire Apparatus and Communications Equipment for Wildland Fires, a water tender sometimes supplies an engine directly from its tank. In other cases, the tender dumps the water into a large-capacity portable drop tank from which the engine drafts water to fight a fire. Many water tenders have 6- to 10-inch (150 mm to 250 mm) quick-dump valves so that their water can be discharged quickly into a drop tank at a fire (Figure 4.9). Dumping the water quickly allows the water tender to return to the water source for another load while the engine pumps the water from the drop tank. Drop tanks may be made of rubber, plastic, or canvas (Figures 4.10 a and b).

Figure 4.9 A water tender quick-dumps into a portable drop tank.

Drafting

Depending upon the situation, an engine or an auxiliary pump (often called a "skid pump") may be used to draft water from a static source (Figure 4.11). In either case, the pumping apparatus is positioned on a solid foundation and as close to the water as safety permits (Figure 4.12). The required noncollapsible intake hose (hard suction) may be

Figure 4.10a A self-supporting frameless drop tank.

Figure 4.10b A collapsible rigid frame drop tank.

Figure 4.11 A skid pump set up to draft from a static source. *Courtesy of NIFC.*

Figure 4.12 The pumping apparatus must be close to the water source.

connected to the apparatus either before or after it is maneuvered into position, but it may be easier to move the apparatus without the suction hose connected. A rope is attached to the strainer on the end of the intake hose to hold it away from the bottom of the water source (Figure 4.13). The rope is then tied to any solid anchor point such as a tree or a part of the apparatus. Another way to keep the strainer off the bottom but at the required depth (at least 2 feet [0.6 m]) is to tie a spare tire or other flotation device to the strainer (Figure 4.14). Yet another method is to lash the hard suction and strainer to a ladder (Figure 4.15).

When drafting with a centrifugal pump, a priming device (usually a small positive-displacement pump) must be used to prime the centrifugal pump. The priming device works by discharging the air from the centrifugal pump and associated piping, creating a partial vacuum. Atmospheric pressure forces water up the hard suction and into the centrifugal pump (Figure 4.16). Being primed, the centrifugal pump then draws (drafts) the water from the source. When pumping from draft, a lift of not more than 10 to 12 feet (3 m to 3.7 m) is optimal. Because of decreasing atmospheric pressure, the maximum possible lift decreases as elevation above sea level increases.

Figure 4.13 The suction strainer should be kept away from the bottom of the water source.

Figure 4.14 A floatation device can be used to keep the strainer off the bottom.

Figure 4.15 Hard suction hose may be lashed to a ladder to keep the strainer up.

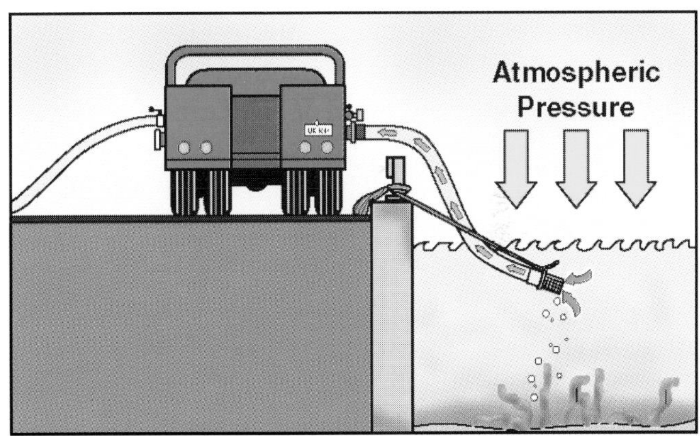

Figure 4.16 Atmospheric pressure forces water into the pump.

The pump manufacturer's instructions should be followed when priming a centrifugal pump. If the pump fails to prime within 30 to 45 seconds at a 10-foot (3 m) lift, disengage the priming device and check the tightness of the caps and connections. Air leaks or pump malfunction may be the cause. When all caps and connections are spanner-tight, try the priming procedure again. If it still doesn't work, use a different pumping apparatus.

Consult the operator's manual for special characteristics of the pump. For more information, see the IFSTA **Fire Department Pumping Apparatus** manual.

Dry-Hydrant Systems

Remotely located farms and other businesses often have ponds, such as settling ponds, for irrigation or for other industrial purposes. These ponds can provide a substantial amount of water for fire fighting if the water can be accessed. Access to these sources of supply can be facilitated if one or more dry hydrants are installed through which engines can draft (Figure 4.17). However, taking water from a dry hydrant does not always require drafting. A hydrant installed at the base of a dam or a large storage tank supplies water under the head pressure created by the height of the water above the hydrant, but it is still considered a dry hydrant because it is not connected to a municipal-type water system. A *dry hydrant* is defined in NFPA 1231 as *"a permanent piping system, normally a drafting source, that provides access to a water source other than a municipal-type water system."*

In general, a dry hydrant consists of a pipe (usually at least 6 inches [150 mm] in diameter) with a strainer on the end permanently submerged in a body of water. This horizontal length of pipe connects to a vertical riser of the same diameter. The riser extends about 2 feet (0.6 m) above grade into a 90-degree elbow. The other end of the elbow is equipped with a strainer inside a threaded fire department connection (Figure 4.18). Dry hydrants installed below elevated water sources must also have a valve to control the flow of water. Dry hydrants (especially the submerged water intakes) should be inspected and tested annually.

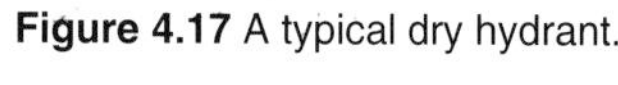

Figure 4.17 A typical dry hydrant.

Figure 4.18 The strainer is visible inside the connection.

Systems/Equipment Compatibility

Regardless of whether the water system is a piped system that supplies water under pressure or an auxiliary source with one or more dry hydrants, the outlet sizes and threads on the hydrants must be compatible with those on the equipment that connects to them. It is also important to identify and rectify any differences in hose coupling threads used by different mutual aid departments *before* a major fire starts (Figure 4.19). One of the best ways to identify such differences is for units from all departments participating in a mutual aid agreement to train together in realistic exercises. When the equipment and systems are not compatible, adapters will be needed to rectify the differences (Figure 4.20).

Figure 4.19 Suctions with sexless couplings will not match the screw threads.

Figure 4.20 A sexless adapter must be used to allow the suction to be connected.

WATER ADDITIVES

Even though water is the most universally used extinguishing agent for both wildland fires and structure fires, its suppression capabilities can be increased two to three times by mixing it with certain chemical additives. But these additives are not magic; it is still the water in the mix that extinguishes the fire. In addition, the class and type of additive must match the fuel on which it is used — for example, Class A foam is ineffective on Class B fuels and vice versa. For more information on the properties of additives used in the wildland, see NFPA 298, *Standard on Fire Fighting Chemicals for Class A Fuels in Rural, Suburban, and Vegetated Areas*. The following types of additives are available for use in fighting wildland fires: *penetrants*, *retardants*, *foaming agents,* and *gelling agents*.

Penetrants

The use of penetrants mixed with water as fire suppressants is widely accepted in wildland fire fighting. Water mixed with a penetrant is often called "wet water" because of its increased ability to soak into fibrous materials. *Penetrants* are surfactants (surface-active agents), which are essentially industrial-strength detergents without the foaming agents and perfumes that are added to ordinary dishwashing liquids. Penetrants work by breaking down the surface tension of water, allowing the water to penetrate (soak into) fibrous materials faster and deeper than plain water. This soaking action reduces the amount of water necessary for initial knockdown and decreases the likelihood of rekindling of fires that had burned in grass, brush, wood, crops, or matted undergrowth. If additional water is applied to the same fuels, the penetrant remaining on the fuel helps the additional water soak even deeper into the material. This deep penetration cools the fuel more effectively than plain water alone. Penetrants make water up to two times more effective as an extinguishing agent, allowing firefighters to cover a larger area with greater fire knockdown than would be possible with untreated water.

Most penetrants are considered nontoxic, but food or water contaminated with penetrants should not be consumed. Waterways and wildlife habitat should be protected from penetrant concentrate spills. As a class, penetrants tend to be corrosive to metals and many metal coatings, but the corrosiveness varies among different products and with the concentration. Penetrants are normally batch-mixed in an engine's water tank because siphoning concentrate directly from a container with an eductor makes the engine too immobile (Figure 4.21). However, when a penetrant is batch-mixed in the water tank of a fire engine, the tank should be thoroughly flushed with plain water after use to remove any remaining penetrant residue unless the tank was specifically designed to hold penetrants.

Figure 4.21 Class A foam should be batch-mixed in the engine's tank.

APPORTIONMENT

Penetrants are usually packaged in a concentrated form, so some type of apportionment (mixing with water in a specific ratio) is required when they are used as extinguishing agents. The two most common methods of mixing are those mentioned earlier — using a siphon eductor or batch-mixing the concentrate with the tank water. When mobility is not a factor, concentrate can be siphoned from its container. The correct eductor setting or concentration is often stamped into the surface of the eductor (Figure 4.22). Otherwise, it can be found in the manufacturer's information on the label of the penetrant container (Figure 4.23). There are two major reasons for maintaining a proper mixing ratio:

Figure 4.22 Most eductors have the correct percentages marked on them.

Figure 4.23 The correct application percentage can also be found on the container label.

- Too much water in the mixture reduces penetrant effectiveness.
- Too little water in the mixture wastes the penetrant.

Proper mixing procedures are outlined in the manufacturer's instructions. The manufacturer often furnishes or markets a mixing device for the concentrate. Following the furnished instructions ensures efficient and economical use of the product with few maintenance problems.

APPLICATION

Wet water is usually applied directly onto the burning fuel with a fog nozzle (Figure 4.24). It must be applied in sufficient volume to allow it to penetrate the fuel without being vaporized by the heat of the fire. Even with the superior extinguishing characteristics of wet water, dense fuels such as logs, fence posts, cow chips, or heavily matted grass may require repeated applications to completely extinguish any deep-seated fire (Figure 4.25).

Figure 4.24 Wet water can be very effective when applied directly onto the fire. *Courtesy of NIFC.*

Figure 4.25 Logs and stumps may require repeated applications. *Courtesy of NIFC.*

Retardants

While penetrant/water mixtures have some fire-retardant effect, they only last until the water content evaporates. At best, they can only be considered very short-term retardants. A true fire retardant, such as diammonium phosphate (DAP), is also delivered in a mixture with water, but its fire-retardant properties remain after the water content evaporates. DAP, which is essentially an agricultural fertilizer, is usually mixed with water and a viscosity agent into a thick mixture called *slurry*. The viscosity agent helps the retardant cling to the surface of the fuels, but it also makes the fuels and the ground around them very slippery. Firefighters can easily slip and fall when walking where fuels have been treated with DAP.

DAP and ammonium sulfate are the two most commonly used wildland fire retardants. They both function by insulating the fuels and by enhancing the dehydration reaction of the cellulose of which the fuels are composed. This reaction increases the ratio of char to tar produced by pyrolysis, which encourages glowing combustion instead of flames. This reduces the fire's intensity and slows its progress. DAP is somewhat more effective than ammonium sulfate but is also more expensive, so the choice of retardant may be made on the basis of cost.

Unlike fire suppressants such as wet water, fire retardants are usually not applied directly onto burning fuels. Retardants are usually applied to uninvolved fuels in the path of a fire. They may also be used for structure protection in the wildland/urban interface or for constructing a wet line (temporary fire-stop made by wetting with water or foam) from which to burn out. While these retardants can be applied with hoselines from an engine, they are most often applied by air drops from tankers or helicopters (Figure 4.26).

Figure 4.26 An air tanker drops a load of retardant. *Courtesy of California Office of Emergency Services.*

Foaming Agents

There are two classes of foam used for fire fighting: *Class A foam* intended for use on Class A fuels such as grass and wood and *Class B foam* intended for use on Class B fuels such as flammable and combustible liquids. Class A foam is a hydrocarbon-based surfactant that lacks the strong film-forming properties of Class B foam but has superior penetrating (wetting) properties. On the other hand, Class B foams are protein- or fluorocarbon-based surfactants with superior film-forming capabilities, but they lack the penetrating capabilities of Class A foam. Because Class A foam is the type used in combating wildland fires, the balance of this section is confined to a discussion of Class A foam and its applications.

CLASS A FOAM

Class A foam is a mechanically generated aggregation of bubbles having a lower density than water. The combination of water and 0.1 to 1.0 percent foam concentrate produces foam solution; adding air to the foam solution (aeration) produces foam. In Class A foam, water constitutes almost 10 percent of the foam mass and is the primary extinguishing agent. Foaming agents can double the effectiveness of water as an extinguishing agent, which increases the amount of fire that can be suppressed with a limited amount of water. Class A foam acts as both a penetrant and a retardant. It acts as a penetrant by decreasing the surface tension of water. It acts as a retardant by insulating the fuel, reflecting heat, suppressing fuel vapors, and excluding oxygen from the combustion process. These characteristics make Class A foam an excellent fire suppressant (Figure 4.27).

Different types of Class A foam are needed for different purposes on the fireground. In this context, "type" refers to the combination of drain time and expansion ratio of a Class A foam. A foam with a fast drain time and a low expansion ratio behaves differently than a foam with a slow drain time and a high expansion ratio.

Drain time. This is the amount of time it takes for a given amount of foam solution to drain from the foam mass, and it is an indication of the foam's longevity. A foam with a fast drain time is best used where quick cooling is needed such as in flame

Figure 4.27 Class A foam is very effective on grass and brush fires. *Courtesy of NIFC.*

Figure 4.28 A firefighter builds a wet line with aspirated foam. *Courtesy of NIFC.*

TABLE 4.1
Expansion Ratios

Category	Ratio
Low	1:1 to 20:1
Medium	20:1 to 200:1
High	200:1 to 1,000:1

Source: NFES 2270, Foam vs. Fire, NWCG.

knockdown and mop-up. A foam with a slower drain time is better suited for exposure protection where more insulating effect is needed.

Expansion ratio. This refers to the amount of foam that is produced from a given amount of foam solution. In other words, it is the volume of foam divided by the volume of foam solution used to produce it. The differences in expansion ratio are due to the characteristics of the foam solution and the means by which air is introduced. Air can be introduced with compressed air foam systems (CAFSs) or air-aspirating nozzles (Figure 4.28). The same concentration of foam solution can produce foam with different expansion ratios depending upon the type of foam generator used. The three common categories of expansion ratios used in wildland fire fighting foam are *low*, *medium*, and *high* (see Table 4.1).

Because of their versatility and greater stream reach, low-expansion foams are the types most often used in wildland fire fighting. In the low-expansion category there are four common types: *foam solution*, *wet foam*, *fluid foam*, and *dry foam*. Their characteristics are as follows:

- *Foam solution*
  - — Is a clear-to-milky fluid
  - — Lacks bubble structure
  - — Is mostly water
- *Wet foam*
  - — Is watery
  - — Has large-to-small bubbles
  - — Lacks body
  - — Has fast drain times
- *Fluid foam*
  - — Is similar to watery shaving cream
  - — Has medium-to-small bubbles
  - — Flows easily
  - — Has moderate drain times

- *Dry foam*
  - Is similar to shaving cream
  - Has medium-to-small bubbles
  - Is mostly air
  - Clings to vertical surfaces
  - Has slow drain times

Because of the variety of wildland fuels, the type of foam used must be adjusted to meet specific fire-suppression situations. Since dry foam coats and adheres well to vertical surfaces and drains slowly, it is well-suited for pretreating exposures (Figure 4.29). Fluid foam adheres well to foliage and drains readily, wetting the foliage in the process. Wet foam penetrates surface fuels rapidly and cools them well. Foam solution should be used as an extinguishing agent only during mop-up when maximum penetration/wetting is needed. Foams delivered by air drops filter through aerial fuels and tend to envelop the fuels on which they settle. Envelopment isolates the combustible vapors that the fuels liberate when heated. The "drizzle" produced by the draining foam decreases the volatility of the fuel vapors and increases the heat energy required to ignite the treated fuels.

Figure 4.29 A firefighter pretreats a structure with a dry foam mixture. *Courtesy of NIFC.*

LINE CONSTRUCTION

If possible, the process of using Class A foam to construct a wet line should be started soon enough to allow time for the fuels to be thoroughly wetted before the fire reaches the line. Initially, mix ratios should be set at 0.3 percent for CAFS or at 0.3 to 0.5 percent for air-aspirating nozzles. These ratios may have to be adjusted, depending upon other variables in the situation. The wet line should be at least two and one-half times as wide as the expected flame lengths. The foam should be applied at close range so that all sides of the fuel are coated (Figure 4.30). Foam can also be used to form an insulating barrier on the surface of brush, tree trunks, and fuel canopies by lofting the foam onto them (Figure 4.31).

Figure 4.30 Class A foam should be applied from close range. *Courtesy of NIFC.*

Figure 4.31 Class A foam should be lofted onto aerial fuels. *Courtesy of NIFC.*

STRUCTURE PROTECTION

The ability of Class A foam to cling to vertical surfaces, and even the horizontal undersides (soffits and eaves), makes it a valuable resource for structure protection. The initial mix ratio of 0.3 percent for CAFS or 0.5 percent for air-aspirating nozzles may have to be adjusted to create the desired consistency. A dry foam should be lofted onto the surface of outside walls, eaves, roofs, columns, and other threatened surfaces (Figure 4.32). Lofting the foam, rather than spaying it directly onto the surface, prevents breakdown of the foam structure due to the force of impact. A foam blanket at least ½-inch (13 mm) thick should be applied because its effectiveness depends on its thickness and moisture content.

In addition to pretreating structures in the path of an advancing wildland fire with low-expansion foam, medium-expansion foam can be used to construct a foam barrier between the structures and the fire. Using air-aspirating nozzles, firefighters can lay a line of foam to intercept the fire front before it reaches the structures. However, if a dry foam mixture is used to construct a foam line, the fuels may not be sufficiently wetted, and the fire can burn under the foam line and resurface on the other side.

Figure 4.32 Dry foam should also be lofted onto structures. *Courtesy of NIFC.*

LONGEVITY OF APPLICATION

Depending upon variables of temperature, humidity, and wind, a blanket of CAFS foam should last about an hour. Foam applied with air-aspirating nozzles normally lasts about half that time. In general, the drier the foam, the more its longevity is affected by wind; the wetter the foam, the more its longevity is affected by temperature and humidity.

Gelling Agents

Other water additives available for use in wildland fires are known as *gelling agents, durable agents,* or *fire gels*. At the time of this manual's publication, these products were relatively new, and the information about them came primarily from the product manufacturers. Agencies considering the purchase of these products are encouraged to research them, and if possible, test them under controlled conditions.

While they are used in the same way as Class A foam, gelling agents are chemically and structurally quite different. Chemically, gelling agents are water-absorbent polymers rather than hydrocarbon-based surfactants like Class A foams. When mixed with water, gelling agents form minute polymer bubbles that are filled with water (hydration); Class A foams form tiny water bubbles that are filled with air (aeration). Therefore, when applied, gelling agents consist primarily of water; Class A foams consist primarily of air. Other major differences between gelling agents and Class A foams are that gelling agents are considerably more expensive, and they make the surfaces to which they are applied *very* slippery. This can create a safety hazard for anyone walking or driving on treated surfaces. Gelling agents are nontoxic and biodegradable.

Gelling agents can be batch-mixed in the water tank of a fire engine or siphoned from the container with an ejector. They can be used for fire extinguishment, line construction, or structure protection. Gelling agents can be applied through hoselines with any standard fire nozzle, applied through master stream appliances, or dropped by air tankers.

FIRE EXTINGUISHMENT

When used for fire extinguishment, gelling agents are normally mixed with water in the tank of a fire engine at a ratio of 1:100, or 1 percent. For every 100 gallons (400 L) of water in the tank, one gallon (4 L) of gelling agent is added. This ratio produces the best results for fire-suppression operations such as mobile attack or pump-and-roll (Figure 4.33). As mentioned earlier, gelling agents can also be dropped directly onto a fire by air tankers.

Figure 4.33 Fire gels can be effective in mobile attack. *Courtesy of Barricade, Inc.*

LINE CONSTRUCTION

Because gelling agents remain hydrated for several hours after application, they are well-suited for constructing wet lines from which to burn out and for pretreating vegetation (Figure 4.34). Gelling agents can also be used to widen and reinforce control lines in indirect fire fighting operations (control lines are located some distance from fire's edge). These agents are normally applied at 1½ to 2 percent for line building.

Figure 4.34 A firefighter pretreats vegetation with fire gel. *Courtesy of Barricade, Inc.*

STRUCTURE PROTECTION

According to manufacturers' information as well as data from controlled tests done by the National Institute of Standards and Technology (NIST), perhaps the most effective use of these agents is in pretreating structures. Applied at a 1½ to 2 percent solution, gelling agents will adhere for several hours to whatever surface they are applied. They cling to wood, aluminum, and vinyl siding. They also cling to the underside of eaves and decks and even to glass (Figure 4.35). Because they remain hydrated for hours after application, more than 24 hours in some cases, gelling agents can be applied well in advance of a fire's arrival, which can significantly reduce the fire risk to those applying the agents. In addition, because these agents can be applied so far ahead of the fire's arrival, the need for applying them from the engine's tank to keep the engine mobile is greatly reduced, so it becomes practical for the agent to be siphoned directly from the container (Figure 4.36).

Figure 4.35 Vertical surfaces pretreated with fire gel are protected for many hours. *Courtesy of Barricade, Inc.*

Figure 4.36 Fire gels may be siphoned from their containers. *Courtesy of Barricade, Inc.*

5 INITIAL FIREGROUND COMMAND

LEARNING OBJECTIVES

This chapter provides information that addresses the following objectives of NFPA 1051, *Standard for Wildland Fire Fighter Professional Qualifications* (1995 edition):

Wildland Fire Fighter I

3-1.1 Prerequisite Knowledge: Fireline safety, use and limitations of personal protective equipment, agency policy on fire shelter use, basic wildland fire behavior, basic wildland fire tactics, fire fighters role within the local incident management system, and first aid.*

Wildland Fire Fighter II

4-1.1.1* *Prerequisite Knowledge:* The Wildland Fire Fighter II role within the incident management system, basic map reading and compass use, radio procedures, and record keeping.

4-1.1.2 *Prerequisite Skills:* Orienteering and radio use.

4.1.2* Lead wildland fire fighters in the performance of a task, given an assignment and performance standards, so that the task is safely completed within the standards.

4-1.2.1 *Prerequisite Knowledge.* Leadership techniques for small groups and recognizing and reacting to unsuitable performance.

4-1.3 Brief assigned personnel, given an assignment, supporting information, and equipment requirements, so that the personnel are informed of specific tasks, standards, safety, operational and special interest area considerations.

4-1.3.1* *Prerequisite Skills.* Briefing skills.

4-5.1 **Definition of Duty.** All activities to confine and extinguish a wildland fire beginning with dispatch.

4-5.2 Select fireline construction methods, given a wildland fire and line construction standards, so that the technique used is appropriate to the conditions and meets agency standards.

4-5.2.1 *Prerequisite Knowledge:* Resource capabilities and limitations, fireline construction methods, and agency standards.

4-5.3 Evaluate the readiness of assigned crew members, given a wildland fire, an assigned task, and agency equipment standards, so that crew members are properly equipped and supplied for suppression duties.

4-5.3.1 *Prerequisite Knowledge:* Agency standards and personnel inspection procedures.

4-5.6* Secure the area of suspected fire origin and associated evidence, given a wildland fire and agency procedures, so that all evidence or potential evidence is protected from damage or destruction and reported to a supervisor.

4-5.6.1 *Prerequisite Knowledge:* Knowledge of types of evidence and the importance of site security and evidence preservation.

4-5.6.2 *Prerequisite Skills:* Evidence preservation techniques and use of marking devices for site security.

Chapter 5
Initial Fireground Command

INTRODUCTION

In some cases, a wildland fire can be controlled with a single engine company; in others, large numbers of firefighters, apparatus, and equipment may be required. Because of this range of resource needs, fireground management can be relatively simple and straightforward or extremely complex. Regardless of whether a wildland fire is relatively small, requiring few resources, or is massive, requiring a multitude of fire fighting resources, one of the keys to successfully handling any wildland fire is preparation. Adequate preparation for handling wildland fires requires the following:

- In-depth pre-incident planning that assesses the fire potential and available resources and serves as the basis for operational plans
- Contingency plans to provide the resources needed for exceptional incidents
- An incident command/management system
- Training at all levels of the organization

This chapter discusses pre-incident planning and the resources needed for major incidents, mutual aid and automatic aid agreements, maps, crew readiness, communications, and record keeping. Also discussed are the principles and functions of developing the fireground organization, extended and major fire attacks, and post-incident critiques.

PRE-INCIDENT PLANNING

Pre-incident planning should take place long before a major fire occurs. A pre-incident plan is quite different from the incident action plan (IAP) that outlines a plan of attack for a fire in progress. However, the IAP may be based on the information gathered during the pre-incident planning process. While an IAP deals with the existing and expected behavior of a fire already in progress, a pre-incident plan attempts to forecast the various fire situations most likely to occur at a particular location, what resources will be needed in each situation, and how those resources may be obtained and deployed. To make the best use of available resources, fire department personnel should develop pre-incident plans for all target hazards (locations with above-average life hazards or fire potential) in their response area and maintain these plans for future use (Figure 5.1). In their assessment of possible fire situations, firefighters should refer to the fire behavior information provided in Chapter 1, Wildland Fire Behavior: Fuel, Weather, Topography.

Figure 5.1 Prefire plans should be developed for all target hazards.

Pre-Incident Operations Plans

Pre-incident plans are guidelines intended to assist fire officers in establishing priorities and making fireground decisions. However, some departments translate the information gathered during prefire planning surveys into detailed operational plans. These plans describe the most likely organizational structures for controlling a range of possible fire scenarios at a given location. They identify needed resources, possible locations for staging areas, most likely access routes, as well as locations of hazardous materials and other extraordinary hazards. The plans outline the most likely deployment of personnel and equipment at a fire scene. The plans are reviewed at least annually and updated as necessary. Prefire operational plans that address initial attack, extended attack, and major fire attack should include the following:

- Organization charts
- Descriptions of organizational functions and responsibilities
- List of types of equipment and number of personnel needed
- Descriptions of support services needed
- Operating procedures
- Radio frequency assignment
- Maps showing topography, possible staging areas, target hazards, access routes, water sources, etc.
- Rosters of other agency resources and ordering procedures
- List of utilities, public works, water, and law enforcement agencies needed
- List of special concerns/needs

Prefire operations plans define the lines of authority and state the duties and responsibilities of all positions within the anticipated organizational structure (Figure 5.2). Wildland fires often involve more than one fire jurisdiction. Therefore, a clear understanding of the fireground organization/management system and the duties and responsibilities of all involved is needed for firefighters from different departments to work together efficiently within the larger structure.

Figure 5.2 Plans should identify needed resources, organizational structure, etc.

These operations plans should also anticipate various ways of deploying personnel and equipment at a fire scene. The plans must anticipate changing fire conditions and make provisions for upgrading response assignments. As mentioned earlier, the plan identifies the priorities, strategies, resources, and possible deployment of personnel and equipment for initial attack, extended attack, and major fire attack (also see Developing the Organization section).

Resources

One of the most important functions of prefire planning is identifying what resources are likely to be needed, what resources are readily available, and how to obtain resources not immediately available. Resources include all fire department personnel and equipment, including those available through mutual aid agreements (see Mutual Aid and Cooperative Agreements section). In addition, many property owners, especially farmers, ranchers, and loggers, can also provide mechanized equipment and other resources for fighting wildland fires. Other resources include privately owned auxiliary water supplies, especially those with dry hydrants to facilitate using the water. In order to have access to these privately owned resources when they are needed, fire departments must get written permission from the owners in advance. Operators of privately owned equipment must also be trained in wildland fire safety procedures and in operating their equipment in fireline conditions. See Appendix B for a sample use and rental agreement.

In addition to lending or renting heavy equipment, local residents can help fire departments through their knowledge of the area. They can supply information on the locations of water supplies, means of access, and other data that may be helpful in controlling a fire. Every fire department should maintain a list of names, addresses, and telephone numbers of individuals and organizations who have resources. However, since people sometimes move away and businesses close or relocate, these lists should be updated at least annually (Figure 5.3).

Figure 5.3 Emergency call lists must be kept up to date.

Mutual Aid and Cooperative Agreements

In addition to identifying the resources available from private sources, the resources available through mutual aid and automatic aid agreements should also be identified during prefire planning. Mutual aid agreements allow fire departments to *request* resources from one or more other departments that are parties to the agreement. These other departments may, *at their option*, honor the request by sending some of their resources into the requesting department's jurisdiction. The mutual aid resources may report directly to the emergency scene or may occupy vacated fire stations to provide protection in case of subsequent alarms. However, if the other departments need to keep their resources within their own boundaries because of emergencies in progress or because of an imminent threat of an emergency developing at the time of the request for mutual aid, they may refuse to honor the request. While mutual aid agreements are often negotiated among adjacent departments and with other departments in the same area, such agreements may also be negotiated on a regional or wider basis such as statewide or provincewide. Units from all departments involved in a mutual aid agreement should periodically train together to identify and rectify any differences in equipment or procedures.

On the other hand, *automatic aid* agreements are usually negotiated only between immediately adjacent jurisdictions. In this type of arrangement, it is not necessary that aid be requested. Both departments *automatically* dispatch resources anytime there is an incident in the geographic area covered by the agreement. This type of agreement is often necessary where district boundaries are unclear. Because they function as one department in the areas covered by the agreement, units from both departments should train together frequently to identify and rectify any differences in equipment or procedures.

Of critical importance in both mutual aid and automatic aid agreements is information regarding the capabilities/limitations of the resources available under these agreements. How the resources are identified — single resources, strike teams (combinations of the same kind and type of resources), task forces (any combination of single resources) — and accessed should be identified. It is also important to know whether communications equipment is compatible with that of the local departments.

Maps

Many different types of maps are available to fire departments. During the pre-incident planning process, it is important to identify where to obtain accurate and easily read maps. Many are available from chambers of commerce, city and county offices, highway departments, and other government agencies (Figure 5.4). Automobile clubs, such as the American Automobile Association (AAA), are other possible sources. It is often advantageous to have different types of maps of the same area. The most common types of maps and the information they contain include the following:

- ***General plot (planometric) maps*** — Show jurisdiction boundaries, distances, roads, communities, and similar features

Figure 5.4 Accurate maps of the district may come from a variety of sources.

- ***Topographic maps*** — Show natural features such as rivers, streams, hills, valleys, open areas, lakes, and ponds (available from the United States Geological Services and local sporting goods stores)
- ***Special maps and aerial photographs*** — Show locations of permanent structures and developed campsites
- ***Rural addressing maps*** — Assign and number residences by the distances in hundredths of a mile (kilometer) from the start of the road (often developed by local utility and service companies)
- ***Orthophoto maps*** — Are aerial photographs corrected to scale so that geographic measurements may be taken directly from the prints

Area Knowledge

Becoming familiar with the actual response area, not just with maps of the area, is an important aspect of planning for wildland fires. In addition to the information available from local residents, firefighters should familiarize themselves with the area for which they are responsible. They should drive the roads and walk the trails to observe the ground cover and make notes concerning terrain, landmarks, trails, access roads, streams, water supply, and structures (Figure 5.5). Data regarding the locations and weight limits of all bridges within the district are particularly important (Figure 5.6).

Figure 5.5 Firefighters should become familiar with their districts. *Courtesy of NIFC.*

Figure 5.6 Bridges with weight limits should be identified.

Crew Readiness

There are a number of different types of crews that may be involved in wildland fire fighting, and they all need to be ready to perform at their best when called. Regardless of the nature of their assignment, firefighters need to be well-trained, properly equipped, physically fit, well-fed and hydrated, and given reasonable rest periods between work cycles in order to perform as required.

Hand crews consist of from 10 to 20 members and have some of the most physically demanding assignments on the fireground. Not only is the process of constructing a fireline with hand tools a very exhausting one, it is often done in the worst atmospheric conditions and terrain possible (Figure 5.7). Another very important aspect of hand-crew readiness is transportation to assignments. If crews do not have to hike for miles (kilometers) to reach their assigned section of fireline, they arrive fresher and capable of being much more productive (Figure 5.8).

Figure 5.7 Hand crews often work under very taxing conditions. *Courtesy of NIFC.*

Figure 5.8 When possible, crews should be transported to their assignments. *Courtesy of NIFC.*

It is no less important for engine crews to be ready to perform when called. While these crews do not have to walk to their assignments, they can be required to perform some very strenuous tasks. For example, setting up a long progressive hose lay in difficult terrain can be very demanding (Figure 5.9).

Figure 5.9 An engine crew begins a progressive hose lay. *Courtesy of Tony Bacon.*

Predesignated Components

Another important aspect of pre-incident planning is to identify and designate certain items as ones most likely to be used during a fire in a given location. Predesignating certain roads or highways for firefighters to access a property is particularly important. The choice of travel routes must take into account other traffic that is likely to be using those routes — normal vehicular and pedestrian traffic, those trying to flee from or return to their threatened property, media vehicles and personnel, and spectators. The plan may predesignate closing certain sections of highway to all but emergency vehicles and owners of threatened properties (Figure 5.10).

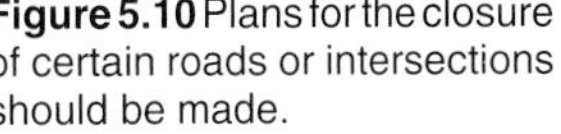

Figure 5.10 Plans for the closure of certain roads or intersections should be made.

Sites that could serve as staging areas should also be identified during pre-incident planning (Figure 5.11). Large, open spaces such as parking lots of schools, churches, and shopping centers should be predesignated as possible staging areas for strike teams and task forces composed of engines or heavy equipment such as bulldozers or tractor-plows (see Chapter 7, Wildland/Urban Interface Fire Suppression). The plan should also note when these areas are not likely to be available — such as when schools and churches are in session and during regular business hours. Those predesignated areas should lend themselves to their intended use — adequate size and configuration, easy ingress/egress, adequate lighting, and available sanitary facilities. Possible staging areas for hand crews should also be identified. They, too, should have the minimum amenities needed — shade, drinking water, and sanitary facilities.

Figure 5.11 Potential staging areas should be identified.

Other sites that should be identified during prefire planning are possible locations for incident command posts, incident bases, helibases and helispots, safety zones, and other Incident Command System (ICS) facilities. All of these functions have different requirements in terms of the amount of space needed, the absence of obstructions, the view provided, and the ease of access. These different requirements should be considered in the selection of possible sites.

Communication System

One of the most important factors in safely and successfully dealing with wildland fires is effective fireground communication. Therefore, one of the most important aspects of pre-incident planning is to identify any differences in various agencies' communications equipment or procedures that would interfere with effective fireground communications. Although modern multichannel radios, many with built-in scanners, have made communications equipment much more compatible, not all agencies have these newer units. Most agencies within a given region are capable of communicating with each other; but when mutual aid units from outside the area (perhaps outside the state or province) are called in, they may not be equipped to monitor or transmit on local frequencies. Some agencies have equipment that uses 800 MHz; others do not. Where equipment incompatibilities are identified, plans must be made to provide portable radios with the appropriate frequencies to those who need them. So that local agencies don't have to buy large banks of portable radios, state or provincial agencies often have caches of multichannel radios available for use during major incidents. The availability of such equipment should be identified during prefire planning. For more information on sources of communications equipment for large-scale emergencies, see Chapter 10, Fire Protection Planning.

In addition to the issue of compatible communications equipment, differences in communications procedures among participating agencies must also be identified and rectified during prefire planning. Agencies that may be involved in large-scale operations must be prepared to forego the use of local or agency-specific radio codes. To be effective, promote fireline safety, and eliminate confusion during multiagency operations, all units must use terminology that promotes clarity and reduces ambiguity during radio communications. One of the best ways of doing this is for all units on a given incident to use *clear text* when transmitting.

While the essence of clear text is replacing the Ten-Code and any other agency-specific codes, acronyms, and abbreviations with plain English, there are certain standard terms and phrases used in clear text. The terms used are mostly resource identifiers that require no explanation — *chief*, *dozer*, etc. However, other terms used in clear text also have specific meanings but are sometimes misused because of regional differences. For example, in clear text a pumper is called an *engine*, not a truck; a *truck* is an aerial device, not an engine or other heavy-duty vehicle; a *tanker* is an aircraft, not a mobile water supply apparatus; and a mobile water supply apparatus is a *water tender*, not a tanker. Some other support vehicles are also called *tenders* — dozer tenders, fuel tenders, etc.

As mentioned earlier, there are certain standard terms and phrases used in clear text. Crew leaders of units dispatched to an incident need only identify their unit and say "*responding*," such as "*Engine 7185 responding*." To announce their arrival at the emergency scene, they again identify their unit and say "*at scene*." If the first-in unit determines upon arrival that no other resources are needed to handle a fire, the crew leader need only say "*can handle*." Once their assignment has been completed, the crew leader announces, "*Engine 7185 available, returning to . . .* (previous assignment, quarters, staging, base, etc.)." In this case, *available* is another of the standard terms in clear text. See Appendix A for a complete list of clear-text terminology.

For fireground communications to be most effective, those transmitting should make their messages as clear and concise as possible. Using the standard clear-text terminology is very helpful. However, one of the best ways of making messages clear and concise is for radio operators to think about both what needs to be said and the best way to say it *before* keying the microphone (Figure 5.12). Another technique that helps improve the quality of radio transmissions is for the operator to demonstrate "command presence." This means carefully choosing the most appropriate terminology and speaking in a calm but authoritative voice.

Organizational Forms

There are a few forms with which all firefighters should be familiar. These forms are an integral part of the incident command/management system, and they provide critical information to all levels of the organization. If these forms are completed properly and submitted promptly, they can assist with personnel accountability, help to identify resource needs, and draw attention to things that need to be done.

Perhaps the most important federal form to firefighters is ICS 201, the *INCIDENT BRIEFING* form. Sometimes called a *tactical worksheet,* this seven-part form consists of the incident name, date prepared, time prepared, a map of the incident, an incident organization chart, a resources summary, and a summary of current actions. An example of a completed Form 201 is shown in Figure 5.13. Also of critical importance to firefighters is ICS Form 202, *INCIDENT OBJECTIVES* (Figure 5.14). These forms may be photocopied from the *Incident Command System* manual published by Fire Protection Publications or obtained from the National Interagency Fire Center in Boise, Idaho.

Figure 5.12 Radio transmissions should be kept short and to the point. *Courtesy of NIFC.*

Other forms used in the command/management of a major fire incident are ICS Form 203, *ORGANIZATION ASSIGNMENT LIST*; Form 204, *DIVISION ASSIGNMENT LIST*; Form 205, *INCIDENT RADIO COMMUNICATIONS PLAN*; and Form 206, *MEDICAL PLAN.* Examples of these incident command/management forms are shown in Appendix C.

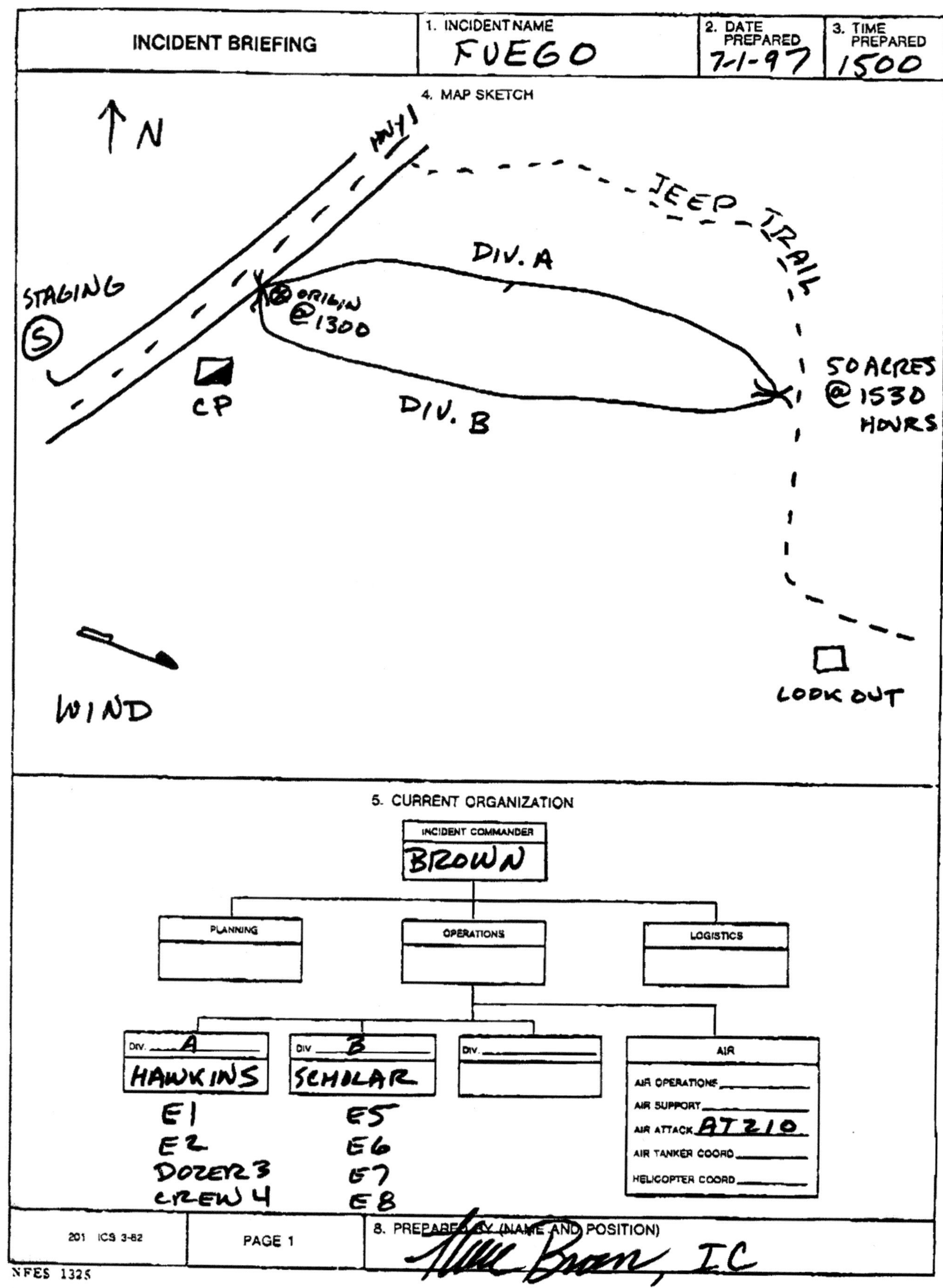
INCIDENT BRIEFING

1. INCIDENT NAME: FUEGO

2. DATE PREPARED: 7-1-97

3. TIME PREPARED: 1500

4. MAP SKETCH

5. CURRENT ORGANIZATION

INCIDENT COMMANDER: BROWN

PLANNING

OPERATIONS

LOGISTICS

DIV. A: HAWKINS — E1, E2, DOZER 3, CREW 4

DIV. B: SCHOLAR — E5, E6, E7, E8

DIV.

AIR
AIR OPERATIONS
AIR SUPPORT
AIR ATTACK: AT 210
AIR TANKER COORD
HELICOPTER COORD

201 ICS 3-82

PAGE 1

8. PREPARED BY (NAME AND POSITION): [signature] Brown, IC

NFES 1325

Figure 5.13 A completed ICS Form 201.

6. RESOURCES SUMMARY

RESOURCES ORDERED	RESOURCE IDENTIFICATION	ETA	ON SCENE	LOCATION/ASSIGNMENT
FIRST ALARM	C 1	1310	✓	IC
	E 1	1310	✓	DIV. A - LEFT FLANK
	E 2	1310	✓	DIV. A - LEFT FLANK
SECOND ALARM	E 5	1320	✓	DIV. B - RIGHT FLANK
	E 6	1320	✓	DIV. B - RIGHT FLANK
ENGINE	E 7	1330	✓	DIV. B - RIGHT FLANK
ENGINE	E 8	1330	✓	DIV. B - RIGHT FLANK
DOZER	DOZER 3	1340	✓	DIV. A - LEFT FLANK
CREW	CREW 4	1340	✓	DIV. A - LEFT FLANK
AIR TACTICS	AT 210	1330	✓	RECON
STAGING MGR	PREV 11	1325	✓	STAGING
DIV. SUP	BATT 9	1325	✓	DIV. A
DIV. SUP	BATT 10	1325	✓	DIV. B

7. SUMMARY OF CURRENT ACTIONS

1310 · INITIAL ATTACK ON LEFT FLANK
1312 · REQUEST SECOND ALARM; OBJECTIVE: CONTAIN FIRE WEST OF JEEP TRAIL
1320 DIFFICULT ACCESS ON LEFT FLANK FOR ENGINES. REQUEST DOZER, CREW, 2 MORE ENGINES, AIR TACTICS FOR RECON, STAGING AREA MANAGER, AND 2 DIVISION SUPERVISORS.
1335 WEATHER FROM BELT WEATHER KIT: CLEAR, 87°, WIND W 5, HUMIDITY 35%
1530 FIRE CONTAINED ON FLANKS AND STOPPED AT JEEP TRAIL; 50 ACRES

201 ICS 3-82	PAGE 2	

Figure 5.13 *Continued*

INCIDENT OBJECTIVES ICS 202

1. INCIDENT NAME	2. DATE PREPARED	3. TIME PREPARED
SQUAW	6·30·96	0600

4. OPERATIONAL PERIOD (DATE/TIME)
6·30·96 0800 - 2000

5. GENERAL CONTROL OBJECTIVES FOR THE INCIDENT (INCLUDE ALTERNATIVES)

1. PERSONNEL SAFETY
2. STRUCTURE PROTECTION - ASTI SUBDIVISION
3. KEEP FIRE EAST OF DUTCHER CREEK RD.
4. KEEP FIRE NORTH OF CANYON RD.
5. KEEP FIRE WEST OF HWY. 101

6. WEATHER FORECAST FOR OPERATIONAL PERIOD
DAYTIME HIGH 95°-100°F, R/H <20%, WINDS NE 05-15, NO FRONTAL ACTIVITY NEXT 48 HRS., CONTINUED HOT & DRY

7. GENERAL/SAFETY MESSAGE

1. KEEP PERSONNEL HYDRATED
2. WATCH FOOTING ON STEEP SLOPES

8. ATTACHMENTS (✓ IF ATTACHED)

- ☑ ORGANIZATION LIST (ICS 203)
- ☑ DIVISION ASSIGNMENT LISTS (ICS 204)
- ☑ COMMUNICATIONS PLAN (ICS 205)
- ☑ MEDICAL PLAN (ICS 206)
- ☑ INCIDENT MAP
- ☑ TRAFFIC PLAN
- ☐ ______
- ☐ ______
- ☐ ______

202 ICS 3-80	9. PREPARED BY (PLANNING SECTION CHIEF)	10. APPROVED BY (INCIDENT COMMANDER)
	J. HAWKINS	S. BROWN

Figure 5.14 A completed ICS Form 202.

DEVELOPING THE ORGANIZATION

Simply stated, an *organization* is a group of people working together to achieve a common objective. To achieve fireground objectives safely and efficiently, the size and complexity of the fireground organization must match the size and complexity of the incident. On small incidents the organization may be small initially, perhaps only an incident commander (IC) (who also acts as the incident safety officer) and members of his or her crew (see Fire Suppression Organization Functions section). If a fire grows, however, the fireground organization must grow with it. As the size and complexity of the organization grow, certain management principles are always applied. The four most important ones are *unity of objective, unity of command, span of control*, and *personnel accountability*.

Unity of Objective

Also called *unity of purpose*, this principle is critical to both safety and effectiveness on the fireground. If everyone in the fireground organization, from the IC down to the individual firefighter, knows what strategic objectives have been established for the fire, it is easier for everyone involved to work toward the same outcomes. Knowing the established objectives helps prevent different units from working against each other and perhaps putting each other in jeopardy. Knowing the established objectives also helps to ensure that the correct types and number of resources are ordered.

Unity of Command

The essence of the unity of command concept is that no firefighter at scene reports to more than one supervisor, regardless of how large and complex the fireground organization is. If everyone at every level in the organization follows this procedure, personnel will not receive conflicting orders from different supervisors. Conflicting orders can cause confusion and may put firefighters in danger. While the IC retains full authority in the fireground organization, he or she must respect the chain of command and issue orders only to direct subordinates.

The only exception to the unity of command principle that is allowed in the fireground organization is when fireground safety is involved. The incident safety officer has the organizational authority to order any officer or firefighter to cease any unsafe activity, regardless of who that individual's supervisor is.

Span of Control

Span of control relates to the number of direct subordinates that one supervisor can effectively manage. Variables such as proximity and similarity of function affect that number. If all subordinates are within sight of the supervisor and are able to communicate effectively with each other, the number of subordinates can be higher than when they are widely separated. Likewise, it is easier to supervise subordinates who are all performing the same or similar functions, so the number of subordinates can be higher than if they are all doing very different tasks. In the ICS, an effective span of control ranges from three to seven subordinates per supervisor, depending upon proximity and functions, with five being the number most often used. If an effective span of control is maintained, it is much easier for supervisors to keep track of their subordinates and to monitor their safety. Unity of objective, unity of command, and span of control all promote one of the most important fireground safety concepts —*personnel accountability*.

Personnel Accountability

Firefighters are at some level of risk whenever they are on the fireground, but wildland fires can be some of the most dangerous environments to which firefighters are assigned. If fire conditions change suddenly and dramatically, firefighters may need to make a quick but orderly retreat. Firefighters usually depend on their supervisors to warn them of impending danger and to order them to leave a dangerous environment in time to escape to safety. However, if firefighters become separated from their unit, they may have to decide for themselves when and in which direction to escape. Those with little fireground experience may be unable to make the decisions needed to save themselves. Therefore, it is critically important that some means be used to keep track of every firefighter on the fireground.

All supervisors in the organization are responsible for keeping track of their subordinates on the fireground.

However, all command/management systems must provide various means of tracking the resources assigned to a given incident. Personnel accountability includes all of the following:

- ***Check-in*** — Requires all responders, regardless of agency affiliation, to check-in to receive their assignments
- ***Incident action plan*** — Identifies incident priorities and objectives, which dictate how tactical operations must be conducted (promotes unity of objective)
- ***Unity of command*** — Ensures that each firefighter has only one supervisor
- ***Span of control*** — Gives supervisors a manageable number of subordinates
- ***Division/group/sector assignment list*** — Identifies resources with active assignments in the Operations section (see Fire Suppression Organization Functions section)
- ***Resource tracking*** — Ensures that each supervisor reports resource status changes as they occur
- ***Resource status unit*** — Maintains status of all incident resources

When personnel at all levels in the fireground organization operate according to these principles and procedures, personnel accountability and safety are maximized. As an incident grows from an *initial attack* to an *extended attack* or to a *major fire attack* incident, these basic principles and procedures must continue to be applied.

Initial Attack

Statistically, a well-planned initial response is adequate to control about 95 percent of all wildland fires. An initial-attack response to a wildland fire may range from a single engine to an assignment that includes a large number of personnel, several pieces of apparatus, heavy mechanized equipment, and aircraft. Its composition is based on prefire plan information and current conditions — the location of the fire, fuel types, values at risk, weather conditions, and available resources.

On small fires, the supervisor of the first-arriving unit assumes command of the incident. This IC personally directs and controls all fire suppression resources and activities unless and until the incident grows beyond his or her span of control (Figure 5.15).

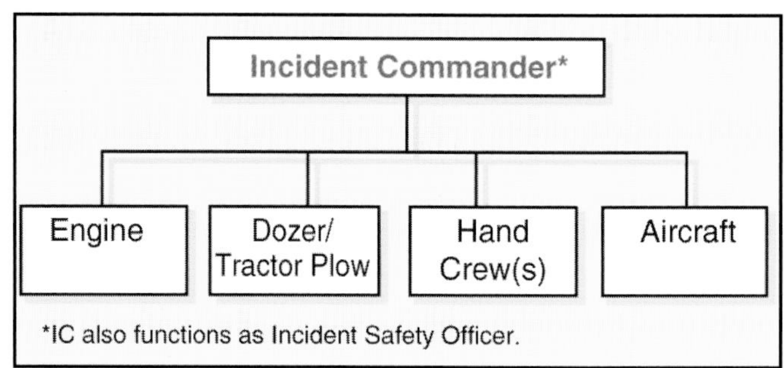

Figure 5.15 A typical initial attack organization.

FUNCTIONS

On an initial-attack fire, the IC may act not only as the incident commander but may also perform all other management functions that the situation requires. One of the most important functions of the IC is to perform a proper size-up and to order any additional resources that are needed. Another of the IC's management functions is that of incident safety officer. The IC does this personally or delegates the authority for this function. Whoever functions as the incident safety officer is also responsible for personnel accountability. However, the primary function of the initial-attack force is to do just that — attack. The personnel and other resources in this attack force are there to achieve the same objectives as any other fireground operation — life safety, property conservation, and incident stabilization.

STAGING

By definition, *staging* is a temporary location where fire fighting resources are gathered. These resources must be available to respond within 3 minutes of being called. The staging function may be slightly different on an initial-attack incident than it is on larger incidents. This is because in the initial-attack incident the exact location of a fire, or the best access to it, may not be known. In this situation, some departments use a procedure called *primary staging* or *level-one staging*. When the exact location of a fire is still unclear, the first-due unit responds directly to the reported location of the incident. If, upon arrival, this unit cannot locate the fire or is unsure of the best access to it, the officer may advise other responding units to stage. The other responding units stage at the last intersection in their response route before the reported location of the fire. They remain there until they receive instructions from the first-due unit.

FIRE SUPPRESSION ORGANIZATION FUNCTIONS

Many wildland fires are detected, reported, and controlled before they become unmanageable for the initial-attack force. When the initial-attack force is unable to control the fire, additional resources must be ordered and the fireground organization expanded to match the size and complexity of the incident. This necessitates the implementation of a larger, more complex fire command/management system.

Only those positions necessary to manage the incident are staffed, and no more.

To be useful in helping to mitigate an incident, an incident command/management system must be structured enough to be consistent from incident to incident, but flexible enough to fit any unique fire situation. The basic structure of a fully developed fireground organization consists of *Command*, *Operations*, *Planning*, *Logistics*, and *Finance / Administration* sections. The command/management functions of each section are discussed in this portion of the chapter along with unified command, extended and major fire attacks, and evidence preservation.

Command

The function of *Command* is to direct the overall management of the incident. The incident commander is responsible for the following:

- Determining overall incident objectives
- Selecting strategies
- Ensuring that tactical activities support the selected strategies
- Approving the incident action plan
- Making maximum use of all assigned resources

For assistance in fulfilling these responsibilities, the IC may appoint one or more members of a *Command Staff*. A full Command Staff consists of a Safety Officer, Information Officer (sometimes called *Public Information Officer* or *PIO*), and a Liaison Officer. Their primary functions are to relieve the IC of various tasks not necessarily directly related to controlling the incident. The functions of the Command Staff are as follows:

- ***Safety Officer*** — Assesses hazardous and unsafe situations and develops measures for assuring personnel safety; has emergency authority to stop and/or prevent unsafe acts; acts as Personnel Accountability Officer for the incident
- ***Information Officer*** — Develops accurate and complete information regarding incident cause, size, current situation, resources committed, and other matters of general interest; acts as point of contact for the media and other governmental agencies that desire information about the incident; gets IC approval of all releases of incident information; coordinates evacuation information broadcast to the public
- ***Liaison Officer*** — Acts as point of contact for representatives of all nonfire agencies assisting or cooperating in the incident (law enforcement, Red Cross, utility companies, etc.)

In addition to the Command Staff, the IC may appoint one or more members of the *General Staff* to assist in managing the incident as it grows in size and complexity. The positions that make up the General Staff are the Operations Section Chief, Planning Section Chief, Logistics Section Chief, and the Finance/Administration Section Chief.

Operations

The primary function of the *Operations* Section is to implement those actions that accomplish the established incident strategic objectives (Figure 5.16). At the direction of the IC, the Operations Chief is responsible for the following:

- Managing all incident tactical activities

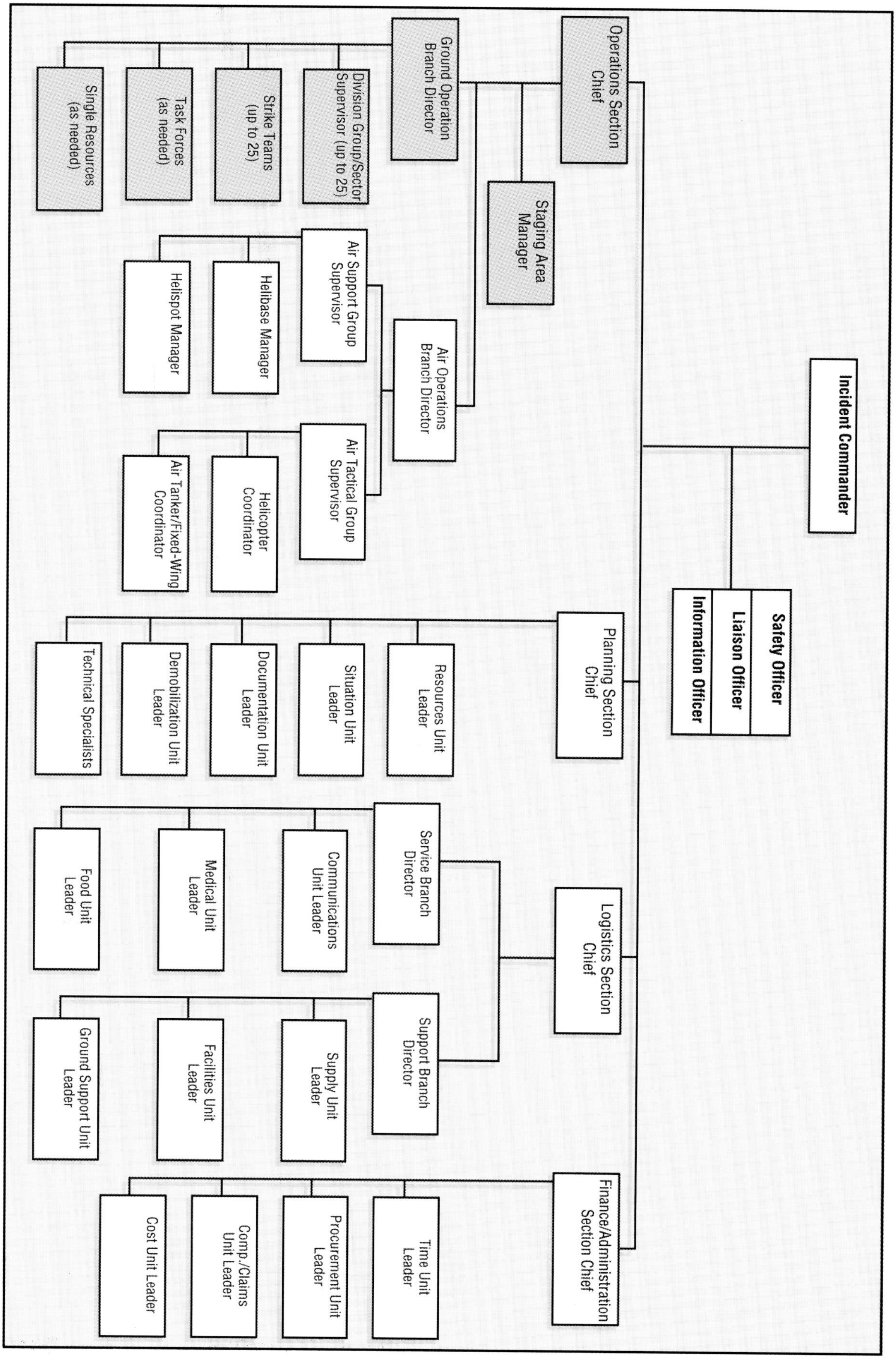

Incident Commander

Safety Officer
Liaison Officer
Information Officer

Operations Section Chief
Staging Area Manager
Ground Operation Branch Director
Division Group/Sector Supervisor (up to 25)
Strike Teams (up to 25)
Task Forces (as needed)
Single Resources (as needed)
Air Operations Branch Director
Air Support Group Supervisor
Helibase Manager
Helispot Manager
Air Tactical Group Supervisor
Helicopter Coordinator
Air Tanker/Fixed-Wing Coordinator

Planning Section Chief
Resources Unit Leader
Situation Unit Leader
Documentation Unit Leader
Demobilization Unit Leader
Technical Specialists

Logistics Section Chief
Service Branch Director
Communications Unit Leader
Medical Unit Leader
Food Unit Leader
Support Branch Director
Supply Unit Leader
Facilities Unit Leader
Ground Support Unit Leader

Finance/Administration Section Chief
Time Unit Leader
Procurement Unit Leader
Comp./Claims Unit Leader
Cost Unit Leader

Figure 5.16 Fireground organization with the Operations Section fully staffed.

- Assisting in the development of the IAP
- Designating incident staging areas (and appointing Staging Area Managers; see following section, Staging Areas)
- Establishing an Air Operations Branch (if needed) (see Air Operations section)

There is only one Operations Chief at any time. However, the Operations Chief may appoint one or more deputies (with qualifications equal to the Operations Chief) to assist in carrying out the responsibilities of this section. The Operations Chief normally functions for one operational period. If fire suppression activities last beyond that, one of the deputies assumes the responsibility for continuing the function. The Operations Chief and each of the deputies assist in the development of the IAP for their operational period.

STAGING AREAS

The Operations Chief designates the locations of any staging areas needed and may relocate or discontinue them as the situation dictates (Figure 5.17). The Operations Chief also appoints a Staging Area Manager for each staging area. As mentioned earlier, all resources in staging must be ready to respond within 3 minutes of being called. The Staging Area Managers keep the Operations Section Chief informed of the number and availability of the resources in staging and request logistical support (food, fuel, portable toilets, etc.) for those resources from the appropriate units in the Logistics section.

Figure 5.17 Staging is where needed resources are gathered. *Courtesy of Monterey County Training Officers.*

AIR OPERATIONS

The Operations Chief also establishes an Air Operations Branch if necessary. Its size, organization, and use depend upon the nature of the incident and the availability of aircraft. The Air Operations Branch may consist of a single aircraft, such as a helicopter, in which case it may be under the direct control of the Operations Chief. In other cases, the Air Operations Branch may consist of a number of fixed-wing and/or rotary-wing aircraft, and the Operations Chief may communicate with an Air Tactical Group Supervisor. The Air Tactical Group Supervisor, in turn, coordinates all airborne activity through a Helicopter Coordinator and/or a Fixed-Wing/Air Tanker Coordinator (Figure 5.18).

Planning

The primary function of the *Planning* Section is to collect, evaluate, and disseminate tactical information about the incident. The section consists of a *resources unit*, a *situation unit*, a *documentation unit*, a *demobilization unit*, and *technical specialists* (Figure 5.19). Under the direction of the Planning Section Chief, the section is specifically responsible for the following:

- Maintaining information on the current and expected situation
- Maintaining information on the status of assigned resources
- Preparing and documenting the IAP
- Preparing incident maps
- Planning demobilization
- Utilizing technical specialists

The Planning Section Chief analyzes incident data and develops tactical alternatives. This person conducts planning meetings, prepares the IAP for each operational period, and may have one or more technical specialists (engineers, meteorologists, chemists, fire behavior experts, water system experts, etc.) to assist with evaluating the situation and forecasting additional resource requirements.

Logistics

The primary function of the *Logistics* Section is to provide all support needs (except aircraft) to the incident. This section may be split into a *Service Branch* (which includes a *communications unit*, a *medical unit*, and a *food unit*) and a *Support Branch* (which includes a *supply unit*, a *facilities*

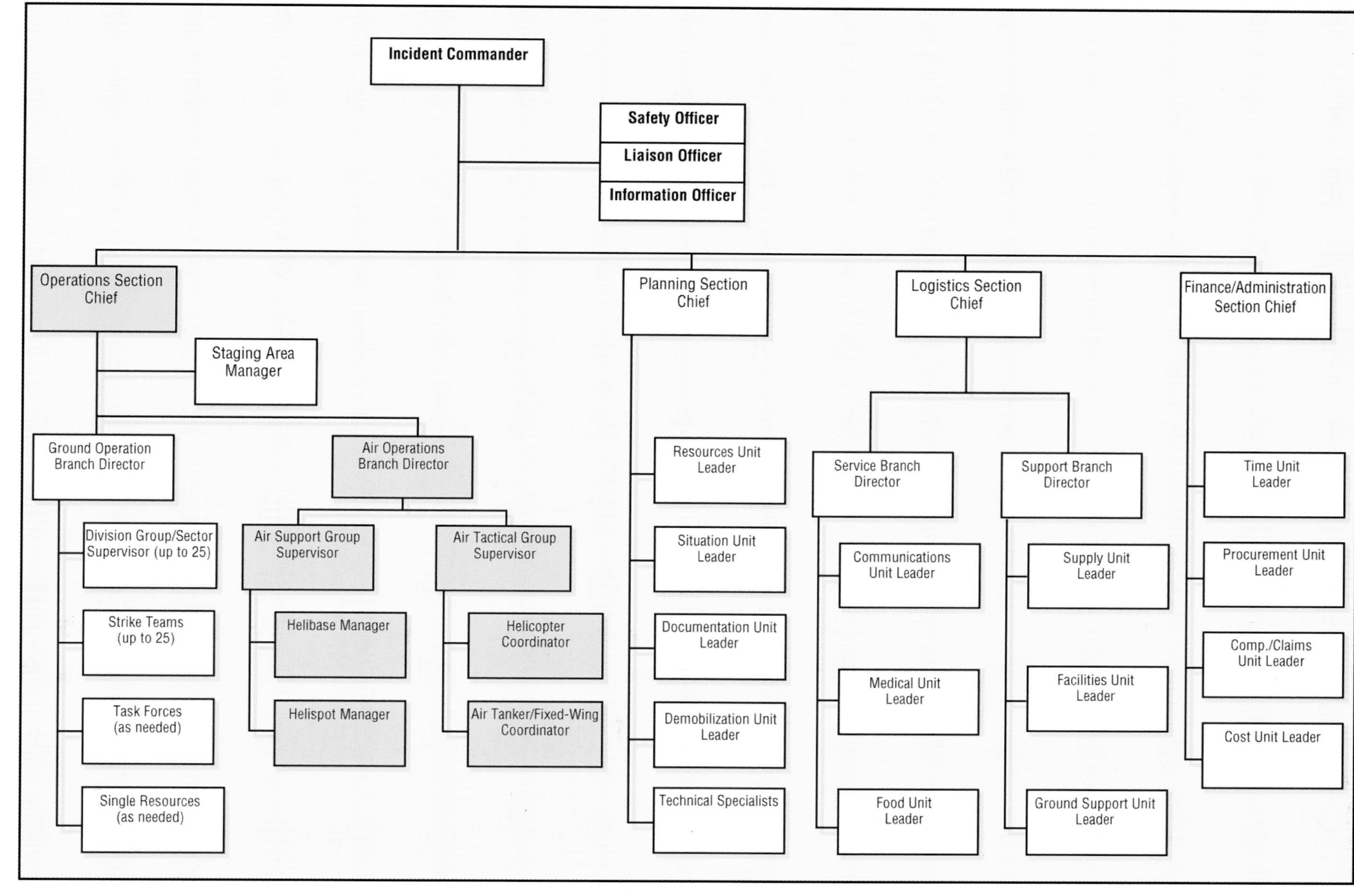

Figure 5.18 Fireground organization including an Air Operations Branch.

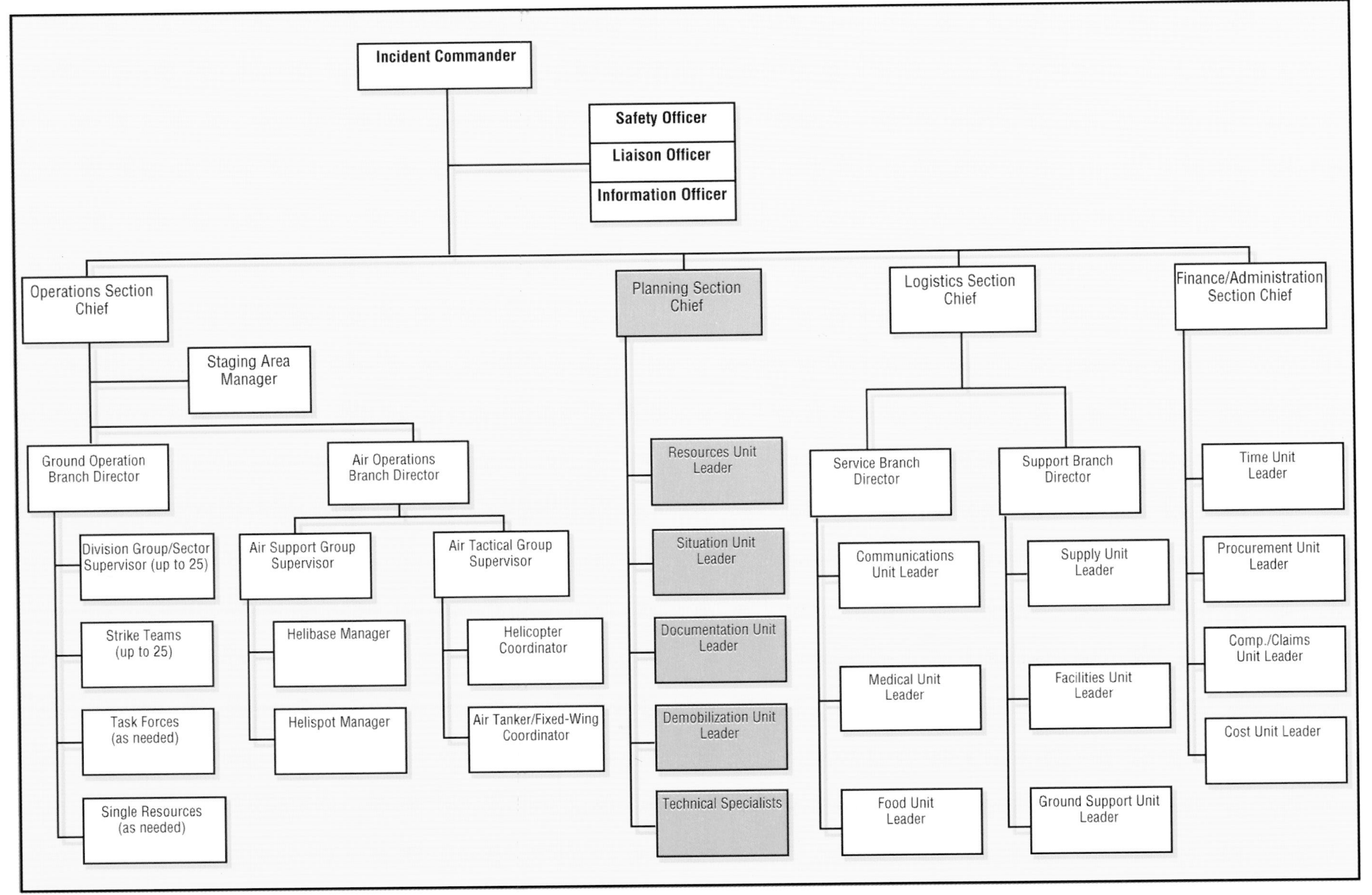

Figure 5.19 Fireground organization including a Planning Section.

unit, and a *ground support unit*) on large incidents (Figure 5.20). Under the direction of the Logistics Section Chief, the section is specifically responsible for the following:

- Ordering fire fighting resources from off-incident locations
- Providing incident facilities and supplies
- Developing the incident traffic plan
- Providing incident transportation, equipment maintenance, and fuel
- Feeding incident personnel
- Developing the incident communications plan
- Providing incident communications services
- Developing the incident medical plan
- Providing medical services to incident personnel (including rehab)

While all of the functions of the Logistics section are important, two of the most important are the services of the communications unit and the medical unit. Considering the vital importance of communications to the safety and effectiveness of those directly and indirectly involved in controlling a fire, it is difficult to overestimate the value of this function. Likewise, the value of rehab stations to firefighters on the line is obvious.

Finance/Administration

The primary function of the *Finance/Administration* Section is to assist those agencies that require cost-recovery and other administrative services. On those incidents where the agencies involved require its services, this section may include a *time unit*, a *procurement unit*, a *compensation/claims unit*, and a *cost unit* (Figure 5.21). Under the direction of the Finance/Administration Section Chief, the section is specifically responsible for the following:

- Ensuring the proper preparation of personnel time-recording documents
- Administering financial contracts with vendors
- Processing compensation-for-injury claims and property-damage claims
- Providing incident cost analysis data

The Finance/Administration section works closely with the Logistics section to see that all incident activities are properly documented. This helps to ensure that all entities requiring compensation or reimbursement for their involvement in the incident are paid fairly and quickly.

Unified Command

When a wildland fire is confined to the jurisdiction in which it started, the command structure is relatively simple and straightforward as described earlier in this chapter. However, many wildland fires extend or threaten to extend across jurisdictional boundaries. Because the legal responsibility and authority of an individual agency is normally confined to its own jurisdiction, a more complex command structure is required when the fire involves more than one jurisdiction. In this situation, representatives from each of the involved or threatened jurisdictions combine to form one *unified command.* These individuals are united by their mutual interest — their unity of purpose — to minimize the negative effects of the fire.

To be successful, those involved in a unified command structure must view the fire as a single incident, regardless of the number of jurisdictions involved. There must be only *one* command post (CP) and only *one* set of incident objectives. The members of the unified command may choose to designate one of their number to be the IC, with the others becoming deputy incident commanders. However, this is not required, and they may choose to work together as a team. Many variables may influence this decision, but the IC is most often the representative in whose jurisdiction the fire is currently burning or in whose jurisdiction there is the greatest amount of fire involvement. The role of IC may transfer from one representative to another if a fire burns out of one jurisdiction into another or if a fire burns for more than one operational period (Figure 5.22). Individually and collectively, Incident Commanders are responsible for the following:

- Determining overall incident objectives
- Selecting incident strategies
- Planning jointly to ensure that tactical activities are in accord with approved incident objectives

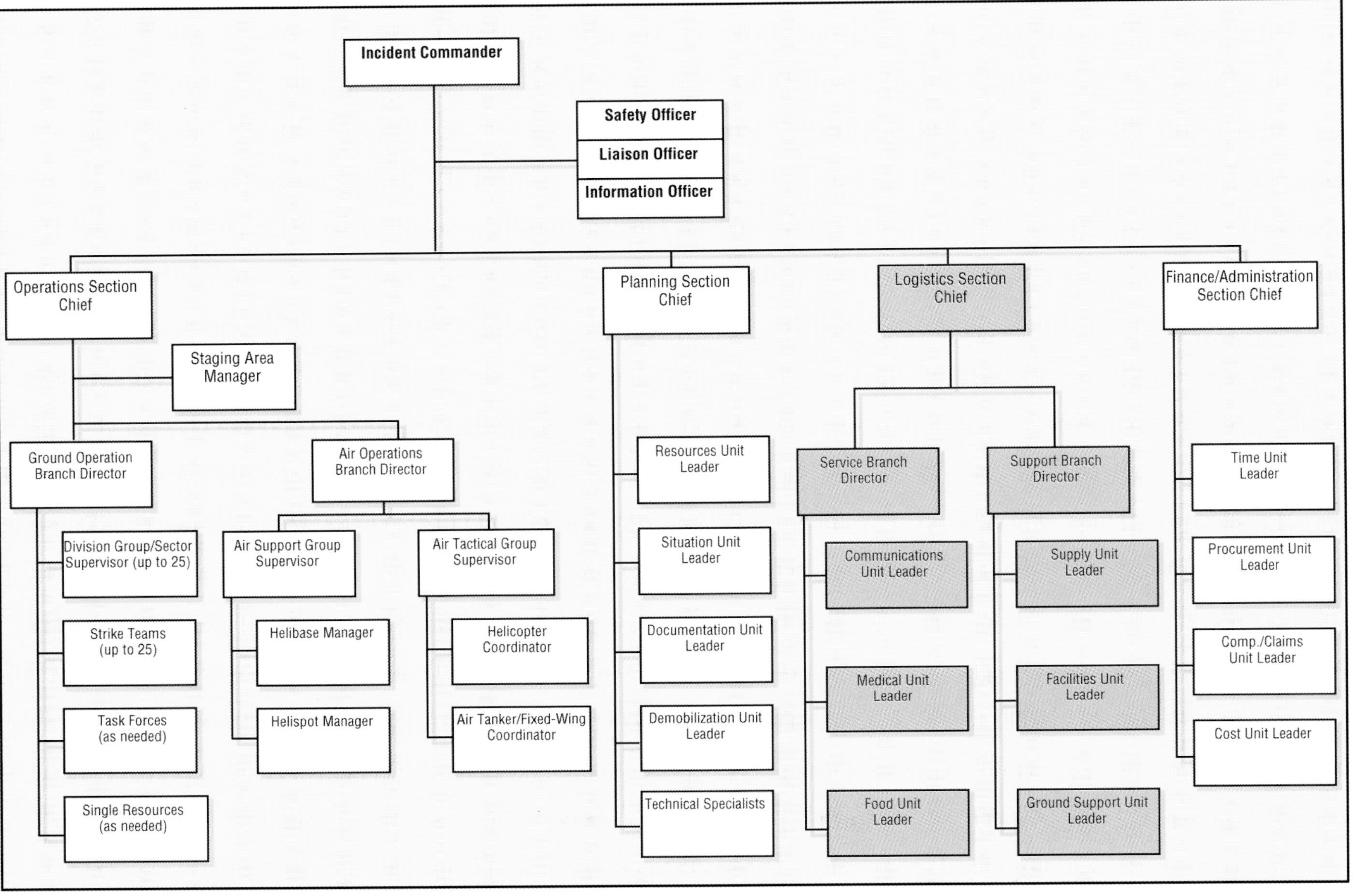

Figure 5.20 Fireground organization including a Logistics Section.

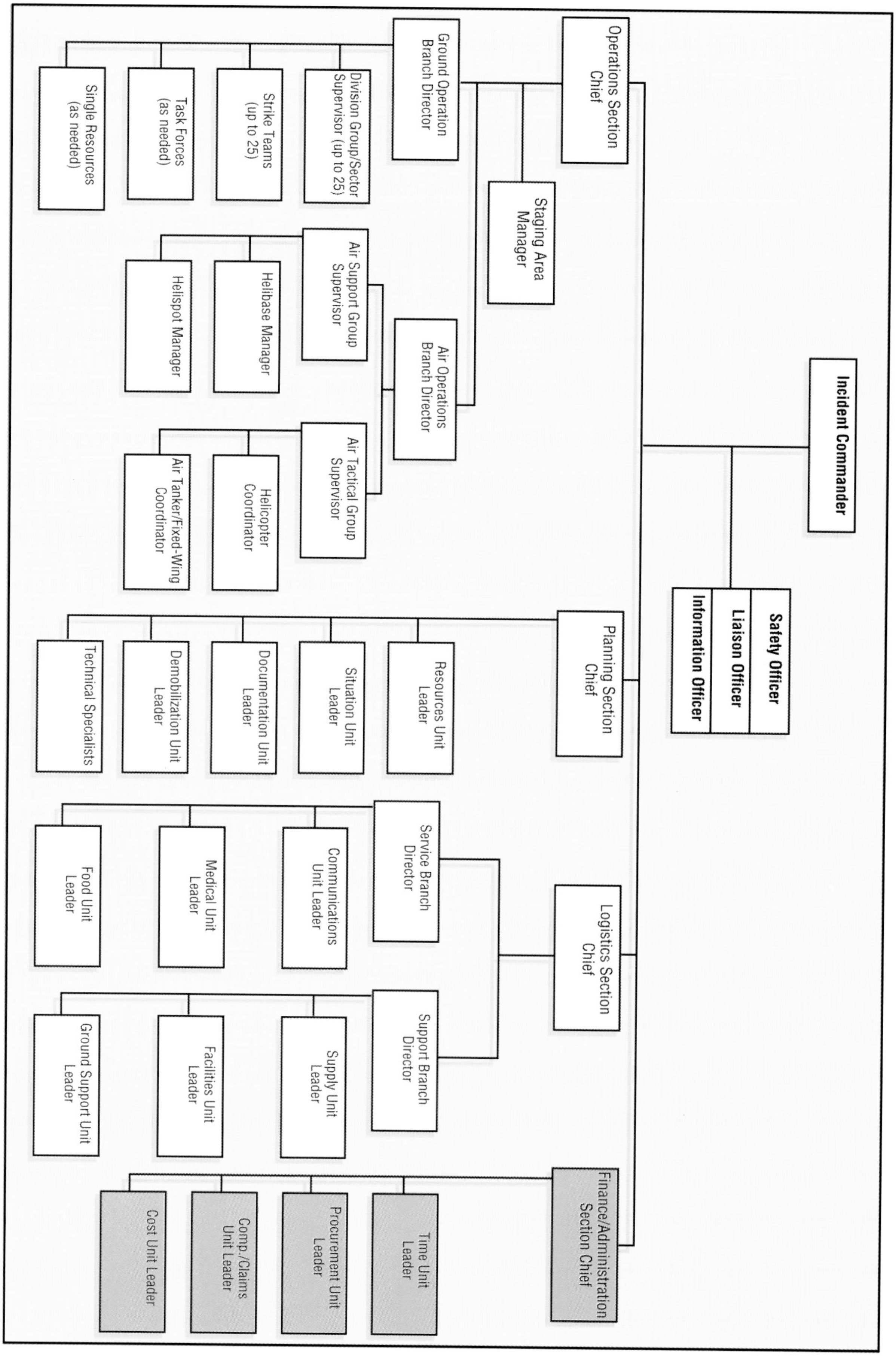

- **Incident Commander**
  - **Safety Officer**
  - **Liaison Officer**
  - **Information Officer**
  - Operations Section Chief
    - Ground Operation Branch Director
      - Division Group/Sector Supervisor (up to 25)
      - Strike Teams (up to 25)
      - Task Forces (as needed)
      - Single Resources (as needed)
    - Staging Area Manager
    - Air Operations Branch Director
      - Air Support Group Supervisor
        - Helibase Manager
        - Helispot Manager
      - Air Tactical Group Supervisor
        - Helicopter Coordinator
        - Air Tanker/Fixed-Wing Coordinator
  - Planning Section Chief
    - Resources Unit Leader
    - Situation Unit Leader
    - Documentation Unit Leader
    - Demobilization Unit Leader
    - Technical Specialists
  - Logistics Section Chief
    - Service Branch Director
      - Communications Unit Leader
      - Medical Unit Leader
      - Food Unit Leader
    - Support Branch Director
      - Supply Unit Leader
      - Facilities Unit Leader
      - Ground Support Unit Leader
  - Finance/Administration Section Chief
    - Time Unit Leader
    - Procurement Unit Leader
    - Comp./Claims Unit Leader
    - Cost Unit Leader

Figure 5.21 Fireground organization including a Finance/Administration Section.

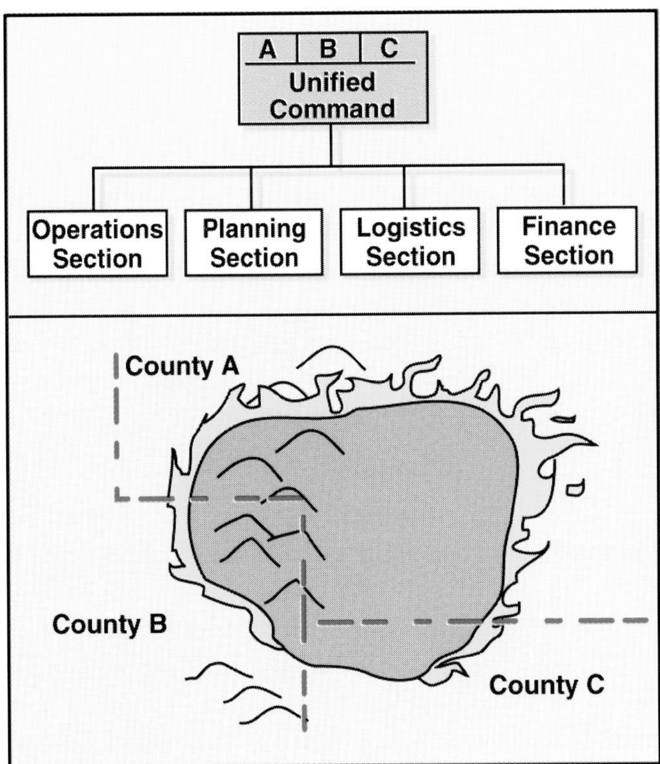

Figure 5.22 A unified command organization.

- Ensuring that integrated tactical operations are conducted
- Making maximum use of all assigned resources
- Providing financial responsibility

Under a unified command, the common objectives and strategies should be in writing because they form the basis for the incident action plan that guides the actions of all involved agencies. In this command structure, the Operations Chief is responsible for implementing the IAP, just as in a single command structure. There is only *one* Operations Chief, and the members of the unified command must collectively agree on which agency provides the individual to fill this role. This decision may be based on greatest jurisdictional involvement, existing statutory authority, or one individual's qualifications.

Extended Attack

If it becomes apparent that an initial-attack force will be unable to control a fire and that it will burn into a second operational period, the IC must request more resources and expand the fireground organization (Figure 5.23). At this point, the IC may decide to appoint an Operations Chief and perhaps one or all of the other members of the Command or General Staffs. Depending upon the number and types of resources included in the initial-attack force and upon the operational procedures of the agency involved, expanding the organization may be done by striking additional alarms or by calling for mutual aid. If the agency is organized to dispatch resources in predetermined increments (alarms), the IC knows how many and what types of resources to expect when each additional alarm is struck. Otherwise, the IC may have to specify exactly what types and numbers are needed in the request for additional resources. Whether the additional resources come from within the agency or from mutual aid sources, the ordering process is simplified if resources are requested in strike teams or task forces (see Chapter 7, Wildland/Urban Interface Fire Suppression).

Figure 5.23 The IC must decide if additional resources are needed. *Courtesy of NIFC.*

Regardless of where the additional resources come from, or in what increments, it is important that the principles of unity of command and span of control be applied as resources are added to the fireground organization. It may be necessary to break down areas of responsibility into geographic divisions/sectors and/or functional groups/sectors. For example, the fire perimeter may be subdivided into divisions/sectors (designated alphabetically) starting at the heel and progressing clockwise around the entire fire (Figure 5.24). Also, certain fireground functions, such as structure protection,

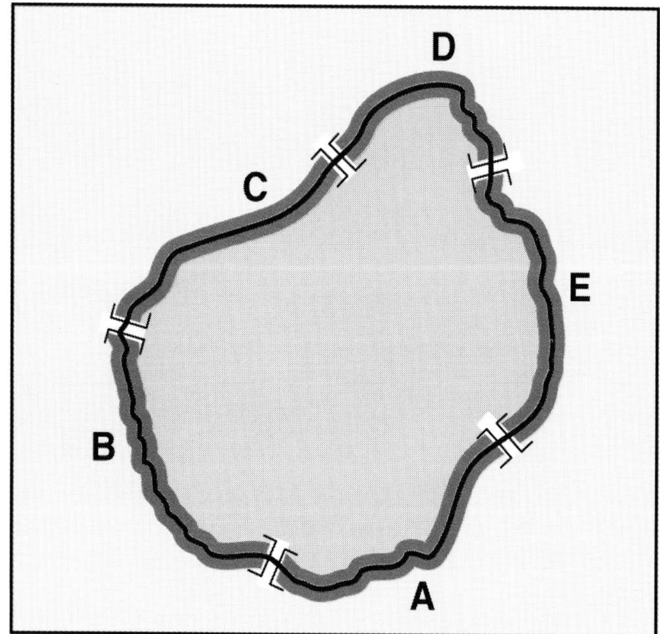

Figure 5.24 A typical fireground organization divided into divisions.

may be assigned as a branch, group, or sector (Figure 5.25).

Another fire management function that becomes operative when the organization is expanded beyond initial attack is staging. As mentioned earlier, the Operations Chief designates how many staging areas are needed, where they are located, and who manages them. The Operations Chief then orders resources from staging and deploys them as needed on the fire. In some cases, prefire planning may have resulted in an operational plan that predesignates certain staging areas, but these are always subject to change by the Operations Chief. Other prefire operational plans may specify that some units report to staging, while others deploy in certain prearranged configurations.

Major Fire Attack

If a fire continues to grow beyond the second operational period, despite the addition of extended attack forces, the IC must increase the number of fire fighting resources assigned to the incident and expand the command/management system. Since the fire has been burning for at least 24 hours, additional resources are needed not only to intensify fire suppression activities but also to provide relief for personnel who have been on the fireline for many hours. Crews coming off the line need food, showers, and sleep (Figure 5.26). Their apparatus and equipment may also need service and repair.

If the full General Staff was not appointed during the extended attack phase, the IC does so now. By delegating the authority for various functions to the General Staff, the IC maintains a manageable span of control. Similar delegation must take place in each of the sections to maintain the span of control of each section chief. One example of a fully developed and staffed major fire organization is outlined in Figure 5.27.

Evidence Preservation

Regardless of whether a fire was controlled during the initial attack or was controlled during the extended attack or developed into a major fire, an important additional duty is determining the fire's cause and origin. It is the duty of every firefighter involved in an incident to preserve any evidence of a fire's cause and origin. Firefighters' responsibilities in this regard actually begin before they arrive at a fire.

While responding to a fire, firefighters should observe things that may relate to the fire investigation such as traffic, access, smoke, etc. However, some of their most important observations and actions take place after they arrive at the scene. Even though most firefighters are eager to carry out their assignments, they should not lose sight of the importance of preserving the scene and any evidence it may contain for an investigator. If firefighters operating near the heel of a fire are not careful when they walk, drive apparatus, or drag hoselines in the black, they can destroy critical evidence of how a fire started (Figure 5.28). Any possible fire sets or other evidence should be protected and preserved for the fire-cause investigator (Figure 5.29). Such evidence may be needed in the prosecution of those responsible for the fire or in recovering the costs of suppressing the fire.

POST-INCIDENT CRITIQUE

Once a fire is extinguished and all resources have returned to quarters and made ready for the next fire, the operation should be analyzed. This analysis, sometimes called a "critique," should be both objective and comprehensive.

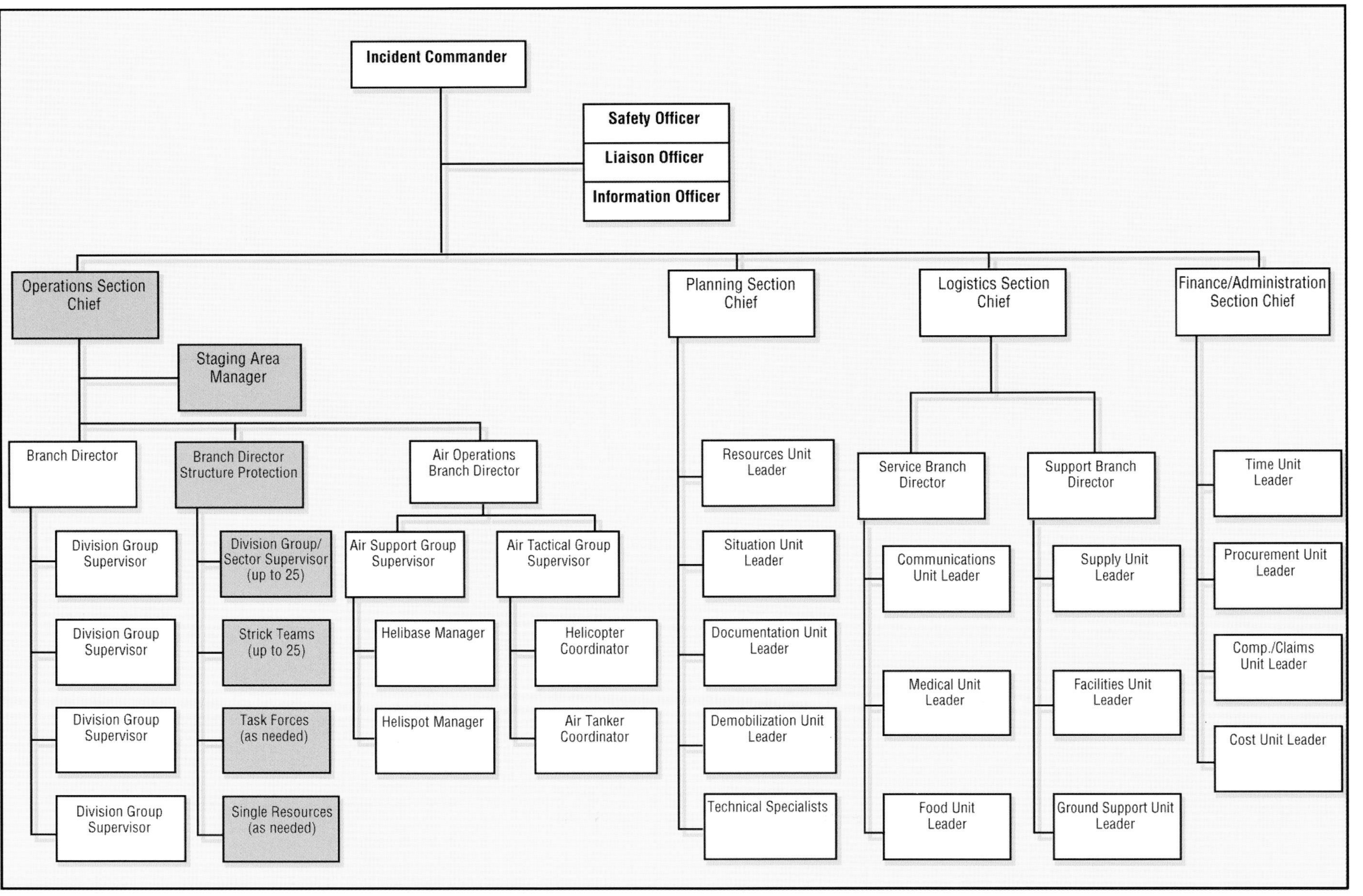

Figure 5.25 Fireground organization including a Structure Protection Branch.

Figure 5.26 Crews rest and replenish at incident base. *Courtesy of NIFC.*

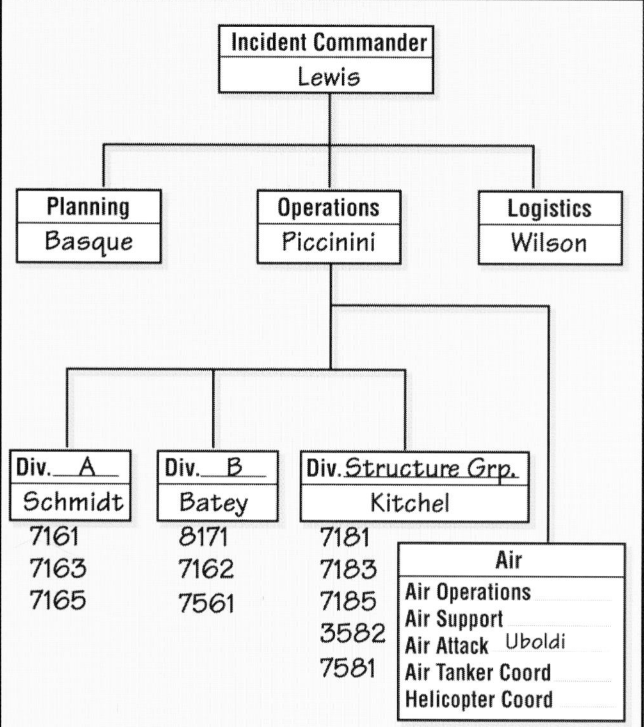

Figure 5.27 A full ICS fireground organization.

To be objective, it must acknowledge the parts of the operation that did not function up to expectations, as well as those that did. In general, more can be learned from mistakes and operational deficiencies than from all the things that went according to plan. However, it cannot be overemphasized that this analysis must not be used to find fault or to blame any individual or group for anything that happened during a fire. This analysis must be used only as a learning opportunity.

For the critique to be comprehensive, every aspect of the operation must be open to scrutiny. The actions taken by everyone from the IC to the

Figure 5.28 Driving apparatus in the black can destroy evidence.

Figure 5.29 Evidence should be protected.

least involved firefighter, and by everyone in between, should be reviewed. If any part of the operation is considered "off limits," the whole process loses its credibility. If the analysis is conducted in an open and matter-of-fact way, no one should feel threatened (Figure 5.30).

Figure 5.30 All major fires should be objectively analyzed.

The entire post-incident analysis process should be used to identify problems that can be corrected — and the needed corrections should be made as soon as possible. If hardware problems emerged, additional and/or newer equipment may be needed, or maintenance procedures may need to be changed. If pre-incident plans were flawed or out of date, they should be changed. If more training on the implementation and use of the incident command/management system is needed, it should be done also.

In this review process, there are a few fundamental questions that must be asked. The nature of these questions reinforces that the focus is on identifying *system failures,* not personnel failures. The main questions to be answered are as follows:

Was the Organization Used Properly?

This is potentially one of the most important questions to consider. Essentially, it asks whether the incident organizational structure was set up and managed correctly. Did the system make everyone's jobs easier or more difficult, and how? Did they have to go outside the system to get something done because the system did not provide for it? Is there a need for additional training or drills?

Were There Safety Issues?

Given that fighting a wildland fire can be dangerous, are there any safety-related issues that need to be addressed? Were any firefighters or civilians injured or killed? If so, have the circumstances been investigated and the results made public? What lessons can be learned? Was there a need for critical incident stress debriefing (CISD), and if so, was it used?

Were There Unity of Command Issues?

Was this critically important principle applied consistently throughout the incident organization? If not, why not? Did everyone know who their supervisor was? Was anyone at any level in the organization required to report to more than one supervisor at a time? If so, how did that happen? Again, is there a need for additional training?

Were Resources Used Properly?

Did those in charge of the various units make the best use of their resources? If not, why not? Were more resources ordered than were really needed? Was there excessive standby time? Is more training needed?

Were There Span of Control Issues?

Did any supervisor in the incident organization experience control difficulty? Were supervisors required to supervise too many subordinates? Were the subordinates separated so widely that effective control was difficult or impossible? Was the nature of the tasks being performed by the subordinates so different from each other that effective control was difficult or impossible? Did these control difficulties compromise firefighter safety?

Were There Personnel Accountability Issues?

Was the status of every individual involved in the incident known at all times? Is there any evidence to the contrary? Is there a need for more training?

Were There Unified Command Issues?

If a unified command was used, how well did it work? Were jurisdictional issues and questions of authority and responsibility resolved in a reasonable and timely manner? What, if anything, would have made the system work better?

Is There a Need for Predesignated Incident Command/Management Teams?

Would incident command/management teams that have worked together on other large-scale incidents have made a positive difference on this incident? Should they be called on future incidents?

Did Everyone Understand the Objectives?

Is there evidence that anyone in the incident organization did not understand the objectives? If so, was this breakdown in communication a system failure or an equipment failure? What might have been done differently?

6 FIRE SUPPRESSION METHODS

LEARNING OBJECTIVES

This chapter provides information that addresses the following objectives of NFPA 1051, *Standard for Wildland Fire Fighter Professional Qualifications* (1995 edition):

Wildland Fire Fighter I

3-1.1* *Prerequisite Knowledge:* Fireline safety, use and limitations of personal protective equipment, agency policy on fire shelter use, basic wildland fire behavior, basic wildland fire tactics, fire fighters role within the local incident management system, and first aid.

3-5.1 **Definition of Duty.** All activities to confine and extinguish a wildland fire, beginning with dispatch.

3-5.2 Assemble and prepare for response, given an assembly location, an assignment, incident location, mode of transportation, and the time requirements, so that arrival at the incident with the required personnel and equipment meets agency guidelines.

3-5.3 Detect potential hazardous situations, given a wildland fire and the standard safety policies and procedures of the agency, so that the hazard is promptly communicated to the supervisor and appropriate action can be taken.

3-5.3.1* *Prerequisite Knowledge:* Basic fireline safety, fire behavior, and suppression methods.

3-5.4 Construct a fireline, given a wildland fire, agency line construction standards, suppression hand tools, and equipment, so that the fireline conforms to the construction standard.

3-5.4.1 *Prerequisite Knowledge:* Principles of fireline construction, techniques, and standards.

3-5.4.2* *Prerequisite Skills:* Proper use of hand tools, fire stream practices, and agent application.

3-5.5 Secure the fireline, given a wildland fire and suppression tools and equipment, so that burning materials and unburned fuels that threaten the integrity of the fireline are located and abated.

3-5.5.1 *Prerequisite Knowledge:* Operational and safety considerations when burning out.

3-5.5.2* *Prerequisite Skills:* Use of basic ignition devices.

3-5.6 Mop up fire area, given a wildland fire, suppression tools, and equipment, so that burning fuels that threaten escape are located and extinguished.

3-5.6.1 *Prerequisite Knowledge:* Mop up principles, techniques, and standards.

3-5.7 Patrol the fire area, given a wildland fire, suppression tools, and equipment, so that control of the fire area is maintained.

Wildland Fire Fighter II

4-5.1 **Definition of Duty.** All activities to confine and extinguish a wildland fire beginning with dispatch.

4-5.2 Select fireline construction methods, given a wildland fire and line construction standards, so that the technique used is appropriate to the conditions and meets agency standards.

4-5.2.1 *Prerequisite Knowledge:* Resource capabilities and limitations, fireline construction methods, and agency standards.

Chapter 6
Fire Suppression Methods

INTRODUCTION

Firefighters use a variety of methods to control and extinguish wildland fires. The most appropriate methods make the best use of the available fire fighting resources and provide for personnel safety. This chapter discusses the parts of a wildland fire, size-up, fire behavior prediction, attack planning, and the most common attack strategies and the circumstances where these strategies are used. Also discussed are the principles of mop-up and patrol.

To make an adequate size-up and initial-attack plan for a wildland fire, firefighters must understand fire behavior as discussed in Chapter 1, Wildland Fire Behavior: Fuel, Weather, Topography. The elements of firefighter safety as discussed in Chapter 8, Firefighter Safety and Survival, are also relevant. To communicate the planning that results from the size-up, they must be able to describe the parts of the fire in terms that are all fireline personnel understand.

PARTS OF A WILDLAND FIRE

To provide a clear report on conditions, give direction, and provide for firefighter safety, firefighters must use standard names for various parts of a wildland fire. The parts are named for their unique characteristics and locations. The most common names are *origin, head, fingers, perimeter, heel, flanks, islands, spot fires, slopover, green,* and *black*. Because it is critical to firefighter safety and survival, Lookouts, Communication, Escape routes, and Safety zones (LCES) is also discussed in this section.

Origin

The *origin* is the area where the fire started (Figure 6.1). It is also the point from which the fire

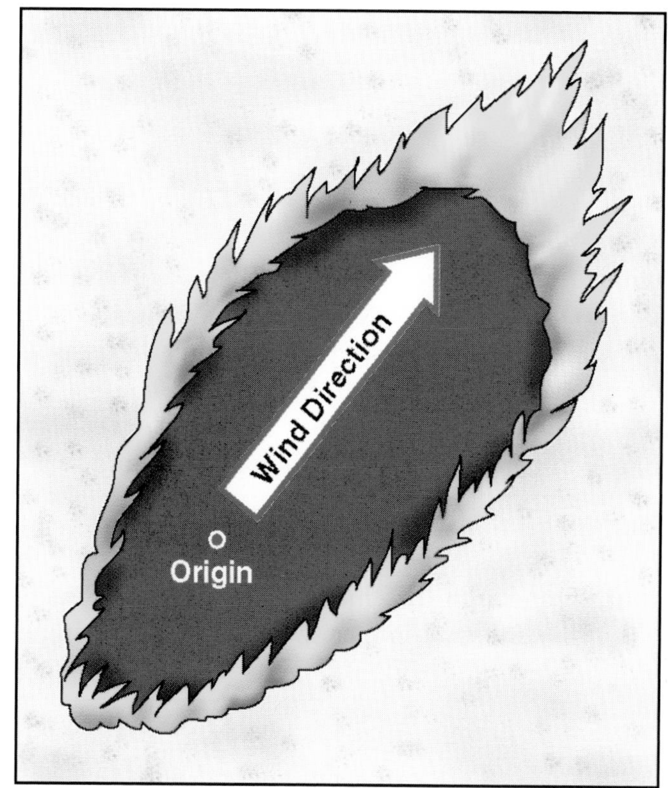

Figure 6.1 The point of origin.

spreads, dependent on the availability of fuel and the effects of wind and slope. The origin is often next to a trail, road, or highway, but it also may be in very inaccessible areas such as those where the fire is started by lightning or campfires. The area of origin should be protected for subsequent investigation of fire cause whenever possible.

Head

The *head* is that part of a wildland fire with the greatest forward rate of spread (Figure 6.2). Because wind and slope affect the rate and direction of spread, the head is normally either on the edge of a fire opposite to the direction from which the

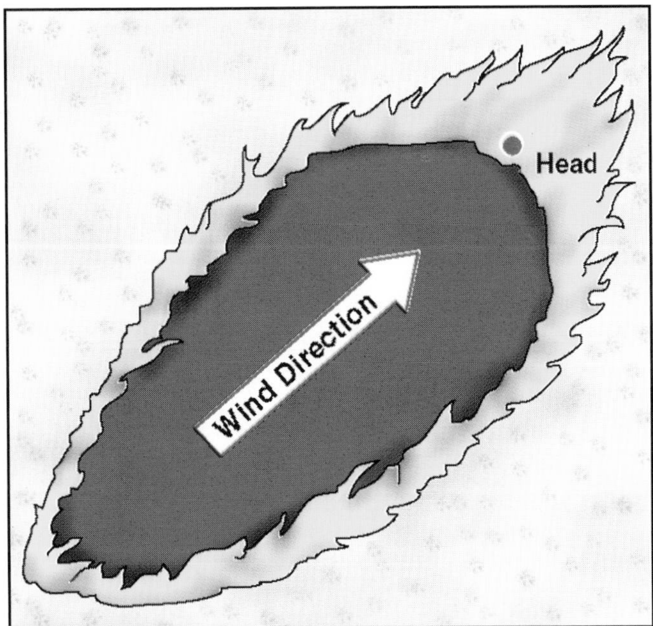

Figure 6.2 The head.

wind is blowing or it is toward the upper part of a slope. The head of a fire often burns intensely and may move with alarming speed. Some large fires may have multiple heads. Ultimately, controlling the head(s) and preventing the formation of new heads are critical to suppressing a wildland fire.

Fingers

Fingers are typically long, narrow strips of fire that extend from the main body of a fire (Figure 6.3). They most often form when a fire burns in an area with both light fuels and patches of heavier fuels. The fire spreads slowly in the patches of heavy fuel, but it continues to spread fast in the light fuels. This inconsistent rate of spread causes "fingers" to form. Fingers can be formed by variations in terrain or wind direction or when the head is split by natural features such as fields, bodies of water, or rock outcroppings.

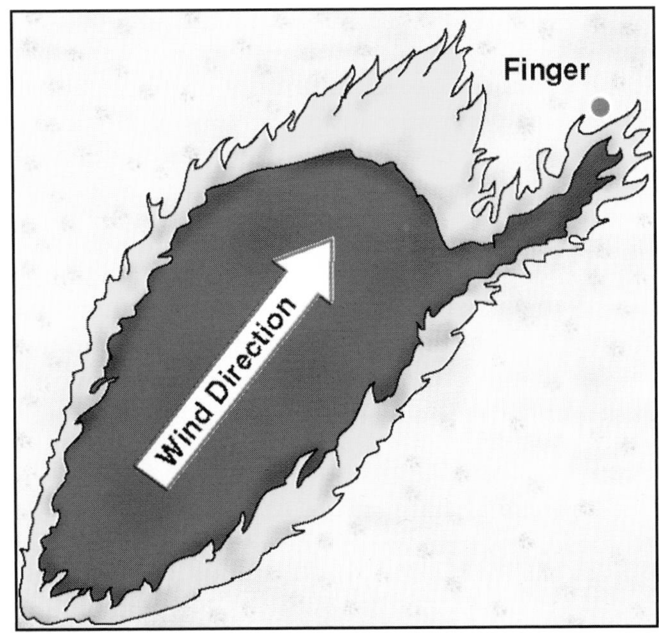

Figure 6.3 Finger.

CAUTION: If fingers are not controlled, they may form new heads.

Perimeter

The *perimeter* is the outer boundary, or the distance around the outside edge, of the burning or burned area (Figure 6.4). Also commonly called the *fire edge*, the perimeter is not necessarily the same as the control line, which may be at another, more convenient, location. Obviously, the perimeter continues to grow until the fire is controlled and extinguished.

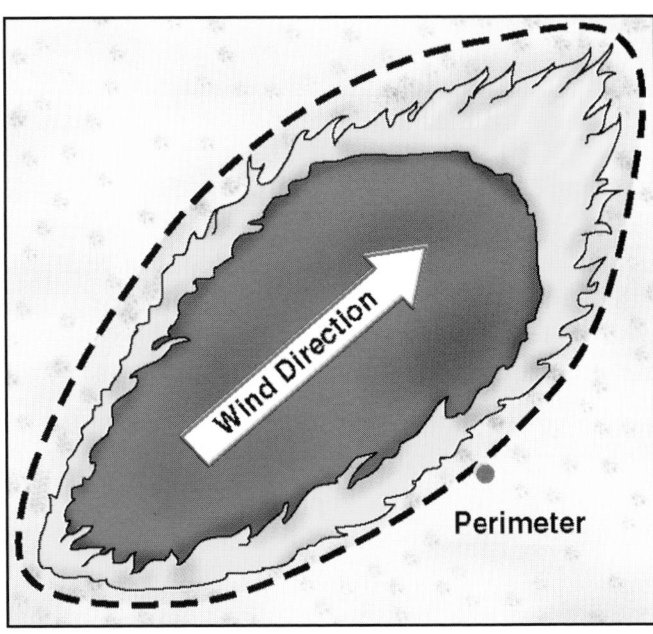

Figure 6.4 The perimeter.

Heel

The *heel*, or *rear*, of a wildland fire is the end opposite the head — that is, relatively closer to the point of origin than to the head. Because fire at the heel usually burns into any prevailing wind, it generally burns with low intensity, has a low rate of spread, and is generally easier to control than the head (Figure 6.5).

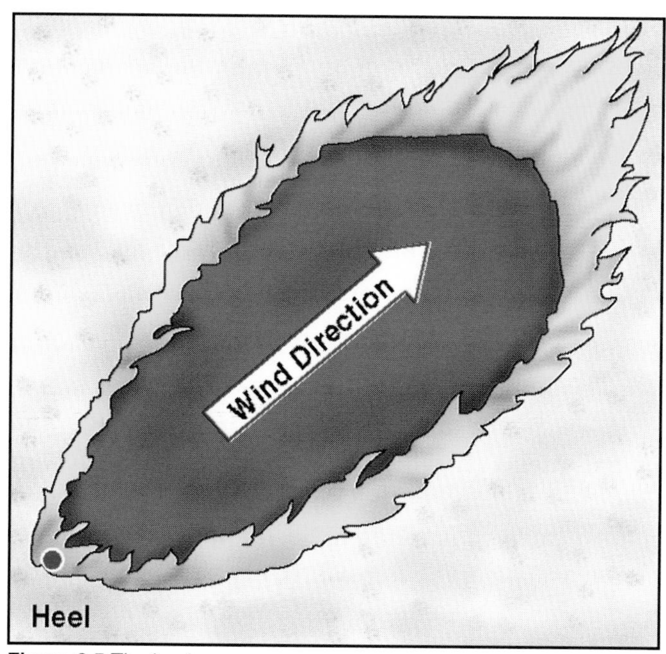

Figure 6.5 The heel.

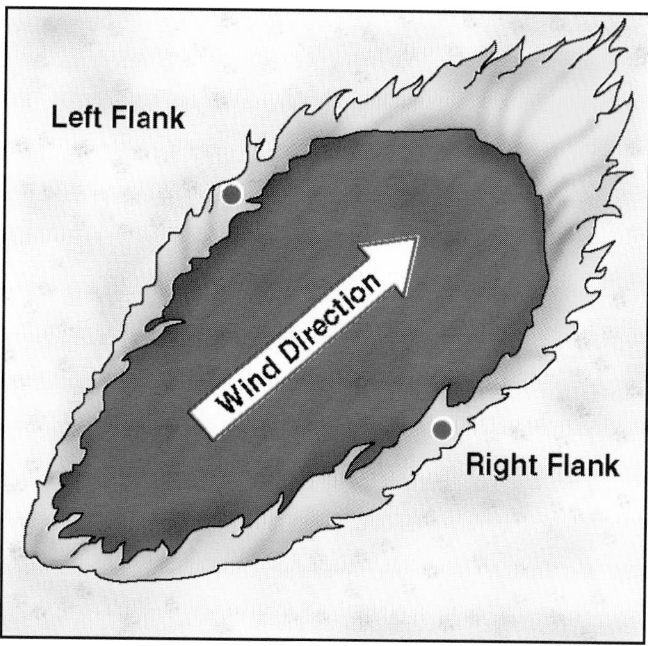

Figure 6.6 The flanks.

Flanks

The *flanks* are the sides of a wildland fire, roughly parallel to the main direction of fire spread (Figure 6.6). Flanks are identified as either left or right (looking from the heel of a fire toward the head). A shift in wind direction may change a flank into a head, and fingers often extend from flanks; therefore, flanks should be controlled as soon as possible.

Islands

Unburned areas inside the fire perimeter (in the black) are called *islands* (Figure 6.7). Because they are unburned potential fuels for more fire, they must be patrolled frequently and checked for any spot fires within (see Spot Fires section). Islands close to a control line *may* flare up later and start spot fires across the control line, so firefighters may burn them out, consuming fuels between the edge of fire and the control line (see Burning Out section).

Figure 6.7 Islands.

Spot Fires

Spot fires are caused by sparks or embers carried aloft by a convection column, blown across a fireline by winds, or rolling downslope into unburned fuels beyond the main fire (Figure 6.8). If spot fires burn unchecked, they may form a new head or another major fire through area ignition (see Area Ignition section in Chapter 1, Wildland

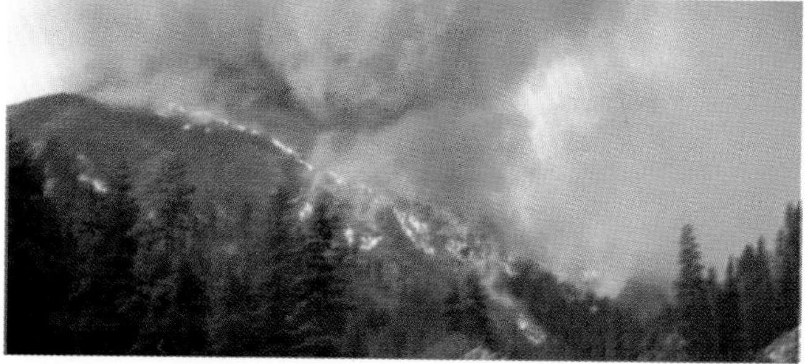
Figure 6.8 A spot fire ahead of the main fire. *Courtesy of NIFC.*

Fire Behavior: Fuel, Weather, Topography). If this happens, personnel and equipment could be trapped between two fires, or the fire may move in an unanticipated direction.

Slopover

Slopover, also known as *breakover,* is when fire crosses a control line or natural barrier intended to confine the fire (Figure 6.9). Slopovers differ from spot fires mainly in their location relative to the control line. A slopover occurs immediately across and adjacent to the control line; a spot fire occurs some distance from it.

Figure 6.9 A firefighter mops up a slopover.

Green

The area of unburned fuels located outside of but adjacent to the involved area is called the *green*. While the term refers to the color of some of the fuels in that area, the green may not be green at all (Figure 6.10). Some fuels in the green may consist of living vegetation with a high moisture content that are relatively slow to ignite and burn.

Figure 6.10 The green.

Others may be living vegetation with a low moisture content that are highly flammable. Still others may be dense, golden-yellow annual grasses and other similar fuels with a low moisture content that may burn vigorously. The term "green" is not meant to define a safe area. It is simply the opposite of "the black," or burned, area. The edge of the green is usually where a control line is constructed.

Black

The opposite of the green — the *black* or the *burn* — is the area (including both surface and aerial fuels) in which the fire has consumed or "blackened," the fuels (Figure 6.11). If it is completely burned over and little, if any, unburned fuel remains, the black is a relatively safe area during a fire; however, the black is not always safe. In steep terrain, where flames from adjacent unburned fuels can lap into it, the black may not be safe. Also, the black can be a very hot and smoky environment with numerous hot spots and smoldering snags (standing dead trees), stumps, and downed trees.

CAUTION: If a surface fire has left aerial fuels more or less intact in the black, or vice versa, a reburn of this area can occur.

Figure 6.11 The black.

LOOKOUTS, COMMUNICATIONS, ESCAPE ROUTES, SAFETY ZONES (LCES)

One vital survival tool for firefighters in wildland fire fighting is commonly called *LCES*. Each initial stands for a critical element in firefighter safety and survival on the fireline. Most fireline injuries and fatalities can be traced to a lapse in one or more of these elements.

LCES
Lookouts
Communications
Escape Routes
Safety Zones

LOOKOUTS

Lookouts are sometimes critical to the safety and survival of crews working on a fireline. Because of the terrain in which firefighters must work and the vegetation covering that terrain, firefighters cannot always see the flame front nearest them. Lookouts serve as the eyes and ears for the rest of the crew when they cannot see a fire and/or their attention is focused on doing their jobs and completing their assignments. Whenever firefighters must work relatively close to a fire that they cannot see, one or more lookouts *must* be posted to observe the fire and inform the rest of the crew of its behavior (Figure 6.12).

Figure 6.12 A lookout monitors a fire.

COMMUNICATIONS

The posted lookouts must be capable of interpreting the fire behavior they observe and communicating quickly and effectively with the rest of the crew (Figure 6.13). Radio communication between the lookouts and the rest of the crew (or *at least* with the crew leader) is essential if crew members cannot see the lookouts. Otherwise, voice communication and/or prearranged signals may be used.

Figure 6.13 Lookout communicates with a portable radio.

ESCAPE ROUTES

Escape routes are the routes firefighters take to reach a safe area or safety zone when they are threatened by an advancing fire. To be most effective, escape routes should be relatively short so that firefighters can reach a safety zone quickly. Firefighters must know where their escape routes are and when to use them. If, "in the heat of battle," firefighters lose track of the nearest escape route, they may not be able to find it when they need it. In addition, firefighters must not become so absorbed in fire fighting activities that they fail to monitor the behavior of a fire. If a fire starts to blow up or make a run toward them, they need to react quickly and retreat. They must also remember that they cannot run as fast as usual in steep terrain when they are fatigued, wearing protective clothing, and carrying tools. At the time of this manual's publication, some agencies were studying the advisability of firefighters abandoning their tools and packs, keeping only their fire shelters, to speed their escape from life-threatening fire conditions (Figure 6.14). Thorough training and adequate fireline communications are essential to firefighter survival in severe fire conditions.

Figure 6.14 A firefighter discards everything but her fire shelter as she escapes.

CAUTION: When fighting a wildland fire, you must always know where your escape routes are and when you should use them.

SAFETY ZONES

Safety zones are areas mostly devoid of fuel that are large enough to assure that flames and/or dangerous levels of radiant heat will not reach the personnel occupying them. Areas of bare ground, burned-over areas, paved areas, and bodies of water can all be used as safety zones. When no natural safety zones are close to firefighters, safety zones have to be constructed. The size of the area needed for a safety zone is determined by its fuel types, its location on slopes, and its relation to topographic features (chutes and saddles), as well as by the observed fire behavior, the number of personnel, etc.

WARNING

Safety zones should never be located in topographic saddles, chutes, or gullies.

High winds, steep slopes, or heavy fuel loads may increase the area needed for a safety zone. An adequate safety zone is large enough that fire shelters are not necessary to prevent injuries to personnel. Most roads are not wide enough to be effective safety zones (Figure 6.15). Natural barren areas such as rock slides or dry stream beds may be used if available (Figure 6.16). When bodies of water are used, the possibility of people drowning must be considered. Burned-over areas (the black) make some of the best safety zones. If fuels are light, the black cools rather quickly; if fuels are heavier, it may take much longer. Once cooled sufficiently, however, the black can provide an excellent area of safety if it is large enough and if firefighters can get into it without having to pass through the flame front.

Figure 6.15 Most roads are not wide enough for safety zones. *Courtesy of Mike Wieder.*

Figure 6.16 Dry creek beds or rock slides make good safety zones.

SIZE-UP

Size-up is an ongoing process of evaluating current and expected fire conditions. It starts before a call comes in from the dispatcher/telecommunicator and continues throughout an incident. All involved fire fighting personnel should constantly monitor fire conditions and make adjustments as needed.

Size-up involves developing a mental picture of the past, current, expected, and potential behavior of a fire. It also involves evaluating the threat to life, property, and resources. The initial size-up consists of gathering information prior to and upon arrival at scene (Figure 6.17). The information needed to make an initial size-up can be broken down as follows:

Figure 6.17 A firefighter sizes up an approaching fire.

Prior to dispatch

- Principles of fire behavior
- Local terrain and fuels
- Fire weather forecasts
- Current/predicted fire danger indices and conditions
- Fire fighting resources available (including backup)
- Pre-incident plans

Figure 6.18 Additional resources must be ordered as soon as possible.

Size-up en route

- Smoke (volume, color, movement)
- Clouds (type, size, movement)
- Time of day
- Weather (temperature, humidity, wind)
- Best access (route in)
- Jurisdiction (government owned or private)
- People leaving the area (vehicles, license numbers, etc.)
- Communications (command, tactical channels assigned while en route)

Size-up on arrival

- What is burning (fire behavior, rate of spread, size, etc.)
- What is threatened (structures, timber, etc.)
- Topography
- Populations at risk
- Access to the fire (vehicles, air, personnel)
- Initial safety zones/escape routes
- Special hazards (snags, hazardous materials, downed wires, etc.)
- Point of initial attack (anchor points, etc. — see Anchor Points section)
- Most likely area of origin
- Resources needed

The information gathered prior to and upon arrival helps the initial incident commander (IC) make one of the first and most important decisions of an incident — are the types and numbers of resources at scene or en route sufficient to control this fire or do additional resources need to be called? If the initial size-up indicates that more resources or different types of resources are needed, then they must be ordered as soon as possible (Figure 6.18). The sooner these units respond, the sooner they will arrive at scene where they are needed. If unsure of what types and numbers of resources are needed, the IC should order any that *might* be needed. Units that prove unnecessary can be held in reserve in staging or returned to quarters.

Size-Up En Route

The smoke from a fire may be visible long before the first units arrive at the fire. The direction of smoke movement can indicate the effect of winds, if any, on the fire (Figure 6.19). The amount of smoke may give a good indication of how big the fire is already. Its color may indicate the types of fuels being consumed — generally, the darker the smoke, the heavier the fuels and the more difficult control and extinguishment will be. The vertical development of the smoke can indicate the intensity with which a fire is burning — generally, the more active

Figure 6.19 Smoke movement may indicate wind direction. *Courtesy of NIFC.*

the smoke, the more intense the fire (Figure 6.20). However, an inversion can cause the smoke column to flatten out and spread laterally (Figure 6.21).

The time of day when the dispatch/telecommunication is received can indicate whether a fire will intensify because of rising atmospheric temperature and decreasing humidity. It also suggests the probable direction of slope winds in the area of a fire — generally upslope in the afternoon and downslope in the evening. The time of day indicates the length of time before a fire is influenced by the moderating conditions normally expected after sunset. The possible impact of any developing inversion layer must also be evaluated.

In general, the weather indicates how quickly a fire is likely to grow and how a fire may behave otherwise. It is important to know the temperature, humidity, and the amount and types of winds expected in the area of a fire. It is particularly important to observe the following:

- Wind speed/direction
- Dust devils
- Weather changes expected such as a cold front passing
- Inversion effect

Based upon what can be seen en route — along with a knowledge of fire behavior, knowing the effects of weather, prefire planning information, and maps of the area — the IC can determine the best route to the fire. A thorough knowledge of the area is useful in choosing a safe route that leads closest to the probable point of initial attack. Knowing the area also helps the IC properly locate staging to make reserve equipment readily available.

Figure 6.20 Smoke movement may indicate fire intensity. *Courtesy of NIFC.*

Figure 6.21 An inversion layer causes smoke to move laterally. *Courtesy of NIFC.*

Size-Up on Arrival

Once at scene, the IC must focus on the situation at hand — where and how to invest the available resources to do the most good until other resources arrive. To do this, the IC should answer the following questions:

- What is the most important work to be done first?
- Where can the most effective work be done?

The IC transmits a clear and concise *report on conditions* for the other incoming units. The report describes (1) what has happened (fire history): "*Fire is 10 acres (4 ha) in size,*" (2) what is happening (current size-up): "*Fire is burning up O'Farrell Hill,*" and (3) what is going to happen (fire behavior prediction without suppression action, initial objective[s], and actions): "*Attacking right flank to protect subdivision.*" The report confirms the address or location of the fire and gives the best means of access to it. Example: "*Ten acres (4 ha) involved, moderate spread up O'Farrell Hill, attacking right flank to protect subdivision. All units enter off Jones Road.*"

The IC may decide that the resources at scene are sufficient to handle the situation and return other responding units. If not, the IC may begin to expand the incident command/management organization, initiate those actions within the capabilities of at-scene units, and assign objectives to those still en route. Example: *"Engine 7182 is O'Farrell Command. Engine 7188 take the left flank, Engines 5660 and 8141 take the head."* These decisions are based on an initial assessment of the following:

- ***Size; perimeter and rate of spread*** — These determine what needs to be done first (evacuation, structure protection, attacking the head, ordering more resources, etc.). (See Pertinent burning conditions bullet.)
- ***Direct or indirect attack needed*** — The attack strategy needed determines how resources can be used to best effect. (See Fire-Control Strategies section.)
- ***Good anchor points*** — Any natural and/or man-made barriers available can be used as anchor points. (See Natural barriers bullet.)
- ***Safety considerations*** — All firefighters must know the locations of escape routes and safety zones (existing/constructed).
- ***Location of head(s)*** — This is generally the most critical part of a wildland fire. Once the head is contained, other parts of a fire are usually less difficult to control.
- ***Pertinent burning conditions*** — Intensity of burning and the rate of spread are related to the following:
  - — *Fuel characteristics* (the fuel burning, the fuel in the path of the fire, fuel moisture, continuity)
  - — *Size of the fire* (the perimeter length or the area involved)
  - — *Special hazards* (spot fires, burning snags, developing fingers, and proximity of flames to flash fuels, sprayed crops, dumps, hazardous materials, etc.)
  - — *Weather factors* (current and expected wind, humidity, and temperature)
  - — *Exposures* (threatened populations and property, for example, structures, fences, utility poles, vehicles, livestock, and valuable timber or crops)
- ***Natural barriers*** — These include streams, roads, cleared fields, swamps, burned areas, and other natural barriers, including the following:
  - — *Access roads and trails* (transportation routes, locations for control lines, and locations for staging and incident command post [ICP])
  - — *Water sources* (the locations of streams, ponds, swimming pools, and wells to supply/supplement water tenders)

NOTE: Natural barriers are possible safety zones or escape routes. They may also provide locations for staging areas.

If a fire appears to be a relatively long-term operation that will involve more than the initial response units, the IC announces the locations of the incident command post and one or more staging areas (Figure 6.22). The IC begins to develop an incident action plan (IAP). Depending upon the potential size and complexity of the incident, and the number and types of units involved, the IC may appoint an Operations Chief, a Safety Officer, and any other members of the Command and General Staffs that may be needed (Figure 6.23).

Figure 6.22 The IC establishes a command post.

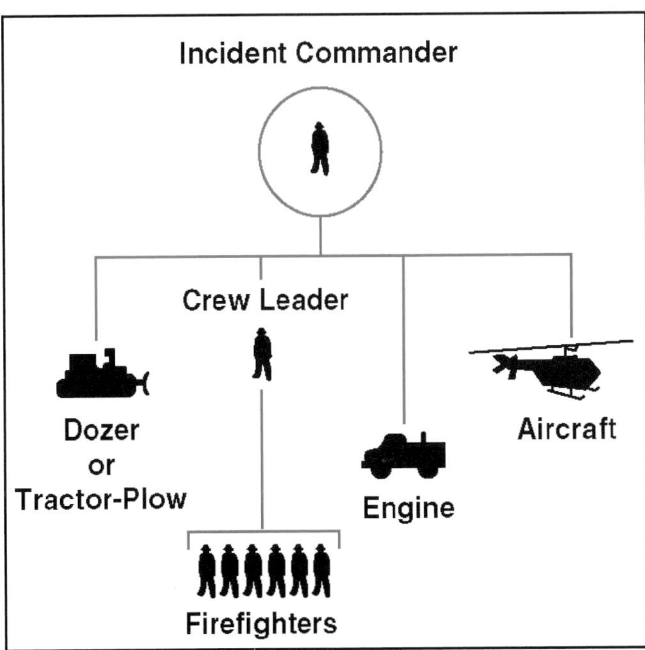

Figure 6.23 A typical initial attack fireground organization.

Forecasting Fire Spread

The next step is to forecast the fire's expected spread based on the initial size-up and any subsequent changes. Wildland fires, like other fires, burn as long as enough oxygen, fuel, and heat are available to sustain the process. Fires behave differently depending upon the weather conditions, the fuel types, and the topography in which they are burning. Some fires generate relatively little heat, burn slowly, and are easy to extinguish. Other fires produce a lot of heat, spread quickly, and may be difficult to control and extinguish.

The terms hot, warm, slow, and fast are vague and subjective. Different individuals interpret these terms differently. What is hot to one person may be warm to another. What is slow to one may be fast to another. Firefighters, therefore, need to use standard terms of measurements to describe fire behavior. Two terms commonly used with wildland fire behavior are rate of spread (ROS) and intensity, normally expressed as flame length(s) (see Chapter 1, Wildland Fire Behavior: Fuel, Weather, Topography).

In addition to the factors considered in the initial size-up, any fire-spread forecast should take the following variables into account:

- Fuel types involved and threatened
- Current and expected weather conditions
- Effects of topography
- Potential fire behavior problems
- Rate of spread (Table 6.1)

One of the most meaningful ways of classifying the rate of spread (increase in the perimeter of the fire) describes the rate at which the head of the fire moves. Table 6.1 shows the rate of spread based on the movement of the head of the fire.

Fire-spread forecasts can also be made by applying some practical guidelines. In general, a fire will continue to spread at a constant rate unless it is affected by changes in fuel, topography, or wind. The fire will usually spread faster in light fuels than in heavier fuels. The rate of spread will double with each 20 percent increase in slope and/or each 10 mph (16 kmph) increase in wind velocity.

Another important factor in predicting fire spread is a reasonably accurate estimate of the acreage (hectarage) already involved. According to the National Wildland Fire Coordinating Group (NWCG) acreage (hectarage) determination chart, a 1-acre (0.4 ha) fire can have a perimeter of from 660 to 1,320 feet (200 m to 402 m), depending upon the shape of the burn. These figures, however, are not very practical for field estimation. A more practical approach is to estimate the area of involvement by picturing each side of a square acre as being about 2/3 the length of an American football field and each side of a square hectare as being roughly the full length of a Canadian football field (Figure 6.24). Once the length of the perimeter has been estimated, the IC can determine the resources required to construct that much control line by applying standard production rates for various resources to the terrain and fuel types involved.

TABLE 6.1
Classifying Rate of Spread

Rate of Movement (Head) (ft/min or m/min)	Rate of Spread
<2 feet per minute (<0.6 m/m)	Slow
2–8 feet per minute (0.6 m to 2.4 m/min)	Moderate
8–30 feet per minute (2.4 m to 9 m/min)	Rapid
>30 feet per minute (>9 m/min)	Extreme

An accurate fire-spread forecast is critically important to the next step in the process — planning the attack.

PLANNING THE ATTACK

The goals of the planning phase are to develop the action plan and to prepare to meet the requirements of that plan. The plan provides for protecting any evidence at the point of origin, matches resource capabilities to the expected rate of spread, and is flexible enough to respond to changing fire conditions. The IC should anticipate that the attack plan may need to be changed and should be prepared to make the necessary adjustments.

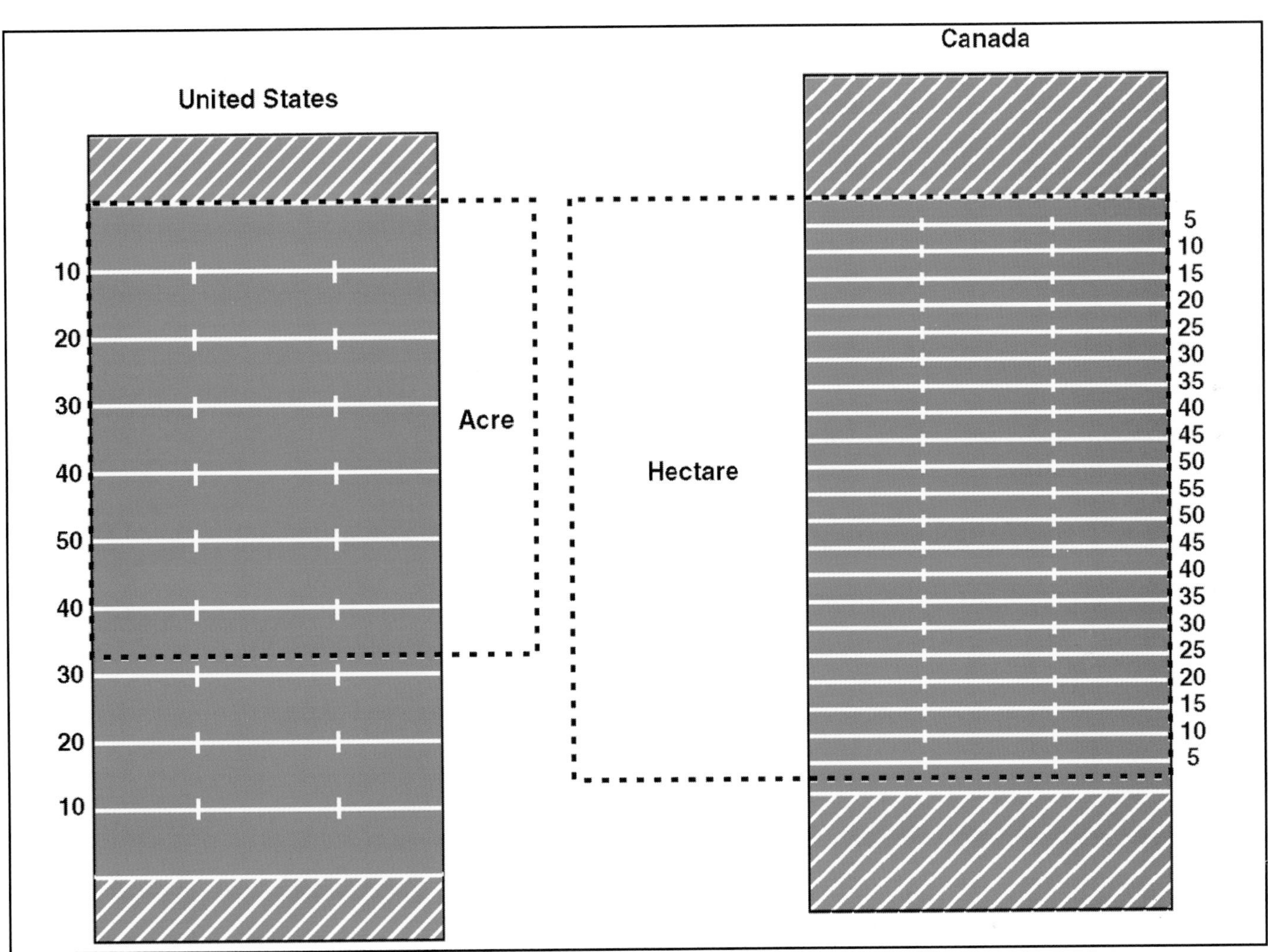

Figure 6.24 A way to visualize the area of an acre or hectare.

A critical initial decision when planning the attack is whether the overall strategy is to be offensive, defensive, or a combination. Unlike structure fire fighting protocols, where the two strategic modes (offensive and defensive) are rarely combined, they are often combined when fighting wildland fires.

A *defensive strategy* is needed when resources at scene are insufficient to extinguish a fire. In this situation, the size of fire limits the actions that the at-scene forces can take. If a fire is beyond the suppression capabilities of the at-scene forces, they may have to use their resources to protect the most valuable exposures while allowing the fire to burn past. This frequently occurs in the wildland/urban interface and in heavy downed fuels. The key is to make the best of a bad situation by using the available resources where they will do the most good.

An *offensive strategy* is the plan of attack when there are sufficient resources at scene to control the fire. If a fire can be controlled and extinguished before it reaches high-value exposures, the exposures will be protected in the process.

When fighting fires in the wildland/urban interface, defensive and offensive strategies are often used in combination. A *combination* strategy may be used when one or more structures are burning while others are merely threatened. The resources at scene may allow for offensive action only on the structure(s) already involved and defensive actions on those nearby.

Regardless of which strategy is used, public and firefighter safety is the first and highest priority. Once personnel are committed to an incident, those resources become the highest value to be protected. Protection of property and natural/cultural resources are secondary to public and firefighter safety. The selected attack plan must provide for the safety of personnel and equipment and for efficient suppression of the fire. The attack plan should follow the Standard Fire Orders (see Chapter 8, Firefighter Safety and Survival) and should be formulated immediately upon arrival at the fire. It is based on the initial size-up of the fire and should be done quickly. The intent is to start work on suppressing the fire as soon as possible. In formulating the plan, determine the following:

- Location of escape routes
- Special hazards such as burning snags, hazardous materials, etc.
- Good anchor points such as roads, burned area, etc. (See Control Lines section.)
- Where to attack fire (head or flank)
- How to attack fire (direct or indirect) (See Fire-Control Strategies section.)
- Type of control line needed (wet lines, scratch lines, bulldozer lines, etc.) (See Control Lines section.)
- Existing barriers that can be used
- When next units will arrive
- How topography will affect fire behavior
- Most likely point of origin

Once the attack has begun, the IC should evaluate it by answering the following questions:

- Is the initial attack plan working? If not, why not?
- Are additional resources needed?
- How much time will it take to construct a control line?
- Will changes in weather, fuel, or topography significantly affect fire behavior prior to control?
- Is rate of spread or intensity higher than expected?

If the initial attack plan is working, it should be continued. If not, it should be changed and the changes implemented. The dispatch/communications center should be notified immediately if a fire exceeds the capabilities of the at-scene forces. *Ask for help if needed!*

The dispatch/communications center should be informed of the progress on the fire and of any significant changes. At the earliest opportunity, the following information should be transmitted to the dispatcher/telecommunicator:

- Fire name
- Location
- Access
- Terrain
- Size of fire
- Anticipated control problems
- Cause (known, suspected)
- Values threatened
- Anticipated time of control
- Weather
- Resources on the fire
- Resources needed if any
- Fire behavior

CONTROL LINES

The terms *control line* and *fireline* are basic to any discussion of attack methods. Their meanings may vary from region to region, and they are often confused with the *fire edge*. The terms as used in this manual are defined as follows:

- ***Control line*** — Any and all constructed or natural barriers and the treated fire edge used to control a fire
- ***Fire edge*** — The boundary of the burned or burning material at any given time
- ***Fireline*** — Part of a control line that is scraped or dug to mineral soil (Figure 6.25); not to be confused with the general use of the term as in "working on the fireline," meaning being in the general vicinity of fire fighting activities

Figure 6.25 A firefighter cuts a fireline. *Courtesy of NIFC.*

Fireline Location

One of the IC's most important initial decisions concerns the location of the fireline in relation to the fire's edge. The location of the control line is related to the method of attack, offensive or defensive. The decision of where to locate the fireline should be based on the following considerations:

- Safety of personnel and equipment must be assured and LCES used where needed. (See safety points in Fireline Construction section.)
- The line should be located far enough from a fire to be completed, burned out, and held before the fire reaches it with predicted rate of spread and fire behavior (see Burning Out section).
- Enough time should be allowed to permit forces not only to build lines, but also to do other needed work, such as felling snags and burning out, in advance of severe burning conditions (see Burning Out section).
- The line should be as short and straight as practical.
- Easiest routes for control should be selected as long as
  - — Line effectiveness is not compromised.
  - — Excessive area or values are not sacrificed.
- Possible hazards should be eliminated from the fire area, and a safe distance should be provided between lines and hazards that must be left in the fire area.
- Mechanized equipment should be used for line construction where possible.
- Areas where spot fires are numerous should be encircled and the unburned fuels burned out (see Burning Out section).
- Undercut lines (lines below a fire on a slope) and sharp turns in the line should be avoided.
- Downhill fireline construction guidelines should be followed (see Downhill Fireline Construction Guidelines section).
- Environmental effects should be minimized and agency policy followed (see Restoring the Environment section).

Anchor Points

All control lines must have an anchor point — a barrier to fire spread where the control action begins. The purpose of an anchor point is to prevent a fire from burning around the end of the control line and possibly outflanking the suppression crews and placing them in danger. An anchor point is usually not a constructed fireline, but it could be a hose lay, aerial retardant drop, road, bare field, stream, cliff, or burned-over section of the fire (Figure 6.26).

CAUTION: Starting fireline construction from an anchor point is critical to firefighter safety.

Figure 6.26 A fireline anchored to a paved road. *Courtesy of John Hawkins.*

Fireline Width

Weather, topography, and the arrangement, volume, and type of fuels all combine to dictate the width of line needed. In sparse surface fuels, such as duff or light grass, the line may only need to be a few inches (millimeters) wide. In heavier fuels and in severe burning conditions, the line must be wider (Figure 6.27). The fuels between the control

Figure 6.27 Fuels and burning conditions dictate line width. *Courtesy of NIFC.*

line and the fire's edge should be burned out, particularly in medium to heavy fuels, to effectively widen the control line.

When a fireline must be constructed by hand crews only, it is important to save time and conserve the firefighters' energy by making the line only as wide as necessary. Most firelines vary in width from a few inches (millimeters) to a few yards (meters); in general, the hotter and faster a fire burns, the wider the line must be. Anything that affects how a fire burns must be considered when deciding how wide the fireline must be. One guideline to determine line width is to figure one and one-half times either the height of the fuels or the height of the flames. The decision regarding line width in any specific situation should be made by a supervisor or experienced firefighter.

The most important factors in determining fireline width are fuel, slope, weather, part of fire, size of fire, and possibility of cooling.

- ***Fuel*** — Fuels that contain flammable oils (chamise, pine, palmetto, juniper, etc.) burn hotter than other fuels, so the line must be wider in these fuels. The higher and denser the fuel, the wider the fireline must be (Figure 6.28). Heavy fuels (logs, snags, etc.) are slower to ignite but burn hotter and much longer than light fuels, so they require a wide fireline. In general, the drier the fuel, the hotter it burns, and the wider the line must be.
- ***Slope*** — For lines constructed above a fire burning on a slope, the steeper the slope, the wider the line must be (Figure 6.29). For lines constructed below a fire burning on a slope, the width of the line is not as critical as the depth of the trench that must be built to catch rolling materials. The steeper the slope, the deeper the trench must be (Figure 6.30).

Figure 6.28 Dozers may be needed to cut fireline in heavy fuels. *Courtesy of NIFC.*

Figure 6.29 A wide fireline is needed above fires on steep slopes.

- ***Weather*** — Temperature, humidity, winds (both natural and convective), and fuel moisture all affect the width of fireline required. In general, the higher the temperature, the lower the humidity will be. The lower the fuel moisture and the stronger the winds, the more intensely a fire burns and the wider the fireline must be.
- ***Part of fire*** — Because fires generally burn with different intensities at the head, flanks, and rear, each part may require a line of different width. Line width must generally be widest at the head, narrower along the flanks, and narrowest at the rear.
- ***Size of fire*** — The amount of heat produced by a large fire is generally greater than that produced by a small fire in the same fuels and burning conditions. A large fire, therefore, usually requires a wider line than a small fire.
- ***Possibility of cooling*** — When plenty of water or dirt is available to cool a fire, a narrow line will usually suffice.

Figure 6.30 Trenches are needed across steep slopes below fires.

Fireline Construction

Firelines are constructed by using hand tools or mechanized equipment, such as bulldozers or tractor-plows, to remove surface and subsurface fuels down to mineral soil. In this process, the continuity of any aerial fuels over the line should also be broken to make them unavailable to a fire (Figure 6.31).

Figure 6.31 A wide fireline that breaks the continuity of aerial fuels. *Courtesy of NIFC.*

Concentrations of surface fuels close to the control line should be broken up and dispersed (Figure 6.32). A pile of brush located inside but adjacent to the control line is a potential source of a hot spot, blowup, or slopover. The same type of pile located outside the line is a good bed of fuel. A spark from a fire could start a rapidly developing spot fire in this pile. Snags and other standing aerial fuels near the line should be cut down, and they must be made to fall inside the control line (Figure 6.33).

Figure 6.32 Fuel should be thrown away from the fireline.

A control line built on a slope below a fire should be trenched or undercut to stop materials such as logs, pine cones, or yucca plants from rolling into unburned fuels and spreading the fire. A good trench has an earthen berm to stop burning materials from rolling downhill and igniting combustibles below the line.

Figure 6.33 Snags should be felled into the black and with the slope. *Courtesy of NIFC.*

To stop a fire that is burning upslope, a control line should be constructed just over the ridge on the other side (Figure 6.34). A line in this location uses the effect of slope on fire behavior to decrease the possibility of the fire jumping the line.

Constructing an ineffective fireline wastes time, energy, and resources. To be effective, a fireline should be constructed according to the following guidelines:

- The line should be wide enough to be effective, but no wider.
- The line should be cleared down to mineral soil where practical.
- Unburned fuel should be discarded and scattered outside the fireline (unless it is needed to help burn out). (See Burning Out section.)
- Charred or burning material should be scattered inside the burned area.
- A line constructed on a steep slope should be trenched to catch rolling material below the fire.
- Adjacent fire should be cooled with water or dirt to increase line effectiveness.
- Rotten logs and stumps near the line should be wet down or covered with dirt.
- Snags near the line should be felled before burning out if time permits.

Figure 6.34 A control line just beyond a ridge can be effective.

- The line should be constructed as close to the fire edge as safety permits.

The fireline should be burned out as control-line construction proceeds. (See Burning Out section.)

DOWNHILL FIRELINE CONSTRUCTION GUIDELINES

When the fire situation requires that a fireline be constructed downhill, the following guidelines should be used:

- Only a competent firefighter who has thoroughly scouted the area should decide to build downhill.
- Downhill line construction should not be attempted when a fire is burning directly below the proposed starting point.
- A fireline should not lie in or adjacent to a chimney or chute that could burn out while crews are in the vicinity.
- Communication must be established between crews working downhill and crews working toward them from below. When neither group can adequately observe the fire, LCES *must* be used.
- Crews *must* be able to reach a safety zone quickly from any point along the line in case the fire suddenly crosses below them.
- Downhill line should be securely anchored at the top. Undercut line should be avoided if possible.
- The line should be burned out as the construction progresses, beginning from the anchor point at the top. The burned-out area provides a continuous safety zone for crews and reduces the likelihood of fire crossing the line.
- Hose lays or air drops may be used to increase safety during downhill line construction.
- Crews must remain aware of and avoid the "Watch-Out!" Situations (see Chapter 8, Firefighter Safety and Survival).
- Crews must comply with the "Standard Fire Orders" (see Chapter 8, Firefighter Safety and Survival).

WARNING

Downhill line construction in steep terrain and fast-burning fuels should be done with extreme caution. Other attack methods should be used whenever possible.

LINE CONSTRUCTION WITH HAND TOOLS

Even though heavy equipment can construct a fireline much faster than hand crews, the use of hand tools for fireline construction is quite common. In many situations, such as in rugged or steep terrain, the use of mechanized equipment is impractical, unsafe, or prohibited by environmental protection regulations. In these situations, hand crews may be the only practical means of constructing a fireline.

When constructing firelines, crew members typically walk and work 10 feet (3 m) apart for safety (Figure 6.35). This spacing helps prevent crew members from being struck by the handles or heads of tools while they carry them or use them in constructing a line. Firefighters should "sound off" (give a loud verbal warning such as "tool coming through!") when they need to pass close to each other on a line, especially if visibility is reduced by smoke or darkness.

Several methods are used to organize hand crews for constructing a fireline. One common method assigns each crew member a few feet (meters) of the line. Each firefighter is responsible for completing that portion of the line before moving to another portion (Figure 6.36). This method is often used to construct a *scratch line,* a narrow line hastily constructed to temporarily stop the spread of a fire. The scratch line can be widened later into a full fireline.

Figure 6.35 Firefighters should maintain a safe working distance from each other.

In another common method known as *progressive line construction* (also called the *one-lick method*), the crew is arranged in a staggered line, and each member remains in position relative to the other members as the line construction progresses (Figure 6.37). Each member takes one stroke ("lick") with the tool before moving one step forward to repeat the action. The crew works in unison until the line is completed. This method of line construction requires teamwork but promotes safety and efficiency because no one passes anyone else on the line.

Regardless of how a crew is organized, they most often work along the line in what is called a

Figure 6.36 Each firefighter is responsible for a portion of the fireline.

Figure 6.37 Firefighters cut a progressive fireline.

typical tool order. The order is dictated by the type of fuel to be cleared. In light fuels, such as pastureland or mowed field crops, it may only be necessary for all members to use scraping tools such as McLeods or swatters (Figure 6.38). In medium fuels, the typical tool order will be cutting tools such as Pulaskis or brush hooks, followed by scraping tools such as McLeods, followed by digging tools such as shovels (Figure 6.39). In heavier fuels, the crew is lead by members with chain saws or axes, followed by brush hooks or Pulaskis, followed by McLeods, followed by shovels (Figure 6.40).

Figure 6.38 All firefighters in the crew may use McLeods in light fuels. *Courtesy of Monterey County Training Officers.*

Figure 6.39 The typical tool order for medium fuels. *Courtesy of Monterey County Training Officers.*

Figure 6.40 The typical tool order for heavy fuels. *Courtesy of Monterey County Training Officers.*

Safe and efficient line construction by hand crews depends on having the right tools for the job and using those tools properly. Some tools are more effective in certain fuel types than others, and knowledge of the fuel types and the topography at the fire helps in selecting the right tools. When tools are issued, crew members should inspect them to be sure they are in good condition (Figure 6.41).

When walking along a line, firefighters should hold tools at their balance points, carry them close to the body at their sides, and never rest them on their shoulders. While walking along a slope, they should carry the tools on the downhill side. Tools should be carried with their heads forward and their cutting edges pointed away from the body or toward the ground (Figure 6.42).

Figure 6.41 Issued tools should be inspected.

Figure 6.42 Hand tools should be carried properly. *Courtesy of NIFC.*

When working with cutting tools such as brush hooks, Pulaskis, or axes, crew members can maintain control of the tools throughout each stroke with stable footing, proper body position, and proper hand placement on the tools (Figure 6.43). Firefighters should be careful that the blade of the tool hits only its intended target. They should be aware of the path of the tool head and watch for obstructions in its path that might deflect the blade into themselves or others. Whenever possible, they should use short strokes when cutting and avoid raising blades above their heads (Figure 6.44). They should also avoid chopping toward their feet or anywhere tools may be deflected toward them. Any crew member who sees a potential threat to the crew such as a rock rolling toward the crew or a snake in the undergrowth should simply yell "*Rock!*" or "*Snake!*" and point toward the threat (Figure 6.45).

Crew members may use a shovel to throw dirt to cool a fire, to scrape, or to cut roots. When throwing dirt, it is important to have stable footing, to have proper body position, and to use a sweeping motion (Figure 6.46). When scraping, use the leg as a leverage point for efficiency (Figure 6.47).

Figure 6.46 A firefighter throws dirt onto a fire.

Figure 6.47 Shovel handle should be braced against the leg when scraping.

Chain saws can make line construction much easier, but they must be used with care. Only trained and experienced personnel wearing goggles, gloves, and protective chaps should operate chain saws (Figure 6.48). Chain saws should not be transported with the motor running. When not in use, the chain should be covered with a safety guard (Figure 6.49). Whenever possible, saws should be turned off and allowed to cool before being refueled.

Figure 6.43 A firefighter in proper stance for using a brush hook. *Courtesy of Monterey County Training Officers.*

Figure 6.44 Firefighters should take short strokes when using an axe. *Courtesy of Monterey County Training Officers.*

Figure 6.45 A firefighter alerts his crew to a snake.

Figure 6.48 Full wildland PPE including chaps should be worn when using a chain saw. *Courtesy of Monterey County Training Officers.*

Figure 6.49 A chain guard should cover the cutting chain when the saw is not in use.

To make a line as effective as possible, do the following:

- Remove all vegetation and debris from the line.
- Clear the line down to mineral soil.
- Widen the line enough to safely burn out from it.
- Throw all burned/charred material into the black.
- Scatter all cut and unburned fuels into the green (unless needed for burning out).
- Remove all branches that hang over the line.

For more information on hand-tool maintenance and use, see Chapter 3, Ground Support Equipment and Personnel Considerations.

LINE CONSTRUCTION WITH MECHANIZED EQUIPMENT

Using mechanized equipment is a fast and efficient way to build a control line if it is available and if the topography allows for its safe and effective operation. Various kinds of mechanized equipment, such as bulldozers, graders (road maintainers), and tractor-plows, are used for constructing control lines. In an emergency, even farm equipment, such as farm tractors with plows or discs, can be effective in situations where they can operate safely.

When conditions allow, mechanized equipment can build a wide control line much faster than hand crews. Bulldozers can cut a one-blade-width control line at about ½ mile (0.8 km) or 880 yards (805 m) per hour, subject to several variables. In general, the newer and larger the bulldozer, the faster it can construct a line. The degree of slope, presence or absence of rocks, fuel type, direction of line construction, operator skill, operator fatigue, availability of lights for night work, and atmospheric temperature all affect the rate of construction.

In areas where wildland fires occur frequently, mechanized equipment specifically designed for building control lines is commonly used. Crawler tractors or all-wheel-drive vehicles with fire plows can build lines quickly and easily if the topography and soil conditions are right for their use (Figure 6.50).

Figure 6.50 A tractor-plow cuts fireline in heavy fuels. *Courtesy of NIFC.*

Bulldozers and other mechanized equipment are typically brought to a scene on low-boy or tilt-bed transports (Figure 6.51). Transporting this equipment to a point close to a line may present problems because of the length and weight of the transport vehicles. In addition, unless fuel tenders accompany the mechanized equipment into remote locations, other provisions must be made to refuel and service the equipment. This may require that fuel and lubricants be flown into the area by helicopter.

Operating heavy equipment on wildland fires requires all of the normal safety procedures for using mechanized equipment plus the additional safety procedures related to fire suppression with hand crews. Even routine preventive maintenance becomes a safety issue if a mechanical breakdown might result in a unit being overrun by a fire (Figure 6.52). One of the most important safety practices around mechanized equipment is maintaining effective communication between the equipment operator and those on the ground. Because it is possible for bulldozers and other equipment to be overrun by a blowup, another important safety feature for the equipment operators is the addition of fire shelters to their personal protective equipment. In some areas, completely enclosed and air-conditioned environmental cabs are used to protect the operator (Figure 6.53). Bulldozers working on slopes frequently dislodge rocks. Firefighters should not work below this equipment without posting a safety watch. If the safety watch sees a rock rolling toward the crew, he or she should immediately yell "*Rock!*" to warn the others.

Figure 6.51 Dozers are transported on low-boy trailers. *Courtesy of John Hawkins.*

Figure 6.52 Dozers may be overrun by fire if they break down. *Courtesy of NIFC.*

Figure 6.53 A dozer with a fully enclosed cab.

Operation of bulldozers or other heavy equipment is usually safer and more efficient if they are used in pairs (Figure 6.54). The operator of the lead bulldozer can rough out a control line and remove some of the fuel. The operator of the second machine can complete the line construction by removing the rest of the fuel down to mineral soil. Operators can also help each other if either machine becomes stuck or stalled. By working together, they can quickly build a safety zone if they are in danger of being overrun by a fire.

Figure 6.54 Dozers may be more effective working in tandem. *Courtesy of NIFC.*

The principles of control-line construction with mechanized equipment are essentially the same as those with hand crews. The procedure involves removing fuel down to mineral soil and disrupting the continuity of canopy fuels. Burning out may still be necessary to widen the line. Proper line locations are also important with mechanized equipment, and a flag person (also called a *swamper* or *line locator*) on foot may be needed to guide an operator in heavy fuels or smoke (Figure 6.55). Control lines constructed with mechanized equipment are usually not built as close to a fire's edge as are hand-built lines. This prevents burning material from being buried by the equipment. Buried burning material may surface later and cause a slopover or complicate the mop-up operation (completing extinguishment after the fire is controlled). Control lines constructed by mechanized equipment must be followed up by hand crews or engine companies. Hand crews are needed to improve and maintain these control lines. Crews equipped with hand tools, water, or both are needed to patrol the lines for spot fires and for doing mop-up (see Mop-up and Patrol section).

Figure 6.55 A line locator may be needed in heavy smoke.

These machines are expensive to operate, but the speed with which they can construct a control line makes their use highly desirable. Mechanized equipment can also construct crude, temporary roads into a fire area to allow access by engines (Figure 6.56). Mechanized equipment, however, can severely disturb the soil, leading to soil erosion later. Also, it may be necessary to restore the area by removing berms or creating water channels after mechanized equipment has been used. Each situation should be analyzed to determine the cost/benefit of any use of mechanized equipment.

Figure 6.56 Dozers may cut temporary access roads. *Courtesy of NIFC.*

In summary, the principles of bulldozer/tractor-plow use are as follows:

- Assure that all personnel know where equipment is working. Post a safety watch if personnel must work on a slope below heavy equipment.
- Be certain all bulldozers/tractors are in good mechanical condition, have approved spark arresters, have safety canopies, and have signed rental agreements if required.
- Take advantage of favorable fuels and topography.
- Consider working equipment in tandem to increase production and safety, especially when working near a fast-moving fire.
- Buck logs and fell trees or snags in the fireline as needed.
- Push flammable material to the outside of the line except on direct attack when hot material should be pushed well inside and scattered (see Fire-Control Strategies section).
- Allow no one other than the operator to ride on equipment.
- During mop-up:
    - — Scatter large logs or hot piles into the burned area.
    - — Scatter unburned piles that are outside of the line.
    - — Construct water bars where necessary to protect against erosion (Figure 6.57).

Figure 6.57 Firefighters construct a water bar to control erosion. *Courtesy of NIFC.*

FIRE-CONTROL STRATEGIES

Fire attack may involve *direct attack, indirect attack,* or both. The two strategies are described as follows:

- ***Direct attack*** — Fighting the fire at its edge or closely parallel to it (Figure 6.58)
- ***Indirect attack*** — Constructing a line some distance from the fire's edge and burning out the unburned intervening fuel (Figure 6.59) (See Burning Out section later in this chapter.)

Figure 6.58 Firefighters in a direct attack. *Courtesy of Monterey County Training Officers.*

Figure 6.59 Firefighters burn out from a road in an indirect attack. *Courtesy of NIFC.*

In this context, a *fireline* is part of a control line constructed very near the fire's edge by clearing a trail of all vegetation. A *control line* is any natural or constructed barrier, including the treated fire edge, used to control a fire. In an indirect attack, the control line may be wider than one at the fire's edge.

While these attacks are presented as separate techniques, most wildland fire-control activities use a combination of strategies to take advantage of varying fuels and topography. The strategy used to control a fire depends on the fire's size, rate of spread, intensity, spotting potential, values at risk, type of available attack resources, and other factors. Control lines should be anchored to an existing barrier such as a road, creek, burned area, etc., to minimize the chance of being flanked by the fire. To select the most appropriate strategy, the IC must understand their advantages and disadvantages and the precautions used with each.

Both strategies of fire attack may involve the use of water or Class A foam in conjunction with other techniques. Using water as part of either direct or indirect attacks can dramatically increase the effectiveness of the attacks. Water applied directly on burning fuel can control and extinguish fires much sooner and with far less effort than other strategies. However, the supply of water in the wildland environment is often limited, so it *MUST* be used sparingly.

The effectiveness of resources, fire-control agents, and attack strategies varies with the intensity of the fire. Table 6.2 illustrates the recommended fire-attack strategies for various levels of fire intensity and flame lengths.

TABLE 6.2
Intensity/Flame Length/Control Strategies

Intensity (Btu/sec/ft or kJ/sec/m)	Flame Length (ft or m)	Control Strategy
<100 Btu/sec/ft (<106 kJ/sec/m)	<4 feet (<1.2 m)	Can be attacked at the head or flanks by crews using hand tools. Handlines should hold the fire.
100–500 Btu/sec/ft (106 to 528 kJ/sec/m)	4–8 feet (1.2 m to 2.4 m)	Too intense for direct attack on the head by hand crews. Heavy equipment and/or aircraft needed. Fire is potentially dangerous to personnel.
500–1,000 Btu/sec/ft (528 to 1 055 kJ/sec/m)	8–11 feet (2.4 m to 3.4 m)	May present serious control problems (torching, crowning, spotting). Attack on the head will probably be ineffective.
>1,000 Btu/sec/ft (>1 055 kJ/sec/m)	>11 feet (>3.4 m)	Crowning, spotting, and major fire runs are probable so attacking the head is likely to be ineffective and dangerous. Indirect attack may be the best alternative.

NOTE: *Torching* is a vertical phenomenon in which a surface fire ignites the foliage of a tree or bush. *Indirect attack* is a method of suppression in which a control line is located at natural or man-made barriers some distance from the fire's edge and the intervening fuel is burned out. *Direct attack* is suppression of a fire by attacking the fire front directly from within or from outside the burned area.

Source: Fire Behavior and Initial Attack, Pacific Northwest Wildfire Conference.

Direct Attack

As the name implies, *direct attack* suppresses a fire by attacking the fire front directly from within or from outside the burned area. A direct attack may involve extinguishing the fire by cooling with water or foam or interrupting the flaming process with chemicals (Figure 6.60). While the use of water makes the attack more efficient, in many cases a direct attack involves smothering the fire with dirt or removing the fuel (Figure 6.61). After a direct attack, crews normally secure the fireline by constructing a control line completely around the fire's edge.

A direct attack is an aggressive, offensive attack at a fire's edge. It is normally used on relatively small fires (flame lengths of no more than 4 feet [1.2 m]) where heat and smoke do not keep firefighters from working at the fire's edge. Whenever conditions permit its use, direct attack is the strategy of choice.

Figure 6.60 A firefighter attacks with water from the black. *Courtesy of Monterey County Training Officers.*

Figure 6.61 A firefighter throws dirt onto a fire.

Like every other tactical decision, the decision to order a direct attack must weigh the risks against the potential benefits. A direct attack is most appropriate on small fires or larger fires of low-to-medium intensity and for hotspotting (knocking down flare-ups [hot spots] ahead of line-construction activities or extinguishing fingers spreading from the main fire). A direct attack may also be used on running (rapidly spreading) fires in light fuels and on fires in high-value property where the amount of area lost to the fire must be kept to a minimum. Mobile attack with engines is a common method of directly attacking small or slow-moving fires (Figure 6.62).

Figure 6.62 A mobile attack from the black. *Courtesy of Monterey County Training Officers.*

Working very near the edge of a fire, as required in a direct attack, can expose firefighters to intense heat and smoke. Firefighters must always know the planned escape routes, have proper protective clothing, stay alert for changes in fire conditions, observe changes in weather, and get weather updates. They must never violate the standard Fire Orders (see Chapter 8, Firefighter Safety and Survival). For a direct attack to be effective, firefighters must be well trained, must be equipped with proper tools (including communications) and protective clothing, and must be in good physical condition.

One primary advantage of a direct attack is that firefighters are close to or working in the burned area, which may be used as a safety zone in the event of a blowup. Other advantages are that firefighters may use burned areas in constructing a fireline (less burning out is required to eliminate unburned fuels than is necessary in an indirect attack) and the need for mop-up and standby is reduced because little unburned fuel remains inside the fireline (see Mop-up and Patrol section).

A major disadvantage of using a direct attack is that firefighters are exposed to heat, smoke, and flame because the fireline is constructed directly adjacent to the edge of the fire. Embers may blow across the line and start spot fires, and constructing firelines with hand tools is physically taxing to firefighters.

In summary, some important things to remember about direct attack are as follows:

- Control efforts, including line construction, are conducted at the fire perimeter, which becomes the control line.

- Direct attack is used when the fire perimeter is burning at low intensity and fuels are light, permitting safe operation at the fire's edge.
- Direct attack is often used where high-value resources and/or improvements are threatened.
- The amount of area burned is kept to a minimum.
- In light fuels, such as grass, attack may be from inside or outside the black.

CAUTION: Working in the green is a situation that shouts "Watch Out!" on a grass fire.

Indirect Attack

Indirect attack sacrifices a certain amount of vegetation or other exposed values. It is used when the intensity of a fire makes direct attack unsafe, when the fire develops long fingers, or when there is not enough time to construct a control line at a fire's edge. In an indirect attack, fire-suppression forces are withdrawn to roads, trails, and other natural fuel breaks or to a constructed control line. Existing safety zones must be identified or new ones constructed. The fuel between these barriers and the fire is burned out or backfired.

While indirect attack does not force firefighters to work as close to a fire front as does direct attack, an indirect attack is not without risk. Firefighters can be in jeopardy during indirect attacks for a variety of reasons. One reason is that while the control line is being constructed, the fire is growing in size and intensity. Because the line is constructed some distance from the fire's edge, firefighters are often unable to directly observe the behavior of the fire. This increases the need for LCES. Another reason is that it is always risky to have firefighters work in an area with unburned fuel between them and the main fire. Whenever firefighters are engaged in constructing an indirect control line, they must be especially mindful of the 18 "Watch-Out!" Situations listed in Chapter 8, Firefighter Safety and Survival.

In an indirect attack, a control line is constructed or located some distance from the edge of the main fire. The distance from the control line to the fire's edge depends on the fire's intensity and rate of spread, the type and volume of fuel, the topography, the wind, and the availability of natural barriers. This attack strategy may be useful if a fire is burning too intensely or spreading too rapidly for firefighters to work safely and effectively at the fire's edge.

A completed control line serves as a barrier to prevent further spread of a fire should it flare up after it has been brought under control. The effectiveness of an indirect attack is determined by the location and construction of the control line. The location selected must be far enough ahead of the fire so that the line can be constructed and burned out or backfired before the main fire reaches it (see Backfiring section). The line must be wide enough to prevent oncoming flames from crossing over it to the uninvolved fuel on the other side. The crews building the line should start from a secure anchor point to prevent the fire from outflanking and potentially surrounding them. Ideally lines should be constructed in light fuels and should take advantage of natural barriers (Figure 6.63). An indirect control line should be no longer than crews can control.

Regardless of the distance between a fire front and the control line, all unburned fuel between these two lines *must* be removed, usually by backfiring or burning out. Removing the fuel creates an effective and safe control line; unburned fuel can allow a fire to threaten the control line.

Figure 6.63 A control line should take advantage of natural barriers.

One advantage of an indirect attack is that it may make better use of natural firebreaks such as lakes, cliffs, or roads. Also, constructing a control line is generally not as physically taxing on the crews because they are working away from the heat and smoke at the fire's edge.

Among the disadvantages of indirect attack is that the fuel left inside a control line can allow a fire to increase and reach such intensity and rate of spread that it could jump the control line. The unburned fuel between the main fire and the control line can also allow the fire to develop a large convection column, increasing the risk of spot fires. This prospect is increased if the winds suddenly pick up after the attack is underway. Changes in wind direction also could change the direction of fire spread, rendering the control line ineffective. These dangers reinforce the importance of continually sizing up and adapting to changing conditions. More fire entrapments and fatalities occur on indirect attack than on direct attack, so indirect attack is generally considered more dangerous.

In summary, some important things to remember about indirect attack are as follows:

- The control line is located along natural firebreaks, along favorable breaks in topography, or at considerable distances from the fire. The intervening fuel is burned out.
- If indirect attack is necessary, the fire may be rapidly growing to the point of requiring extended attack (needing more resources). Be alert to this possibility.
- Indirect attack may be used on crown fires, on steep terrain, on fast moving ground fires too intense for hand crews, or in areas with constructed or natural barriers.
- Indirect attack can involve downhill line construction.

USE OF WATER

As discussed in Chapter 4, Water Supply, water, foam, and other additives have a variety of uses on the fireline. In addition, many of the fire-control tactics discussed in the balance of this chapter may use water (with or without additives) as the primary extinguishing agent. The tactical use of water usually involves one or more of the following basic applications:

- Making the initial attack
- Cooling hot spots to permit a direct attack
- Holding or strengthening lines for burning out
- Extinguishing spot fires
- Protecting exposed structures
- Conducting mop-up operations

While a full study of field hydraulics and pump operation is beyond the scope of this manual, firefighters need to understand the basic mechanics of moving water from its source to the point of application. Excluding the water delivered by aircraft, firefighters most often apply water through hoselines supplied by engines or portable pumps.

Basic Hydraulics

To be useful in fighting a fire, water must be delivered to the fire in sufficient quantity and at sufficient pressure to be applied effectively. This usually requires a pump at the water source (plus additional pumps as needed), hose and various fittings and appliances, and a nozzle (Figure 6.64). The pump creates the pressure needed to move the water through the hose to the nozzle at the desired working pressure. To do this, the pump must overcome the resistance to the flow of the water through the hose. The resistance comes from two sources: (1) friction between the moving water and the inside of the hose, fittings, and appliances (*friction loss*) and (2) the force of gravity if the water must

Figure 6.64 A basic water delivery system.

be pumped uphill (*elevation loss, back pressure,* or *head*). In addition to overcoming this resistance, the pump must supply the water to the nozzle at its designed operating pressure (*nozzle pressure*). The aggregate of these demands for pressure from the pump is called *pump pressure* or *engine pressure.*

The amount of friction loss in a given hose lay depends upon a number of variables: hose size (inside diameter), length, and lining material; the number and types of fittings and appliances through which the water must flow; the volume of water (gpm or L/min) flowing; and the desired nozzle pressure. In general, the smaller the hose diameter and the longer the hose lay, the higher the friction loss. Friction loss in rubber-lined fire hose is less than in unlined forestry hose. Every tee, wye, or other appliance (including the nozzle) through which the water must flow adds friction loss. If a hose lay is long enough and/or the volume of water being flowed is high enough and/or the elevation of the nozzle above the pump is high enough, more pressure may be lost than the pump can produce. The amount of pressure lost may even exceed the safe working pressure of the hose. In this situation, one or more additional pumps may have to be installed at intervals in the hose lay. In wildland fire fighting situations, this is usually done by *relay* pumping (Figure 6.65).

The amount of back pressure (if any) in a given hose lay is dependent upon the number of vertical feet (meters) the water must be pumped. In round figures, for each foot (0.3 m) of elevation, gravity produces a back pressure of about 0.5 psi (3.5 kPa). If the nozzle is lower than the pump, gravity *reduces* the pressure needed by the same amount. So, if a hose were laid up one side of a hill and down an equal vertical distance on the other side, the back pressures would cancel each other, and there would be no net elevation loss (Figure 6.66). The pump would still have to overcome the friction loss in the lay and supply the required nozzle pressure.

If the water source is below the level of the pump, *suction* or *lift* must be considered. In general, pump suction is a function of atmospheric pressure. That is, when the priming device discharges the air within, creating a partial vacuum in the pump and suction hose, atmospheric pressure forces water up into the suction hose (Figure 6.67). Since atmospheric pressure decreases about 0.5 psi (3.5 kPa) for each 1,000 feet (305 m) of

Figure 6.65 Basic relay pumping.

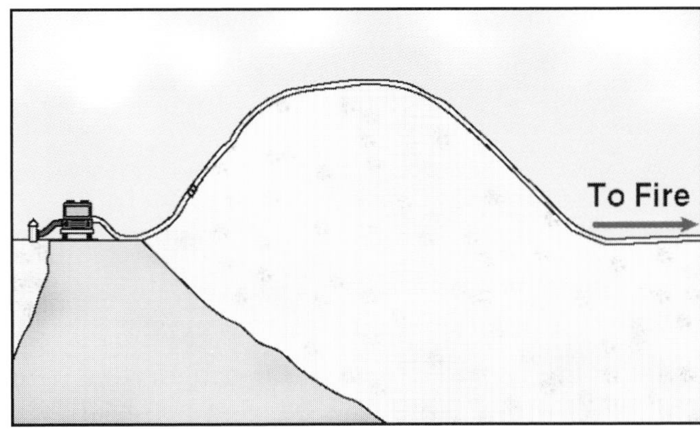

Figure 6.66 Elevation loss is equaled by gain on the downhill side.

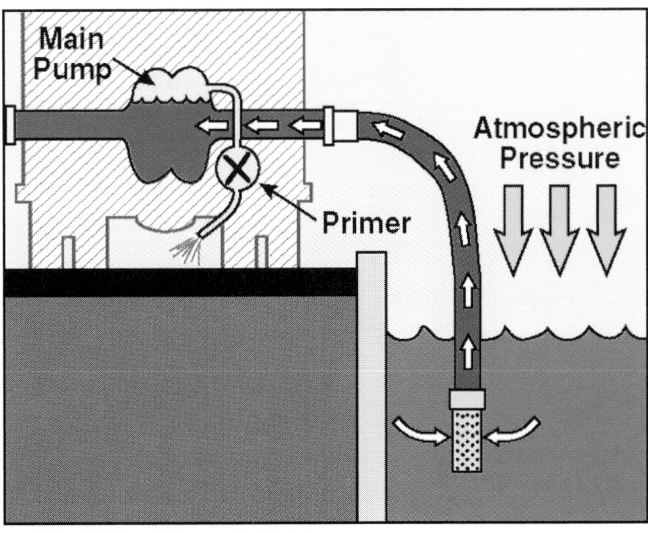

Figure 6.67 The basics of suction lift.

elevation above sea level, the maximum effective lift is about 20 feet (6 m). If a primed pump cannot lift water or performs poorly, the pump may have to be relocated closer to the water level, the fittings may have to be tightened, or both.

Principles of Water Use

- Use water sparingly, especially when it is in short supply.
- Conserve water by shutting off nozzles while traveling between hot spots.
- Direct water at base of flame (at what is burning).
- Have hand-tool personnel train with nozzle personnel to ensure the most effective use of water, especially during mop-up operations.
- Maintain good communications between nozzle personnel and water-source personnel.
- Plan for ample water supply in advance, including possible use of water tenders.
- Coordinate so all units do not run out of water at the same time during critical periods.
- Do not block roads with hoselines or apparatus.
- Keep engines positioned in the direction of exit for quick escape if necessary.
- Cut a fireline to mineral soil around the entire fire where appropriate after direct attack with water.
- Provide eye protection for nozzle operators.
- Consider using foam or other water additives to increase effectiveness and save water.
- Maintain a minimum nozzle pressure of 25 psi (175 kPa).
- Avoid scattering burning fuel with the hose stream.
- Apply water as close to the burning fuel as possible, but use the stream reach when needed.
- Reserve the last 100 gallons (400 L) of water for firefighter/engine protection.

Obviously, some hose lays take longer than others to set up, even with well-trained crews. Depending on the particular fire fighting situation, however, the time spent setting up an effective hose lay and pumping operation may be more than offset by the increase in fire-suppression capability and firefighter safety. For more information on fireground hydraulics and pump operations, see the IFSTA **Fire Stream Practices** manual.

FIRE-CONTROL TACTICS

The following fire-control tactics are forms of initial attack, either direct or indirect. They may be used singly or in combination. The decision regarding which tactic(s) to use in any situation should take into account the actual or potential life hazards, exposures (property or resources threatened), the fuel in which the fire is burning, current and expected weather (especially wind), topography, and anchor points. Flank, pincer, and frontal attacks can all include aircraft as one of the attack resources.

Flank Attack

A *flank attack,* sometimes also referred to as a *flanking attack* or *flanking the fire,* is used for moderately hot fires moving at a moderate rate of spread. The attack is started at a secure anchor point on one or both flanks of a fire and works toward the head (Figure 6.68). The flanks may be attacked simultaneously or successively, depending on fire conditions and resources available. The attack may be either direct or indirect, and the distance of the control line from the fire edge usually depends on fire intensity. The strip of unburned fuel between the line and the fire's edge is burned out as soon as possible during fireline construction. Whenever possible, firefighters work in the black for safety.

Pincer Attack

A *pincer attack* is a simultaneous attack on two sides of the fire from a secure anchor point, such as the heel or point of origin, working toward perimeter control. A pincer attack is similar to a flank attack and requires two or more crews working in a highly coordinated effort (Figure 6.69). That coordination demands effective communication between the participating units, especially if there is a mix of hand crews and mobile units. Again, whenever possible, crews work in the black for safety.

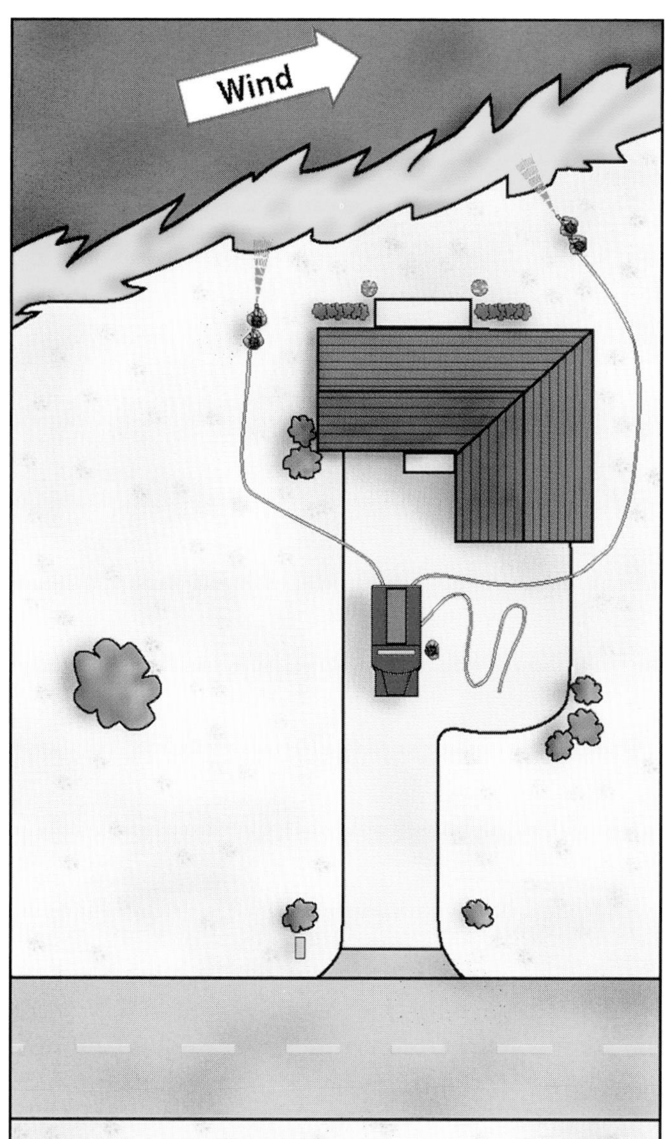

Figure 6.68 Firefighters may attack from the green if lives or structures are at risk.

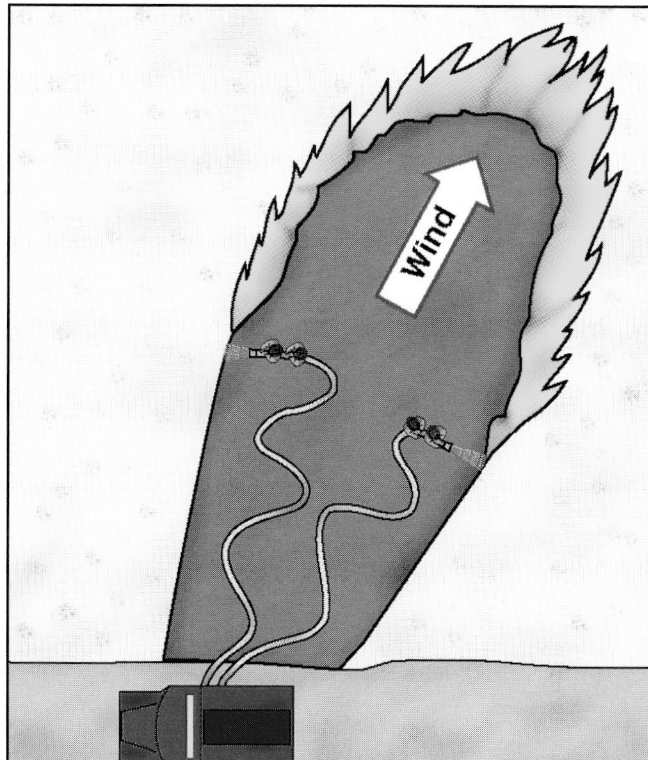

Figure 6.69 A typical pincer attack.

Figure 6.70 A frontal attack may be needed to protect structures.

Frontal Attack

A *frontal attack* attacks the head of a fire or the head of fingers extending from the main fire. Because this attack is made without an anchor point, *frontal attacks can be very dangerous*. While attacking the fingers of a fire is always a high priority in order to slow the rate of spread, it is usually done from within the black for safety. The frontal attack starts at or near the head of a fire and then proceeds to the flanks (Figure 6.70). A frontal attack does limit the spread of a fire; however, such factors as fire intensity, fuel type, wind, or topography usually make a frontal attack too dangerous to attempt except by aircraft (Figure 6.71). As previously shown in Table 6.2, flame lengths in excess of 8 feet (2.4 m) normally preclude frontal attack with ground resources. If a fire is too intense for a safe and effective frontal attack with ground resources, a flank attack may be another alternative.

Mobile Attack

Mobile attack (also known as *pump and roll*) is a fast and efficient method of extinguishing wildland fires when conditions allow its use. Primarily

Figure 6.71 An aerial frontal attack.

used on grass fires, it involves employing fire apparatus in a fast-moving direct attack with water or foam as the primary extinguishing agent (Figure 6.72). The major requirement for this method of attack is that the fire be in terrain that the apparatus can safely negotiate. Mobile attack is started from a secure anchor point; ground sweep nozzles are used or hoselines are kept as short as possible to allow the engine maximum mobility.

Figure 6.72 An engine crew in mobile attack. *Courtesy of Monterey County Training Officers.*

While a mobile attack is sometimes made with the vehicles in the green, the safest method is for the apparatus to work inside the black. In either case, the attack should always start from a secure anchor point. Typically, one or more attack vehicles enter the black at the heel and attack the fire along the burning edge from within (Figure 6.73). This allows the attack to be made directly on the head of a fire from the relative safety of the black. Because the majority of fuels within the black have already burned, there is less danger of crews being overrun by fire. Working in the black, crews are generally exposed to less heat and smoke. Apparatus operators can see obstacles, such as logs, stumps, and ditches, that might otherwise be hidden if they were in high grass or brush. Although the prevailing wind tends to blow the smoke away from firefighters, smoldering material within the burn can still produce enough smoke to obscure vision. Operators need to turn on vehicle lights and roll up windows. The smoke can also cause respiratory difficulties, so crews may have to be rotated more frequently. When operating within the black, operators must be careful not to stop their vehicles on smoldering or burning materials that can severely damage tires or undercarriages.

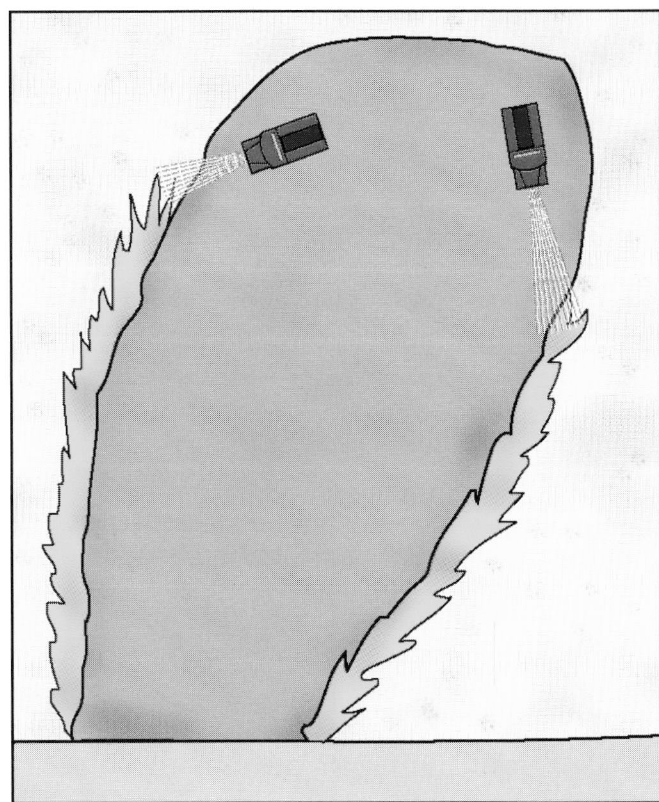

Figure 6.73 Mobile attack should be made from the black for safety.

Combination nozzles, those capable of producing both a fog stream and a straight stream, are the nozzles of choice for mobile attack. The attack lines and their nozzles must have a high enough output to absorb the amount of heat produced by the fire. While "hard lines" (booster lines) are very durable and highly maneuverable, their relatively high friction loss (¾ inch [19 mm] is about 13 times that of 1½-inch [38 mm] lines) means they may not be able to flow the required quantity of water. In most cases, larger lines should be used (Figure 6.74).

Figure 6.74 Attack lines should be of sufficient size for the fire conditions. *Courtesy of Monterey County Training Officers.*

CAUTION: Never ride on the outside of a moving fire apparatus, especially during a mobile attack. Major injuries may result.

Tandem Attack

A *tandem attack* is a direct attack along a part of the fire perimeter by engines, mechanized equipment, hand crews, or aircraft working together to achieve greater effectiveness. A tandem attack may be made by aircraft and bulldozers, bulldozers and engines, engines and hand crews, or any combination of two resources working in a coordinated effort. The first units do a quick knockdown of the fire; the other units follow closely and do a more thorough job of extinguishment and mop-up.

TWO ENGINES

When two engines attack in tandem, the first can move along a fireline at a relatively fast pace, knowing that the second engine will be hotspotting and securing the line behind it. The second engine moves more slowly, making sure that the fireline and any hot spots along it are completely extinguished (Figure 6.75). The first engine completes its assigned portion of the fireline before the second engine. The first engine can then either reverse its direction and do a more thorough extinguishment as it works back toward the other engine or begin to extinguish hot spots burning in the black (Figure 6.76).

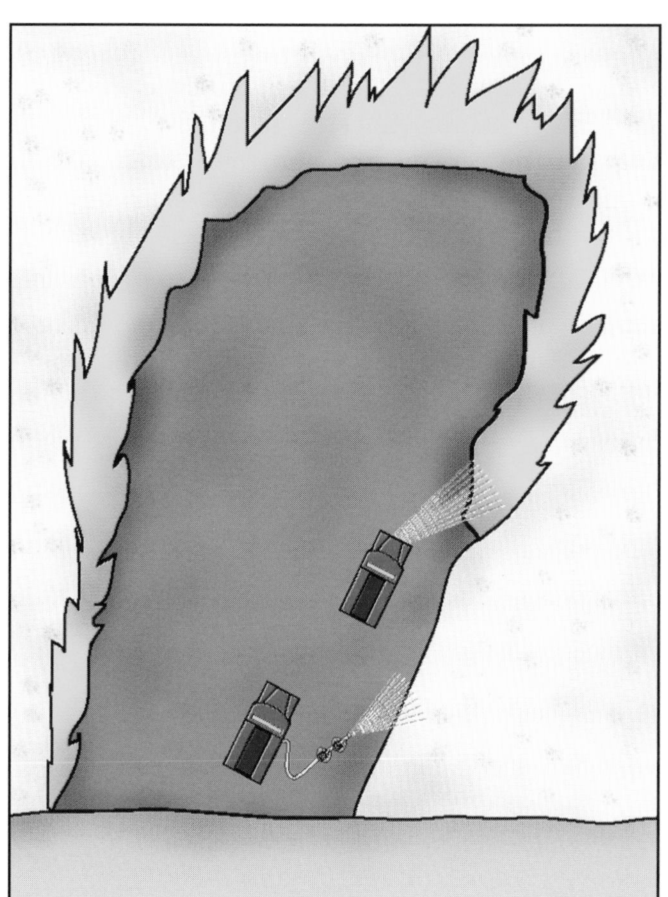

Figure 6.75 Engines working in tandem.

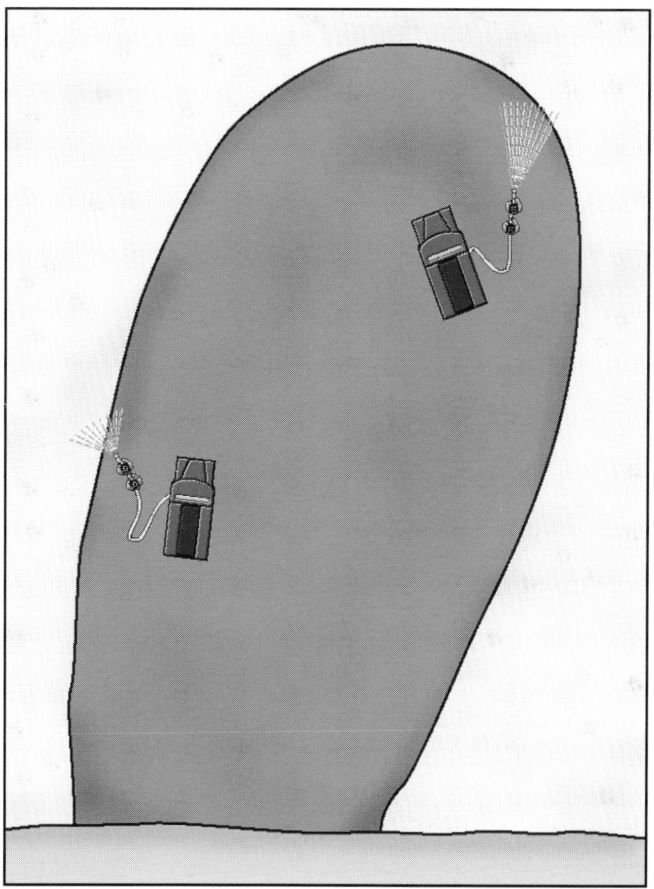

Figure 6.76 The first engine works back toward the other.

ENGINE AND AIRCRAFT

An engine can also mount an effective attack in tandem with either an air tanker or a helicopter. The engine stands by as the aircraft drops retardant or an extinguishing agent on the fire (Figure 6.77). Because an air attack is not effective in the long term, the engine immediately follows up to complete the extinguishment and secure the control line.

Figure 6.77 An engine waits for a helicopter to complete its drop. *Courtesy of Monterey County Training Officers.*

HAND CREWS

Two hand crews can work effectively in tandem in the same way as two engines (Figure 6.78). Although crews may progress slower than engines, they are no less effective. Because the work of the lead crew may be more physically demanding, crews should alternate taking the lead.

Figure 6.78 Hand crews also work in tandem.

MECHANIZED EQUIPMENT

Bulldozers, tractor-plows, and other earth-moving equipment can work in tandem to construct an effective control line very quickly. Working as close to the fireline as safety allows, the first unit makes a single pass. The second unit widens and secures the line behind the first unit (Figure 6.79). This works particularly well when using private equipment in tandem with fire service equipment. The fire service equipment leads while the private equipment follows.

Figure 6.79 Dozers working in tandem. *Courtesy of Monterey County Training Officers.*

Parallel Attack

The parallel attack fire-control tactic is used whenever a fire is too intense for direct attack or when a fire's edge is so irregular that direct attack would result in an excessively long control line. In a parallel attack, a control line is constructed as near to the fire's edge as possible while still allowing enough time to complete the line before the fire front arrives. For this reason, mechanized equipment such as tractor-plows and bulldozers is often used in a parallel attack (Figure 6.80). The area between the control line and the fire is usually burned out as the work progresses, or firefighters patrol the line to ensure that it is not breached when the main fire reaches it.

The parallel attack is similar to an indirect attack, but the control line is constructed much closer to the fire's edge, typically within 100 feet (30 m). The area sacrificed to the fire is smaller than in an indirect attack.

Figure 6.80 Fuel is burned out behind a dozer in a parallel attack.

The parallel attack reduces the labor of hand crews because a line can be constructed straight across indentations of the fire front, thus shortening the line (Figure 6.81). This method can also be used to keep the fire away from heavy fuels or to encircle spot fires.

CAUTION: Any time you are making any form of attack from the green, you must always be aware of the need for a planned escape route and/or safety zone.

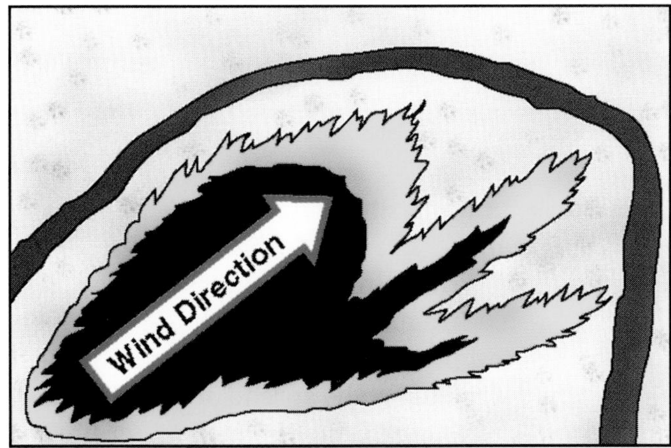

Figure 6.81 Building a parallel line across the ends of fingers saves time and effort.

Hotspotting

Hotspotting is one of the more dangerous fire fighting tactics, primarily because it is not initiated from an anchor point. Hotspotting involves making a rapid attack on hot-burning points on a fire's edge — fingers that are developing rapidly or active parts of a fire that are threatening life or high-value property — until control lines are constructed. Since this is a potentially dangerous tactic, only experienced firefighters should be assigned to hotspotting. Hotspotting is usually done by hand crews or smoke jumpers and should be done from within the black whenever possible.

Backfiring

Backfiring and burning out both eliminate unburned fuel between a fire's edge and a control line, but they are used under different conditions. Backfiring is an offensive tactic used to stop the spread of a very intense wildland fire and is usually done on a much larger scale than burning out. Backfiring is done from the downwind side of a large fire (the active fire edge) and takes advantage of the natural indraft of the main fire to draw the backfire toward the main fire (Figure 6.82). Although backfiring does widen the defense perimeter, it may also be used to change the force of the convection column. Except for rare circumstances meeting specific criteria, backfiring should be executed only by experienced firefighters and only after being approved through the chain of command.

CAUTION: Backfiring must be done only by experienced personnel because it involves extensive burning under adverse conditions.

Figure 6.82 When backfiring, smoke usually blows *away* from the control line. *Courtesy of NIFC.*

Judgment and experience must be relied upon to decide when conditions would make a backfire impractical or unsafe. Some of the most important factors in successful backfiring operations are the following:

- Obtaining permission from the operations chief or the incident commander
- Ensuring personnel safety
- Establishing communications
- Avoiding saddles and chutes
- Obtaining current weather data
- Obtaining current fire behavior predictions
- Establishing an anchor point for the operation
- Using a wet line, road, stream, or constructed control line (see Line Construction With Water section)
- Ensuring that all personnel are aware of the backfiring operation (communications)
- Patrolling the control line to check for spotting across it
- Observing the principle of Lookouts, Communications, Escape routes, Safety zones (LCES)

Burning Out

Burning out within the control line is intentionally setting fire to uninvolved fuels between a control line and an approaching (inactive) fire front to strengthen the fireline (create a black line) (Figure 6.83). Burning out is done to widen a control line, to eliminate islands of unburned fuel, and/or to create escape routes and safety zones. It must be done far enough away from a main fire so that there is no indraft effect from it. Burning out is usually done on a smaller scale and for a different purpose than backfiring. Burning out also reduces the danger of fuel near a line burning when no one is around or when conditions are such that flare-ups near the line could spot across it. Burning out is used to consume fuel that would otherwise be available to feed the main fire. Burning out around structures can be a very effective tactic in the wildland/urban interface.

The effectiveness and safety of burning-out operations are affected by the fuel, the weather, and the topography. Light, patchy fuels may not burn cleanly and may leave islands of unburned vegetation. Heavy, highly compact fuels may burn too intensely and spread too slowly to create the kind of firebreak needed to stop the main fire. Burning out in heavy fuels may also generate more heat than available resources can handle. This can cause a fire to spot across a line, doing more harm than good. High winds can also drive a backfire away from the main fire, but there are no established rules for determining the maximum safe wind speed for backfiring or burning-out operations.

Whenever possible, burning out should be done from the top of a slope toward the bottom (Figure 6.84). A fire started at the top of a ridge has to burn against both the slope and the wind, so it spreads slowly toward the main fire.

Figure 6.83 When burning out, smoke usually blows *toward* the control line. *Courtesy of Tony Bacon.*

Figure 6.84 Burning out from the top of a slope can be effective.

Line Construction With Water

Water can be used in either a direct or an indirect attack. In a direct attack, water is applied directly onto a fire either through hoselines operated by ground personnel or by air drops. In most cases, this attack halts the spread of a fire but does not achieve complete extinguishment; therefore, a direct attack with water must be followed by hand crews who construct a control line and complete the extinguishment.

In an indirect attack, water or foam may be used to construct a temporary fire-stop (wet line) by wetting down fuels. Water, foam, or retardant from hoselines or air tankers is used to wet a strip of fuel sufficiently to prevent ignition for a few minutes (Figure 6.85). The adjacent fuels, if any, are backfired or burned out to widen the line. Just as with a direct attack, a wet line is followed by hand crews who construct a control line.

Figure 6.85 A firefighter creates a wet line with foam. *Courtesy of NIFC.*

WARNING

A wet line by itself will not reliably stop an advancing fire. It is not a control line. To be effective, wet lines must be burned out to a regular control line or followed up with a fireline cut to mineral soil.

Since water alone may not sufficiently wet fine fuels and grasses, a common practice is to use water in conjunction with penetrants to wet the fuel better. Penetrants also help to extend the water supply. Class A foam is excellent for laying a wet line from which to start a backfire or burn out. One advantage of foam is that firefighters can clearly see where the line is (Figure 6.86). For more information on Class A foam, see Chapter 4, Water Supply.

The location of a wet line is important to its effectiveness. The distance required between the main fire and the wet line depends on fire behavior, fuel types, topography, and weather conditions. The line must be far enough ahead of the flame front to be sufficiently burned out and create an effective line before the fire reaches it.

When selecting nozzles and patterns for wet-line construction, the type of fuel, flame lengths, thermal output of the fire, and the available water supply should all be considered. For instance, when the fire is in light fuel and flame lengths are short, a narrow fog pattern from a nozzle delivering 6 to 15 gpm (22 L/min to 56 L/min) may be sufficient (Figure 6.87). In heavier fuels, when flame lengths are higher and thermal output is greater, a slightly wider fog pattern at 23 to 95 gpm (87 L/min to 360 L/min) may be needed (Figure 6.88). To knock down hot spots, a straight stream or a narrow fog pattern may be used if they will not scatter burning fuel across the control line (Figure 6.89).

Figure 6.86 Firefighters burn out from a wet line of foam. *Courtesy of NIFC.*

Figure 6.87 Low output lines and nozzles may be used in very light fuels. *Courtesy of NIFC.*

Figure 6.88 Large lines and nozzles must be used in heavy fuels. *Courtesy of Monterey County Training Officers.*

Figure 6.89 Firefighters must be careful not to scatter burning material with straight streams. *Courtesy of Tony Bacon.*

WARNING

An adjustable fog nozzle is essential to provide a protective pattern for the nozzle operator in case of a flare-up, and the gpm (L/min) output must exceed the thermal output expected on the fireline.

HOSE LAYS

A hose lay used either to make a direct attack or to support an indirect attack can be an important element in controlling wildland fires. A *hose lay* consists of lengths of hose and accessories laid from a pump and extending to the point of water delivery. A hose lay can be used either as the primary means of suppression or as one of several means in the overall suppression effort. The variables discussed earlier — fuel, weather, and topography — determine the techniques used.

A booster line or hard line may be used for attacking a small, very low-intensity fire; however, because friction loss in a booster line is high, it delivers a relatively limited volume of water. These hose lays should be no more than a few hundred feet (meters) long and should not be used on large or intensely burning fires where fire-stream effectiveness and personnel safety require the use of larger diameter hose. On actively burning wildland fires, or where long hose lays are needed, at least 1½-inch (38 mm) attack hose must be used. As mentioned earlier, the hoselines must deliver enough water (in gpm or L/min) to absorb the heat generated by the fire and/or to protect the crew in case of a flare-up. To provide adequate personnel protection, attack lines must not be extended with smaller hose such as 1-inch (25 mm) forestry hose (Figure 6.90). These smaller hoselines are adequate for mop-up within the black, but create too much friction loss to be safe and effective for fire attack.

Figure 6.90 Large attack lines should *not* be extended with smaller ones.

One very effective technique for extended hose-lay operations, a *progressive hose lay,* is used primarily for a quick attack on a fire from a secure anchor point, usually a road. It allows a fast, aggressive attack while maintaining a continuous water supply without having to take an engine off the road. A progressive hose lay consists of advancing a hoseline from an engine along the fire's edge, knocking down fire as the hose is extended, connecting another length of hose from a hose pack, and then advancing and extinguishing more fire. Most hose packs contain two, 100-foot (30 m) lengths of 1½-inch (38 mm) hose with lightweight couplings and one water thief (1½- by 1-inch [38 mm by 25 mm] gated tee). Some departments include a 100-foot (30 m) length of 1-inch (25 mm) forestry hose in their hose packs; others keep the 1-inch (25 mm) hose on their engines until needed for mop-up (Figure 6.91).

Figure 6.91 Some departments carry their 1-inch (25 mm) hose in compartments.

Progressive hose lays require coordination and teamwork to be effective. The hose lay is usually started at the heel, attacking the most active flank (Figure 6.92). When the hose is fully extended, the stream is used to extinguish fire as far ahead as it will reach (Figure 6.93). The nozzle is then shut down, the hose is clamped, and the nozzle is removed (Figure 6.94). An additional length of hose is rolled out and attached to the end of the first hose (Figure 6.95). A water thief may be installed at any point where two lengths of hose are connected (every 200 feet [60 m] is recommended) if significant

Figure 6.92 An engine crew starts a progressive hose lay. *Courtesy of Tony Bacon.*

Figure 6.93 Stream reach is used when the hose is fully deployed. *Courtesy of Tony Bacon.*

Figure 6.94 Hose is clamped as another section is pulled from a pack.

mop-up is anticipated or there is a fire threat to the hose lay. The nozzle is screwed onto the new length of hose and the hose clamp is released, charging the additional hose (Figure 6.96). This process is repeated either until the hose is fully extended and another length added or until the fire is completely encircled. Additional personnel may be needed to help pull hose, to extinguish hot spots, and to help

Figure 6.95 An additional length is connected to the original lay.

with mop-up. As a rule, it takes about 5 minutes to extend each 100 feet (30 m) of 1½-inch (38 mm) hose in a progressive hose-lay operation while fighting fire in medium brush.

Figure 6.96 The hose clamp is released to charge the line.

As attack lines are extended around the perimeter of a fire, 1-inch (25 mm) lateral lines can be connected to the water thieves (Figure 6.97). Working in the black, the laterals are used primarily for mop-up, but they can also be used to protect the attack line from minor flare-ups. The laterals can be as long as 300 feet (90 m), but 200 feet (60 m) is recommended. Especially when working in the black, firefighters must not allow the forestry hose to contact flames or glowing embers because this hose quickly burns through and fails (Figure 6.98).

Figure 6.97 Lateral mop-up lines are connected to the water thieves.

Key points to remember about progressive hose lays:

- Be sure sufficient hose and appliances are available.

Figure 6.98 Water sprays from burn-damaged hose. Courtesy of Monterey County Training Officers.

- Choose the engine with the greatest pressure pumping capability when pumping uphill.
- Use pumping engines with the largest booster tanks when possible.
- Assume 5 minutes per 100 feet (30 m) of 1½-inch (38 mm) line for deployment.
- Use a minimum of nine personnel for progressive hose lays over 1,000 feet (305 m) in length.

As mentioned earlier, Class A foam and other additives can be mixed with the water delivered through hoselines to enhance its effectiveness. For more information on the application of these chemicals, see the Water Additives section of Chapter 4, Water Supply.

AIR OPERATIONS

Air operations should be considered when they are most likely to be effective and can be applied safely. In a reconnaissance role, aircraft can locate exposures and access roads and provide the IC with extremely valuable information about fire activity not clearly visible from the ground such as spotting (Figure 6.99). Aircraft can also be used in both direct and indirect attacks on a fire. In the indirect mode, aircraft can apply fire-retardant chemicals to uninvolved fuels in the path of a fire (Figure 6.100). In a direct attack, aircraft can apply water and other suppressants directly onto a fire (Figure 6.101). The tactical decisions regarding when, where, and how to use aircraft must take weather, topography, fuels, and fire behavior into account.

Figure 6.99 Air reconnaissance can provide the IC with critical fire-behavior information. *Courtesy of California Office of Emergency Services.*

Figure 6.100 An air tanker pretreats fuels in the path of a fire. *Courtesy of NIFC.*

Figure 6.101 An air tanker drops water directly onto a fire. *Courtesy of New Jersey Forest Fire Service.*

The use of aircraft as part of a coordinated fire attack must also be safe for both the aircraft and the firefighters on the ground. Safety for the aircraft depends upon favorable flying conditions such as relatively stable air over the fire, relatively consistent winds, ample daylight, and smoke conditions that do not obscure the pilots' vision. It is critically important for aircraft safety to clearly identify physical hazards such as electrical transmission towers, power lines, and tall timber or snags in the drop zone or approach (Figure 6.102). Firefighters on the ground must be aware of an impending air drop so they can get clear of the drop zone or protect themselves if they cannot get out of the way.

Figure 6.102 Pilots must be warned of hazards in the drop zone. *Courtesy of NIFC.*

As described in Chapter 2, Fire Apparatus and Communications Equipment for Wildland Fires, a variety of aircraft are used to help ground forces control wildland fires. In general, air-attack operations involve the following types of aircraft:

- Rotary-wing aircraft (helicopters) equipped with buckets or tanks
- Fixed-wing aircraft, including tankers capable of dropping 100 to 3,000 gallons (400 L to 11 400 L) of retardant, water, or foam; those used for reconnaissance; and those used to transport cargo

Helicopters may be used to transport firefighters and supplies. They may also be used to drop suppressants or retardants in either a direct or an indirect attack on a fire. Helicopters may apply water or foam directly onto a fire in a salvo drop (dropping the entire load), which is often done to extinguish spot fires and hot spots (Figure 6.103). Some helicopters have a limited trail-drop (dropping in sequence) capability, so they can be used to close gaps in a retardant line (Figure 6.104). If helicopters have a nearby water source, they can deliver more agent per hour onto a fire than air tankers even though their tank capacity is relatively limited.

Figure 6.103 A helicopter makes a salvo drop onto a fire. *Courtesy of NIFC.*

Figure 6.104 A helicopter starts a trail drop. *Courtesy of NIFC.*

Many fixed-wing air tankers have more than one agent tank, so they can deliver their loads in a variety of ways. They can do a trail drop by opening their tanks in sequence. They can do a split drop by dumping part of their load in one place and another part somewhere else. Tankers can also drop their entire load at once in a salvo drop (Figure 6.105). Some air tankers with computer-controlled tanks can drop loads in a more precise pattern.

Figure 6.105 An air tanker makes a salvo drop. *Courtesy of NIFC.*

How aircraft are used on any given fire depends on the conditions where the drops are needed and the objectives. Air drops are always used in conjunction with ground attack. Under ideal circumstances, a fire may actually be contained by air drops alone, but air drops without ground support cannot be depended upon to completely control a fire.

Tactical-Use Criteria

When the situation indicates a need for fire suppressants and/or retardants to be dropped by rotary- or fixed-wing aircraft, the IC should use the following criteria:

- Call for retardant early if it is needed, and follow up with aggressive ground suppression action.
- Consider the following:
  - — Will air drops be effective?
  - — Will air drops be safe for ground personnel?
  - — Can the mission be accomplished during daylight?
  - — What is the fuel type?
  - — Are wind conditions favorable (normally not over 30 mph [48 kmph])?
  - — What is the fire behavior?
  - — Will firefighters be able to follow up with ground action?
  - — Will the terrain allow effective air drops?
  - — Can the pilots see the targets?
- Suspend drops when they are no longer effective or essential.
- Notify pilots of physical hazards in the approach or drop zones such as utility lines, towers, trees, other aircraft, etc.
- Consider the use of helicopters if conditions or terrain make the operation of fixed-wing air tankers unsafe or ineffective.

Ground forces and aircraft working a fire must be coordinated to be effective. On fires where more than one aircraft are used, an Air Tactical Group Supervisor (Air Tac) in a light plane coordinates all air-attack operations. While observing the fire from the air, the Air Tac consults on a separate radio channel with the ground officer designated by the IC (IC, operations chief, etc.) to determine priorities and to formulate an air-attack plan. Then on the air-to-air channel, the Air Tac advises the tankers or, if available, the Air Tanker Coordinator/Fixed-Wing Coordinator and/or the Helicopter Coordinator where and how the drops are to be made (Figure 6.106). The Air Tanker Coordinator/Fixed-Wing Coordinator may lead the air tankers through the drop zone. On small, initial-attack fires, the aircraft usually contacts the IC on the fire directly.

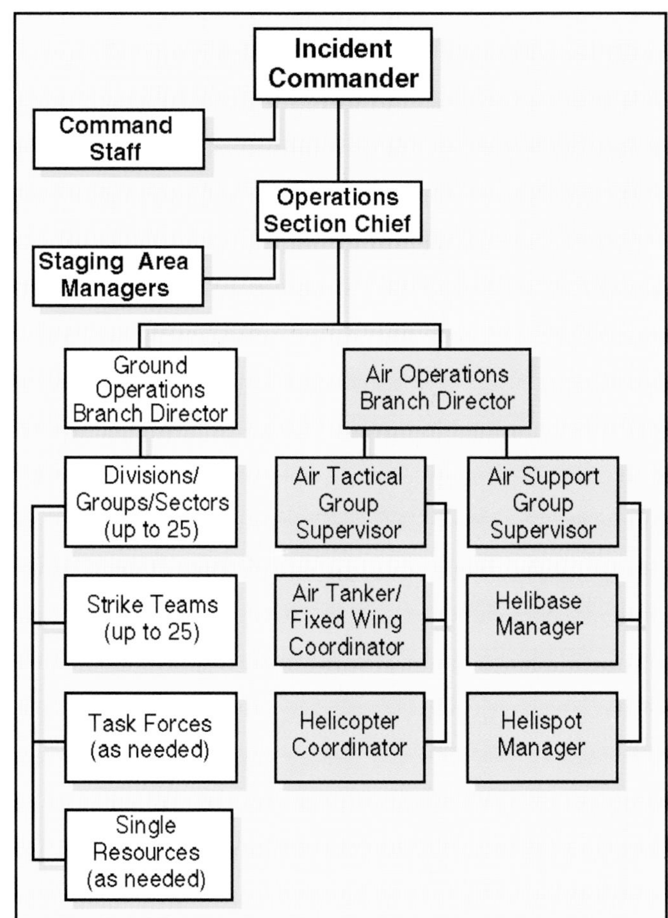

Figure 6.106 The Air Operations branch.

Incident commanders who are relatively unfamiliar with deployment of aircraft for fire fighting should rely on and work with the Air Tactical Group Supervisor. This person is an experienced ground-attack firefighter who is knowledgeable in the application of rotary- and fixed-wing aircraft to wildland fire fighting. The key is for the Air Tactical Group Supervisor to know the objectives and plans for the incident (Figure 6.107).

Figure 6.107 An Air Tac prepares for takeoff.

Direct Air Attack

A direct air attack involves either applying suppressants directly onto a fire's edge to cool the fire or applying retardants in a fire's path to slow the rate of spread. Cooling the head may allow ground forces to contain and control that part of a fire. Slowing the rate of spread can give ground forces time to construct a control line around a fire. In the most common forms of direct attack, drops are made across the head, on the most active flank, or on the flank with the most fire potential (Figure 6.108).

Attacking the head can significantly reduce a fire's intensity and allow ground crews to build a line that might not otherwise hold if the fire were allowed to burn unchecked. Attacking the flanks can keep a fire from spreading or from forming additional heads. In mountainous terrain, a direct attack can slow a fire sufficiently to allow ground crews time to reach the fire.

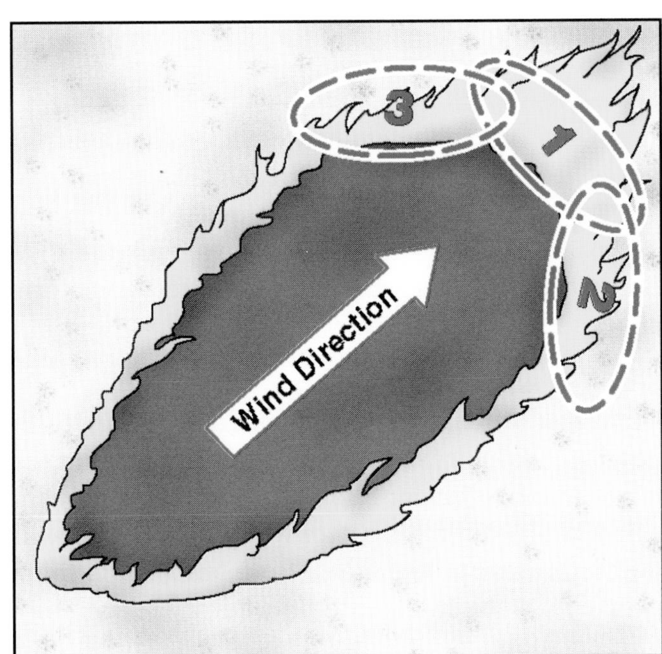

Figure 6.108 Direct attack from the air.

Indirect Air Attack

An indirect attack on a fire may also be supported by air drops. Application of retardants may slow the rate of spread sufficiently to allow time for construction of a secure control line. In mountainous terrain, air drops can be used to pretreat ridges to form effective control lines (Figure 6.109). Air drops can be used to build a wet line from which burning out can be started or to widen a hand-built control line. As mentioned earlier, cooling the fire with air drops may allow crews to contain a fire with a control line that would not otherwise have been sufficient. Drops may also be used to protect structures and other high-value exposures (Figure 6.110).

Figure 6.109 An air tanker pretreats a ridge. *Courtesy of NIFC.*

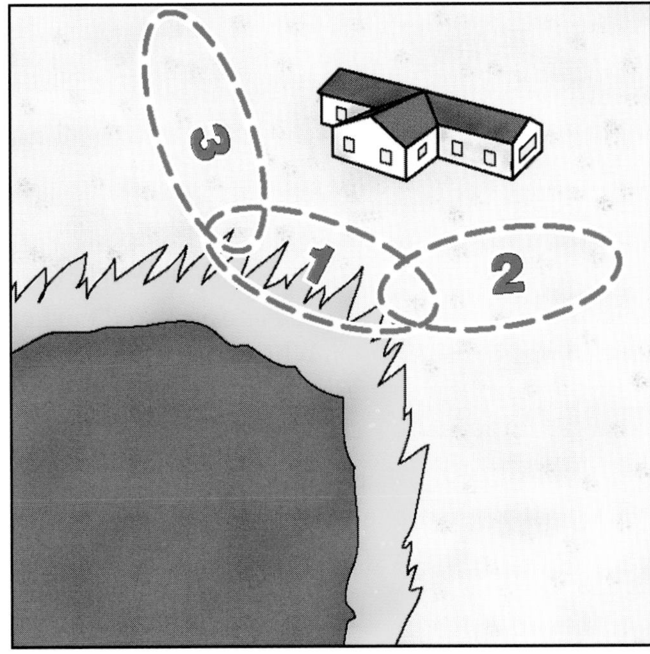

Figure 6.110 Structure protection from the air.

Aerial Ignition

One tactic involving helicopters is dropping burning material onto uninvolved fuels (Figure 6.111). This technique may be used for both backfiring and burning out. As part of an indirect attack, this technique may be used to initiate or assist with burning out from a control line or to burn out islands that could otherwise flare up later.

Figure 6.111 A typical helitorch operation. *Courtesy of NIFC.*

Air Tanker Retardant Drops

When deciding whether to request an air tanker retardant drop, the IC should consider the fact that air turbulence near hills, in canyons, and along ridge lines may make it extremely dangerous for the aircraft to approach a fire, as well as the fact that air drops are very expensive. For these reasons, air tanker retardant drops are used very selectively. The tactical-use criteria listed earlier should also be considered. When the decision to use retardant drops is made, the following principles of retardant application should be applied:

- Determine tactics of direct/indirect attacks based on fire size-up and resources available.
- Establish an anchor point and work from it.
- Use the proper drop height.
- Apply proper coverage levels.
- Drop downhill and away from the sun.
- Drop into the wind for best accuracy.
- Maintain honest evaluation and effective communication between the ground and the air.
- Use direct attack only when ground support is available to complete extinguishment (Figure 6.112).
- Avoid dropping on personnel and vehicles whenever possible.
- Plan drops so that they can be extended or intersected effectively.
- Monitor retardant effectiveness and adjust its use accordingly.
- Suspend air drops when wind speed exceeds 25 mph (40 kmph).

Figure 6.112 A ground crew waits to follow up an air drop. *Courtesy of NIFC.*

Water/Chemical Applications

Many types of chemical or water solutions are applied by aircraft. Some situations lend themselves to the use of water, others to the use of retardants or other agents. These agents are generally classified according to their effects on a fire: short-term or long-term.

SHORT-TERM AGENTS

Water and foam are considered short-term agents. Foam creates small bubbles that have a lower surface tension than water; thus, they penetrate and cover fuels better while resisting breakdown from wind and heat. Both water and foam provide a 5- to 15-minute knockdown, depending on the fuels. Because of this short knockdown period, ground crews must reach the drop area quickly to prevent a fire from flaring up again. When this level of coordination is impossible or impractical, the drops may be ineffective.

As discussed in Chapter 4, Water Supply, the foam used on wildland fires is known as Class A foam because it is intended to be used on Class A materials (ordinary combustibles such as grass and wood). These agents are mixtures of water and a foam concentrate. Foam is delivered in various expansion ratios, depending upon the application. The ratio of concentrate to water for aerial application is typically in the low-expansion range, from 0.1 to 1.0 percent.

The foams delivered by aircraft are all in the low-expansion range because they are short-term agents. These agents are usually applied as suppressants directly onto a fire or immediately in its path. To be most effective, these agents are applied to low-intensity fires in light fuels with little or no canopy. High-intensity fires may dehydrate the agent before it settles onto the fuel, and a canopy may prevent the agent from reaching burning surface fuels.

LONG-TERM AGENTS

Long-term retardants usually contain water and chemical salts such as diammonium phosphate (DAP) or ammonium sulfate. The heat from an approaching fire evaporates the water content and releases ammonia and carbon dioxide from the retardant. This process cools the uninvolved fuels and thereby retards pyrolysis, which slows the progress of the fire. These retardants also leave a salt residue on the surface of the fuels after the water evaporates. Depending on the fuel type and the level of coverage, this residue normally gives ground crews up to 1½ hours to reach treated areas before the agent loses its effectiveness and the fire flares up again. The sooner ground personnel back up air drops, the more likely they can achieve total extinguishment.

Smoke Jumpers/Helicopter Rappellers

Smoke jumpers/helicopter rappellers are firefighters specially trained for making an initial attack on wildland fires in remote areas. They also perform rescues and emergency medical care and sometimes construct helispots (temporary landing areas for helicopters) and paracargo drop zones. Smoke jumpers parachute from fixed-wing aircraft when they are over the incident scene (Figure 6.113). Rappellers reach the scene by rappelling from helicopters to the ground (Figure 6.114). Smoke jumpers and rappellers can often reach and extinguish fires while the fires are still small. These highly trained firefighters may also reinforce other units on large wildland fires.

Figure 6.113 A smoke jumper nears the ground. *Courtesy of NIFC.*

Figure 6.114 A rappeller descends from a helicopter. *Courtesy of NIFC.*

Safety In and Around Aircraft

Most air operations at wildland fires are relatively safe when common sense and established procedures are used. Personnel working in and around operating aircraft must know the applicable safety procedures and follow them without exception.

AIR BASE SAFETY

Observe the following safety procedures when working in and around fixed-wing aircraft:

- Do not smoke within 100 feet (30 m) of aircraft.
- Wear both eye and hearing protection around operating aircraft.
- Attempt to maintain eye contact with the pilot.
- Be alert for unexpected forward movement when propellers are spinning.
- Remain at least 15 feet (4.6 m) from the front of a spinning propeller.

- Avoid the area directly behind operating engines.
- Avoid working within the arc of a propeller until the engine has cooled at least 20 minutes.
- Avoid touching hot wheels, and do not attempt to cool them with water.
- Seek and wait for permission from the control tower before crossing runways.
- Remain alert for aircraft movement on taxiways.

HELIBASE/HELISPOT SAFETY

Observe the following safety procedures when working in or around helicopters:

- Keep landing zones clear of loose equipment, objects, and unauthorized personnel.
- Avoid shining lights in the direction of operating helicopters aloft or on the ground.
- Show wind direction to an approaching helicopter with flag, dust, or other visual indicators.
- Follow instructions of the helicopter crew at all times. The helicopter crew should provide detailed instructions on safety procedures to all personnel.
- Stay at least 100 feet (30 m) from an operating helicopter unless authorized to approach by the pilot or crew.
- Stay well clear of operating tail rotors.
- Approach or leave in the pilot's field of vision.
- Stoop when approaching or leaving an operating helicopter.
- Approach or leave an operating helicopter on the downhill side only.
- Wear both eye and hearing protection around operating helicopters.
- Carry (rather than wear) helmets that do not have chin straps to and from operating helicopters.
- Carry all tools horizontally, close to the side when approaching or leaving helicopters.
- Do not smoke within 100 feet (30 m) of a helicopter, fuel storage, or fueling equipment.
- Avoid standing directly beneath a hovering helicopter unless making a sling load hookup.
- Keep safety harness fastened when aboard a helicopter until directed to release it.
- Have an engine with Class B foam stand by whenever possible.

INCIDENT SAFETY

Firefighter safety associated with aircraft is also extremely important on the incident site — the fireline. Both fixed- and rotary-wing aircraft operating in the fire area produce turbulence that can cause erratic fire behavior. Turbulence from fixed-wing aircraft and rotor downwash from helicopters can cause significant localized changes in fire behavior (Figure 6.115).

Figure 6.115 Helicopters can create significant turbulence near the ground. *Courtesy of NIFC.*

DROP-ZONE SAFETY

The impact of agents dropped by aircraft can injure fireline personnel. If possible, personnel should leave the target area before a drop is made. However, fireline personnel should not attempt to run unless escape is assured. If they are unable to retreat to a safe place, the safest procedure is the following:

- Stay away from dead snags, tops, and limbs in the drop area.
- Hold hand tools away from the body.
- Lie face down with legs spread, helmet on, and head toward oncoming aircraft.
- Grasp something secure to prevent being carried or rolled about by the impact of the dropped agent.
- Walk carefully because of slippery surfaces in any area covered by wet retardant.

MOP-UP AND PATROL

Mop-up is the process of completing the extinguishment of a wildland fire after it has been controlled. It involves making sure that the fire along the edge is completely out and locating and extinguishing remaining hot spots (especially those close to the line) such as snags, logs, cow chips, and similar materials that may continue to smolder within the burned area (Figure 6.116). It also involves making sure that the control line is scraped completely down to mineral soil and is wide enough to keep any burning materials from being blown across it into unburned fuels (Figure 6.117). Complete extinguishment of all material burning within the black may be impractical in large fires; however, all smoldering material within 100 feet (30 m) of the control line should be extinguished. Mop-up must be thorough, because a small spark or flame left along the line could rekindle hours or days later, starting another and perhaps larger fire. If available, an infrared device may help locate hidden hot spots (Figure 6.118).

Figure 6.116 Hot spots may smolder long after the fire is controlled. *Courtesy of NIFC.*

Figure 6.117 A hand crew widens a control line. *Courtesy of NIFC.*

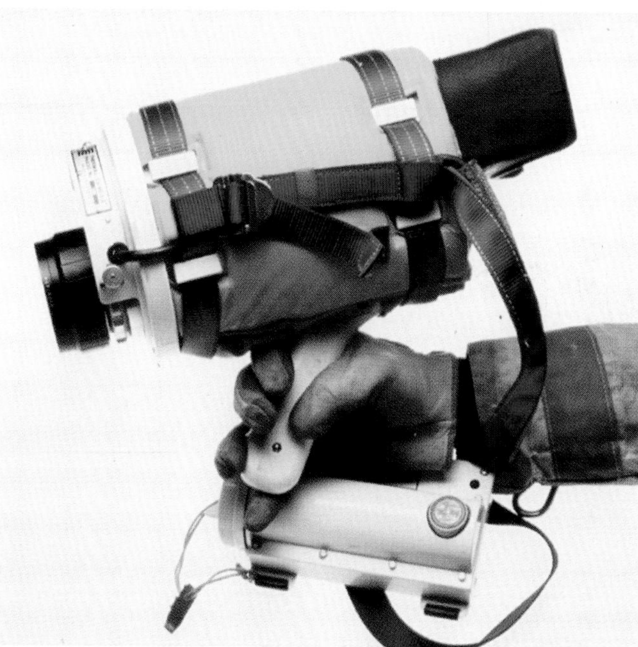

Figure 6.118 A typical infrared heat detector. *Courtesy of NIFC.*

WARNING

You must continue to wear full protective gear and remain alert for changes in fire behavior and weather conditions during mop-up.

Principles of Mop-Up

When assigned to mop-up after a wildland fire has been controlled, firefighters should be guided by the following principles:

- Start mop-up as soon as line construction and burnout are complete. Mop up the most threatening areas first.
- Allow fuel to be consumed if it will do so quickly and safely.
- Mop up entire area on small fires if practical.
- Mop up on large fires far enough into the black to be certain that no fire can blow, spot, or roll over the fireline under the worst possible conditions anticipated.
- Fell all snags that could cause spotting or fire spread across the line.
- Search for smoldering spot fires.
- Consider potential for problems from snags, punky (rotten) logs, and fuel concentrations outside the control line.

- Search for and dig out burning roots and stumps near the line.
- Cut unburned or partially burned brush near the line.
- Scatter concentrations of burning fuels to reduce heat and danger of spotting.
- Trench below, block, or turn heavy logs, stumps, or similar material so they cannot roll.
- Feel with back of hands without touching for possible smoldering spots close to the line (use care, go slow).

  CAUTION: When using your hands during mop-up, place the back of your bare hand close to but not in contact with the fuel to test for heat.
- Use water in conjunction with hand tools where possible or practical. In dry mop-up, stir and mix hot embers with dirt.
- Use water sparingly, but use enough to do the job. Match the amount of water to the job.
- Scrape or stir the fuel while applying water when moping up deep-burning fuels such as peat, duff, or needles.
- Add wetting agents or foam to water to increase its effectiveness, especially in deep-burning fuels.
- Cold trail (expand/augment fireline) where applicable (see Cold Trailing section).

The following paragraphs expand on some of the items in the foregoing list. In particular, they describe the usual methods of dealing with materials that are still burning and unburned fuels close to the fireline.

Burning logs too large to move. Work from the cool edges of the logs to avoid burning the feet. Remove hot coals and ashes from beneath logs. Cool logs with dirt or water (Figure 6.119). Cover hot ground with cool earth. Alternate use of water or dirt and scraping until the fire is out. If a chain saw and an operator are available, cut logs to separate burning and unburned sections.

Figure 6.119 A firefighter soaks a smoldering log.

Figure 6.120 Smoldering material should be thrown well away from the fireline.

Burning material near fireline. Scatter fuel well back from the line into the black and allow it to burn (Figure 6.120). Cool heavy burning material with dirt or water. Then complete mop-up of individual logs and stumps as soon as they are cool enough to handle.

Fuels outside of but adjacent to fireline. Remove and scatter unburned fuels a safe distance from the fireline. If removal is impractical, dig a trench and roll the material into it. Cover the trench with a layer of clean dirt (Figure 6.121). This is safer than simply attempting to cover the fuels on the ground with a mound of earth.

Burning snags. For safety, post a lookout to watch for and warn personnel of falling tops or limbs. If a snag is burning above the reach of firefighters on the ground, scrape away all hot materials from the base of the snag and cool the hot ground with water or cover it with cool dirt. If the

Figure 6.121 Smoldering logs may have to be covered with dirt.

Figure 6.122 Snags should be felled into the black with the slope.

lean of a snag allows, fell it into the black with the slope, not across it (Figure 6.122). If the lean does not allow a snag to fall into the black, scrape away all combustible fuels from the space where the snag will fall. When the snag is down, scrape all burning material from its surface and extinguish with water or dirt.

CAUTION: Beware of snags and trees that could fall and of material that could come loose and roll downhill and hit you during mop-up.

If a snag is burning at its base only, scrape away hot coals and ashes and cool the hot ground with water or cool dirt. Chop or scrape burning portions of the snag with an axe or shovel. Peel off loose bark as high as can be reached. Extinguish burning material, including sparks or embers in cracks, with water or dirt. Scatter burned material from the snag into the black.

Stumps/logs near fireline. Stumps or logs burning near the fireline must be extinguished with water or dirt. In addition, burning portions of any roots that extend under the fireline must be uncovered and removed (Figure 6.123). This is necessary to prevent fire from burning under the fireline and coming up in unburned areas. Dig a trench below any stumps or logs burning on steep

Figure 6.123 Burning roots must be uncovered and removed.

slopes (Figure 6.124). For safety, work uphill of logs that must be moved. If the logs are moveable, separate and move them to lie with the slope, not across it, and trench below them where necessary (Figure 6.125). Scrape off hot coals and extinguish them with water or dirt.

Figure 6.124 A firefighter trenches below a burning log.

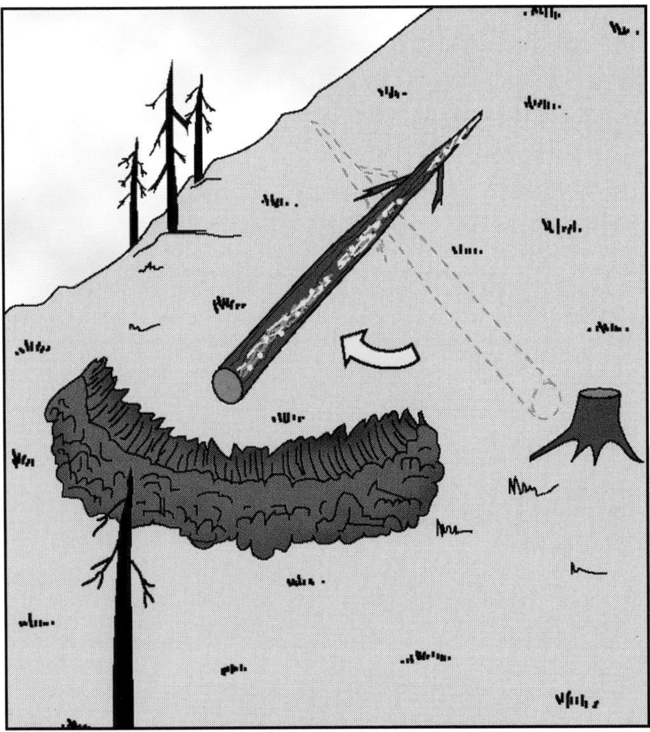
Figure 6.125 Logs should be moved to lie with the slope.

Burning chunks, limbs, and small logs. On a steep slope, dig a trench close to but downhill from the burning materials and place them in the trench to burn. Systematically work the entire fire area using water or dirt to cool burning material. Test limbs and other materials for heat by passing the back of an ungloved hand along their surfaces with the hand close to but not in contact with the material (Figure 6.126). If no heat is detected this way, pick up the material and feel with the hands. If the material is no longer burning, spread it in an area within the black that has been cleared of all burning or hot fuels. This is called *boneyarding.*

Figure 6.126 A firefighter checks a limb for heat.

Patrol

Patrol is an important phase of the mop-up operation. It is the part of mop-up that includes watching for spot fires and preventing hot spots from breaking over the control line. Each firefighter in a crew is usually assigned a certain section of line to watch during patrol. Within the assigned area, any smoldering material is extinguished or thrown into the black and allowed to burn (Figure 6.127). As many patrols as necessary are made until there is reasonable assurance that the fire edge is completely out. Depending upon the size and amount of fuel still smoldering, patrolling may last from a few minutes to a few weeks.

Figure 6.127 Smoldering material should be thrown farther into the black.

Cold Trailing

Cold trailing is a method of improving a control line by checking for hot spots and widening the line or cutting a new one if necessary. It involves carefully inspecting and feeling with the hand to detect any fire, digging out every hot spot, and trenching any hot items on a slope (Figure 6.128).

Figure 6.128 A firefighter checks a log for heat.

Restoring the Environment

Even when a wildland fire is completely extinguished, the firefighter's job may not be finished. Firefighters should attempt to prevent or reduce any additional damage that may result from the fire or fire fighting operations. Fences that were cut to provide access to the fire must be repaired to prevent livestock from wandering (Figure 6.129). Where it is necessary and practical to do so, any soil that was disturbed by mechanized equipment should be restored to prevent erosion (Figure 6.130). Openings in structures that were damaged by the fire should be covered to protect them from vandalism or from damage by the elements.

Figure 6.129 Barbed-wire fences that were cut must be repaired.

Figure 6.130 Dozer tracks can lead to soil erosion if not smoothed out.

WILDLAND/URBAN INTERFACE FIRE SUPPRESSION 7

LEARNING OBJECTIVES

This chapter provides information that addresses the following objectives of NFPA 1051, *Standard for Wildland Fire Fighter Professional Qualifications* (1995 edition):

Wildland Fire Fighter I

3-5.1 **Definition of Duty.** All activities to confine and extinguish a wildland fire, beginning with dispatch.

3-5.2 Assemble and prepare for response, given an assembly location, an assignment, incident location, mode of transportation, and the time requirements, so that arrival at the incident with the required personnel and equipment meets agency guidelines.

3-5.3 Detect potential hazardous situations, given a wildland fire and the standard safety policies and procedures of the agency, so that the hazard is promptly communicated to the supervisor and appropriate action can be taken.

3-5.3.1* *Prerequisite Knowledge:* Basic fireline safety, fire behavior, and suppression methods.

3-5.4 Construct a fireline, given a wildland fire, agency line construction standards, suppression hand tools, and equipment, so that the fireline conforms to the construction standard.

3-5.5 Secure the fireline, given a wildland fire and suppression tools and equipment, so that burning materials and unburned fuels that threaten the integrity of the fireline are located and abated.

3-5.5.1 *Prerequisite Knowledge:* Operational and safety considerations when burning out.

Wildland Fire Fighter II

4-5.1 **Definition of Duty.** All activities to confine and extinguish a wildland fire beginning with dispatch.

4-5.2 Select fireline construction methods, given a wildland fire and line construction standards, so that the technique used is appropriate to the conditions and meets agency standards.

4-5.2.1 *Prerequisite Knowledge:* Resource capabilities and limitations, fireline construction methods, and agency standards.

Chapter 7
Wildland/Urban Interface Fire Suppression

INTRODUCTION

A growing number of fire departments face the problem of protecting the wildland/urban interface. Scenic beauty and changing lifestyles have motivated people to purchase and develop homesites and businesses in once pristine areas. People and their homes and workplaces worsen the wildland fire problem in a number of ways. Fire protection planners must be concerned with the increased life hazards the larger populations in these areas create. The expanded human activity multiplies both the ways in which fires can start and the total number of fires in these areas. Structures in the wildland greatly raise the dollar value of property to be protected (Figure 7.1).

Patterns and types of development also influence the fire environment and the requirements of effective protection. The sometimes limited ability and motivation of local governments to provide the needed infrastructure of roads, water, and fire stations leave many developments with less than adequate fire protection. Minimal planning and building-code enforcement often result in a random pattern of structure types in heavy vegetation and on terrain that often provide breathtaking scenic vistas but also make the structures vulnerable to fire (Figure 7.2). When these factors are combined with minimal fire fighting resources, a single fire can quickly destroy literally hundreds of homes.

Because of increased human habitation in the wildland/urban interface and the resulting growth in the values at risk and the number of fires, departments that never expected to function at wildland fires have been asked to respond to mutual aid calls in the interface. Therefore, it is critically important for all firefighters — from those in single-station volunteer companies to those in large metropolitan departments with hundreds of stations — to be able to perform safely and effectively in these fires.

This chapter reviews the elements of size-up as they apply to the wildland/urban interface, and it discusses developing an incident action plan based

Figure 7.1 Many homes in the wildland/urban interface are very expensive.

Figure 7.2 Homes in such situations are difficult to protect from fire. *Courtesy of John Hawkins.*

on that size-up. Other topics covered are the report on conditions, techniques for organizing and assigning resources, defending structures in the wildland, offensive/defensive operations, traffic management, communications, and interface safety considerations.

SIZE-UP

As discussed in Chapter 6, Fire Suppression Methods, size-up is an ongoing process of evaluating current and expected fire conditions that starts before the call comes from the dispatcher/telecommunicator and continues throughout the incident. The same is also true of size-up in the wildland/urban interface. Some of the most important information gathered during the pre-incident size-up is a knowledge of the area and the expected fire behavior.

Utilizing Area Knowledge

A pre-incident size-up can be especially important in the wildland/urban interface. Because of the greater life hazard and the relatively high values of the properties at risk, planning how to best protect a particular development is very important. Planning how to protect residents and their properties involves knowing the topography, where development is concentrated, where the heaviest fuels are located, where water sources are located, the best/worst access routes, and the local weather normal for the season. The best way to gather this data is by touring the area, making firsthand observations and talking to local residents (Figure 7.3). This survey, sometimes referred to as "pre-incident triage," helps firefighters make a preliminary determination of which houses have the most/least potential to be saved (see Property Conservation section later in this chapter).

Figure 7.3 Firefighters should talk to local residents to learn more about the area. *Courtesy of NIFC.*

NOTE: During actual fire incidents, it is often helpful to use local firefighters or others (sometimes called *bird dogs*) to assist mutual aid resources with directions and information about local conditions.

Another important source of pre-incident size-up information is the fire history of the area as reflected in reports, maps, and records of past fires and fire seasons (Figure 7.4). One of the most important factors in the pre-incident size-up process is the consideration of the expected fire behavior.

Figure 7.4 Fire department reports and records may reveal valuable information about the area's fire history.

Expected Fire Behavior

Based on the data gathered while studying the area, its weather patterns, and its fire history, one or more models of expected fire behavior should be developed. Models based on a number of possible weather patterns and other variables can be developed using very sophisticated computer-based fire modeling programs or by manually plotting various fire-spread possibilities on maps of the area

(Figure 7.5). The data on which a fire behavior model is based can come from one or more of the following sources: personal experience and observations, reading the local paper or watching the local weather channel, and monitoring fire-weather broadcasts that include burning indexes and components. Regardless of how these data are gathered or what fire behavior model is used, the resulting pre-incident plans(s) should be written and include appropriate maps (Figure 7.6). For information on specific fire behavior models, see Chapter 10, Fire Protection Planning.

Figure 7.5 A firefighter uses a computer-based fire behavior model.

Figure 7.6 A firefighter compares the map to the area. *Courtesy of NIFC.*

Size-Up on Arrival

Just as in any other type of fire, proper size-up on arrival at a fire in the wildland/urban interface is critical to saving lives and minimizing loss and damage. As described in Chapter 6, Fire Suppression Methods, life safety is the first and most important consideration in the size-up at any fire. In addition to the elements of size-up described in Chapter 6, the size-up on arrival at interface fires must include the following considerations:

- ***Distance from the fire*** — Knowing how far away a fire front is and how fast it is moving helps approximate the time firefighters have to prepare (Figure 7.7). There may be time to evacuate occupants, move combustibles away from structures, deploy hoselines around them, and/or burn out around them.

Figure 7.7 A firefighter checks the progress of an approaching fire.

- ***Construction features*** — Combustible roofs, eaves, siding, and decks are vulnerable to ignition from burning embers, radiated heat, and direct flame impingement (Figure 7.8). If there is time and if Class A foam or gelling agents are available, these exposed parts of structures can be pretreated.

Figure 7.8 Given the setting and the construction materials, this home is vulnerable to fire. *Courtesy of John Hawkins.*

- ***Terrain*** — Structures on steep slopes or in saddles or chimneys where fire spread will be fastest are the most difficult to defend (Figure 7.9). They may sometimes have to be left undefended in order to maximize the use of available resources on structures more likely to be saved.

Figure 7.9 This home would be very difficult to defend.

- ***Spacing between buildings*** — Structures that are close together are potential exposure threats to each other (Figure 7.10). In general, buildings separated by a distance of less than 50 feet (15 m) are mutually exposed. The larger the structure, however, the greater the exposure potential, so large buildings may be considered exposures even with greater separation.
- ***Accessibility*** — Structures that are accessible to fire apparatus from only one direction are at much greater risk than those that can be reached several ways. Long, narrow driveways or dead-end roads make it difficult for fire apparatus to reach structures because safety dictates that the apparatus must be backed in from the last known turnaround to allow rapid escape if necessary (Figure 7.11).
- ***Power lines*** — Overhead power lines close to structures may fall on the structures or the firefighters or both. Fire apparatus should never be parked under power lines in fire situations (Figure 7.12).

Figure 7.10 Homes situated close to each other increase the exposure risk. *Courtesy of Tony Bacon.*

Figure 7.11 Engines should be backed down narrow driveways.

Figure 7.12 Apparatus should not be parked under overhead power lines.

- ***Wind*** — The direction and speed of the prevailing wind may seriously alter the situation (Figure 7.13). Obviously, if a strong wind is blowing a fire toward the structures, the increased rate of spread will allow less time to prepare.

The more items that are in the firefighters' favor, the better the chances of stopping a fire's spread and minimizing loss. These factors are also major life-safety considerations.

Figure 7.13 A wind-driven fire threatens a structure. *Courtesy of California Office of Emergency Services.*

Report on Conditions

One of the most important parts of the initial size-up on arrival is the report on conditions. This information is transmitted to other incoming units and the communications center, and it is the basis for many of the decisions that set the tone for the rest of the operation. The report should be organized and transmitted in a way that describes the essential elements of the situation in the most concise and meaningful terms possible. While there may be minor differences in the information needed on different incidents, in general the following should be included in the report on conditions:

- Reporting unit and location (by address or landmarks)
- Size of the fire (acres [hectares] involved or length of fire front)
- Type of fuel involved (light, medium, or heavy)
- Rate and direction of spread (toward or away from structures)
- Fire behavior (crowning, surface, etc.)
- Best access (road name, property name, etc.)
- Life safety (evacuation needed, etc.)
- Exposures threatened
- Hazards (power lines down, bridges that might not support engines, etc.)
- Additional resources needed

It also helps other responding units to hear the overall strategic objectives for the incident as part of the report on conditions. This creates unity of objective among the responding resources. This statement could be as simple as "*we will try to confine the fire between Jones Road, Smith Ridge, and Interstate 80.*" The IC should also consider updating the other responding units regarding weather and additional resources available (aircraft, hand crews, etc.).

Once units are at scene and operating, they should give an initial update on their progress. In particular, they should report any problems impeding their progress so that more resources can be ordered or the action plan adjusted if necessary. Their initial update should be followed by periodic updates as needed to keep the IC informed.

ACTION PLAN

Every incident requires its own action plan. All incident action plans (IAPs) have certain common elements, but because the factors they must address vary in importance from one fire to the next, every IAP is unique. On small, relatively routine incidents, the plan need not be in writing, but there ***must*** be a plan. It is important for the first-in officer to avoid the *candle / moth syndrome* — being drawn to the first flame seen, while ignoring the rest of the situation. The first company on the scene may be more productive by assuming command and developing a plan rather than attacking the fire. On larger, more complex incidents, the plan should be in writing. Since every interface fire poses the possibility of subsequent litigation, a written record of the incident becomes even more important.

While individual plans vary, they all have the same overall priorities:

- Life safety
- Incident stabilization
- Property conservation

These three priorities guide the development of the incident action plan. Deciding how to meet them dictates the strategic objectives for the incident. Examples of some common strategic objectives are as follows:

- Provide for the safety of firefighters, residents, and others.
- Contain the fire to a specified geographic area.
- Save structures exposed to the fire.
- Extinguish the fire.
- Determine fire cause/origin.

Life Safety

Life safety is always the first and highest priority in any incident action plan. As always, life safety includes the lives of firefighters as well as the civilian population. Protecting firefighters can take many forms — from properly sizing-up a fire and correctly estimating its potential, to providing sufficient resources to safely implement the plan, and to making sure that all personnel wear the safety gear that they have been issued. Protecting residents of the wildland/urban interface can also take a variety of forms. It can range from eliminating the threat by quickly extinguishing the fire, to locating the residents and informing them of the situation, to assisting in their evacuation to a place of safety, and to coaching them on how to protect themselves if they decide to stay with their properties.

The following sections focus on the safety of the residents of the wildland/urban interface; firefighter safety is covered in Chapter 8, Firefighter Safety and Survival. The three most common ways of protecting the lives of those in the path of an advancing wildland fire are *evacuation, shelter in place,* and removal to a *safe refuge*.

EVACUATION

Requesting that law enforcement authorities remove civilians from the path of an advancing wildland fire is one option available to the IC. The decision to evacuate should be based on a size-up of the current and expected fire behavior, the number of people to be evacuated, the resources required, the adequacy of evacuation routes, and the resources available. If a large number of people have to be displaced, one or more relocation centers should be established as close as possible to, but outside of the threatened area. Schools, churches, and other large public buildings are often used for this purpose because they provide shelter, sanitary facilities, and sometimes kitchens (Figure 7.14). In some areas, the American Red Cross (ARC) may allow the fire department to authorize temporarily housing displaced persons at ARC expense. Check with your local ARC chapter.

Figure 7.14 Large churches and other public buildings can serve as relocation centers.

In some evacuations, all that is necessary is to notify residents of the need to evacuate and they can do the rest. Law enforcement personnel can, and usually do, notify residents, so there is little if any direct impact on fire fighting resources (Figure 7.15). In other situations, firefighters must make the notifications and, in some cases, provide needed transportation. Evacuating residents who are bedridden or attached to various life-support devices may require an extraordinary commitment of resources.

Evacuation may be desirable to clear the area for fire fighting operations and to minimize the risk to citizens. Anyone who will not be directly involved in defending their homes should be asked to leave. In most jurisdictions, however, firefighters can merely ask people to evacuate; only law en-

Figure 7.15 Evacuation notification should be done by law enforcement personnel.

forcement officers have the power to make them leave. Little is gained by arguing with someone who will not leave. The IC should be advised of the locations of those not evacuating.

Evacuees should be advised to take a minimum of belongings with them and to close but not lock their residences. Locking the residence makes it difficult for firefighters to quickly enter to save the house or take refuge. Evacuees should be advised of the appropriate route to their gathering place and to watch out for incoming fire equipment. The evacuating population will also increase the amount of traffic on these sometimes narrow country roads.

Under the best of circumstances, large-scale evacuations take time. An automated notification system can save time, as can a standard marking system to indicate that the occupants of a structure have been notified (Figure 7.16). This keeps other personnel from duplicating effort by searching empty residences or talking to residents who have decided not to leave.

Figure 7.16 The tape indicates that these residents have been notified to evacuate.

Some residents will be ready to leave at the first sign of smoke, others will stay with their homes until the fire overruns them. Those who stay may be helpful to firefighters if they know the locations of other structures, water sources, access routes, hazards, etc. Residents can help prepare their homes before the fire arrives and may be able to assist firefighters when it does. Residents who remain should be advised on water conservation and basic safety considerations. They should also be advised to stay out of the way of fire equipment, stay out of unburned fuel, know their escape routes and safe zones, remain calm, and follow instructions from emergency response personnel.

SHELTER IN PLACE

When residents choose to remain with their homes, the firefighters' responsibility is to provide advice and direction (Figure 7.17). Firefighters should advise residents on how to help the most and interfere the least and how to do it safely. This tactic is sometimes called *shelter in place*.

In this tactic, residents do what they can to protect their properties before the flame front

Figure 7.17 Firefighters should advise residents about defending their homes.

arrives and then stay in the structures until the fire burns past. In most cases, structures can provide shelter to occupants while the flame front passes, even if a structure catches fire and eventually burns to the ground. Once the flame front has passed, occupants can get out of the structures and attempt to extinguish the fire, leave the area, or take whatever other action they feel is required and within their means.

Those residents who decide to shelter in place rather than evacuate should be advised to do the following:

- Wear boots, jeans, long-sleeved cotton or wool shirts, hats, and sturdy gloves — not shorts, tank tops, and flip-flops (Figure 7.18).
- Increase the water available by connecting garden hoses to faucets and filling wading pools, hot tubs, or other large containers (Figure 7.19).
- Fuel, test, and make auxiliary or floating pumps ready for use.
- Turn off gas to the structure at the gas meter or liquefied petroleum gas (LPG) tank, but leave electricity on (Figure 7.20).
- Move combustible materials such as wooden patio furniture or firewood away from the structure (Figure 7.21).

Figure 7.18 A properly dressed resident prepares to defend her home.

Figure 7.19 Residents should fill every available large container with water.

Figure 7.20 Residents should turn off the gas supply.

Figure 7.21 Firewood and other combustibles should be moved away from the home.

- Place a ladder for access to the roof (Figure 7.22). The most likely source of ignition will be burning embers landing on a combustible roof or deck.
- Back vehicles into the garage and close the garage door (Figure 7.23). Disconnect the garage-door-opener mechanism because it will not work if power fails (Figure 7.24). Place valuables, prescription medicines, and irreplaceable photographs in a vehicle, and roll up all vehicle windows.
- Close or cover all windows, doors (including interior doors), and other openings such as attic vents with heavy material such as plywood. Close heavy drapes, but remove lightweight combustible curtains from windows.
- Leave lights on so structures are easier to find in the dark (Figure 7.25).

Residents should be instructed to use garden hoses or buckets to get water to roofs and/or to

Figure 7.22 A ladder should be placed against the eaves of the roof.

Figure 7.23 Vehicles should be backed into the garage.

Figure 7.24 Automatic garage door openers should be disconnected.

Figure 7.25 The porch light should be left on.

wooden decks to extinguish any embers that land there. They should not waste water by wetting down structures beforehand because the water usually evaporates before it does any good (Figure 7.26). Putting lawn sprinklers on roofs rarely helps, and may waste limited water supplies. Residents should also be cautioned not to attempt to burn out around structures or attempt to extinguish the flame front when it arrives.

Figure 7.26 Wetting down the roof before the fire arrives wastes water.

SAFE REFUGE

The concept of *safe refuge,* which is very similar to sheltering in place, involves placing residents outside structures but somewhere that affords them protection from the flame front — by definition, an area where they will not suffer direct flame impingement. If a structure is indefensible and evacuation is impractical, moving residents into a safe refuge may be the best option. An example of a safe refuge would be inside a vehicle (with the windows rolled up and vents closed) parked in a clearing devoid of vegetation with no overhead power lines. A large, barren parking area, golf course, school athletic field, or a wide road might also provide safe refuge.

Incident Stabilization

Incident stabilization generally involves containing or controlling a fire so that the area involved is no longer increasing and structures in the interface are no longer threatened. To achieve this level of stabilization, firefighters must determine the number and types of resources needed and use one or both of two operational modes — offensive and defensive.

OPERATIONAL MODE

Operational mode (offensive/defensive/combination) refers to the type of attack to be made on a fire. The operational mode is determined by the types and number of resources available and the size and behavior of the fire. Tactical objectives can be developed using either operational mode or a combination of both.

- If ample resources are readily available, an offensive mode utilizing direct or indirect attack may accomplish all established strategic goals.
- If resources are limited, a defensive mode may minimize losses and accomplish some high-priority strategic goals until enough resources arrive to control the incident.
- A combination of offensive and defensive modes may be used.

The offensive mode is directed toward attacking and containing a fire. It requires that sufficient numbers of resources be available soon enough to arrest the progress of a fire and extinguish it. On the other hand, the defensive mode attempts to defend exposures where possible and to continue containment efforts until additional resources are available or the fire burns itself out. The defensive mode may be used when the number of resources available is insufficient to contain or control a fire. It is not uncommon to use a combination of these two modes at various times and in various locations on a major wildland fire.

RESOURCES NEEDED

Estimating the number and types of resources needed to stabilize an incident is difficult, and the ability to do so usually comes only with training and experience. If the IC is unsure about the number and/or types of resources needed, too many should be ordered rather than too few. In general, the resources needed directly reflect the strategic goals and tactical objectives of the IAP. These goals and objectives are broken into specific tasks that are assigned according to priority and resource capability.

To determine the specific resources needed, the IC must consider a number of variables. The most common are as follows:

- ***Populations at risk (firefighters/residents)***
  - — Is evacuation needed?
  - — Do fire fighting resources need to assist in evacuation?
  - — Is livestock control needed?
  - — Is law enforcement available?
- ***Structures and improvements***
- ***Fuels***
- ***Weather***
- ***Topography***
- ***Fire behavior***
- ***Response times***
  - — How soon can additional resources be expected?
  - — How much will the fire spread before additional resources arrive?
- ***Kinds and types of resources needed***
  - — Will hand crews be effective on this fire?
  - — Is air support needed, and will it be effective?
  - — Are bulldozers or tractor-plows needed, and will they be effective?
  - — Are engine strike teams or task forces needed, or are single resources enough (Figure 7.27)? (See Strike Teams and Task Forces sections.)
  - — Are off-road engines needed (Figure 7.28)?

One of the most important decisions the IC makes is whether the fire fighting resources at scene or en route are sufficient to complete the assignment. Whether the assignment is rescue, fire control, structure protection, or evacuation, if

Figure 7.27 A Type-1 strike team also includes 21 personnel.

more resources are needed, they must be ordered as soon as possible (Figure 7.29). While agency protocols must always be followed, the first-in officer should have the authority to order whatever is needed. Additional resources should generally be ordered in groups organized as strike teams or task forces (discussed later in this chapter; see Standard Techniques section). Other resources such as air tankers, helicopters, water tenders, or ambulances may be ordered individually.

Figure 7.28 A typical Type-3 strike team. *Courtesy of NIFC.*

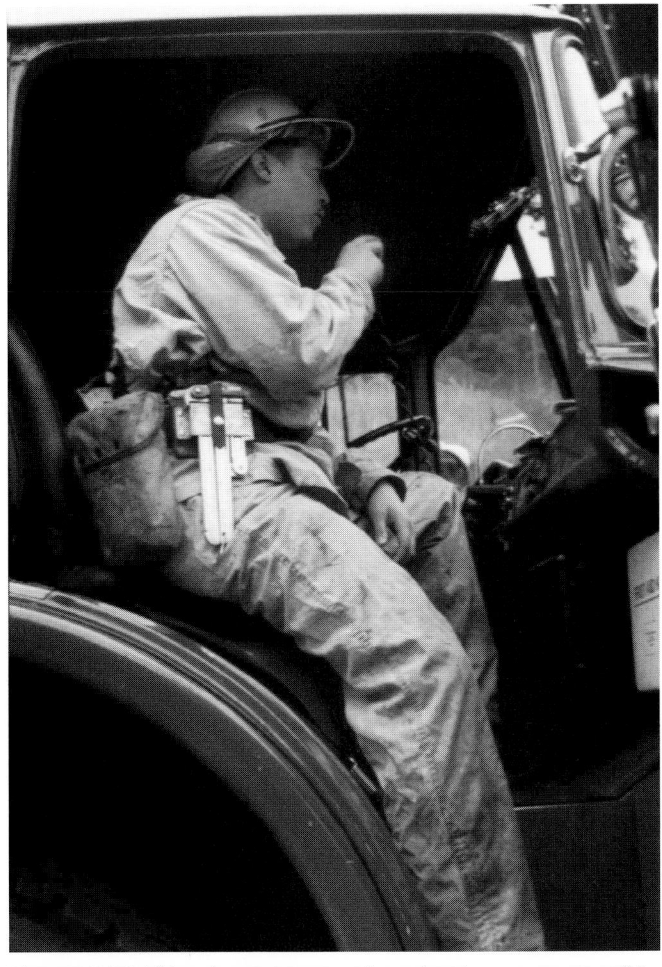

Figure 7.29 Additional resources must be ordered as soon as possible.

Experience over many years and in countless interface fires has produced the following general guidelines to use when estimating minimum resource needs for structure protection:

- For widely separated structures surrounded by wildland fuels, order one engine per structure.
- For a row of structures less than 50 feet (15 m) apart, order one engine for every two structures.
- To calculate the number of water tenders needed, use estimated water consumption, tank capacity, and the travel time to and from the water source.

Property Conservation

One major element of the action plan addresses property conservation. Property conservation focuses on protecting the property at risk, rather than on suppressing the fire. To decide how best to protect the structures in the path of a fire, it is necessary to identify and classify them. This process is commonly called *structural triage.*

STRUCTURAL TRIAGE

The French word *triage* means "sorting" or "prioritizing." *Structural triage* is the process of categorizing threatened structures according to their defensibility. Its goal is to determine how to do the most good with the resources and time available. Structural triage may be done at two different times:

- Before the fire front reaches the structures (perhaps weeks or months before)
- When the structures are immediately threatened or already involved in fire (Figure 7.30)

In both cases, each structure or group of structures is triaged (sized-up) in terms of its susceptibility to ignition and the resources needed to defend it. As part of pre-incident planning, triage provides information on which to base decisions about possible future resource needs. If triage is done during a fire but before the arrival of the fire front, there may be time for measures to reduce a structure's susceptibility to ignition. If firefighters and/or residents can take some of the

Figure 7.30 Structural triage would classify this structure as indefensible.

steps described earlier in this chapter, an indefensible structure can sometimes be made defensible. If the flame front reaches the structures before triage can be done, the system of classification is slightly different. Firefighters must decide which structures to defend and which to let burn. As difficult as these decisions are, under these conditions firefighters must be coldly analytical and must resist the urge to attempt the impossible. Trying to save indefensible structures may spread the available resources too thinly, resulting in the loss of those that could have been defended. Assuming that there are not enough resources at scene to reasonably defend all structures, the priorities used in this very difficult decision-making process are as follows:

- If both a defensible structure and an indefensible structure are threatened but not yet involved, protect the defensible one.
- If two structures are both burning, attack the fire in the structure with the least involvement.

WARNING

If fire threatens your escape route, abandon all structures and GET OUT!

When structural triage is done during an incident, it helps firefighters to classify threatened structures as follows:

- Needing little or no attention (at least for now)
- Defensible, but needing protection
- Indefensible (hopeless)

Theoretically, any structure can be defended if enough resources are available; however, since the quantity of fire fighting resources is usually limited, the following variables must be considered:

- Firefighter safety
- Structural characteristics
- Fuels
- Fire behavior
- Available resources

Firefighter safety. Since protecting the lives of area residents was discussed earlier in this chapter, this section focuses on the safety and survival of the firefighters charged with defending the structures in the interface. The most common considerations are as follows:

- ***Ingress/egress routes***
  - — Two ways out are safer than one way out.
  - — Heavy adjacent fuels are dangerous.
  - — Canopy fuels threaten safety.
  - — Steep grades can slow egress.
- ***Power lines***
  - — Roads may be blocked by downed power lines.
  - — Electrocution is possible.
- ***Smoke***
  - — Drivers' vision can be impaired.
  - — Ability to observe fire behavior can be reduced.
- ***Hazardous materials***
  - — Smoke may be toxic.
  - — Other airborne contaminants may be present.
  - — Runoff water may be contaminated.

- ***LPG and aboveground fuel storage***
  - — Boiling liquid expanding vapor explosion (BLEVE) is possible.
  - — Engines should not be parked near fuel storage sources.

Structural characteristics. The structure's susceptibility to ignition is another important variable when a wildland fire is approaching. The following structural characteristics should be evaluated:

- ***Roof***
  - — Is it combustible (wooden shakes/shingles, tar paper, etc.)?
  - — Is it noncombustible (tile, metal, fiberglass, etc.)?
  - — Is the pitch safe to operate on?
  - — Do gutters hold flammable debris?
- ***Siding***
  - — *Combustible (wood)* — Wood siding ignites easily when exposed to direct flame impingement.
  - — *Noncombustible (metal, brick, etc.)* — Aluminum siding melts away quickly, exposing flammable materials underneath. Brick, stone, stucco, or corrugated steel siding are much more fire resistant.
- ***Windows***
  - — Double pane windows are better than single pane.
  - — Large windows allow in more radiant heat.
- ***Heat traps***
  - — Open gables may allow fire into attic.
  - — Vents without fire-resistant screens may allow fire into interior spaces.
  - — *Overhanging decks* — Wooden decks are susceptible to ignition by embers from above, and those with exposed wooden underpinnings are susceptible to ignition by direct flame impingement from below (Figure 7.31). Because of their physical arrangement, wooden-deck supports burn vigorously. Decks that have steel or masonry supports or enclosed wooden underpinnings are much less susceptible to ignition from below, but spaces beneath decks are often used for covered storage of flammable materials.

Figure 7.31 This wooden deck is vulnerable to ignition from above and below. *Courtesy of NIFC.*

- ***Size of structures***
  - — Large structures require more resources.
  - — Large structures have greater heat-trap potential.
- ***Position on slope*** — Structures on steep slopes or in chimneys or saddles are difficult if not impossible to defend because they are in areas where fires will be most intense (Figure 7.32). Those on the flat are much easier to defend.

Figure 7.32 Homes in saddles or on steep slopes are difficult to defend. *Courtesy of California Office of Emergency Services.*

Fuels. Another very important variable in the wildland/urban interface is the fuel surrounding and immediately adjacent to the structures.

- ***Surrounding fuels*** — Vegetation within 30 to 100 feet (9 m to 30 m) of a structure increases the chances of fire spreading to the structure (Figure 7.33).
- ***Size and arrangement***
  - — What is the Btu output potential?
  - — Interlocking crowns spread fire quicker.
- ***Age*** — Older fuels tend to have more dead woody material.
- ***Proximity*** — What is the radiant heat spread potential?
- ***Loading*** — How much fuel is actually available to burn?
- ***Types***
  - — Are they resistant or flammable?
  - — Are they landscape/ornamental shrubs?
  - — Is it grass, brush, timber, or exotic (palmetto, etc.)?
  - — Are there wood piles?
- ***Improvements***
  - — Is there wooden fencing?
  - — Are railroad ties used?
  - — Is there a tool shed?
- ***Defensible space access***
  - — Is it accessible by firefighters?
  - — Is it accessible by fire apparatus?

Figure 7.33 Houses surrounded by dense vegetation and dry grass are vulnerable to fire. *Courtesy of NIFC.*

- ***Yard accumulations***
  - — Are they flammable?
  - — Are they composted?
- ***Flame or heat duration***
  - — Heat from light fuel passes quickly.
  - — Heat from heavy fuel lingers.
- ***Explosives***
  - — Are there LPG tanks?
  - — Are there aboveground fuel tanks?

Fire behavior. Some of the most important variables in the fire situation are the following elements of the current and expected fire behavior:

- Rate and direction of spread
- Topographic influence
- Weather influence
- Flame length
- Spotting/fire embers
- Time of day
- Time of year (season)
- Natural or other barriers

Available resources. Having sufficient resources available can mean the difference between success and failure in protecting structures threatened by an approaching wildland fire. Important resource considerations are as follows:

- ***On-site private resources*** — Water, pumps, tools, equipment, ladders, etc.
- ***Kind, type, and amount of fire department equipment*** — Engines, bulldozers, tractor-plows, aircraft
  - — From home agency?
  - — By mutual aid?
- ***Fire department personnel resources available***
  - — From home agency?
  - — By mutual aid?
- ***Where resources are***
  - — Dispatched
  - — En route
  - — In staging

- ***When resources will be available*** — Response time
- ***Capabilities and limitations***
  - — Equipment's mobility
  - — Water/foam aboard
  - — Retardants aboard
  - — Pumping capacity
  - — Reliability
  - — Communications

TYPES OF STRUCTURES

All types of structures in the wildland/urban interface should be triaged before and/or during a fire. The most common types are farm/ranch structures, commercial structures, and residential structures. When fire fighting resources are limited, as they often are, the triage priorities are as follows:

- Life safety (including pets and livestock)
- Residences
- Commercial or government buildings
- Unoccupied outbuildings

Farm/ranch structures. Farm/ranch houses, barns, and other outbuildings are typically wood-frame structures with wooden siding and either wood shingle or composition roofing (Figure 7.34). While a growing number of these structures are being made of corrugated metal, many of the traditional wooden buildings remain. These traditional buildings are made of wood that tends to be old and dry. Consequently, they ignite easily and burn vigorously. They also are often many miles (kilometers) from the nearest fire station. Their vulnerability is compounded because if they do catch fire, they may go unnoticed for some time, so the fire department's response may be delayed.

Figure 7.34 A typical wooden farm structure.

Commercial structures. For many of the same reasons that people build homes in the wildland/urban interface, many businesses also locate office buildings and other business structures there (Figure 7.35). Sawmills in rural areas are often surrounded by huge log decks and large inventories of cut lumber (Figure 7.36).

Figure 7.35 Many office buildings are located in the wildland/urban interface.

Figure 7.36 Sawmills concentrate massive amounts of combustible materials. *Courtesy of Kirk Hale.*

Residential structures. There are many isolated residences, cabins, hunting lodges, and other residential/recreational structures in wildland areas (Figure 7.37). The number of residential structures in the wildland has increased enormously in recent years, and many of these homes are very expensive (Figure 7.38). Regardless of age or market value, many homes (both old and new) in the wildland would be untenable if they were in the path of an advancing wildland fire. Structural triage would classify many of them as indefensible.

A growing number of mobile homes and manufactured homes are also located in wildland areas (Figure 7.39). These structures are manufactured of lightweight materials that make them susceptible to early ignition and intense burning. Unless they have an adequate water supply and are located in a large cleared area, many of these structures are indefensible.

Figure 7.37 A typical wilderness cabin. *Courtesy of NIFC.*

Figure 7.38 Many homes in the wildland are extremely expensive. *Courtesy of NIFC.*

Figure 7.39 Mobile homes and manufactured homes are common in the wildland. *Courtesy of John Hawkins.*

Natural Resource Conservation

Protecting natural resources must also be considered in developing the incident action plan. In many cases, the highest values at risk in the wildland/urban interface are the structures, but the commercial value of timber, crops, and pasture must also be considered (Figure 7.40). Nor should the value of undeveloped scenic areas such as watershed and wildlife habitat be overlooked (Figure 7.41). These resources contribute significantly to our economic well-being and quality of life. Because of their intrinsic value, these resources are considered part of the overall property conservation element of the action plan and may well be more valuable than the structure(s) at risk. For example, the commercial value of a single mature redwood tree exceeds the market value of any mobile home.

Figure 7.40 A harvest-ready wheat crop is extremely valuable.

Figure 7.41 The scenic beauty of wilderness areas is difficult to calculate.

During a fire, protecting natural resources by immediately attacking the flanks and/or head of a fire almost always also reduces the potential for loss of structures. First alarms should always include adequate resources to attack the flanks and protect life and property simultaneously.

STANDARD TECHNIQUES

Standard techniques are used to implement the action plan. They have been developed over many years of fireground experience by many different fire fighting agencies. When all units use the same techniques in the same ways, each unit understands what the others are doing. That promotes safety and effectiveness and makes outcomes more predictable.

Organizing

Standard techniques are especially important in organizing individual fire fighting resources into more effective units. Most fire fighting agencies now operate under ICS or a similar command/management system as described in Chapter 5, Initial Fireground Command. To maximize resource effectiveness and to help maintain span of control, ICS organizes basic resources into two clearly defined units: *strike teams* and *task forces*.

Both strike teams and task forces are assigned specific missions, such as structure protection or hot spotting, during a major incident. The difference is that once its assignment is finished, a task force is disbanded, while a strike team remains together as a unit throughout the incident, available to be reassigned or demobilized. While individual units within a strike team may be temporarily assigned to different locations, the assignments must be relatively close together because strike teams are intended to stay together and function as teams. Individual units should not be so widely dispersed that they are beyond the effective control of the strike team leader.

STRIKE TEAMS

A *strike team* consists of specified combinations of similar resources with common communications and a leader. Strike teams are most often composed of engines, hand crews, or bulldozers; however, they may be composed of any like resources (Figure 7.42).

Strike team leaders must have their own vehicles and must be able to communicate with their immediate supervisor and all units within their strike team. This may necessitate strike team leaders having more than one radio. If they don't have multichannel radios, they may have one radio on the incident command channel and another radio on a tactical channel. If staffing levels permit, each strike team leader should have an assistant to operate radios and help with paperwork, act as a safety officer, and train to become a strike team leader (Figure 7.43).

TASK FORCES

A *task force* consists of a group of resources with common communications and a leader temporarily assembled for a specific mission. Unlike strike teams, task forces may include a variety of individual resources. For example, a task force might consist of a bulldozer, a bulldozer tender, a couple

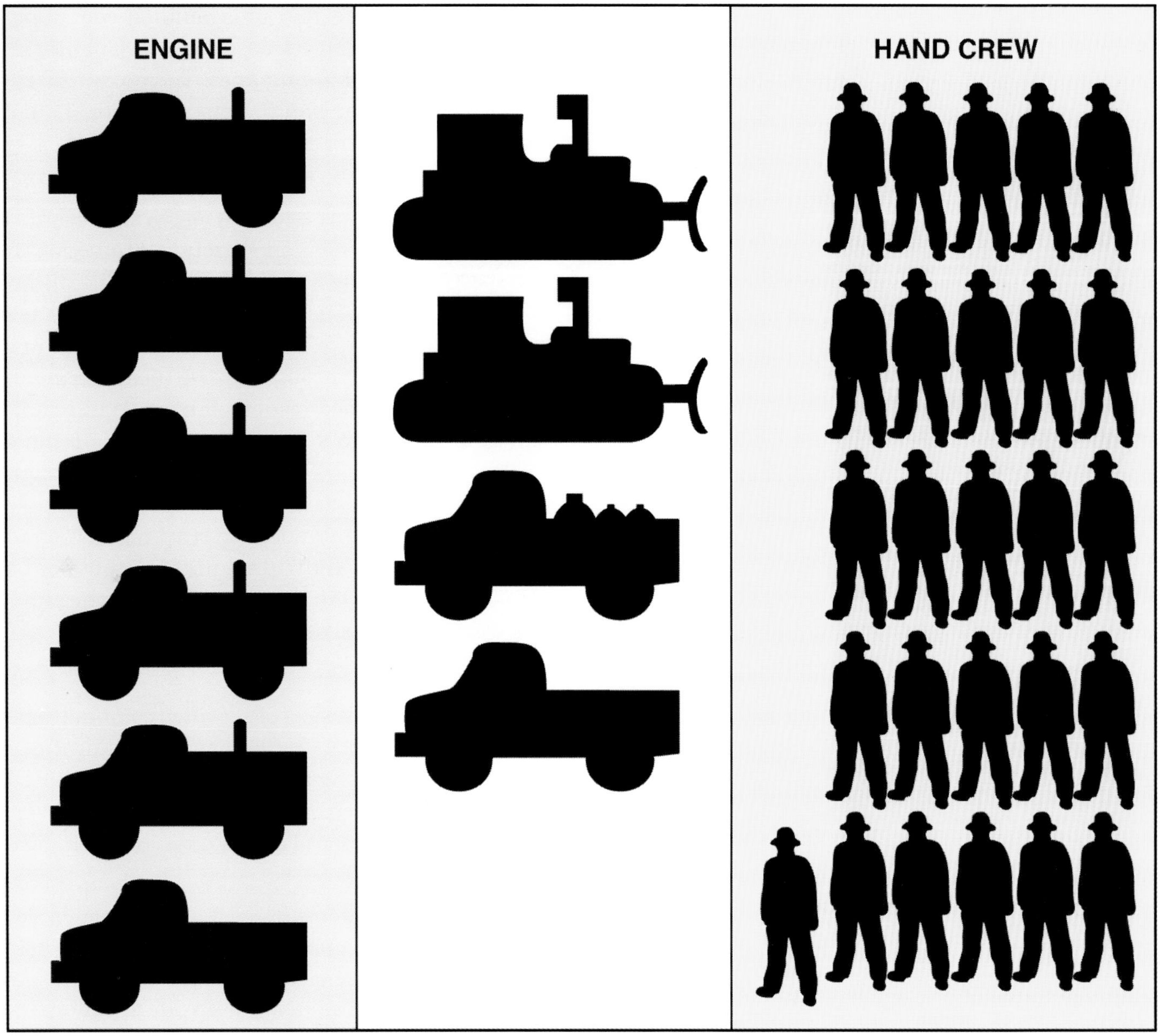

ENGINE

HAND CREW

Figure 7.42 Engine, dozer, and hand crew strike teams.

Figure 7.43 Strike team leaders should have assistants.

of hand crews, an engine, and a leader (Figure 7.44). Task forces can be composed of any kinds of available resources, depending upon the mission.

STRUCTURE PROTECTION GROUP/SECTOR

One very common use of strike teams and task forces, especially in the wildland/urban interface, is for structure protection. Depending upon the number and spacing of structures to be protected, strike teams and task forces (or individual resources) may be combined into a structure protection group/sector. This organizational structure also helps in maintain span of control by allowing the group/sector supervisor to deal only with the

Figure 7.44 A typical dozer task force.

strike team/task force leaders in the group/sector. Each unit in the group/sector may be assigned to protect a structure or a group of structures. For example, an engine might be assigned to protect a single residence or two residences if they are within 50 feet (15 m) of each other. A strike team or task force might be assigned to protect all houses on a particular road. A single engine might be assigned to protect a farmhouse and its outbuildings.

Safety

As always, safety is a primary consideration in assigning and using resources in the wildland/urban interface. When firefighters are actively trying to reduce a structure's susceptibility to ignition, their attention may be diverted from the fire. The concept of Lookouts, Communications, Escape routes, and Safety zones (LCES) is extremely important (see Chapter 8, Firefighter Safety and Survival). As with any wildland fire, safety in the wildland/urban interface begins with the tailboard briefing. The briefing includes the following:

- Checking all personal protective equipment (PPE) including fire shelters and communications equipment
- Reviewing assignments and answering questions (Figure 7.45)
- Reviewing LCES (Figure 7.46)

STRUCTURE DEFENSE IN THE INTERFACE

Structure defense often involves equipment and personnel normally assigned primarily to either wildland fire fighting or structure fire fighting. Structure defense, however, is unique; it is neither wildland fire fighting nor interior structure fire fighting. It is most akin to exposure protection.

Figure 7.45 Crew leader makes sure everyone understands the assignment. *Courtesy of Monterey County Training Officers.*

Figure 7.46 Crew leaders review the LCES plan. *Courtesy of Tony Bacon.*

Some things that work well in conventional structure fire fighting do not work in structure defense. Even those structures that do catch fire do not burn in the usual way, but conventional structure fire fighting equipment and some standard techniques can still be used.

Fire-control actions are dictated by the approaching fire. Staying mobile, using water efficiently, and using effective wildland fire-control methods are commonly required. In spite of all efforts to provide coordination and organization, resources may still be overtaxed. In the midst of a rapidly changing, often poorly understood situation, resources must be ready to take immediate and decisive action. Resources defending structures must be mobile, resourceful, and self-reliant.

Initial Operations

Often units moving into position for structure defense will be the first fire fighting equipment at scene. This is equally true of initial response units and those deployed under an action plan. These units commonly encounter the following problems:

- Traffic congestion
- Concerned and even panicky citizens
- Lack of information on access, number of structures, and their locations
- Deciding which structures to defend

Engine crews, hand crews, and heavy equipment crews, as well as the incident commander or other command officer, can all face such problems. Whoever is in charge must be prepared to take action. Fire-control actions must often be delayed until these problems have been handled.

In addition to the steps residents can take to protect their structures, firefighters assigned to structure protection can also act to defend the structures. If time allows, firefighters should do the following:

- Survey the situation.
- Prepare the site.
- Prepare the structure.
- Construct wet line/treat structure.
- Construct fireline.

SURVEY THE SITUATION

Firefighters must first familiarize themselves with the structure and its immediate surroundings. This will alert them to hazardous situations and help them determine the most effective ways to defend the structure. Their survey should include the following:

- ***Surrounding yard and wildland fuels***
  - — Proximity of wildland fuels to structure
  - — Clearance around structure
  - — Accumulations in yard
  - — Narrow gates
- ***Structure***
  - — Combustible roof, siding, etc.
  - — Unscreened vents
  - — Open windows (left open or broken by heat)
  - — Holes between shingles or tiles
  - — Space under skirting of mobile homes
  - — Open doorways or breezeways
  - — Crawl spaces under porches or decks
  - — Large windows facing fire approach
- ***Hazards***
  - — Holes/drop-offs, septic tanks, leach fields, soft ground
  - — Wire fences, low wires or cables
  - — Overhead power lines
  - — Agitated or protective animals
  - — Hazardous materials (LPG tanks, pesticides, explosives, solvents)
- ***On-site resources***
  - — Materials for covering openings (plywood, boards, sheet metal)
  - — Hammers, saws, nails, etc., for securing coverings
  - — Ladders
  - — Rakes, brooms, blowers, etc., for removing leaves, needles, or grass
  - — Chain saws, trimming saws, axes, shovels, etc.
- ***Water sources (even small ones)***
  - — Hydrants
  - — Pools or tanks (if accessible quickly)
  - — Irrigation systems
  - — Garden hose connections
  - — Ponds
  - — Wells
  - — Portable irrigation pumps
- ***Adjacent resources*** **—** Contact adjacent fire fighting resources in order to work effectively together. This may include the following:
  - — Sharing information
  - — Identifying mutual boundaries
  - — Adjusting and equalizing workloads
  - — Identifying the best access routes for mutual assistance

PREPARE THE SITE

Before they prepare the structure itself, firefighters should do as much as time allows to reduce the fire's impact on the things outside the structure.

- ***Clearance***
  - — Fuels should be cleared away from the structure at a distance of at least two to three times expected flame length.
  - — Slope or winds increase clearance needed.
- ***Removing/trimming fuels***
  - — Not all plants must be removed/trimmed.
  - — Flame-producing potential determines clearance (greater potential requires more space).
- ***Hazardous materials***
  - — Turn off LPG tanks and apply foam blanket.
  - — Apply foam to pressure vessels (even if contents are not flammable).
  - — Protect vehicle components (shocks, tanks, mounted tires, drivelines, etc.).
  - — Apply foam to aboveground gasoline or diesel tanks.
  - — Remove or protect pesticides, fertilizers, explosives, paints, etc.
- ***Private vehicles*** — Private vehicles left on-site are often relatively high-value property that should be protected if possible. However, they should not be left where they might interfere with fire fighting operations. Private vehicles should be parked as follows:
  - — In sheltered location (away from heat and fire embers; not in dry vegetation)
  - — Where they will not interfere with the movement of fire apparatus
  - — In the direction of egress, with keys in the ignition
  - — With doors and windows closed, but unlocked
- ***Pets and livestock*** — Animals that are free to move around will generally avoid being burned by a wildland fire. However, those that are restrained by fences or chains may need to be freed (Figure 7.47). Animals that are agitated or frightened by the fire and the sounds of fire fighting equipment and operations are easily panicked and can be dangerous to firefighters trying to lead them to safety.
  - — Pets should be placed inside a garage or residence if possible.
  - — An animal control agency should be called if large numbers of pets and/or livestock must be relocated.

Figure 7.47 Livestock may need to be evacuated.

PREPARE THE STRUCTURE

In this context, preparing a structure refers to reducing its susceptibility to ignition and limiting fire spread. These ends can be achieved in a number of ways; the most common are as follows:

- ***Clear the roof and rain gutters of accumulated combustibles*** (Figure 7.48).
  - — Remove dry leaves.
  - — Remove pine needles.
  - — Remove trash and debris.
- ***Cover openings and potential openings*** (Figure 7.49).
  - — Nail or lean plywood or metal over unscreened vents or ducts.
  - — Close windows and cover with plywood or metal.

Figure 7.48 Debris should be cleared from the roof and rain gutters.

Figure 7.49 Vent openings should be covered.

— Cover doorways or other large openings with sheets of plywood or tarps.

— Cover evaporative (swamp) coolers with wood, metal, or canvas; if they cannot be covered, turn off blowers and turn on the water.

• ***Prepare the interior*** (Figure 7.50).

— Remove lightweight curtains and other easily ignited materials.

— Close nonflammable window coverings (drapes, shades, blinds, etc.).

— Close interior doors.

— Turn off fans or coolers (anything that blows air).

— Turn off flammable gas (natural or LPG) at the source.

— Leave electricity *ON* (unless interior of structure becomes involved).

— Leave a porch light and a central interior light on to make the structure easier to locate in the dark.

— Make sure essential doors can be opened.

— Leave a note for the resident describing the changes that have been made.

• ***Set up ladders*** — Whenever possible, the resident's ladders should be used instead of fire department ladders. Some extension ladders can be separated into two or more single ladders that can be used at different points on the same structure or on different structures. The ladders should be placed where they will do the most good. For example, the roof should be laddered because it provides a good vantage point and because quick access to the roof may be needed if burning embers land on it (Figure 7.51). Position ladders as follows:

— On the sheltered side of the structure (away from the approaching fire)

— Where they will be most visible to anyone needing access to the roof

— As close as possible to the engine assigned to the structure (to keep hose lays as short as possible)

Figure 7.50 Heavy drapes should be closed. *Courtesy of Tony Bacon.*

Figure 7.51 The homeowner's ladder should be placed against the roof.

CONSTRUCT WET LINE/TREAT STRUCTURE

The techniques for using Class A foam for wet-line construction and structure protection are discussed in Chapter 4, Water Supply. However, when applying Class A foam for either of these purposes, *timing is critical.* Unlike gelling agents, which remain effective for up to several hours, Class A foam must be applied shortly before the fire's expected arrival. While Class A foam applied through a compressed-air foam system (CAFS) may last more than an hour under ideal conditions, it is always best to apply any Class A foam within 15 minutes of the time the fire is expected to arrive. Class A foam should also be applied to any exposed LPG tanks or aboveground fuel storage tanks.

Wet-line construction. When insufficient time and/or resources prevent constructing a fireline between the structure and the approaching fire, a wet line can be constructed as an initial fire-stop or as a line from which to do burning out. The wet line should be at least two and one-half times as wide as the expected flame lengths. The foam should be applied at close range so all sides of the fuel are coated (Figure 7.52). Foam can also be lofted onto the surfaces of brush, tree trunks, and fuel canopies to form an insulating barrier (Figure 7.53).

Figure 7.53 Foam should be lofted onto canopy fuels. *Courtesy of NIFC.*

Burning out. Under some conditions, burning out the fuel around a structure may be the best tactic. It should be considered when existing control lines might not hold and the structure would be threatened, when there is no time to wait for the fire front to arrive, or when burning out on the surface would prevent a more destructive crown fire. Burning out should be done only by qualified personnel and only after the division/group/sector supervisor gives permission. Adjacent units (including air attack) should also be informed of the impending operation. Before starting to burn out, crews must construct a wet line or scratch line from a natural or constructed anchor point. Burning out should be started as soon as the control line is finished and should not be delayed until the fire front creates an indraft. The fire started for burning out must be small enough that the resources on hand can control it.

Treating a structure. A structure's susceptibility to ignition can be reduced by applying Class A foam or gelling agent to the roof and the walls facing the oncoming fire front (Figure 7.54). Properly applied, these agents provide some protection from both burning embers and radiant heat. Under

Figure 7.52 Firefighters build a wet line with foam. *Courtesy of NIFC.*

Figure 7.54 Structures should be pretreated with foam or gelling agents. *Courtesy of NIFC.*

ideal conditions, a blanket of Class A foam or gelling agent may last up to several hours; however, it may last as little as 15 to 30 minutes.

CONSTRUCT FIRELINE

Unlike a wet line, which is a temporary firestop, a fireline is a control line that is cut down to mineral soil. While both wet lines and firelines may serve as a point from which to burn out fuels between the line and a fire, a fireline is more likely to stop an advancing wildland fire, either with or without burning out. The fireline should be located in fuels and terrain where the main fire or the burning-out operation can be controlled with the resources available. It should be located as close as possible to natural or man-made barriers such as green lawns, other green landscaping, or parking areas devoid of vegetation. The aim is to eliminate all flammable fuels between the line and the structure if possible. The fuels, the slope, the wind, and the resources available to construct the line determine its width.

Remove intermediate fuels. Intermediate fuels are those left between the fireline and the structure. When ignited, they may produce embers or radiant heat that could threaten the structure. Examples of common intermediate fuels are as follows:

- Wood piles (firewood, lumber, fencing materials) (Figure 7.55)
- Wooden fences attached to the structure (Figure 7.56)
- Attached decks and combustible awnings (Figure 7.57)
- Combustible lawn furniture (Figure 7.58)

Figure 7.55 Wood stacked next to a structure may provide more fuel to a fire. *Courtesy of NIFC.*

Figure 7.56 Attached wooden fencing can help a fire spread to the structure.

Figure 7.57 Exposed wooden deck supports are vulnerable to ignition.

Figure 7.58 Wooden lawn furniture can provide more fuel to a fire.

Remove yard accumulations. In addition to the obvious flammable/combustible items that can directly threaten the structure, common objects that have value may be scattered about the yard and should be protected. Items such as boats, inoperative vehicles, small trailers, and stacks of building materials can create problems by providing more fuel to the fire and by impeding firefighters' efforts to deploy hoses to protect the structure (Figure 7.59). Flammable/combustible items in contact with the structure should be moved away from it. If moving these items is impractical, the flammable vegetation immediately around them must be cleared or burned out.

Engine-Based Operations

This section addresses engine-based operations in interface fires. The principles and techniques

Figure 7.59 Abandoned vehicles may also provide additional fuel.

discussed apply equally whether an engine is assigned to protect a single structure or a pair of closely adjacent structures or remains mobile. To apply these principles and techniques, crews should have familiarized themselves with the area as described earlier.

ACCESS AND POSITIONING

Many structures threatened by a wildland fire are not on wide, paved streets. They are often at the ends of long, narrow driveways opening off rural lanes. They are also often surrounded by dry, flammable vegetation.

Negotiating the access. While getting into and out of a structure's location can be easy under normal conditions, it may be much more difficult if the driveway is completely obscured by smoke. For safety, engines should be backed in from the last known turnaround, and the crew should note the locations of landmarks along the way (Figure 7.60).

Figure 7.60 Engines should be backed down narrow driveways.

Positioning the engine. The engine should be positioned so that it is safe and convenient to work from. This can be accomplished by the following:

- Park off the roadway to avoid blocking other fire apparatus or evacuating vehicles.
- Scrape away fuel if necessary to avoid parking in flammable vegetation.
- Park on the lee side of the structure to minimize exposure to heat and blowing fire embers.
- Park near (but not too close to) the structure so that hoselines can be kept short.
- Keep cab doors closed and windows rolled up to keep out burning material.
- Place engine's air-conditioning system (if so equipped) in recirculation mode to avoid drawing in smoke from outside.
- Do not park next to or under hazards such as the following:
  - — Power lines
  - — Flammable trees or snags
  - — LPG tanks or other pressure vessels
  - — Structures that might burn

HOSELINES AND NOZZLES

The hoselines and nozzles selected for structure defense must be large enough to handle the expected fire intensity; however, they should not be so large that they are difficult to maneuver or might waste water. The lines should be as short as possible but long enough to adequately protect the structure, including the roof. A short line should always be maintained exclusively for protecting the crew and the engine (Figure 7.61). Where available, hydrants should be used to keep engine water tanks full.

Figure 7.61 A charged engine/crew protection line should always be ready.

Hoselines. Unless anticipated fire conditions dictate otherwise, 1½-inch (38 mm) or 1¾-inch (45 mm) hoselines with combination nozzles are usually a good choice (Figure 7.62). These hoselines provide an adequate stream while being reasonably easy to handle. Smaller 1-inch (25 mm) or ¾-inch (19 mm) lines may be used if it is *certain* that they will be sufficient to handle the highest anticipated fire intensity (Figure 7.63). Even though these smaller hoselines are more maneuverable, they provide much less protection for firefighters in case of a sudden increase in fire intensity. Lightweight or single-jacket hose is often used for exterior structure defense because it is easy to maneuver.

In addition to the engine/crew protection line, two attack hoselines should be deployed from each assigned engine. One line should be laid around each side of the structure being protected. The lines must be long enough to meet behind the structure but should not be more that 200 feet (60 m) long. Deploying the lines this way allows the following:

- Two streams are available to direct at hot spots and flare-ups.
- A backup line is available in case one line fails.
- Hoselines are between the fire and the structure — the most effective position.
- The fire front can split and burn around the structure.

Additional attack lines may be needed for spot fires on the roof or for fighting fire inside the structure. If possible, these lines should be set up in anticipation of their need. The roof hoseline should be flaked out at the base of a ladder to the roof, and the interior line(s) flaked out in front of the entry door(s) (Figure 7.64). Lines intended for interior fire fighting must be at least 1¾ inches (38 mm) and should be 1½ inches (45 mm) if available. Lines intended for roof protection can be hard lines or even garden hoses (Figure 7.65). If the roof becomes well involved, larger lines will be needed.

Figure 7.62 Attack lines should be of sufficient size. *Courtesy of Bill Lellis.*

Figure 7.63 Booster lines are intended primarily for mop-up.

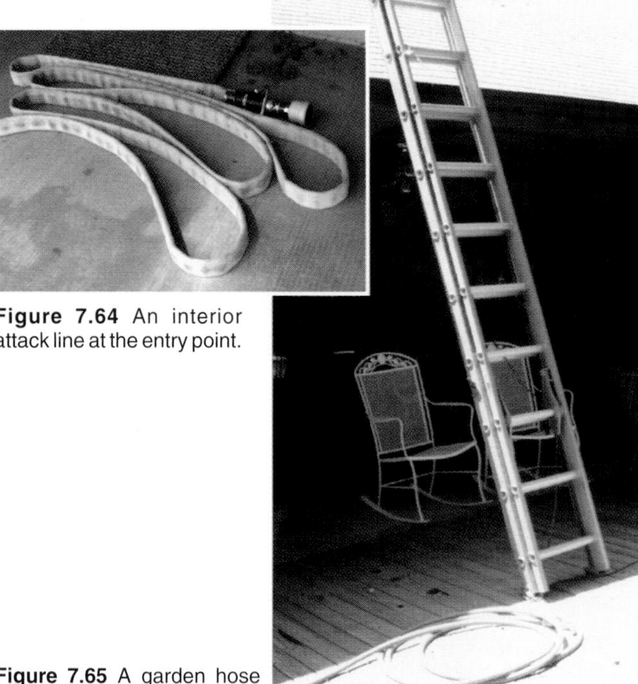

Figure 7.64 An interior attack line at the entry point.

Figure 7.65 A garden hose may be used for roof protection.

As mentioned earlier, an additional hoseline for protecting the engine and crew should also be ready in case the fire overruns the position or if the engine must be driven through the fire to escape. The protection line should be long enough to reach all the way around the engine, and it can be loosely coiled or flaked out on top of the hose bed ready for use (Figure 7.66).

The hoselines should all be connected to the same side of the engine, preferably the pump panel side, and only to valved discharge outlets (Figure 7.67). This makes it easier to disconnect the lines in a hurry if a retreat becomes necessary. For the same reason, the lines should be extended to the rear of the engine, not around to its front, so they won't become entangled in the wheels if the engine must be moved quickly (Figure 7.68).

Hoselines are sometimes abandoned in place when an engine retreats or is reassigned. These lines may be useable by another incoming engine. Flagging tape may be used to mark their locations if they are not obvious (Figure 7.69). When hoselines are abandoned, the ends of the hose should be draped over a fence, mail box, tree limb, etc., to keep incoming vehicles from running over them (Figure 7.70).

Figure 7.66 A typical engine/crew protection line. *Courtesy of Tony Bacon.*

Figure 7.67 Lines should be connected on the pump panel side. *Courtesy of NIFC.*

Figure 7.68 Lines should *not* be deployed around the front of the engine.

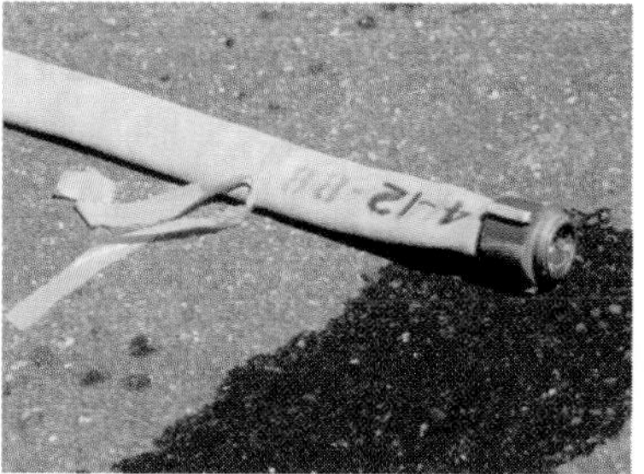

Figure 7.69 Abandoned hose may be marked with tape.

Figure 7.70 Abandoned hose may be draped over a fence or mail box.

Nozzles. In most cases, adjustable fog nozzles are recommended. Most of these nozzles provide a variety of stream patterns and flow rates, and most importantly, they provide a wide-angle fog pattern for operator protection if needed (Figure 7.71). In high winds, solid-stream nozzles can provide better reach, but they are safe only for mop-up and very small spot fires (Figure 7.72).

Figure 7.71 Fog nozzles can provide a crew protection pattern.

Figure 7.72 Solid-stream nozzles are safe only for mop-up and small nuisance fires. *Courtesy of NIFC.*

Hydrant use. If a hydrant is available, a short length of hose and/or a manifold or gated wye can be attached to the hydrant to facilitate its use (Figure 7.73). Except to refill their tanks, engines usually should not be connected directly to a hydrant. Operating from a hydrant prevents an engine from staying mobile and moving quickly when necessary. Another option is to connect a 2½-inch (65 mm) or larger hose to the hydrant and use it to supply the engine near the structure. Most municipal systems, however, are not designed for dozens of hydrants to be used simultaneously, and severe water system damage may occur. In the absence of a hydrant, water tenders can be used to supply the engines with water.

CAUTION: If operating from the engine's water tank, the last 100 gallons (400 L) of water in the tank should always be reserved for engine/crew protection.

Figure 7.73 A hydrant ready for use.

Contain the Flanks

If a well-developed head is approaching the structures and air drops are not immediately available, the actions just described are probably the most appropriate. The crews in the structure protection group/sector assume a defensive posture and defend the structures as the fire burns past. In other cases, however, the structures will not be threatened by the head of the fire but by one of the flanks. In these situations, the group/sector crews may be able to act more aggressively and do more than wait for the fire to arrive.

If a flank of a fire is approaching but is still some distance away, crews may be able to contain the flank before it ever really threatens the structures. One way they can do this is by constructing a wet line or bulldozer line near the structures and burning out the fuel between the line and the advancing fire (Figure 7.74). Another way is to move the engines into position for a direct attack on the fire, with or without the aid of air drops.

Dealing With the Homeowner

Firefighters may have to deal with angry or panicky homeowners. If any homeowners or occupants are out of control, abusive, or hysterical, law enforcement should be called to deal with them. The more rational homeowners should be advised to evacuate the area until the danger has passed. If they choose not to leave, advise them on what they can do to defend their properties without putting themselves in too much danger. The actions they can take were described earlier in this chapter.

Figure 7.74 Burning out from a dozer line can protect threatened structures.

Rescue and Safety

Homeowners' justifiable concern for their property, pets, and livestock sometimes exceeds their good judgment, and they can put themselves in serious jeopardy. If they cannot extricate themselves from a serious predicament, firefighters may have to rescue them. These rescue situations can be as simple as removing someone from a roof after a ladder has blown down, or it may involve stabilizing someone who has been injured in a fall or other accident and moving the victim to a safety zone. In the stress of a major fire and in their zeal to protect their properties, it is not uncommon for homeowners to suffer life-threatening medical emergencies such as heatstrokes or heart attacks.

However, homeowners are not the only potential victims in a major fire in the wildland/urban interface. Firefighters are also vulnerable to accidents and injuries. The most common threats to firefighter safety are discussed in Chapter 8, Firefighter Safety and Survival. But one of the most significant threats to firefighter safety in interface fires is that firefighters often have to fight the type of fire for which they have the least training and with which they have the least experience. In other words, structure firefighters may have to fight a major wildland fire, and wildland firefighters may have to fight structure fires. This can sometimes lead to firefighters improperly sizing-up a situation or unknowingly using unsafe tactics. The obvious answer to this dilemma is to provide cross training for both groups, but this is not always possible.

Firefighters may suffer excessive heat stress if they wear structure turnouts while fighting a wildland fire, and they risk burn injuries if they attempt to fight interior structure fires in wildland gear. Crew leaders and supervisors must critically evaluate their crews' capabilities and limitations and not hesitate to ask for advice and assistance — firefighter safety is at stake. Ideally, structure crews working in the wildland should be under the supervision of an experienced wildland firefighter and vice versa. In addition, firefighters should have both wildland and structure fire PPE on their engines at all times.

Structures Ahead of Fire

Depending upon the resources available at the scene and the amount of time remaining before a fire front reaches a structure, firefighters may use both offensive and defensive strategies. If a fire front is still some distance away from a structure, they may have time to deploy hoselines to attack the fire before it reaches the structure. They may also use the time to construct a bulldozer line or a wet line and burn out the fuel between the line and the fire front (offensive mode). But if an intense fire front is moving rapidly toward a structure, firefighters may have to curtail their fire fighting efforts and take refuge inside the structure until the fire front passes. Once the front has passed, firefighters can attempt to save any property that was ignited by the fire front (defensive mode).

Fire at the Structure

Confronting the fire at the structure is often a complex process that involves consideration of the fire situation, the structure, and the status of adjacent fire fighting resources. Most often, crews will find themselves in one of four general situations:

- Where spotting-zone fires are possible
- Where full containment is possible
- Where only partial containment is possible
- Where containment is impossible

SPOTTING-ZONE FIRES POSSIBLE

The spotting zone is generally the area immediately downwind of the fire front (Figure 7.75). Even though the main fire may never reach the structure(s), the threat of new fires starting in this zone may last for hours. Fire crews must remain vigilant and stay mobile enough to reach any point within their areas of assignment quickly. If spot fires start, crews should attack them while they are still small. Spot fires should be fully extinguished or fully contained within an effective control line (Figure 7.76).

FULL CONTAINMENT POSSIBLE

In this situation, firefighters can stop the fire before it reaches the structure. They can accomplish this by constructing a control line completely around the structure or tying it to adjacent control lines (Figure 7.77). Otherwise, if time allows, firefighters can simply wait for the fire to arrive and extinguish it if it is not too intense. If there isn't time to wait for the fire or it is expected to be too intense, then burning out from the control line is the best option.

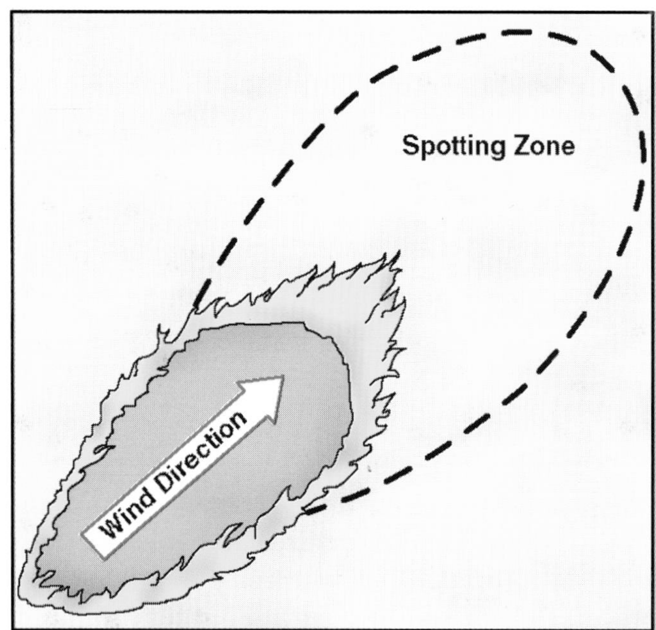

Figure 7.75 The spotting zone.

Figure 7.76 Firefighters extinguish a small spot fire. *Courtesy of Monterey County Training Officers.*

Figure 7.77 A fireline may have to completely surround a structure.

ONLY PARTIAL CONTAINMENT POSSIBLE

In this situation, firefighters may be able to modify or diminish the fire as it burns by the structure. They may be able to backfire from a short segment of the control line to mitigate the most threatening part of the fire front. Or they may use hoselines to knock down that part of the fire front that is moving directly toward the structure (Figure 7.78).

After the main fire passes, the crews must quickly check the most vulnerable points of the structure for fire. If there is no structure involvement, they can then extinguish any fire remaining on the fringes of the control line.

Figure 7.78 Firefighters may have to protect threatened structures with hoselines.

CONTAINMENT IMPOSSIBLE

In this situation, the wildland fire will burn by the structures essentially unchecked. The fire will arrive before the crews can do anything to stop it, or it will be too intense for a direct attack when it does arrive. Fire crews must concentrate first on surviving the passage of the fire front and then attacking any fire in or on the structure. Firefighters may be able to protect the structure from ignition by spraying water or foam on it as the fire passes (Figure 7.79). But if the situation becomes untenable, they should abandon their hoselines and retreat to a safety zone until the flame front passes. They can then return and extinguish any remaining fire.

Staying Mobile

Engine crews should stay mobile in order to save as much property as possible, especially when resources are limited and a fire front is moving through a development. To do this they may have

Figure 7.79 Water should be applied to exposed structures as the fire passes. *Courtesy of Tony Bacon.*

to use hit-and-run tactics. This involves making an initial knockdown on a fire and then, without taking time to do a thorough overhaul (structure fire mop-up), moving on to the next involved structure. Crews should return as soon as fire conditions allow to complete the extinguishment and to salvage what can be saved.

Structure-protection crews should avoid becoming tied down by lengthy supply lines and attack lines. In general, attack lines should be no longer that 200 feet (60 m). When not in use, they may be tossed on top of the hose bed in loose folds or coils (Figure 7.80). In addition, crews should remove as little equipment as possible from their engines. Crews should remain close together and in direct communication with their leader to facilitate a rapid retreat if that becomes necessary.

Figure 7.80 The attack line may be coiled on top of the engine hose bed. *Courtesy of Tony Bacon.*

If crews must relocate in a hurry, they may abandon their hoselines, taking only their nozzles and appliances with them. When the situation allows, they can return to pick up their hose. Even if retreat becomes necessary because of the intensity of the fire, crews must still be careful as they leave. They must watch for hazards along their route, remain calm, and resist becoming excited or careless.

When the worst of the fire has passed, crews may be able to return to the structures to which they were assigned; however, the fire may have created new hazards. These new hazards may include the following:

- Downed power lines
- Burning snags that can fall or drop large pieces without warning
- Rocks, logs, or other debris on the roadway
- Objects (burning and otherwise) that may roll downslope
- Hot spots next to the road
- Heavy smoke
- Weakened bridges or cattle guards

When crews have returned to their assigned structure, they must save as much property as possible by extinguishing any fire that may have started on or in the structure. But in addition to extinguishing the obvious fires, they must do as thorough an overhaul as time allows. Except for reasons of life safety, it is both unprofessional and irresponsible for a fire crew to leave a structure before *all* of the fire has been found and extinguished.

Water Usage: Supply and Application

Water usage depends primarily upon water availability. Regardless of the total quantity of water available for fire fighting, it must be used when and where it will do the most good.

WATER SUPPLY

In addition to the various sources of water and the ways in which it is supplied to the fireline as discussed in Chapter 4, Water Supply, a water tender may be assigned to each engine strike team. This combination is called an *enhanced* or *reinforced* strike team (Figure 7.81). When reinforced strike teams are available, the water tenders can be coupled with the engines assigned to protect the most vulnerable structures. This makes the greatest quantity of water available to the engines that need it most. However, to be effective as a team, the water tenders must have enough speed to keep up with the engines they are assigned to support. Class A Foam, gelling agents, or other water additives can also be used to multiply the extinguishing capability of the water and thereby effectively extend the supply of water available.

Figure 7.81 An enhanced engine strike team includes a water tender.

Firefighters should know the characteristics of whatever water supply is available to them on the fireline. The most important of these characteristics are as follows:

- ***Capacity*** — What is the total quantity available; how long will it last?
- ***Flow rate*** — Is gpm (L/min) continuous/intermittent?
- ***Pressure*** — Is pressure adequate to run hoselines directly?
- ***Reliability*** — Will pressure (gravity or pumped) drop because of use elsewhere?

WATER APPLICATION

Regardless of how much water is available, this precious commodity should never be wasted. Nozzles should be turned off when water is not being applied to extinguish fire, construct a wet line, or protect exposed structures. Water may be used to construct a wet line from which fuel can be burned out, but it should not be wasted by wetting vegetation or exposed parts of structures in the path of an advancing fire. The heat, low humidity, and winds associated with most wildland fires will usually evaporate the water before the fire arrives.

Water can be effective against radiant heat and helps to prevent ignition if it is applied to an uninvolved structure *while* the fire is threatening the structure, not before.

In general, water should not be used to pretreat exposures, but to extinguish fire. As the fire approaches, water can be used effectively to extinguish spot fires on the roofs of structures or on wooden decks. The rate at which water is applied depends on the situation — the more fire there is, the more water is needed to extinguish it. Small spot fires can usually be extinguished with booster lines (hard lines) flowing about 25 gpm (100 L/min). But as mentioned earlier, 1½- or 1¾-inch (38 mm or 45 mm) lines capable of flowing at least 60 gpm (228 L/min) are recommended for most wildland fires. It is also important that the nozzles be adjustable so that the flow rate can be varied to match the fire conditions. Adjustable fog nozzles with a flow of at least 95 gpm (360 L/min) are needed for structure protection. Inside a structure, both safety and effectiveness dictate that nothing smaller than 1½-inch (38 mm) hoselines be used for attack lines.

Water can also be used effectively to reduce or limit the potential heat buildup from the heat wave created by an approaching fire. This can be done by extinguishing any spot fires and fires in light, surface fuels that might otherwise spread to available aerial fuels. It can also be done by keeping the fire out of concentrated fuels such as woodpiles or patches of brush. If the radiant heat is too intense for hose streams to be effective when the peak of the heat wave arrives, firefighters should take shelter behind a solid object until the heat wave passes. When the heat begins to subside, they can step out and apply the water where it will do the most good.

Hazardous Materials

Dealing with hazardous materials in the wildland/urban interface usually calls for a defensive strategy. In most cases, it involves firefighters recognizing that a material is hazardous, isolating it, controlling the scene, and calling for the specialized resources needed to mitigate the problem (Figure 7.82). Regardless of what the material is or how it came to be where it is, firefighters must first protect themselves and others from exposure. If the material is contained, it may only be necessary to cordon off the area and wait for the hazardous materials team. When a toxic or potentially explosive material is in the path of an advancing fire front, it may be necessary to evacuate the area immediately. If a toxic material is uncontained, firefighters can avoid exposure by staying as far away (uphill and upwind) from the material as possible. If the material is involved in fire, firefighters must protect themselves from the smoke by wearing SCBA or staying upwind of the fire. If water is used to extinguish a fire in a hazardous material, the runoff may be contaminated and must be prevented from entering streams, rivers, or other bodies of water (Figure 7.83). If contaminated runoff soaks into the ground, the area must be cordoned off and access denied pending soil analysis and/or removal.

Figure 7.82 Many firefighters carry flagging tape for cordoning off areas.

Figure 7.83 Runoff water may have to be contained. *Courtesy of Tony Bacon.*

Aircraft Operations in Structure Defense

Air tankers, helicopters, and small fixed-wing aircraft can be used effectively on fires in the wildland/urban interface. As described in Chapter 6, Fire Suppression Methods, small fixed-wing aircraft and helicopters can be used in a reconnaissance and coordination role. Air tankers and helicopters can drop water, foam, or retardants to knock down hot spots threatening crews or structures (Figure 7.84). However, these resources must be used wisely and coordinated with ground units, otherwise their efforts can be wasted or even counterproductive. The possible damage to structures and vehicles by retardants may affect how fire-control aircraft are used.

Figure 7.84 Helicopters can make very precise drops. *Courtesy of John Hawkins.*

Figure 7.85 The IC may request a "no divert" if the situation warrants.

In general, fire-control aircraft are dispatched to the fire from which the most recent request for them came. This means that aircraft currently en route to or working on a particular fire may be diverted to another fire in order to keep the new fire as small as possible. If the initial attack has not brought a fire under control, the IC may request a "no divert" to ensure that the aircraft will not be diverted to another fire. However, a no-divert request will only be allowed if there is an imminent threat to life and valuable property (Figure 7.85).

COORDINATION

For reasons of both safety and effectiveness, close coordination between air and ground operations is critical. The IC must prioritize when and where to use these air resources to maximize their effectiveness and their value to the ground operations. For example, pilots must be aware of any planned firing operations so they won't interfere with these intentional fires. Even though drops from fixed-wing aircraft are not usually made any lower than 150 feet (46 m) above the vegetation, ground crews must alert pilots to potential hazards (tall trees or snags, transmission towers, or power lines) in the approach or drop zones (Figure 7.86). When identifying landmarks for pilots, ground personnel must realize that objects easily seen from the ground may not be clearly visible from the air. Anything smaller than a water tower, transmission tower, or the roof of a building may be too small to use as a landmark or reference point.

Figure 7.86 Pilots should be notified of tall obstacles. *Courtesy of Tony Bacon.*

CAPABILITIES

While the specific capabilities of the most common types of rotary- and fixed-wing fire-control aircraft were described in Chapter 2, Fire Apparatus and Communications Equipment for Wildland Fires, and in Chapter 6, Fire Suppression Methods, the following is a brief discussion of how these resources may be used in the wildland/urban interface:

Air tankers. Within standard guidelines for the length of effective fireline created by retardant drops, air tankers can lay a strip of retardant to extinguish fire or pretreat fuels. As mentioned earlier, all fixed-wing aircraft must maintain a safe minimum altitude above the vegetation, but there are differences in how low various models and types of aircraft can safely operate (Figure 7.87).

Figure 7.87 A MAFFS C-130 drops retardant on a fire. *Courtesy of NIFC.*

Helicopters. Helicopters are excellent in a reconnaissance role on wildland fires. When used in an attack role, they commonly drop water or foam, rather than retardant. While they cannot deliver as much agent per drop as most air tankers, they can often deliver more total agent per hour than fixed-wing aircraft because they are able to refill from water sources much closer to the drop zone. In addition, helicopters can drop from a very low altitude and on a specific point. This makes them especially good for working localized hot spots and in close support of ground crews (Figure 7.88).

Figure 7.88 A helicopter makes a pinpoint drop on a fire. *Courtesy of John Hawkins.*

AGENTS

As described in Chapter 4, Water Supply, the agents used to control fires in the wildland/urban interface, whether delivered by rotary- or fixed-wing aircraft, are either *short-term* or *long-term agents*. As mentioned earlier, none of these agents can be relied on for total extinguishment when applied from the air. Their effectiveness depends on prompt follow-up by ground forces.

Short-term agents. Water and Class A foam are the most common short-term agents used in the interface. Since their effectiveness lasts only as long as it takes for their water content to evaporate, these agents are generally not used as retardants to pretreat fuels or structures. To take advantage of their quick knockdown capability, they are most often dropped directly onto burning vegetation when a structure is immediately threatened. One exception to this generalization is that Class A foam can be effective as a retardant when dropped directly onto a structure to give ground forces time to reach it.

Long-term agents. Long-term retardants, such as diammonium phosphate (DAP) and ammonium sulfate, knock down fire well and have a residual retardant effect for several hours after application. However, these agents can severely damage a structure aesthetically if they are dropped directly on it, so this is not generally recommended.

HAZARDS AND LIMITING CONDITIONS

As described in earlier chapters, the use of aircraft for fire-control operations is potentially hazardous for flight crews and ground forces alike. Also, because of the potential hazards to the aircraft and their crews, there are certain limitations to the safe operation of aircraft under fire conditions.

Hazards. Both rotary- and fixed-wing aircraft can generate vortices that reach the ground. They often take the form of strong, turbulent winds that can cause a fire to flare-up or cross a control line (Figure 7.89). Retardant drops that hit ground personnel or equipment can cause injuries or damage. Crews should stay out of the center of a drop pattern and protect themselves if being hit is unavoidable. For more information on drop-zone safety, see Chapter 6, Fire Suppression Methods, and Chapter 8, Firefighter Safety and Survival.

Limiting conditions. As mentioned earlier, trees, poles, power lines, and other obstacles are potentially hazardous to aircraft and their crews operating on a fire. Such obstacles are very common in the interface. Because these hazards are sometimes difficult to see from the air, ground crews should always alert aircraft personnel to their presence. High winds, often turbulent just above tree tops and topographic features, can also be hazardous to low-flying aircraft. Winds in excess of 30 mph (48 kmph) are potentially dangerous. Smoke can reduce visibility to the point where air operations are unsafe.

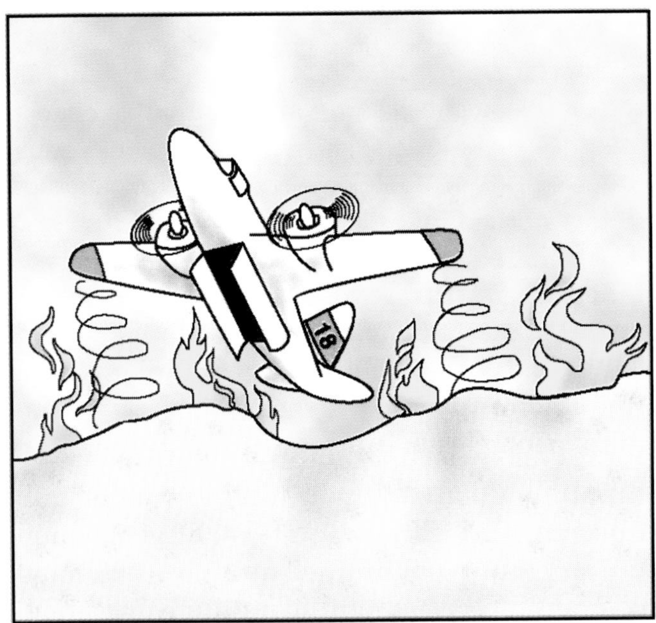

Figure 7.89 Wingtip vortices can cause erratic fire behavior.

Heavy Equipment Operations in Structure Defense

Heavy equipment can be very effective in structure defense in the interface. These machines can do some tasks that cannot be done any other way and can do others much faster than by any other means. Heavy equipment is most effective if used before a fire reaches the interface. If time allows, firelines can be constructed in advance to stop or slow the fire when it arrives (Figure 7.90). These lines may be used in conjunction with backfiring or burning-out operations. When used in conjunction with water or retardant applications, these lines can be extremely effective.

Figure 7.90 A tractor-plow cuts a fireline. *Courtesy of New Jersey Forest Fire Service.*

Among the most common pieces of heavy equipment used in structure protection are various types of bulldozers as described in Chapter 2, Fire Apparatus and Communications Equipment for Wildland Fires. They may be used to construct firelines, clear heavy fuels, construct roads, build safety zones, and clear helicopter landing areas (Figure 7.91). However, bulldozers and other types of heavy equipment can be very dangerous. When firefighters are assigned to work around heavy equipment, they should apply the safety points described in Chapter 8, Firefighter Safety and Survival.

Figure 7.91 A dozer constructs a safety zone. *Courtesy of NIFC.*

TRAFFIC MANAGEMENT

Because there may be a significant number of residents in the wildland/urban interface, the roads leading into and out of the area may be full of vehicles (Figure 7.92). Fire apparatus trying to get into the area may be impeded by civilian vehicles moving in and out of the area. Some vehicles may contain residents trying to return to their property to protect it or to gather belongings; others may contain residents trying to leave. Other vehicles may contain the curious who have been attracted by the smoke or news coverage of the fire. Still others may contain those who would take advantage of the residents' absence to loot their homes.

Regardless of why numerous vehicles are on the road, they can seriously delay the arrival of badly needed fire fighting resources. Law enforcement should be called very early in the incident to assist with traffic control and/or evacuation of residents (Figure 7.93). The IC should provide law enforcement personnel with very specific instructions regarding what interface areas should be closed, what roads should be blocked, and who should be allowed in and out of the controlled areas. However, law enforcement personnel should not be sent into areas where their safety may be in jeopardy; they have no hoselines or fire shelters with which to protect themselves.

Figure 7.92 Roadways may become congested with traffic. *Courtesy of California Office of Emergency Services.*

Figure 7.93 Law enforcement can assist with traffic control and evacuation. *Courtesy of NIFC.*

Routing Traffic and Maintaining Access

The following list includes the actions most often necessary to establish and maintain access into and out of interface fire areas. These initial actions may have to be taken in an order other than as listed.

- Devise a traffic plan that identifies routes into and out of the area. Where roads are too narrow to allow two-way traffic, a one-way loop may have to be established. When possible, post signs or individuals (possibly civilian) to indicate the routes and directions.
- Advise incoming units of routes and conditions. This reduces confusion and expedites the arrival of other resources.
- Assign traffic control to law enforcement. The ranking law enforcement officer needs to know what has been done and what traffic patterns have been established.
- Identify narrow access roads that may already be filled with and even blocked by

local traffic and devise a means of gaining access to threatened structures.

- Clear existing traffic and keep it moving to make way for fire equipment. Alternatively, civilian traffic may be pulled over to the roadside until fire equipment has passed and then allowed to continue.
- Leave a clear path for other incoming units. Identify weight limits or bottlenecks that may limit some equipment. To reduce confusion, mark routes and intersections with flagging tape or other signs.
- Make use of staging areas so uncommitted fire fighting resources are not clogging roads.
- Identify preferred ingress and egress routes, post signs on them, and broadcast this information over radio and television using the Emergency Alert System.

Smoke

Smoke drifting or blowing across roads and highways can be very dangerous (Figure 7.94). Driving any vehicle through blinding smoke puts both drivers and pedestrians at risk, whether on or off a roadway. Hand crews and others on the ground, such as photographers and other members of the media, are also at risk of being run over by vehicles driving through heavy smoke. Fire apparatus can be accidentally driven into holes, gullies, or creek beds unless a scout is used. Even if the driver or other crew members are not injured in such an accident, it could damage the apparatus, perhaps disabling it and putting it at risk of being overrun by a fire. On roads obscured by smoke, civilian vehicles should be stopped and escorted through or detoured to another road if possible.

When fire apparatus must be driven in smoke, all vehicle lights (including emergency lights) should be on. A scout should walk ahead of any vehicle that must be driven through smoke off a surfaced road. Fire apparatus drivers must keep their scouts in view because the scouts are in danger from the fire, downed power lines, and other unseen hazards. Drivers must also drive their vehicles slow enough to allow them to stop within the driver's range of vision. If the vehicle is surrounded by heavy smoke, all windows should be rolled up and vents closed. If the vehicle has air-conditioning, it should be set on *recirculate* only. All personnel should stay in the vehicle until ordered out. This not only maintains crew integrity, it protects them from possibly being run over by another vehicle, stepping on downed power lines, or being overrun by a fire.

Two Ways In and Out

Whenever possible, firefighters should not be sent into an area where there is only one way in and out. If time allows, bulldozers can be used to cut a secondary access road (Figure 7.95). As mentioned earlier, if engines must be positioned at the ends of narrow roads or long driveways, they should be backed in from the last known turnaround so they can be driven out quickly if necessary (Figure 7.96). Even where there are two ways in and out, it

Figure 7.94 Smoke may create traffic hazards if it blows across a highway. *Courtesy of NIFC.*

Figure 7.95 Dozers may be used to build secondary access roads.

Figure 7.96 Engines should be backed down narrow roads and driveways.

may be necessary to designate one of the roads for entry and the other for egress if they are very narrow.

COMMUNICATIONS PLAN

One of the most important elements of an incident action plan is the communications plan. On small incidents, all units normally operate on a common frequency (channel). On larger incidents, a formal communications plan is necessary to prevent the chaos that might develop when units from many different agencies and jurisdictions attempt to communicate on a single channel.

To prevent any channel from being overloaded with radio traffic and to keep channels clear for emergency traffic, different parts of the fireground organization are assigned different channels. A separate channel is dedicated for each incident function: command, tactical, support, air-to-air, and air-to-ground. Multiple tactical channels may be assigned on more complex incidents. There may also be a separate mutual aid channel. Because mutual aid units may not be equipped with radios that have the necessary incident channels, it is sometimes necessary to issue portable radios to strike team leaders when they check in (Figure 7.97). This allows the strike team units to communicate among themselves on their mobile radios and the strike team leader to also communicate with the division/group/sector supervisor on the issued portable radio.

One of the most effective ways that fire fighting units operating in the interface can help keep tactical channels from becoming overloaded is by using direct, face-to-face communication whenever possible (Figure 7.98). If the assigned tactical channel does become overloaded, the crew leader should request a different channel assignment. For their own safety and effectiveness, firefighters who are issued radios should test them before entering any hazardous structure-

Figure 7.97 Radios may have to be issued at check-in.

Figure 7.98 Face-to-face conversation will help keep radio channels clear. *Courtesy of NIFC.*

protection situation (Figure 7.99). For more information on fireline communications and communications equipment, see Chapter 2, Fire Apparatus and Communications Equipment for Wildland Fires. For information on obtaining multichannel radios for large-scale emergencies, see Chapter 10, Fire Protection Planning.

Figure 7.99 Firefighters should test any issued radios before they need them.

INTERFACE SAFETY CONSIDERATIONS

Firefighter safety is covered in Chapter 8, Firefighter Safety and Survival; however, the items discussed in this section apply specifically to safety in the wildland/urban interface. Many of these safety points are equally applicable to other wildland fire situations, but given the importance of firefighter safety, they bear repeating here.

General Safety Considerations

When assigned to protect structures in the wildland/urban interface, firefighters and their crew leaders should remember and apply the following:

- Make sure all crew members have fire shelters and know how to use them.
- Always wear and use appropriate personal protective clothing.
- Protect the engine as well as the structure.
- Park the engine facing the direction of exit. Don't block escape routes. Back into narrow driveways and access roads.
- Avoid allowing an engine to idle with lights and radios on unless adequate rpm can be maintained using the hand throttle. Excessive idling can deplete the battery, stall the engine, make the radio inoperative, and put the engine and crew in jeopardy.
- Keep headlights and all emergency lights on when driving in heavy smoke.
- Keep a length of charged 1½-inch (38 mm) hose looped on top of the hose bed for the protection of the engine and crew.
- Reserve the last 100 gallons (400 L) of water in the engine's tank for protection of the engine and crew. Never pass up an opportunity to refill the tank if it is less than full.
- Never leave the engine unattended unless it is parked in a safe area (in the black, on a gravel or paved road, etc.).
- Maintain control of fire fighting crews. Crew leaders must keep calm, display a positive attitude, and maintain communications.
- Stay out of possible lethal areas:
  - — Saddles
  - — Chimneys
  - — Chutes
  - — Areas of heavy fuel buildup
  - — Structure collapse zones
  - — Any other area that doesn't look/feel safe for crews or the public

IF TRAPPED BY FIRE

If a sudden increase in fire intensity has cut off the planned escape route, firefighters must be prepared to protect themselves. Some of the actions that have proven to be effective in past fires are as follows:

- Take refuge in a structure. It usually burns slowly and provides protection from the fire outside. Park the engine in as safe a place as possible, perhaps in the garage.

- Take refuge in the engine. If it is in a good location, stay there. If not, even though visibility is poor, keep it moving and seek a place where the fire is less intense.
  - Keep the pump running and use the 1½-inch (38 mm) protection line to spray a fog pattern over the cab of the engine.
  - Take SCBA, if available, into the cab for protection from the smoke.
  - Use fire shelters or salvage covers for protection from radiant heat coming through the windows.
  - Request air drops.
  - Stay inside the cab. If the motor has stalled, it was probably from oxygen deficiency — continue to use SCBA. If the engine catches fire, stay in the cab because it may provide protection until it begins to cool down outside. Do not leave the cab until it is safe to do so.

HAZARDOUS MATERIALS SITUATIONS

Firefighters assigned to defend structures in the interface should expect hazardous materials to be present and be prepared to deal with them. They should pay particular attention to the color of smoke being produced by fires because that can be an early indication of hazardous materials being involved. In addition to fighting these fires from upwind, SCBA may be necessary, especially during overhaul (mop-up). The most common types of hazardous materials to be expected in fires in the interface are as follows:

- ***Residences***
  - Chemicals (garden fertilizers, pesticides)
  - Paints
  - Cleaning materials
  - Fuel cans (tanks on lawn mowers/weed cutters, etc.)
  - Synthetics (plastics, man-made fabrics/building materials, etc.)
  - Ammunition/reloading supplies
  - Vehicles (automobiles may contain as much as 300 pounds [135 kg] of plastic)
- ***Farms and ranches*** — Includes all items listed for residences, plus the following:
  - Bulk chemicals (fertilizers, pesticides, etc.)
  - Fuel storage tanks (gasoline, diesel, safety cans, tanks on farm machinery, etc.)
  - LPG tanks
  - Explosives (dynamite, blasting caps, etc.)

Structural Situations That Shout "Watch Out!"

- Wood construction (shake/shingle roofs, elevated decks, etc.)
- Poor access (narrow, one-way roads)
- Inadequate water supplies
- Flammable vegetation within 30 feet (9 m) of structures
- Extreme fire behavior
- Strong winds (in excess of 25 mph [40 kmph])
- Emotional residents (angry, panicky, hysterical)
- Structures situated in chimneys or box canyons, on slopes in excess of 30 percent, in flashy fuels, etc.
- Weak and/or narrow bridges
- Open eaves and soffit vents

Dos and Don'ts in Protecting Structures in the Interface

- Do wear full protective clothing and equipment.
- Do reserve at least 100 gallons (400 L) of water for engine/crew protection.
- Do have a charged line for engine/crew protection.
- Do back engines into position for quick exit if necessary.
- Do use at least 1½-inch (38 mm) hoselines whenever possible.
- Do post lookouts whenever necessary.
- Don't park in saddles or chimneys.
- Don't attempt interior structural fire fighting unless trained and equipped to do so.

Don'ts When Working Around Power Lines

- When a downed wire is on vehicle:
  - Don't leave vehicle until power is turned off by utility personnel.
  - Don't step from the vehicle if fire forces you to abandon it before the power can be turned off. Jump clear and hop away with feet together.
- Don't operate heavy equipment under overhead power lines.
- Don't use transmission line rights-of-way as jump or cargo drop zones.
- Don't drive under overhead power lines in vehicles equipped with long antennas.
- Don't refuel vehicles under overhead power lines.
- Don't stand near power lines during a retardant drop.
- Don't park apparatus under overhead power lines.
- Don't apply streams to energized power lines.

FIREFIGHTER SAFETY AND SURVIVAL 8

LEARNING OBJECTIVES

This chapter provides information that addresses the following objectives of NFPA 1051, *Standard for Wildland Fire Fighter Professional Qualifications* (1995 edition):

Wildland Fire Fighter I

3-1.1* *Prerequisite Knowledge:* Fireline safety, use and limitations of personal protective equipment, agency policy on fire shelter use, basic wildland fire behavior, basic wildland fire tactics, fire fighters role within the local incident management system, and first aid.

3-1.2* *Prerequisite Skills:* Basic verbal communications, the use of appropriate personal protective equipment.

3-3.1 **Definition of Duty.** Activities in advance of fire occurrence to ensure safe and effective suppression action.

3-3.2 Maintain assigned personal protective equipment, given the standard equipment issue, so that the equipment is serviceable and available for use on the fireline and defects are recognized and reported to the supervisor.

3-3.2.1 *Prerequisite Knowledge:* Maintenance of personal protective equipment including inspection, the recognition of unserviceable items, and proper cleaning procedures (including manufacturer's and authority having jurisdiction's recommendations).

3-3.3* Maintain assigned suppression hand tools and equipment, given tools and equipment, agency maintenance specifications, supplies and small tools, so that assigned equipment is safely maintained, serviceable, and defects are recognized and reported to the supervisor.

3-3.3.1 *Prerequisite Knowledge:* Inspection of tools and assigned suppression equipment, the recognition of unserviceable items, and safe maintenance techniques.

3-5.2.1* *Prerequisite Knowledge:* Equipment requirements, agency time standards and special transportation considerations (weight limitations), agency safety, and operational procedures for various transportation modes.

3-5.3 Detect potential hazardous situations, given a wildland fire and the standard safety policies and procedures of the agency, so that the hazard is promptly communicated to the supervisor and appropriate action can be taken.

3-5.3.1* *Prerequisite Knowledge:* Basic fireline safety, fire behavior, and suppression methods.

3-5.4.2* *Prerequisite Skills:* Proper use of hand tools, fire stream practices, and agent application.

3-5.5.2* *Prerequisite Knowledge:* Operational and safety considerations when burning out.

Wildland Fire Fighter II

4-3.1 **Definition of Duty.** Responsibilities in advance of fire occurrence to ensure that tools, equipment, and supplies are fire ready.

4-3.2* Maintain power tools and portable pumps, given agency maintenance specifications, supplies, and small tools, so that equipment is safely maintained, serviceable, and defects are recognized and repaired.

4-3.3 Inspect tools and equipment, given agency specifications, so that availability of the tools and equipment for fire use is ensured.

Chapter 8
Firefighter Safety and Survival

INTRODUCTION

The fact that this chapter on firefighter safety and survival appears near the end of the manual does not reflect its relative importance in wildland fire fighting. It is necessary to discuss wildland fire behavior, apparatus, equipment, and fire-suppression methods to establish a frame of reference for this chapter. Safety is an integral part of every chapter in the manual, and the material in this chapter is intended to reinforce and expand on the safety practices discussed in the other chapters. No grass, brush, timber, or structure is worth a firefighter's life — only a reasonable chance to save another human life is worth that risk.

Fighting wildland fires is inherently dangerous, and firefighters risk injury or even death in these operations. There are firefighter fatalities in the wildland nearly every year. As recently as 1994, as many as 34 died in one year. In addition to the danger from the fire itself, the need to use cutting tools, mobile apparatus, heavy equipment, and aircraft add to the risk involved. If firefighters know how to recognize potentially hazardous situations and how to mitigate them, they can reduce or eliminate much of that risk. The first principle of safety is to avoid high-risk situations that might result in injury or death. In a wildland fire, this may not be easy to do because appearances can sometimes be deceiving. Apparently innocuous fire conditions can quickly develop into extremely hazardous situations.

This chapter builds on the information contained in the previous chapters to explain the principles of firefighter safety and survival in all types of wildland fires, including those in the wildland/urban interface. The chapter includes sections entitled the Common Denominators of Fire Behavior on Tragedy Fires, the Ten Standard Fire Fighting Orders, Eighteen Situations that Shout "Watch Out!," Structural "Watch-Out!" Situations, and Look Up, Look Down, Look Around. In addition, downhill fire fighting, personal protective equipment, safety briefings, fire shelters, respiratory protection, personnel monitoring and firefighter rehabilitation, and safety around fire apparatus and mechanized equipment are discussed. Also covered are air-operations safety, tool safety, electrical safety, personal survival techniques, first aid, hazardous materials situations, working in burned-over areas, backfiring and burning-out safety, traffic hazards, smoke, and fireground communications.

While reading this chapter, keep the *remember and follow* and *supervisors' safety responsibilities* in mind.

Remember and Follow

All fireline personnel, regardless of rank or assignment, must remember and follow these guidelines:

- Standard Fire Orders
- Watch-Out Situations
- Common Denominators of Fire Behavior on Tragedy Fires
- Downhill Fire Fighting Guidelines

Supervisors' Safety Responsibilities

In addition to their other duties, crew leaders and other fireground supervisors are responsible for the following:

- Evaluating firefighters' physical and mental qualifications — assigning to fire duty only those who are trained, equipped, and physically fit for the job

- Analyzing work problems to eliminate or avoid hazards
- Discussing safety at the beginning of each shift
- Becoming immediately involved whenever injury occurs by investigating the accident with persons involved
- Inspecting work to be sure it is done safely and efficiently
- Providing leadership in applying corrective action aimed at eliminating causes of accidents
- Setting a personal example of safe behavior and enforcing safety practices and procedures

COMMON DENOMINATORS OF FIRE BEHAVIOR ON TRAGEDY FIRES

The U.S. Forest Service (USFS) has identified several common fire-behavior characteristics that have resulted in firefighter fatalities or near misses. These characteristics are called the Common Denominators of Fire Behavior on Tragedy Fires. The study emphasized that firefighters must remain alert for potentially life-threatening situations, even when a fire does not appear to be dangerous. Firefighters on the line must monitor both the fire's behavior and the factors that could modify the fire's behavior. For example, if the fire is burning in light fuels, they must remember that these fuels are more responsive to changes in weather conditions than are heavier fuels. Wind shifts or changes in topography can cause rapid changes in the behavior of any wildland fire. Firefighters must watch for signs of these situations developing and be prepared to modify plans accordingly.

Common Denominators of Fire Behavior on Tragedy Fires

- **Most incidents happen on the smaller fires or on isolated portions of large fires.**
- **Most fires are innocent in appearance before the "flare-ups" or "blowups." In some cases, tragedies occur in the mop-up stage.**
- **Flare-ups generally occur in deceptively light fuels.**
- **Fires run uphill surprisingly fast in chimneys, gullies, and on steep slopes.**
- **Some suppression tools, such as helicopters or air tankers, can adversely affect fire behavior. The blasts of air from low flying helicopters and air tankers have been known to cause flare-ups.**

LOOKOUTS, COMMUNICATIONS, ESCAPE ROUTES, AND SAFETY ZONES (LCES)

LCES is an extremely important safety tool on the fireline. It serves to remind crew leaders and supervisors of the essential elements involved in providing for the safety of their crews working on the line. When tactical plans are made, they should be checked against LCES to make sure that these safety considerations are included. If not, the validity of the plan should be questioned, and it should not be implemented until these safety provisions are included.

LCES

LOOKOUTS
COMMUNICATIONS
ESCAPE ROUTES
SAFETY ZONES

Lookouts

A *lookout* is a member of a fire crew who is assigned to observe a fire and warn the crew when there is danger from a fire. When crews are assigned to work in drainages or other areas where they cannot see the fire, but the fire front is relatively close by, one or more lookouts who can see both the crew and the fire should be posted to monitor the progress of the fire (Figure 8.1). The lookout should be able to see both the fire and the firefighters and un-

Figure 8.1 A lookout watches a wildland fire's progress.

derstand the fire behavior they are seeing. If there is a sudden or unexpected change in fire behavior that might threaten the crew, the lookout's function is to notify the crew in time for it to retreat to a safety zone or leave the threatened area. However, it is the responsibility of every crew member to watch for hazards, such as blowups, rolling rocks, and falling snags, and to warn the others.

Communications

Some form of absolutely reliable communication between the lookouts and the crew leader is imperative. This is most often done by radio, but radios are not always reliable in some areas such as mountainous terrain. If radios are not effective or not available, another form of communication must be devised. In these cases, direct voice communication is best, but using a system of hand signals or signal flags can work if the signals used are clearly understood (Figure 8.2).

Figure 8.2 The lookout uses hand signals to alert the crew.

Escape Routes

An *escape route* is a path by which crew members can rapidly leave an area of danger and find an area of safety (Figure 8.3). Escape routes should be identified in advance of their need, and all members of the crew should be informed of where the routes are located. Escape routes *must* be positively identified whenever firefighters are required to work in the green near the fireline, especially if there is unburned fuel between them and a fire. However, escape routes should not normally be located above a fire burning uphill on a slope. The routes may lead into the black or farther into the green away from the fire such as to a safety zone (see following section). As a fire front progresses, the adequacy of previously identified escape routes changes also. New escape routes may have to be selected from time to time, and the location of these new routes communicated to the crew.

Figure 8.3 A firefighter checks an escape route. *Courtesy of NIFC.*

Safety Zones

A *safety zone* is an area into which firefighters can escape if their line is outflanked or if spot fires render the line ineffective or unsafe. When crews must work in heavy fuels, a safety zone should be maintained as close by as possible in case of a blowup. A safety zone must be large enough for all crew members to deploy their fire shelters safely if necessary. However, if a safety zone is large enough and the fuels light enough, firefighters may not need to deploy their fire shelters within the zone. To be effective, safety zones should have a radius equal to at least double the height of the adjacent fuel or the flame length, whichever is larger. Depending on the situation and the resources available, the larger the safety zone the better it is.

NATURAL AND CONSTRUCTED

Safety zones are open areas characterized by an absence of fuel available to the fire. They may be naturally barren areas, such as rock slides or cliffs, or they may be streams or other bodies of water (Figure 8.4). Safety zones may also be previously burned-out areas. In the absence of natural safety zones, they may be constructed by scraping away surface fuels down to mineral soil. In heavy fuels, mechanized equipment may be needed to construct safety zones, but in light fuels hand tools may be used. Safety zones can also be created by burning out. However, caution must be used because burning out can create additional safety problems.

Figure 8.4 A stream bed is a natural safety zone.

SUBSTANTIAL STRUCTURE

In the wildland/urban interface, a substantial structure may provide an area of refuge. The level of protection provided by a structure can be enhanced by burning out around it. Even if the structure ignites and is eventually lost to the fire, it can provide a refuge for firefighters long enough for the vegetation fire to burn past before they have to leave the structure (Figure 8.5).

Figure 8.5 Even though seriously damaged, a structure may provide refuge for fire crews. *Courtesy of NIFC.*

ENGINE

If firefighters are about to be overrun by an intense and fast-moving wildland fire, they can take refuge in the cab of their engine. The engine should be positioned in an area where the least amount of vegetation is available as fuel to the fire. Once in the cab, they should close all vents and roll up the windows. They should also cover the inside of the windows and windshield with their fire shelters (Figure 8.6). If they have self-contained

Figure 8.6 Firefighters cover the windows of their engine with fire shelters. *Courtesy of Tony Bacon.*

breathing apparatus (SCBA), they should use them when the smoke reaches the vehicle. The engine should be left in road gear with the motor running —this may allow a rapid escape from the area if the opportunity arises. They *should not* take a charged hoseline into the cab with them — this would provide an opening for smoke and hot embers to enter the cab, and the water could cause steam burns to the occupants.

TEN STANDARD FIRE FIGHTING ORDERS

The fire-behavior characteristics listed earlier are not the only fire conditions that are dangerous to fire personnel, only the most common ones. Other situations can put firefighters at risk. Studies of firefighter deaths led to the development of the Ten Standard Fire Fighting Orders. In every case where a firefighter was killed while fighting a wildland fire, it was shown that one or more of the Ten Standard Orders had been ignored. Violating one or more of these orders may result in firefighter deaths.

The orders are guidelines that help firefighters identify and avoid high-risk situations. Every firefighter should know and follow them. Being able to recite them is commendable, but putting them into practice is more important. Each order should be considered separately so firefighters will recognize when it applies during a fire and respond correctly. A description of the orders follows:

FIRE ORDERS

- **Fight fire aggressively but provide for safety first.**
- **Initiate all action based on current and expected fire behavior.**
- **Recognize current weather conditions and obtain forecasts.**
- **Ensure instructions are given and understood.**
- **Obtain current information on fire status.**
- **Remain in communication with crew members, your supervisor, and adjoining forces.**
- **Determine safety zones and escape routes.**
- **Establish lookouts in potentially hazardous situations.**
- **Retain control at all times.**
- **Stay alert, keep calm, think clearly, act decisively.**

1. *Fight fire aggressively but provide for safety first.*

Aggressive action is the key to a successful fire-suppression operation; however, safety is the first priority. Crew leaders should analyze the situation and make an attack consistent with accepted practices and methods under the existing conditions.

2. *Initiate all action based on current and expected fire behavior.*

Every action that firefighters take should be based on what a fire is doing and what they think it might do. They must make an educated guess. A few of the things to consider are as follows:

- What is the fire doing now? Can it be controlled with the resources at scene or en route? Do more resources need to be requested?
- What is the fire likely to do later? Can we flank it? Should we back off for an indirect attack?
- What action is being taken now? Are we using equipment effectively? Are we making the attack at the right spot?
- What is the weather in the fire area? What is the weather likely to do?
- What type of fuel is burning? Can we build control lines fast enough? Is the fuel a type that can cause spot fires?
- What type of fuel is the fire heading toward? Do we have the proper tools? Will we need aircraft or mechanized equipment?

Every fire must be approached differently because of the varying conditions that may be encountered. While all fires behave according to the same principles, no two fires are exactly alike because of all the possible variations in fuel, weather, and topography.

3. *Recognize current weather conditions and obtain forecasts.*

Weather is a major factor in fire behavior, and it is essential that firefighters stay informed of weather conditions and forecasts. However, firefighters must use their senses and every other means available to them to answer the following questions:

- What is the weather in the fire area? Is the weather against us? Is this the critical burning period?
- What is the weather likely to do? Are we going to have winds, high temperatures, or low humidity? Is there a chance of wind from a passing weather front or wind associated with cumulus or cumulonimbus clouds?

Firefighters must not ignore what their senses tell them about current and expected local weather conditions. Portable/belt weather kits also give data on weather conditions.

- ***Feeling*** — Even without the aid of a weather station, a firefighter can feel whether the temperature is increasing or decreasing. The time of day is also a good indicator of whether the temperature is likely to increase or decrease.
- ***Sight*** — Wind vanes and anemometers give precise wind information, but they are not necessary to determine whether the wind is blowing. The direction and velocity of the wind can be determined by looking at the vegetation, a flag, the movement of the clouds, or the drift of smoke or dust (Figure 8.7).

Figure 8.7 A firefighter checks wind direction by pouring sand from his hand.

- ***Hearing*** — Firefighters should listen to fire-weather forecasts, but they should also listen to people who are familiar with the area and ask their opinions of what to expect from local weather patterns. For example, locals can confirm whether gravity/gradient winds are common in the area at a particular time of year.
- ***Portable/belt weather kits*** — Portable or belt weather kits enable firefighters to measure relative humidity, wind speed and direction, and temperature (Figure 8.8).

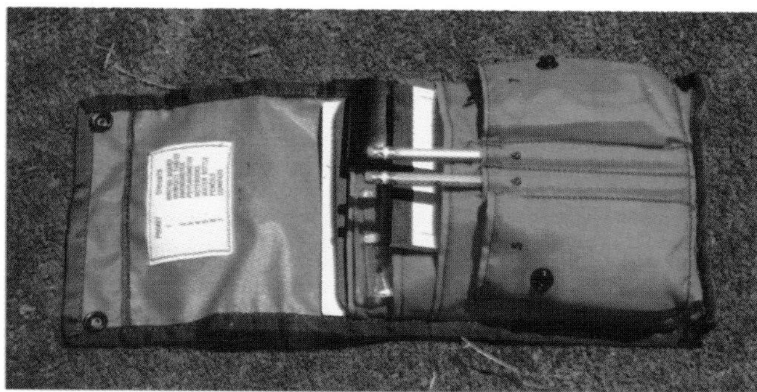

Figure 8.8 A typical belt weather kit.

4. Ensure instructions are given and understood.

The likelihood of accidents occurring is increased if vague or ambiguous oral instructions are given instead of concise, written instructions. Oral instructions are sometimes incomplete and are more likely to be misinterpreted or forgotten. Crew leaders must verify the following when given an assignment:

- What to do (objective)
- How to do it
- Where to go
- Where to finish
- When to finish
- Whom to coordinate with
- Identity of the supervisor on the line
- Identity of the relieving person
- Expected duration of attack
- Available transportation to and from the fireline
- Other pertinent information (emergency procedures and safety considerations)

5. Obtain current information on fire status.

Many small fires become large ones, sometimes very quickly. Firefighters' lives have been lost because they failed to properly size up and evaluate what confronted them based on *current* information or because they failed to observe the whole situation.

- Personally observe the fire from a vantage point.
- Personally scout ahead.
- Send out reliable scouts (who then report back).
- Observe the fire from a helicopter or other aircraft if available.
- Inform all crew members of the current situation.

6. Remain in communication with crew members, your supervisor, and adjoining forces.

Adequate communication is essential to fireline safety (Figure 8.9). Good communication must be maintained within each unit, between the unit and any scouts or lookouts, and with other fire fighting units. Fireground communication may be face to face or by radio, telephone, or any other reliable means.

Figure 8.9 Firefighters must be able to communicate under adverse conditions. *Courtesy of Tony Bacon.*

7. Determine safety zones and escape routes.

This order must be followed at all times, but it is especially important that it is followed when crews are traveling cross-country to a fire that has no control lines or to an area with which they are unfamiliar. Some good areas to select are as follows:

- ***Burned area*** — Make sure it is close enough to be reached.
- ***Safety islands*** — Identify or construct them, and let their locations be known to all.
- ***Escape route*** — *Have one,* most important, even if it must be constructed. Mark a safe route into the burned area, or cut one if the brush is too thick for travel.
- ***Natural barriers*** — Locate rocky areas, riverbeds, streams, lakes, and slide areas; let others know about them.

Some cautions with regard to escape routes are as follows:

- Firefighters must be cautious of areas that are not completely burned out.
- Once escape routes are selected, crew members must be told where they are, how to travel to them, and what to do when they get to a safety zone.
- As a fire progresses, it may be necessary to change the escape route, perhaps more than once.

8. Establish lookouts in potentially hazardous situations.

Lookouts must be in constant communication with firefighters or supervisors. If radios are not available, a system of visual signals may have to be used. Also, enough lookouts must be used to maintain visual contact with both the fire and the crew. Some situations that warrant a lookout are as follows:

- When the head of the fire is not visible to the crew
- When felling snags
- When personnel and mechanized equipment are working close together
- When falling rocks could strike someone or burning material could cross the control line
- When an obvious hazard such as a snag cannot be felled

9. Retain control at all times.

The capabilities and limitations of crew members must be considered when making work assignments. Are they rested and ready to work, or are they already exhausted from a previous assignment? Crew leaders should also inspect tools, coordinate work with available equipment, and make provisions for safety.

10. Stay alert, keep calm, think clearly, act decisively.

When faced with a possible life-threatening problem, crew members and leaders must keep calm and analyze what is happening. Panic can endanger the leader as well as the crew. After evaluating the situation, they should decide how to deal with it, and ***do it!*** Staying calm is not the same as moving slowly — do whatever you must to save yourself and your crew, and do it quickly!

EIGHTEEN SITUATIONS THAT SHOUT "WATCH OUT!"

Experience over many years and in countless wildland fires has produced a list of fireline situations that resulted in firefighters being put in serious jeopardy. To avoid finding themselves in the same danger, firefighters must be familiar with that list of situations. Each of the following situations has resulted in at least one firefighter death:

"WATCH-OUT!" SITUATIONS

1. Fire not scouted and sized-up
2. In country not seen in daylight
3. Safety zones and escape routes not identified
4. Unfamiliar with weather and local factors influencing fire behavior
5. Uninformed on strategy, tactics, and hazards
6. Instructions and assignments not clear
7. No communication link with crew members or supervisor
8. Constructing line without safe anchor point
9. Building fireline downhill with fire below
10. Attempting frontal assault on fire
11. Unburned fuel between you and fire
12. Cannot see main fire, not in contact with anyone who can
13. On a hillside where rolling material can ignite fuel below
14. Weather becoming hotter and drier
15. Wind increases and/or changes direction
16. Getting frequent spot fires across line
17. Terrain and fuels make escape to safety zones difficult
18. Taking nap near fireline

The "Watch-Out" list identifies some specific situations where the fire-related risk is higher than normal. Any firefighter in such a situation must be especially alert for changes in fire behavior that might increase the danger. The listed items are expanded as follows:

1. ***Fire not scouted and sized-up*** — If crews are assigned to fires or parts of fires where they cannot see the entire perimeter and have not had the opportunity to adequately perform a size-up, they are in danger. To provide any degree of safety, crews must be aware of where the fire is and what it is doing.
2. ***In country not seen in daylight*** — When crews are in an area that was not seen in daylight, they don't have the topographical information needed to work in an adequate degree of safety. Being unable to see the shape of the land, the density of the vegetation, and the distances between points all make the situation unsafe.
3. ***Safety zones and escape routes not identified*** — If crews or individual firefighters find themselves on the line not knowing where the safety zones and escape routes are, they should ***stop what they are doing*** until they find out where these critical safety features are located.
4. ***Unfamiliar with weather and local factors influencing fire behavior*** — The crew may find the local microclimate or burning conditions different from those in other areas. Safely coping with different conditions may require a conscious change in strategy or tactics.
5. ***Uninformed on strategy, tactics, and hazards*** — Not knowing what the plan of attack is, and how they fit into it, can place crews in serious jeopardy. They may find themselves in the path of danger such as air drops or backfires.
6. ***Instructions and assignments not clear*** — The results can be both unproductive and dangerous. Once a firefighter or a crew is on the fireline, it may be too late to get orders clarified.
7. ***No communication link with crew members or supervisor*** — This can result in critical, life-saving information not being passed up or down the chain of command. Maintaining reliable communications between all levels is an absolute necessity.
8. ***Constructing line without safe anchor point*** — While this may have to be done under some extraordinary circumstances, it is a dangerous practice that risks the crew being flanked and possibly surrounded by a fire. The decision to take this risk must be made only by a very experienced firefighter who has carefully weighed the benefits against the risk and established LCES.
9. ***Building fireline downhill with fire below*** — Building a line downhill (or making a hose lay in the green) toward a fire below is very hazardous. Fire normally burns faster uphill than downhill, and there is a greater risk of the fire flanking the crew working downhill. Also, convected heat, smoke, and flame rising upslope make it difficult for firefighters to breathe or see clearly, and they are likely to have very poor footing.
10. ***Attempting frontal assault on fire*** — A crew attempting a frontal assault on a fire (from the green) is in a dangerous position, especially if it has too few hoselines or hoselines that are too small. The fire may overrun the firefighters or spot behind them.
11. ***Unburned fuel between you and fire*** — It is dangerous for firefighters to be in any type of ground cover with unburned fuel between them and a fire. They are actually in the green and are susceptible to being overrun by the fire as they attempt to move through the unburned fuel to reach the burned area. Having readily available escape routes and safety zones is critical.
12. ***Cannot see main fire, not in contact with someone who can*** — If firefighters are working out of sight of the fire and they are not in contact with anyone who can see the fire, an unseen blowup can put them in danger of being overrun. Post one or more lookouts.
13. ***On a hillside where rolling material can ignite fuel below*** — If rolling materials start spot fires below, a new fire may

run upslope toward the crew. Since fire can spread rapidly upslope, firefighters are not likely to be able to outrun the fire.

14. ***Weather becoming hotter and drier*** — When the weather becomes hotter and drier, fires become increasingly active. New smoke may appear within the burn, smoldering duff supports visible flame, and upcanyon winds may start to blow through the ravines and across control lines. Spot fires may increase in number, and smoldering spot fires may be fanned back into life. Pyrolysis increases, and fuels become more susceptible to ignition. The likelihood of crowning increases, even during mop-up.

15. ***Wind increases and/or changes direction*** — Be aware of winds increasing or changing direction. Wind flattens out the flames, which results in the ignition of new fuels and increases the rate of spread. Wind blowing from the green into the black may suddenly reverse direction and blow hot materials or flames into new fuels.

16. ***Getting frequent spot fires across line*** — If spot fires across the line are numerous, it may be difficult to reach them while they are still small. However, if they are not controlled, they can combine into area ignition. Spot fires can also develop into separate major fires, and crews can suddenly find themselves and their equipment between two or more fires.

17. ***Terrain and fuels make escape to safety zones difficult*** — A crew some distance from the burned area or another safety zone can be in terrain or cover that makes travel difficult and slow. Slopes present hazards such as falling rocks and the potential for slipping. Irregular terrain can put firefighters out of sight of the fire and other personnel. Heavy cover may also restrict their ability to see the fire and may obscure escape routes.

18. ***Taking nap near fireline*** — A sleeping firefighter can be overrun by a fast-moving fire. Crew members should be allowed to sleep on the fireline only in safety zones and only when lookouts have been posted. Drowsiness may also be an indication of carbon monoxide poisoning.

STRUCTURAL "WATCH-OUT!" SITUATIONS

Just as there are situations that have resulted in firefighter fatalities in the wildland, there are also situations in structural fire fighting that can place firefighters in serious jeopardy. The most common of these Structural "Watch-Out!" Situations are as follows:

- Electrical lines (down or overhead)
- Wooden construction; shake roofs; overhanging eaves, porches, and decks; and large windows
- Poor access, narrow one-way roads
- Inadequate water supply
- Vegetation within 30 feet (9 m) of a structure
- Extreme fire behavior
- Strong winds (25 mph [40 kmph] or more)
- Evacuation needed
- Structures located in chimneys, in box or narrow canyons, on slopes of 30 percent or more, and in continuous, flashy fuel types
- Propane tanks and other aboveground fuel storage

The safety hazards that exist when assigned to protect structures from wildland fires can be significant. In addition to applying the Standard Fire Orders and avoiding the "Watch-Out" Situations, good judgment and planning are extremely important because of the presence of homeowners and their families, the media, pets and livestock, traffic, and unfamiliar combustibles. The safety considerations in these situations are as follows:

- Be aware of possible toxic fumes, and stay upwind and out of the smoke.
- Wear full structural protective clothing.
- Keep at least 100 gallons (400 L) of water reserve in the engine tank.
- Have a charged protection line for the crew and engine.

- Back the engine into position; firefighters may need to leave quickly.
- Use 1½-inch (38 mm) lines for all interior fire fighting.
- Use foam, if available, to coat the structure if time permits.
- DO NOT park under power lines, in saddles, or in chimneys.
- DO NOT enter a burning structure unless properly trained and equipped for interior fire fighting.

NOTE: If a fire threatens to overrun your position, the safest place for a short period of time (until the flame front has passed) may be inside the structure.

LOOK UP, LOOK DOWN, LOOK AROUND

Continually checking their environment is a habit that all firefighters should develop. It is one of the best ways that firefighters can protect themselves in hazardous situations. Some of the more common safety hazards that may be seen in this way are described in the following sections.

Look Up

When firefighters look up, they may see any number of potential hazards (Figure 8.10). When crews or equipment are working upslope, firefighters below must beware of rocks that may dislodge and roll down the hill toward them or burning materials that may roll into unburned fuel below them. Overhead power lines may burn through and fall on anyone below. Staying aware of what aircraft overhead are doing can reduce the chances of being hit by an air drop. During mop-up, firefighters must be wary of tree limbs weakened by fire and burning or burned-out snags that may fall without warning.

Figure 8.10 Overhead power lines are but one of the hazards that firefighters may see when they look up.

Look Down

Firefighters should always be careful about where they step (Figure 8.11). Stepping on a downed power line can be fatal. Poor footing, especially on steep slopes, can cause firefighters to fall. Unseen ditches, holes, or drop-offs can injure or kill firefighters who walk into them. In the wildland, snakebite is a possibility for the unwary.

Figure 8.11 Firefighters must pay attention to where they step as they walk.

Look Around

In the sometimes noisy environment of the fireline, firefighters may not hear signs of danger approaching. Someone operating a chain saw may not hear a vehicle approaching. Looking around can alert firefighters to approaching vehicles as well as to changes in the fire's behavior that might threaten them.

USING SAFETY AND SURVIVAL GUIDELINES

Any fire department with wildland responsibility, either within its jurisdiction or through mutual aid, should teach the Common Denominators, the Standard Fire Fighting Orders, LCES, and the "Watch-Out" Situations to all personnel. It is especially important for structure firefighters who may have little contact with wildland fires to know and understand these guidelines. Together these lists provide a formula for firefighter survival and are used as general guidelines for field operations.

The National Wildland Fire Coordinating Group (NWCG) has developed an interactive video course to teach the Eighteen "Watch-Out" Situations and the Ten Standard Fire Fighting Orders. For more information, contact the National Interagency Fire

Center (NIFC), 3833 S. Development Avenue, Boise, ID 83705. These guidelines are also included with this manual in the form of a pocket-size card that is ideal for review while waiting for deployment in staging areas or while en route to a fire.

PERSONNEL ACCOUNTABILITY

Keeping track of all personnel operating on a fire is a major priority and one of the most important safety functions. Properly applied, a personnel accountability system assists in planning rest and rehabilitation (rehab) for fireline crews and reduces the chances of a firefighter or a crew becoming lost or isolated during a fire. Personnel accountability, an integral part of incident command/management, involves a very structured process by which individual firefighters and crews check in and out with their supervisors when starting and finishing each assignment.

An on-duty list or other system, such as T-cards or a tag system, and frequent head counts are used to track and account for all personnel on the fireground (Figure 8.12). Supervisors are responsible for keeping track of where their subordinates are and what they are doing at all times. Using on-duty lists allows managers and supervisors to know exactly who is operating on the fireground, but crew leaders must be in constant contact with their personnel visually or by radio. Each crew leader keeps a copy of the crew's on-duty list. Leaders of engine crews and other apparatus-based crews keep a second copy in the apparatus. Frequent head counts reduce the possibility of someone becoming separated from the rest of the crew.

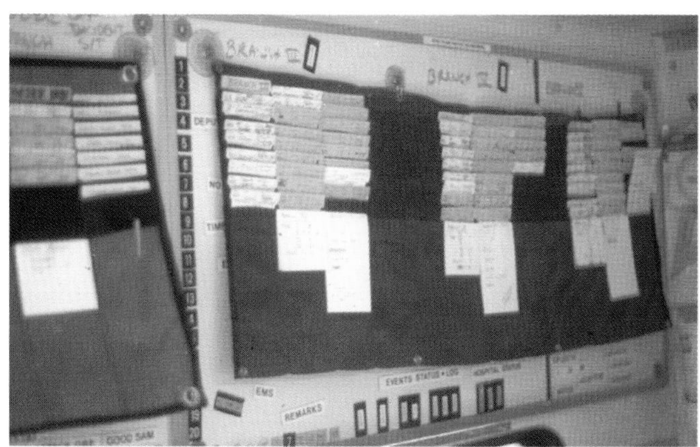

Figure 8.12 A T-card system helps to track personnel and other resources on the fireground. *Courtesy of Monterey County Training Officers.*

DOWNHILL FIRE FIGHTING SAFETY

Whenever possible, a direct attack should be used, but it is sometimes necessary to construct a control line or extend a hose lay downhill toward a fire. This is very dangerous in fast-burning fuels and steep terrain because the fire may burn across the slope below the crews and sweep uphill to trap them. However, if it is decided that this attack must be made to protect life or valuable property and the benefits outweigh the risks in this particular situation, the following guidelines should be used:

- The decision should be made by a competent firefighter only after thoroughly scouting the situation.
- Downhill line construction or hose lay should not be attempted when an intense fire is present directly below the proposed starting point.
- An air drop to knock down the fire should be considered along with keeping a loaded air tanker or helicopter circling overhead to protect the crews.
- The fireline should not be constructed in or adjacent to a chimney or chute that could burn out while a crew is in the vicinity.
- A downhill line or hose lay should be securely anchored at the top. Avoid undercut line whenever possible.
- Blacklining (burning vegetation adjacent to a control line to widen and strengthen it) should be done as the line progresses, beginning from the anchor point at the top. The burned-out area provides a continuous safety zone for the crew and reduces the likelihood of fire crossing the line.
- Communication must be established between the crew working downhill and crews working toward them from below. When no crew can adequately observe the fire, communications must be established among the crews, supervisors, and lookouts who are posted where the fire's behavior can be seen.
- The crew must be able to rapidly reach a safety zone from any point along the line in case the fire suddenly crosses below them.

- A downhill hose lay must be made with 1½-inch (38 mm) hose or larger. Hose must provide sufficient water flow to contain the worst-case fire conditions expected.
- Personnel must be aware of and avoid the WATCH-OUT SITUATIONS.
- Full compliance with the STANDARD FIRE ORDERS must be assured.

Line Scouting

To increase crew safety when hand crews are concentrating on constructing a line in brushy terrain, the area should be scouted.

- When scouting or working ahead of a crew in brushy terrain, carry a cutting tool and clear any vegetation that might hamper escape.
- A lookout should be posted to warn of danger when personnel are scouting in unburned areas of dense vegetation.

Line Construction/Hose Lays

When constructing line or laying hose, firefighters must remain alert for changing fireline conditions that may threaten their safety.

- Firefighters must make sure their footing is secure and working positions are safe. Walk, DON'T run (except to an escape route or safety zone).
- Personnel or machines should not work directly above each other or close together when working on steep slopes.
- When there is a danger of rolling rocks and logs, the crew leader should do the following:
  - — Post a lookout to warn the crew.
  - — Space crew members more than 10 feet (3 m) apart.
  - — Stagger crew members so they are not working or walking directly above or below each other.
- Firefighters should be briefed on what to do when a warning is given:
  - — Face the danger and do not move until the danger is seen.
  - — Quickly move behind the protection of the nearest large tree or other stable barrier.
  - — Quickly move into an opening offering maximum upslope visibility, and stand facing the oncoming danger, prepared to react instantly, if suitable protection is not nearby.
- Loose rocks along bulldozer breaks should be stabilized before a crew works below them.
- Firefighters should pass a burning or fire-weakened tree only on the uphill side, or above the lean, and then watch it closely.
- In fast-burning fuels, firefighters should watch out for fast fire runs in any direction, at any time of day or night. If cutting a line or laying hose across the head means a slow access and retreat, they should use a flank attack, starting at a safe anchor point.
- Firefighters should patrol below for spot fires from hot materials rolling down slopes.
- Firefighters should stay calm but act decisively — panic is dangerous.
- Firefighters should never try to outrun the head of a fast-moving fire. They should try to get to the flanks or into the burned area.

PERSONAL PROTECTIVE EQUIPMENT

The safety of firefighters is significantly enhanced if they wear the proper protective ensemble for the type of fire being fought. Personal protective equipment (PPE) for structural fire fighting and that for wildland fire fighting is quite different, except for the fire-resistant materials of the protective clothing. It is not recommended that structural gear be worn when fighting a wildland fire and vice versa. Civilian volunteers (Red Cross, Salvation Army, etc.) and contract workers must also be provided with protective clothing if they are to be outside the incident base (Figure 8.13). No one should be allowed near a fire without proper protective clothing for the job they are assigned. All personal protection items must meet applicable local, state/province, and national standards.

Figure 8.13 Support personnel must wear full wildland PPE when they work outside the incident-base area.

Structural Gear

Specifications for personal protective equipment used in structural fire fighting are contained in several NFPA standards: 1971 (protective clothing), 1972 (helmets), 1973 (gloves), 1974 (protective footwear), and 1404 (self-contained breathing apparatus) (Figure 8.14). This gear is designed to be functional in the activities associated with fighting structure fires but is generally too bulky, too hot, and too heavy to be practical for use in wildland fires. However, if they are trained in its use, wildland firefighters should have structural turnout gear available to them when assigned to structure protection in the wildland/urban interface.

Figure 8.14 A firefighter in full structural gear and SCBA.

Wildland Gear

Specifications for wildland fire fighting PPE are contained in NFPA 1977, *Standard on Protective Clothing and Equipment for Wildland Fire Fighting*. This equipment includes fire shelters, helmets, goggles, protective footwear, brush jackets/pants or one-piece jumpsuits, and gloves (Figure 8.15). Different forms of respiratory protection for wildland firefighters are available (Figure 8.16).

Figure 8.15 Firefighters in typical wildland PPE.

When assigned to the fireline, firefighters should wear the following protective clothing and equipment:

- Fire shelter
- Helmet or hard hat (with shroud)
- Lace-up leather boots with slip-resistant soles
- Flame-resistant clothing (no synthetic clothing — even undergarments)

Figure 8.16 Various forms of respiratory protection may be used in wildland fire fighting.

- Leather gloves
- Eye protection
- Hearing protection (when working near noise-producing equipment, such as helicopters, chain saws, or pumps)
- Chaps (when operating chain saws)
- Dust/smoke mask (when in smoky or dusty environments)

FIRE SHELTERS

Fire shelters are extremely valuable pieces of safety equipment that have saved many lives and prevented many serious burn injuries. In one fire alone (Butte Fire, Salmon National Forest, 1985), 73 firefighters survived a high-intensity crown fire in the Wallace Creek drainage by deploying their fire shelters. Five of the firefighters suffered heat exhaustion and dehydration, requiring hospitalization overnight for observation; the others were uninjured. However, fire shelters should be viewed only as a last resort when there is no way out of a life-threatening situation. Firefighters must also guard against becoming overconfident because they have a shelter and must not place themselves in danger unnecessarily. Fire shelters do not reduce the need for proper scouting, posting lookouts, and establishing escape routes and safety zones.

Every firefighter must have his or her own shelter, and shelters should not be shared except to save the life of someone who has no shelter. To keep fire shelters readily available for use, and to make their deployment easier and faster, fire-shelter pouches should be worn in a position that keeps the pouch visible to the wearer at all times

(Figure 8.17). An extra fire shelter should also be stored in the cab of each engine to cover the windows in case the engine is about to be overrun by fire.

Figure 8.17 The fire shelter should be visible to the wearer at all times.

Fire shelters are small fold-up tents made of fiberglass cloth bonded to aluminum foil with a nontoxic, high-temperature adhesive (Figure 8.18). They protect by reflecting radiant heat, and this gives occupants cooler, more breathable air.

Figure 8.18 A fully deployed fire shelter.

The unreflected radiant heat is absorbed by the shelter, which can cause the temperature inside to gradually rise. However, these elevated temperatures are survivable, and it is important for firefighters to remain calm and stay inside their fire shelters until the flame front has passed. Because of their importance to firefighter survival, shelters must be treated with reasonable care:

- Inspect shelter according to manufacturer's instructions — unbroken seal, smooth edges.
- Keep shelter away from sharp objects that may puncture it.
- Do not place heavy objects on top of a shelter.
- Avoid rough handling of a shelter if possible.
- Do not lean a shelter against objects.
- Do not sit on a shelter or use it as a pillow.

Effective fire-shelter deployment depends on recognizing the need for the shelter soon enough to allow time to get to a safety zone and deploy it. In most cases, the crew leader or supervisor decides when and where crew members deploy their shelters. If firefighters are not part of a crew or if they become separated from their crew, they must make these decisions for themselves.

Fire shelters work best in areas where fuels are light and the flame front passes quickly. Natural firebreaks (creek beds, depressions in the ground, rock slides, and the lee side of ridges and hills) can be used if they are available. Wide control lines such as bulldozer lines and burned areas may also be used.

Because fire consumes anything that burns, the following items should be kept well away from the spot where shelters are to be deployed:

- Fusees
- Gasoline cans
- Supply boxes
- Packsacks
- Other combustible fire fighting gear

When selecting a spot for fire-shelter deployment, avoid areas with heavy brush, trees with low branches, and logs and snags. Also avoid narrow draws, saddles, chimneys, and similar terrain features. Once a specific spot for a shelter is selected, an area about 4 by 8 feet (1.2 m by 2.4 m) (larger if time allows) should be scraped free of vegetation down to mineral soil (Figure 8.19). Burning out light fuels may be easier and faster than scraping.

Figure 8.19 A firefighter scrapes a spot for fire-shelter deployment.

The spot should be as free of fuels as possible because grass and duff inside a shelter can smolder or ignite, filling the shelter with smoke. A clean area also minimizes flame contact with the shelter.

If the flame front arrives before an area is cleared, firefighters can still get under their shelters in a matter of seconds, although shelters are more difficult to deploy in high winds. After the shelters are out of the package and opened, firefighters stand inside their opened shelters with their backs toward the oncoming flame front (Figure 8.20). They then lie face down with their feet toward the fire (Figure 8.21). The foot end will be the hottest spot in the shelter but is easily held down with boots. Firefighters should take only their protective clothing (including helmet, shroud, and gloves), portable radios, and canteens into their shelters. Their boots, gloves, and helmets help protect them from conducted heat as they hold down the edges of their shelters (Figure 8.22).

Figure 8.20 The firefighter stands with his back toward the approaching fire.

Figure 8.21 The firefighter drops to his knees and falls forward.

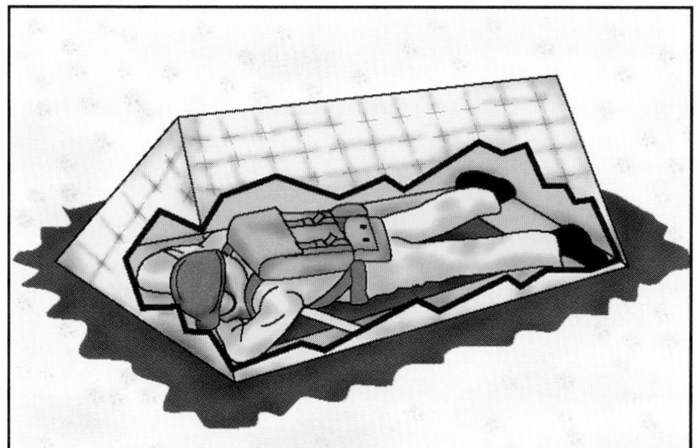

Figure 8.22 Cutaway illustration of a firefighter inside a fully deployed fire shelter.

Their portable radios allow them to contact others for help and communicate with each other about when it is safe to leave their shelters. If they are in the shelters for a prolonged period, they can sip water from their canteens.

CAUTION: Firefighters should never wet themselves down before or after entering a fire shelter. The moisture greatly increases the likelihood of their being burned.

While fire shelters are one of the most significant developments in wildland firefighter safety, they do have limitations. As mentioned earlier, they function primarily by reflecting 95 percent of the radiant heat. However, a fire in dry chaparral can burn at more than 1,000°F (538°C), and this can raise the temperature inside a shelter by more than 50°F (10°C) *above* the atmospheric temperature. In addition, a fire could consume the majority of the available oxygen, leaving little if any for the firefighters in their shelters.

HELMETS

Hard hats or helmets with chin straps must be worn for head protection. Lightweight wildland helmets are preferred to structural helmets, although some helmets are designed to be used in both types of fires (Figure 8.23). They should be equipped with a protective shroud for face and neck protection, but firefighters in some municipal fire departments use their Nomex®/PBI® hoods to provide face and neck protection on wildland fires (Figure 8.24). Goggles with clear lenses should also be worn (Figure 8.25).

Figure 8.23 A typical combination helmet.

Figure 8.24 Firefighters sometimes wear Nomex® hoods with combination helmets.

Figure 8.25 Most wildland firefighters wear goggles with clear lenses.

PROTECTIVE FOOTWEAR

What is deemed acceptable in footwear for wildland fire fighting varies in different geographical regions, but some standard guidelines apply in all areas. Lace-up leather safety boots with lug or grip-tread nonslip soles are used most often. The laces should be of leather to be as durable as the boots. Boots should be at least 8 to 10 inches (203 mm to 250 mm) high to support the ankle and protect the lower leg from burns, snakebites, and cuts and abrasions (Figure 8.26).

Figure 8.26 A typical wildland boot.

PROTECTIVE CLOTHING

The protective clothing worn by wildland firefighters, often called "brush gear," consists of unlined pants and jacket or a one-piece jumpsuit. In either case, the cuffs of the sleeves and the pants legs can be closed snugly around the wrists and ankles. The fabric is treated cotton or some other inherently flame-resistant material such as Nomex. Underwear of 100 percent cotton, including a long-sleeved T-shirt, should be worn under brush gear. Socks should be made of wool or other natural fiber.

CAUTION: Never wear synthetic materials at a fire. These materials will melt when heated and stick to your skin, and this greatly increases the likelihood of burn injuries.

GLOVES

Wildland fire fighting gloves are made of leather or other suitable materials and must provide wrist protection (Figure 8.27). They should be comfortable and sized correctly to prevent abrasions and blisters.

Figure 8.27 Wildland firefighters' gloves should provide wrist protection.

Agency Policy

Every fire department should have established policies and procedures relating to the maintenance and use of personal protective equipment, including fire shelters. These policies and procedures should be in compliance with applicable laws and standards, such as NFPA 295, *Standard for Wildfire Control*, and NFPA 1051, *Standard for Wildland Fire Fighter Professional Qualifications*, and with accepted safe practices within the fire protection community. Firefighters are responsible for knowing and complying with their agency's policies and procedures.

One important part of agency policy regarding fire shelters should be a requirement for thorough training in their deployment and use. Agencies should provide shelters that are used for training only so that every firefighter can practice deploying them. Shelters used for training or demonstration should not be used on the fireline. Firefighters should be required to practice shelter deployment until *mastery* is achieved. Nothing less is acceptable. In some agencies, mastery is defined as correctly deploying a shelter within 25 seconds. Firefighters should practice shelter deployment under a variety of conditions — in the wind, on steep terrain, while running, etc. For more information on fire shelters and their use, a videotape entitled *Your Fire Shelter* is available for purchase from the National Interagency Fire Center in Boise, Idaho.

Inspection

Wildland PPE is designed to provide a high degree of safety if properly worn and maintained. To be sure that it provides the intended level of protection, firefighters must regularly inspect it. Inspection is normally done at the beginning of each work shift, but it may be required more frequently under heavy use during a fire. Firefighters should follow departmental guidelines and apparel/equipment manufacturer's recommended inspection procedures.

Maintenance

Personal protective equipment should be cleaned periodically and after every major fire in which it was used. Improper cleaning can reduce the flame-retardant properties of some materials, so all maintenance should be done according to the manufacturer's specifications.

Wearing Gear Properly

For personal protective clothing to provide its designed level of protection on the fireline, firefighters must not only wear the proper gear, they must wear the gear properly. This means that all buttons should be buttoned and all zippers zipped (Figure 8.28). Gloves must be on, and goggles and helmet shrouds must be in place.

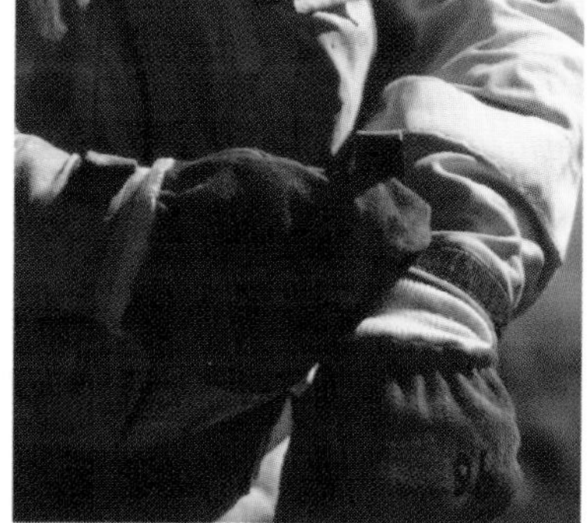
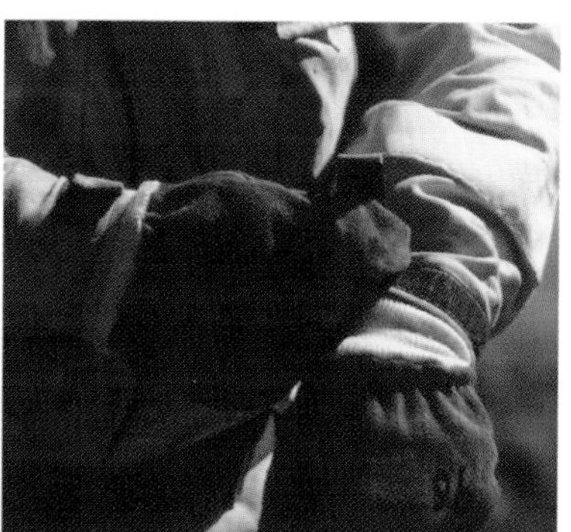

Figure 8.28 All closures should be fastened before work is begun.

Canteens

Carrying water is vital to firefighters being able to stay sufficiently hydrated (Figure 8.29). Therefore, canteens should be refilled at every opportunity during a fire. They should be kept as clean and sanitary as possible under the conditions, and they should not be shared among crew members.

Figure 8.29 Each Firefighter should carry two canteens of water. *Courtesy of NIFC.*

Web Gear

Web gear refers to the traditional web belt with shoulder straps or the more recent backpacks that firefighters wear in the wildland. Various tools and canvas pouches are carried in or attached to the gear. Items typically carried are a canteen (or two), small hose clamp, weather kit, fire shelter, and first-aid kit (Figure 8.30). There is often a pouch for holding a portable radio (Figure 8.31). Also carried in the web gear are fusees used for burning out (Figure 8.32).

Figure 8.30 A wildland firefighter in typical web gear.

Figure 8.31 Web gear usually has a pouch for a portable radio. *Courtesy of NIFC.*

Figure 8.32 Firefighters often carry fusees in their web gear.

Fusees

From a safety standpoint, the main reason firefighters carry fusees with them is to be able to burn out an area to create a safety zone. If there isn't time to create a safety zone any other way, burning out a small area may be the only option. While fusees can enhance a firefighter's level of safety, they can also be a safety hazard under some circumstances. Fusees are designed to burn, so if there is any danger that they will be exposed to direct flame impingement, they should be discarded. For example, fusees should never be taken into a fire shelter.

SAFETY BRIEFING

Prior to an engine or a hand crew being committed on a fire, the crew leader or supervisor should gather all members together for a safety briefing.

Sometimes called "tailgate safety sessions," these briefings give everyone an opportunity to obtain information that may prove critical to their safety and survival once they are on the line.

During these sessions, the crew is informed of what is known about the incident in general, including the weather forecast, and its assignment in particular. Crew members are informed of how they are to reach their assigned work area and of the location of escape routes and safety zones (Figure 8.33). They are informed of their assigned radio frequency and reminded to test their radios. Crew leaders inspect all tools and PPE and make repairs or replacements.

Figure 8.33 A crew gets a preassignment briefing. *Courtesy of NIFC.*

PERSONAL SURVIVAL TECHNIQUES

By staying alert for hazardous situations developing, keeping the Eighteen Watch-Out Situations in mind, following the Ten Standard Fire Fighting Orders, and remembering LCES and the Common Denominators of Fire Fatalities, firefighters can reduce the risk of being trapped by a fire. However, it is not possible to completely eliminate that risk. Therefore, every firefighter must know and apply the techniques for safety and survival in wildland fires.

When they are threatened with being overrun by a rapidly advancing flame front, crews should stay together, stay calm, and follow the orders of the crew leader. Crew leaders must remain calm but act decisively. Every crew member should know the locations of escape routes and safety zones, but they should resist the urge to try to outrun a fire, especially uphill. If they must run at all, they should run laterally or downhill. Especially in thick brush, trying to run from a fire will almost certainly not work. In addition, firefighters may take foolhardy risks because heat and smoke can impair their judgment. Poor judgment has been the cause of many firefighter fatalities in wildland fires. If ordered to take shelter, firefighters should follow the procedures described in the Safety Zones section earlier in this chapter.

Fluids

Fighting a wildland fire can be very hard work, and firefighters often must work in conditions of high temperatures and low humidity. As a result, they perspire profusely (Figure 8.34). It is possible for firefighters to lose as much as 8 quarts (7.6 L) of water in one hour to perspiration. This rate of fluid loss cannot be sustained, but a rate of 1 quart (0.95 L) per hour over a 12-hour work shift is not uncommon. To maintain their ability to work and to avoid developing heat cramps, heat exhaustion, or heatstroke, it is critical that firefighters replace lost water, salt, and potassium.

Figure 8.34 Firefighters perspire profusely when working hard in hot, dry conditions.

REPLACING LOST WATER

To replace the water lost to perspiration, firefighters must drink lots of fluids, but water is best (Figure 8.35). Coffee and carbonated soft drinks are not recommended. To avoid dehydration, they should drink 8 to 16 ounces (240 ml to 480 ml) of water or juice before starting a work shift. They should carry canteens of water with them on the fireline and should drink at every break (whether they feel thirsty or not) and at every meal break (Figure 8.36). They should not drink stream water unless they have an approved purification method.

Figure 8.35 Water is the best fluid replacement for firefighters.

Figure 8.36 Firefighters should drink water at every opportunity.

REPLACING LOST SALT

To replace salt lost through perspiration, most physically fit and acclimated firefighters need only to lightly salt their food at mealtimes. Unacclimated firefighters and those in less-than-top condition may have to add slightly more salt to their food to replace what was lost. However, firefighters should *not* take salt tablets because they are not needed and may result in excessive salt intake. This may cause an elevated core temperature that can lead to the heat-related illnesses described later in this chapter. An excess of salt can also cause stomach distress, muscle soreness, fatigue, impaired heart function, high blood pressure, mental confusion, and a loss of potassium.

CAUTION: Firefighters should not take salt tablets because an excess of salt can interfere with the body's ability to regulate its core temperature.

REPLACING LOST POTASSIUM

Potassium is vital for adequate muscle function, but it can be depleted through perspiration caused by working in the heat for extended periods. To replace this loss, potassium-rich foods like bananas and other fruits should be eaten at frequent intervals during fire fighting operations (Figure 8.37). Lemonade, tomato juice, and electrolyte replacement fluids ("sports drinks") can also help to replace lost potassium (Figure 8.38).

Figure 8.37 Firefighters should eat fruit to replace lost potassium.

Figure 8.38 Sports drinks can also help replace lost potassium.

Food in Vehicles

Because firefighters may be working miles (kilometers) from the incident base and hot meals may not always be available on the line, firefighters should keep food in their vehicles whenever possible. Foods that do not require refrigeration, such as fresh fruit and packaged snacks, can provide the necessary energy boost that firefighters need during a long shift on the line (Figure 8.39).

Figure 8.39 Packaged snacks can help maintain energy levels.

First Aid

A knowledge of first aid and how to recognize and treat those injuries and illnesses common to the wildland environment is an important survival skill for firefighters. Because they can be life-threatening, the heat-related illnesses discussed in the following sections are among the most important to firefighters in wildland fire fighting. A knowledge of cardiopulmonary resuscitation (CPR), how to render aid for those suffering from strains, sprains, and fractures, and how to properly treat burn injuries are also important to all firefighters (Figure 8.40). Firefighters in the wildland are vulnerable to snakebites and a variety of insect bites.

Figure 8.40 All firefighters should have basic first-aid training.

They may also be exposed to certain plants, such as poison oak and poison ivy, that can produce some uncomfortable and debilitating effects, especially if their oils are inhaled in smoke. For information on treating fireground medical emergencies, firefighters should refer to their agency's medical first responder text.

On large-scale incidents, the Medical Unit (when activated) is responsible for handling all medical emergencies for personnel assigned to the incident

(Figure 8.41). First-aid facilities, supplies, and treatment should be made available on the line and at all incident facilities. Prompt first aid should be given for all injuries. In addition to their first-aid training, firefighters should be trained in recognizing and treating those conditions resulting from heat stress.

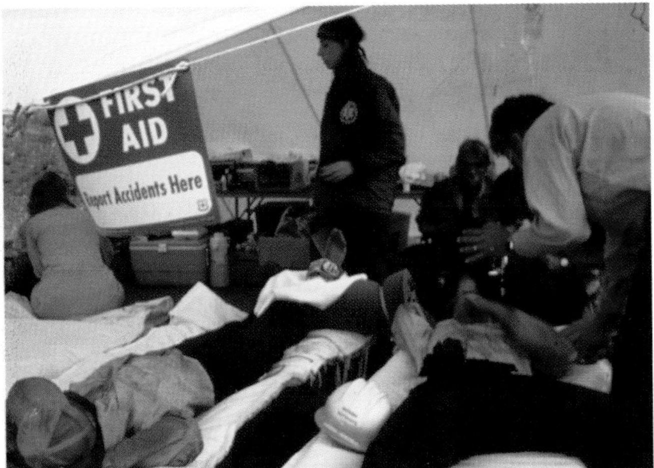

Figure 8.41 The Medical Unit treats incident personnel injured on a fire.

HEAT STRESS

The atmospheric conditions that favor combustion — high temperature and low humidity — combined with the heavy manual labor involved in fighting fire over an extended period create conditions in which firefighters become vulnerable to heat-related problems such as heat exhaustion and even heatstroke. On the fireline, firefighters must pace themselves, take periodic short rest breaks, drink plenty of water, and monitor each other for signs of physical distress. When firefighters anticipate being on the fireline on very hot days, they can reduce the likelihood of suffering heat-stress disorders by drinking at least 2 quarts (1.9 L) of water before noon (prehydrating) and continuing to drink water or sports drinks throughout the day. Heat-stress disorders are divided into three categories: heat cramps, heat exhaustion, and heatstroke.

Heat cramps. This sometimes painful and debilitating condition is caused by not replacing salt lost through perspiration. The symptoms of heat cramps are simply muscle cramps in the legs and abdomen, but they can be quite painful. The victim should be encouraged to drink electrolyte replacement fluids.

Heat exhaustion. This condition is caused by a loss of both water and salt due to perspiration. The symptoms are general weakness and fatigue, an unstable gait, and pale, cool, clammy skin. The victim may faint briefly but quickly regain consciousness. The victim should be moved into the shade, have outer clothing removed, and be given water or electrolyte drinks if tolerated. The victim should lie down with the feet slightly elevated.

Heatstroke. This is a life-threatening condition caused by a total collapse of the body's temperature-regulating mechanisms. The symptoms of heatstroke are hot, red, dry skin, with a body temperature in excess of 105°F (41°C) and delirium or loss of consciousness.

WARNING

Heatstroke victims must be immediately and aggressively cooled with cool water and transported to a medical facility as soon as possible. Brain damage or death can result if treatment is delayed!

SERIOUS INJURY/FATALITY PROCEDURES

Every fire protection agency should have policies regarding how to deal with personnel who have been seriously injured or killed. Firefighters should know and follow their agency's policies. Agency policy may include the following suggested procedures:

Serious injury

- Give first aid — call for medical aid and transportation.
- Do not release victim's name, except to authorities, nor use it on the radio.
- Do not allow unauthorized picture-taking of victim or release of pictures.
- Notify incident commander, who will do the following:
  - Assign someone to lead evacuation, if necessary, and stay with the victim until under medical care. In rough terrain, as many as 15 workers may be needed to transport a litter.

— Assign someone to get facts and witness statements and preserve evidence until investigation can be taken over by the safety officer or appointed investigating team.

— Notify the agency administration.

Fatality

- Do not move body except as necessary to establish positive identification.
- Do not release victim's name, except to authorities, nor use it on the radio until next of kin is notified.
- Do not allow unauthorized picture-taking of victim or release of pictures.
- Notify incident commander, who will do the following:
  - — Assign someone to conduct investigation until relieved by agency-investigating team.
  - — Notify agency administration and report essential facts. The agency administration will notify proper authorities and next of kin as prescribed by agency regulations.
  - — Assist authorities, if requested, in transporting remains, marking location of body on the ground, photographing the scene, and noting the location of tools, equipment, or personal gear.

FIREFIGHTER REHABILITATION

Crew leaders are responsible at all times for knowing the location and status of every member of their crew on the fireline, but firefighters also must operate as a team and look out for each other. Firefighter fatigue compromises safety because fatigued firefighters have less control over their tools, and they may become less alert to changing fire conditions.

Firefighters must be in good physical condition to do the arduous manual labor required. They must drink plenty of water or other recommended fluids and stay alert for signs of heat stress in themselves and other crew members, especially during the hottest part of the day. Crew leaders must set and maintain a work pace in which their personnel are productive but not exhausted before their assignments are completed.

Firefighters in the wildland typically work 12- to 24-hour shifts on the fireline, usually in high atmospheric temperature, sometimes in high humidity, and sometimes at high elevations. Just as there is less oxygen available to support a fire at higher elevations, there is less oxygen for hard-working firefighters to breathe. In addition, they may have had to reach that elevation on foot and may have had to construct a scratch line (unfinished preliminary control line) along the way. The firefighters may arrive already physically exhausted. These conditions — in addition to the normal fatigue that results from doing hard work under adverse conditions — combine with the lack of oxygen to create conditions in which the firefighters become vulnerable to the heat-related problems described earlier in this chapter. These conditions take a terrific toll on even the fittest firefighters, and those who are in less-than-ideal condition suffer even more. On large-scale incidents, the IC should consider the need for one or more firefighter rehabilitation stations, usually called "rehab" in fireline terminology, to be set up to reduce these effects and to allow firefighters to continue to perform their duties (Figure 8.42). These stations provide firefighters with shade, water and other fluids, and first aid if needed. However, under these typical fireline conditions, firefighters should also pace themselves, take frequent short breaks, drink plenty of water, and monitor each other for signs of physical distress.

Figure 8.42 Firefighters can rest and replenish in the shade of a rehab station.

Monitoring Personnel

NFPA 1500, *Standard on Fire Department Occupational Safety and Health Program*, says that crew supervisors must continuously monitor the condition of their personnel and provide adequate fluid replenishment and work relief to maintain their health and safety. These general requirements are translated into specific actions on the fireline. In addition to general rehab procedures, crew leaders must know what the incident action plan specifies and must weigh the effects of temperature, humidity, topography, and workload on their personnel to establish appropriate intervals for rest and rehydration.

Individual firefighters have a responsibility to monitor their own condition and that of their co-workers on the line. Is anyone in the crew sweating more than the others? Is anyone behaving abnormally? Is anyone dizzy, chilled, or nauseated? Is anyone showing signs of excessive fatigue? If the answer to any of these questions is "yes," the affected personnel should immediately be taken off the line, evaluated, and treated as described earlier.

Work Pace

Whenever possible, firefighters should adjust the pace of their work to fit the conditions. They should do the heaviest work during the coolest hours of the shift. They should change tools or jobs occasionally to either work different muscles or work the same muscles in different ways. Firefighters working on the line should take frequent short breaks to catch their breath and take a drink of water.

Rest Periods

Crew leaders or supervisors should schedule work breaks for their crews as the work and working conditions dictate. However, the harder the work and the hotter the weather, the more frequent the rest periods should be. The length of the breaks is also a matter of judgment, but personnel should be allowed to rest, in the shade if available, at least long enough for their heart rate (pulse) to return to normal. In general, the work/rest ratio should be 2:1; that is, 16 hours of work should be followed by 8 hours of rest. During long-duration assignments, it is recommended that firefighters be given a minimum of 1 day off in 14 days and/or 2 days in 21.

MONITORING FIRELINE CONDITIONS

Changing fire conditions can put firefighters in mortal danger if they do not monitor their situation carefully enough to recognize what is happening and react quickly and appropriately. Firefighters can protect themselves from a sudden blowup if they are alert to changing conditions, avoid known hazardous situations, and have a plan of action if they do get into a life-threatening situation. Following are some safety practices that firefighters should use in situations with blowup potential.

Monitoring the Weather

When working on or near an active fireline, firefighters must constantly monitor the weather for signs of change. Some of the indicators of sudden weather changes are as follows:

- Trees crowning-out inside fireline
- Smoldering fires intensifying over large areas
- Approaching thunderheads with dark clouds beneath
- Presence of dust devils and whirlwinds
- Increased spotting
- Sudden calm
- High clouds moving fast in direction different from surface wind

Monitoring Working Situations

Some working situations on the fireline involve more potential risk than others. Firefighters should monitor the following working situations closely:

- Building fireline downhill into an area where an intense fire is burning
- Building fireline on hillside below a fire
- Building fireline through heavy cover some distance from a fire
- Building fireline in country not seen in daytime

Planning for Potential Blowup Conditions

In situations that have a potential for blowups, crew leaders must have a plan for protecting their personnel. To increase the safety of fireline personnel, crew leaders should do the following:

- Advise personnel of escape routes, and make necessary provisions to ensure the route is clear for foot travel.
- Give crew frequent rest periods.
- Verify location of rock slides, open hillsides, streams, etc.

- Post lookouts to give alarm when firefighters are working where they cannot directly observe danger points.
- Consider possibility of retreating into burned area.
- Have personnel protect their faces when crossing fire edge into burned area.
- Do not travel in direction of fire spread unless it is certain a safe spot can be reached.
- Carry matches and/or fusees to burn out safety zones.

RESPIRATORY PROTECTION

Considering the toxic products of combustion of many wildland fuels, such as carbon monoxide, nitrogen oxides, sulfur oxides, and reactive aldehydes, there is always some need for respiratory protection for firefighters in the wildland. The widespread use of pesticides and herbicides on crops and rangelands has increased this need. As mentioned earlier, several types of respiratory protection devices are available for use in wildland fire fighting, but many are heavy and may increase heat stress. However, if firefighters in the wildland are faced with having to suppress a fire in any known or suspected concentration of pesticides or other toxic chemicals, such as in old dump sites, they should wear structural SCBA despite its weight and relatively short duration. If the situation permits, an indirect attack may reduce the need for respiratory protection.

In many dusty and/or smoky situations, firefighters can protect themselves from inhaling some of the airborne particulates by using some form of filter mask. Some firefighters carry surgical-type filter masks, and even a bandanna helps filter out some of these irritants (Figure 8.43).

Figure 8.43 Some firefighters wear bandannas as filter masks. *Courtesy of NIFC.*

APPARATUS OPERATIONS SAFETY

Wildland fire apparatus safety consists of considerations for the safety of the personnel operating on and around the equipment and those for the protection of the apparatus from abuse and exposure to the fire. It can also be broken down into the following broad categories: general guidelines, off-road guidelines, engine operations and safety, power-winch operations and safety, and personnel transport.

General Guidelines

When responding to, working on, or returning from a fire, apparatus operators are responsible for the safe operation of the vehicle and for the safety of the personnel on and around the vehicle, including pedestrians. Firefighters must ride inside the vehicle (never on the tailboard, running boards, bumpers, or fenders) and wear safety belts and protective clothing at all times (Figure 8.44). The headlights on fire apparatus should be on, day or night, whenever the vehicle is in motion (Figure 8.45).

Figure 8.44 Firefighters should ride inside the vehicle at all times.

Figure 8.45 For safety, headlights should be on whenever apparatus is in motion. *Courtesy of NIFC.*

When the vehicle operates under conditions of reduced visibility because of smoke or darkness, it should be driven at an appropriately reduced speed. A spotter (scout) may be needed to walk ahead of the vehicle to help locate and avoid obstacles such as logs, stumps, rocks, low-hanging limbs, ditches, and gullies. Spotters should be equipped with reliable hand lights, wear highly visible clothing, and stay within the driver's field of view at all times (Figure 8.46).

Figure 8.46 Spotters should wear highly visible clothing. *Courtesy of Monterey County Training Officers.*

When the apparatus is operated in a stationary position, such as when pumping from draft, it should be positioned for maximum protection from heat and flames. Natural or man-made firebreaks such as streams or roads can be used. Driver/operators should consider potential hazards such as falling trees, rolling rocks, incoming air drops, and heavy equipment building control lines when selecting a position for the apparatus. A short 1½- or 1¾-inch (38 mm or 45 mm) line should be deployed and charged for protection of the apparatus (Figure 8.47). Whenever the apparatus is parked, the wheels should always be chocked (Figure 8.48). The apparatus should be parked heading in the exit direction.

Off-Road Guidelines

Fighting wildland fires often requires apparatus to leave the roadway to reach the fire. The

Figure 8.47 A charged protection line should be immediately available.

Figure 8.48 Apparatus wheels should be chocked when the vehicle is parked.

weight, low ground clearance, and large turning radius make some structural fire apparatus unsuited for off-road use. Even with apparatus specifically designed for off-road use, apparatus operators must know their units' capabilities and limitations and operate them accordingly. Damage to the vehicle and injury to personnel can result from mishandling or overtaxing the vehicle. The speed at which a vehicle is driven should reflect the driving conditions (Figure 8.49). The apparatus and crew could be in jeopardy if a breakdown occurs.

Figure 8.49 Apparatus should be driven at a speed appropriate for the conditions. *Courtesy of Monterey County Training Officers.*

On steep hillsides, loose or unstable ground can cause the apparatus to slide or overturn, especially if it has a relatively high center of gravity (Figure 8.50). Even on level terrain, a vehicle can become mired in soft ground, sand, or mud, leaving it vulnerable to being overrun by a fire (Figure 8.51). Do not drive apparatus across a bridge unless the bridge is known to be strong enough to support the vehicle's weight (Figure 8.52). Do not attempt to ford streams with apparatus that is not designed to do so.

Figure 8.50 Wildland fire apparatus should not be driven beyond its limitations. *Courtesy of NIFC.*

Driving fire apparatus on the shoulders of railroad roadbeds can result in tire damage from the coarse, angular rock of which the roadbeds are made (Figure 8.53). In addition, this rock has relatively little cohesion, so apparatus may be in danger of sliding and/or rolling over on these steep inclines.

Figure 8.51 Even "Hummers" can get stuck. *Courtesy of NIFC.*

Figure 8.52 Apparatus should only be driven across bridges that are strong enough to support the weight of the vehicle.

Figure 8.53 Apparatus should not be driven on the slopes of railroad roadbeds.

When apparatus is used in a mobile attack, hoselines should be kept as short as possible (Figure 8.54). This helps to keep the apparatus mobile because shorter hoselines are less likely to become looped around stumps or other objects and are easier to disentangle if they do get caught. Except when operating in the black, the last 100 gallons (400 L) of water in the vehicle's tank should be reserved for protection of the apparatus. When progressing along the fire's edge, the crew should make sure that the fire is out. To ensure complete extinguishment, engines can work in tandem or a single engine can work with a hand crew (Figure 8.55). The second engine or the hand crew mops up and patrols the line, making sure all fire is extinguished.

Figure 8.54 Mobile attack lines should be kept as short as possible. *Courtesy of Monterey County Training Officers.*

Figure 8.55 A hand crew follows an engine to complete extinguishment.

Engine Operations and Safety

Experience using engines in the wildland environment has resulted in certain basic safety procedures being developed. These general engine-operation safety procedures are as follows:

- All engines responding to wildland fires should stop for traffic lights and stop signs (even when using emergency warning lights, sirens, and air horns), and then proceed when safe to do so.

- Engines parked on a roadway or shoulder at fires should be marked with traffic cones front and rear to warn motorists of the presence of apparatus and personnel (Figure 8.56).
- Engines should be positioned on the side of the road (not blocking it) to protect the vehicles and to allow room for other vehicles to pass.
- Engines should be parked in a safety zone and should not be left unattended at fires.
- Effective communication/coordination with the rest of the fireground organization is critical for safe and effective engine operations.
- Engines should operate in the black whenever possible (Figure 8.57).
- Headlights should be on whenever the engine is running.
- Engines should be backed into one-way roads and driveways facing the escape route (Figure 8.58).
- All windows should remain rolled up to prevent burning embers from entering the cab of the vehicle (Figure 8.59).
- A driver/operator, a nozzle operator, and at least one backup are needed for effective, safe use of engines.
- Nozzle operators should wear eye protection (Figure 8.60).
- When a fire spreads rapidly upslope, it is safer to draw back to the flanks than to attempt a frontal attack.

Figure 8.56 Traffic cones make parked apparatus more visible.

Figure 8.57 Apparatus should operate in the black for safety.

Figure 8.58 Apparatus should be backed down narrow driveways.

Figure 8.59 Windows should remain closed to keep out sparks and embers.

Figure 8.60 Nozzle operators should wear eye protection. *Courtesy of NIFC.*

Power-Winch Operations and Safety

Power winches found on some fire apparatus and dozers have many uses, especially when used with pulleys or snatch blocks to increase mechanical advantage (Figure 8.61). Winches use a steel cable wound onto a rotating drum that is geared to give maximum pulling power. Winches may be powered by the apparatus engine or an electric motor. Most vehicle-mounted winches are operated with controls located adjacent to the winch or remotely by means of a long electrical cord. A drum brake prevents the drum from overrunning when the winch clutch is disengaged and the cable is being unwound.

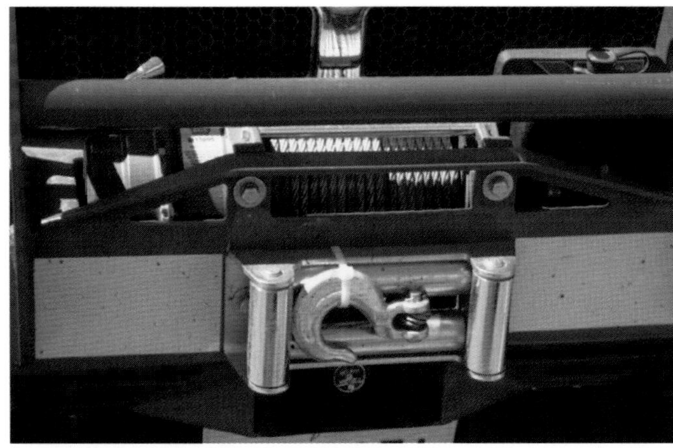

Figure 8.61 A typical bumper-mounted winch.

While these devices can make firefighters' work easier and safer, there are some important safety procedures that must be followed. When using a power winch, firefighters should remember the following points:

- Set the vehicle's parking brake and chock the wheels.
- Pull no more than 2 degrees to either side of the center line of the drum.
- Do not overheat the winch motor with prolonged use.
- Post a lookout to watch the drum and cable.
- Wear gloves when handling the cable.
- Secure pulleys or snatch blocks if used.
- Stay away from the cable a distance equal to the amount of cable between the winch and the load when a winch cable is under tension (Figure 8.62).
- Use a heavy stick or bar to guide the cable onto the drum (Figure 8.63).

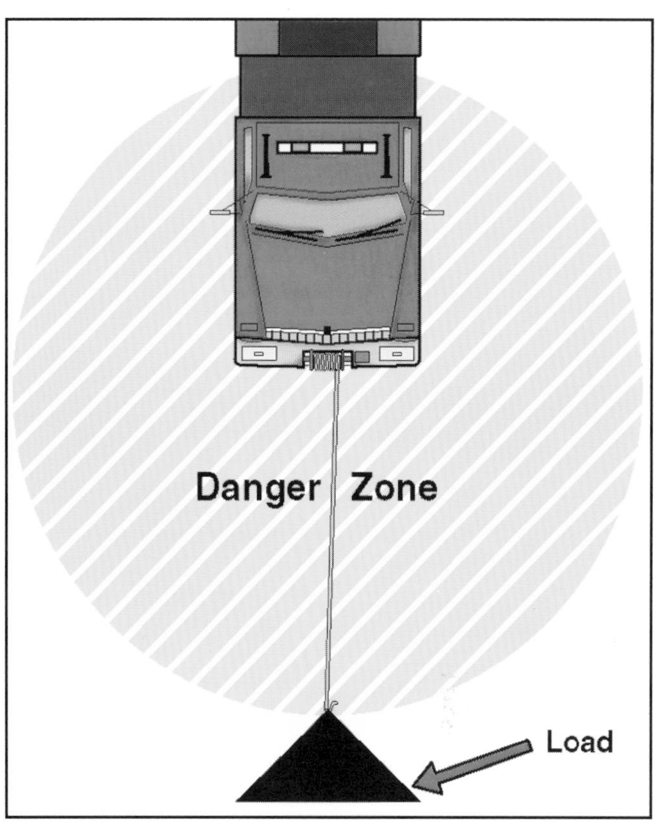

Figure 8.62 Personnel should stay out of the danger zone when a winch is in use.

- Do not allow the cable to kink.
- Replace damaged cable before the winch is used.

A bumper-mounted winch may be used to pull the vehicle on which it is mounted if the vehicle gets stuck (Figure 8.64). However, a winch should never be used in an attempt to move any object that is beyond the rated capacity of the winch.

Figure 8.63 A firefighter guides cable onto the cable drum. *Courtesy of Monterey County Training Officers.*

Figure 8.64 A vehicle can pull itself out of a ditch with its own winch. *Courtesy of Monterey County Training Officers.*

Personnel Transport

Firefighters are often transported from an incident base to a forward staging area or from one section of fireline to another in some sort of vehicle. The type of vehicle can be an engine, a helicopter, or a personnel transport unit (Figure 8.65). The basic safety rules that apply to passengers in any emergency vehicle also apply to those being transported to or from their assignments in a wildland fire. All passengers must ride inside the vehicle (never on the tailboard, running boards, or fenders), and seat belts and any other safety restraints provided must be used until directed otherwise. Whoever is responsible for the vehicle is in charge, and that person's instructions must be followed by all passengers and crew (Figure 8.66). Transporting personnel with land-based vehicles can be a relatively safe operation if the following guidelines are used:

Figure 8.65 Buses are often used to transport fireline personnel. *Courtesy of NIFC.*

Figure 8.66 A helicopter pilot signals ground personnel to approach.

- Drivers must be qualified for the vehicle and operating conditions. If not, they must not be permitted to drive.
- Drivers' shifts should not exceed 12 hours with no more than 8 hours of actual driving.
- Drivers should walk around vehicles to make sure all is clear before departure.
- Crew leaders should ride in the cab with the drivers when transporting personnel whenever possible.
- All passengers should be seated with arms and legs inside the vehicle.
- Personnel and loose tools should not be transported together.
- All traffic signals, safe speed limits, and safety rules should be observed when traveling to a fire.
- Drivers are responsible for the safe operation of their vehicles, including the use of wheel chocks if provided.
- Drivers should make daily mechanical checks of vehicles before driving. Unsafe equipment should be removed from service and reported to the Ground Support Unit.
- Drivers should use spotters outside of vehicles when backing or turning around.
- Vehicles should be operated with headlights on at all times.

HEAVY EQUIPMENT OPERATIONS SAFETY

All mechanized equipment and mobile apparatus used on wildland fires (including contract equipment) must be equipped with proper rollover protection, lighting, back-up alarms, fire curtains, fire shelters, and personnel protective gear before being placed in service (Figure 8.67). Operators must be trained in the use of this safety equipment and required to use it. They must also be trained in wildland fire operations with their type of equipment. Operators should be required to check all safety gear and safety features at least daily, usually during a regular morning inspection, and when changing shifts.

When working around heavy equipment, personnel must be aware of the dangers of being run over by the equipment or of being struck by debris

Figure 8.67 A properly equipped fire fighting bulldozer.

Figure 8.69 Dozers often operate in tandem.

dislodged upslope. Because of the noise produced by operating heavy equipment and by the fire, most equipment operators must rely primarily on their vision to locate hazards. The equipment must have lights for night operations, and ground crews working around the equipment must have headlamps and reflective clothing to increase their visibility. Night or day, ground personnel should not approach mechanized equipment from the sides or rear because the operator's vision is limited in these areas. They should approach from the front after being signaled to do so by the operator (Figure 8.68).

Figure 8.68 Personnel should approach an operating bulldozer only when the operator signals for them to do so.

Operations

Operation of bulldozers or other heavy equipment is usually safer if they are used in pairs (Figure 8.69). Operators can help each other should either machine become stuck or stalled. Working together, they can build a safety zone quickly if they are in danger of being overrun by a fire. Heavy equipment operations can be made safer by using aircraft to support them. Air drops of water or retardants ahead of heavy equipment reduces the fire to which the equipment is exposed and increases the equipment's rate of progress.

The safe operation of heavy equipment in wildland fires requires all of the normal safety procedures for using mechanized equipment and the additional safety procedures related to fire suppression. Even the normal operating procedures must be followed more closely under fire conditions. For example, maintenance is more critical if a mechanical breakdown might result in the unit being overrun by a fire.

Communication with heavy equipment operators is essential for safe and efficient operations. Operators must be in contact with both supervisors and ground crews to integrate the equipment activities into the overall strategy. Most important, operators must receive information for the safe operation of the equipment. Because of the noise of the equipment, radio communication usually requires that operators wear radios built into their hearing protection (Figure 8.70). Voice communication may be difficult or impossible except when the unit is stopped and throttled down. Using either radio communication or hand signals saves time and reduces the chances of misunderstanding. A few simple hand signals agreed upon in advance are useful to aid the spotter or line supervisor in directing the activity of earth-moving equipment (Figure 8.71).

Figure 8.70 Heavy equipment operators wear head gear with built-in communications.

HAND SIGNALS

Stop
Back and forth, waist high, swinging motion.

Come Ahead
Up and down in front of spotter, from waist to arm's length above.

Turn
Swing flag or light on side to which operator is to turn.

Reverse Or Backup
Full circle in front of spotter.

Caution
Wave flag or light in half circle at arm's length above head.

Attract Operator's Attention
May also use one blast on a police whistle or suitable substitute.

DOZER OPERATOR SIGNALS

Can't See Spotter
Gun motor twice.

Want Dozer Helper To Come To Dozer
Gun motor once.

Figure 8.71 Hand signals can be used to communicate with equipment operators.

Safety Procedures

Heavy equipment is typically slow moving and is therefore vulnerable to being overrun by a fast-moving fire. Operators and their support personnel must pay close attention to a fire's behavior and be prepared to withdraw into safety zones if the fire threatens their positions. Periodically, the equipment should be used to develop safety zones by clearing fuel from certain areas (Figure 8.72).

When firefighters are assigned to work in close proximity to mechanized equipment on the fireline, they are exposed to an additional set of risks. The possibility of being struck or run over by a piece of heavy equipment is always present. The following safety procedures should be used by firefighters and heavy equipment operators:

- Park heavy equipment in a safety zone — create one if necessary.
- Load/unload heavy equipment onto or from a transport on a level, stable surface.
- Work from an anchor point when heavy equipment is used to construct safety zones.

Figure 8.72 A dozer clears a safety zone.

- Do not sit or bed down on or near heavy equipment.
- Walk around heavy equipment before starting and moving it to be sure that it is safe to do so.
- Lower dozer blade and/or fire plow to the ground when the equipment is idling or stopped.
- Stay at least 50 feet (15 m) behind and at least 100 feet (30 m) in front of operating bulldozers or tractor-plows when working as part of a ground crew.
- Do not approach heavy equipment until it has stopped and the operator has signaled that it is okay to approach.
- Stay in full view of the heavy equipment operator at all times.
- Do not use a bulldozer/tractor-plow without a canopy, brush guard, and radio communications.
- Make sure each bulldozer/tractor-plow crew consists of *at least* two people.
- Do not allow anyone except the operator to ride on heavy equipment.
- Do not try to get on or off heavy equipment while it is in motion.
- Make sure heavy equipment has both front and rear lights on when operating, especially at night or in heavy smoke.
- Make sure ground personnel wear helmet lights and reflective vests to make them more visible to equipment operators when they are working near heavy equipment at night.
- Make sure the equipment operator is aware of the location of ground personnel.
- Be aware of the fire-behavior characteristics of the fuel types in which ground personnel and equipment are operating.
- Assess the safety of the situation continually.
- Watch out for steep slopes, rocks, ditches, or other obstacles that might disable the heavy equipment in the path of a fire.
- Shout "*Rock!*" if rocks or other materials are dislodged by heavy equipment working upslope and anything starts to roll downhill toward firefighters.
- Be aware of the work area. Watch out for wetlands, steep slopes, rocks, ditches, and other obstacles that might stop the equipment.
- Make sure operators have PPE including a fire shelter available.
- Do not let heavy equipment get too far ahead of crews during backfiring operations.
- Start the control line at a secure anchor point to prevent a fire from burning around the line.
- Take heavy equipment with a hand-operated clutch lever out of gear any time the operator gets on or off while the engine is running.
- Conserve water and let the equipment do as much line construction as possible when heavy equipment and engines are working together. Engine crews use water on hot spots, providing safety for themselves and the equipment.

ROLLING MATERIALS

When heavy equipment is operating on hillsides, a variety of objects can be dislodged.

Firefighters working on the hillside below are at particular risk. Large rocks can be dislodged and roll down the hill with sufficient speed to injure or kill any firefighter in their path. Burning pine cones, logs, and other objects can roll down the hill and start fires below firefighters on the hillside.

FIRE-RESISTIVE CURTAINS

Bulldozers and other heavy equipment used on wildland fires should be equipped with fire-resistive curtains or fully enclosed safety cabs to protect the operators (Figure 8.73). These protective measures also increase the productivity of the units by allowing them to operate longer in the fireline environment.

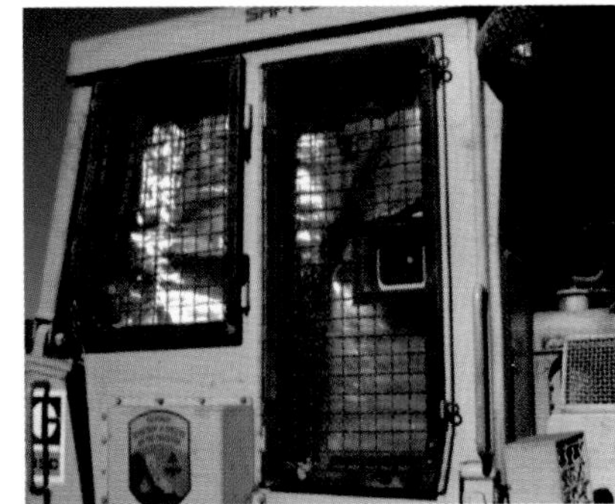

Figure 8.73 A dozer with both a fully enclosed cab and fire-resistive curtains.

OTHER SUPPLIES

Each piece of heavy equipment should also have a fire shelter for the operator, as well as a radio and a pressurized water fire extinguisher. Even though heavy equipment that is not turbocharged must be equipped with spark arrestors on their exhaust pipes, they can still start small fires during operation so fire extinguishers are often needed (Figure 8.74).

Figure 8.74 Heavy equipment should be equipped with a fire extinguisher.

BACKFIRING AND BURNING-OUT SAFETY

It may be advantageous or necessary to backfire or burn out certain areas during a fire. *Backfires* are used as a way of attacking a very intense fire. *Burning out* is used to widen a control line by eliminating unburned fuels between the control line and an advancing fire front. Regardless of the purpose and which tactic is chosen, certain basic safety procedures are followed in backfiring and burning-out operations because any additional fire will somewhat increase the risk to life and property. When backfiring or burning out is to be done, the following safety guidelines should be used:

- Backfiring must be done by personnel trained to do so and only after being authorized by the incident commander or operations chief.
- Only approved equipment and certified personnel are used when backfiring or burning out with aircraft.
- Constant radio communications must be maintained so backfiring operations can be coordinated with other fireline operations.
- Drip torches are preferred to using fusees because it is easier for the operators to remain erect and observe what is gong on around them.
- Escape routes and safety zones must be clearly identified before starting backfiring or burn-out operations.
- Personnel must always be assigned to watch the area behind the drip torch operator to look for spot fires that might trap the backfire crews.
- If backfiring or burning out becomes too intense for the crews to control, it must be stopped immediately.
- If backfiring or burning out is not igniting all available fuel along the control line, supervisors should be consulted before continuing.

AIR OPERATIONS SAFETY

Most air operations at wildland fires are relatively safe when common sense and established procedures are used. Personnel working in and around operating aircraft must know the applicable safety procedures and follow them without exception.

Helicopter Safety

There are certain well-established safety procedures that relate to the use of helicopters (Figure 8.75). When working in or around helicopters, ground personnel should do the following:

- Keep landing zones clear of loose equipment, objects, and unauthorized personnel.

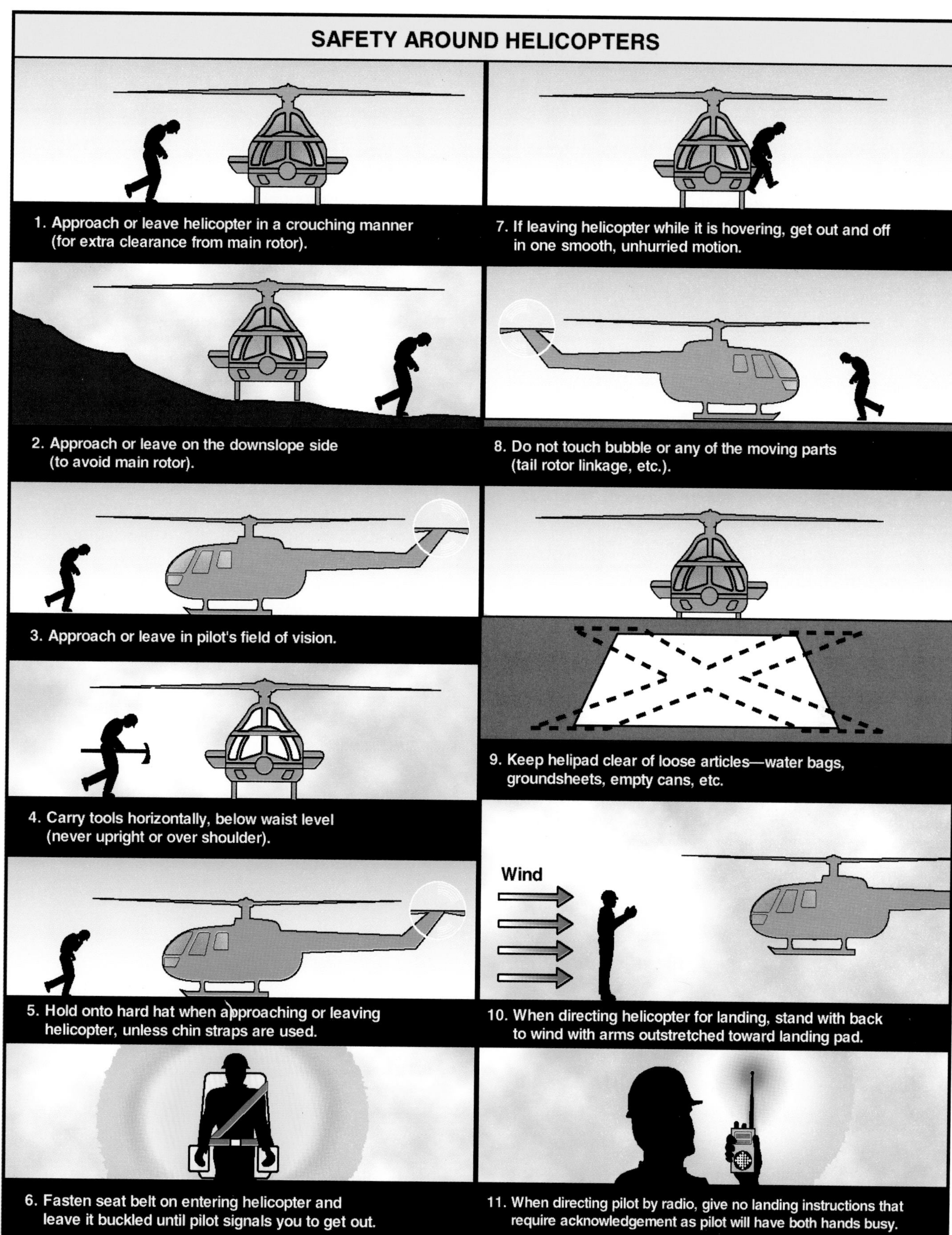

Figure 8.75 Safety procedures for working around helicopters.

- Avoid shining lights in the direction of operating helicopters aloft or on the ground.
- Show wind direction to approaching helicopter with flag, dust, or other visual indicators.
- Follow instructions of helicopter crew at all times.
- Stay well back (at least 100 feet [30 m]) from an operating helicopter unless authorized to approach by the pilot or crew.
- Stay well clear of operating tail rotors.
- Approach or leave in full view of pilot.
- Stoop when approaching or leaving an operating helicopter to avoid the main rotor until at least 100 feet (30 m) away.
- Approach or leave an operating helicopter on the downhill side only.
- Wear both eye and hearing protection around operating helicopters.
- Carry hard hats that do not have chin straps to and from operating helicopters.
- Carry all tools horizontally and close to the side when approaching or leaving helicopters.
- Do not throw anything out of a helicopter.
- Do not slam aircraft doors.
- Prohibit smoking within 100 feet (30 m) of helicopter, fuel storage, or fueling equipment.
- Avoid standing directly beneath a hovering helicopter unless making a sling load hookup.
- Keep safety harness fastened when aboard a helicopter until directed to release it.
- Get briefing on safety procedures from helicopter crew.

Fixed-Wing Aircraft Safety

Observe the following safety procedures when working in and around fixed-wing aircraft:

- Do not smoke within 100 feet (30 m) of aircraft.
- Wear both eye and hearing protection around operating aircraft.
- Attempt to maintain eye contact with the pilot.
- Be alert for unexpected forward movement when propellers are spinning.
- Remain at least 15 feet (4.6 m) from the front of a spinning propeller.
- Avoid the area directly behind operating engines.
- Avoid working within the arc of a propeller until the engine has cooled at least 20 minutes.
- Avoid touching hot wheels, and do not attempt to cool them with water.
- Seek and wait for permission from control tower before crossing runways.
- Remain alert for aircraft movement on taxiways.

Drop-Zone Safety

Under ideal conditions, retardants or suppressants dropped from air tankers disperse in the air and fall to earth straight down in a fine mist. However, conditions in the drop zone are often less than ideal, and the agent may reach the ground at a very shallow angle, at high speed, and in a cohesive mass. Personnel in the drop zone may be seriously injured if they are hit by a concentrated mass of agent or by rocks, limbs, or other debris sent flying by the force of the impact (Figure 8.76).

Figure 8.76 Personnel can be injured if hit by a low-altitude retardant drop. *Courtesy of NIFC.*

If possible, ground forces should be out of the drop zone when a drop is made. If they cannot get out of the drop zone in time, they should get behind a solid object such as a boulder or large, live tree. If caught in the open, they should try to find a place where there is as little loose rock and surface litter as possible. They should wear their helmets and goggles and lie face down with their heads toward the approaching aircraft. They should hold the helmets or chin straps in place with one hand and hold their hand tools to the side of their bodies with the other hand. Their feet should be spread apart for stability (Figure 8.77).

Figure 8.77 The recommended position for personnel who may be hit by an air drop.

Even if the retardant has stopped all forward motion when it falls to earth, personnel in the drop zone can be covered with retardant. While these agents are nontoxic, they can be irritating to the skin. They also make everything very slippery, so footing can be very unsure. These agents should be rinsed from personnel and equipment as soon as possible.

Aircraft Turbulence

Firefighters near a drop zone can also be adversely affected by changes in fire behavior because of turbulence created by the aircraft making drops. In fixed-wing aircraft, the wingtip vortices (eddies formed at the wingtips) can create very erratic winds that can cause a fire to blow up or increase in intensity. The same is true of the downdraft or rotorwash from low-flying helicopters (Figure 8.78).

Figure 8.78 Wingtip vortices can cause erratic fire behavior.

TOOL SAFETY

The fact that firefighters have to use various types of cutting tools makes their jobs more dangerous. Obviously, some tools are more dangerous than others, and those using them (or working in close proximity to others who are using them) are at increased risk of injury. In general, however, if fire fighting tools are well maintained and used according to manufacturer's recommendations and accepted practice, they are not unsafe to use. Regardless of the type of cutting tool being used, one with a sharp blade is easier and safer to use than one with a dull blade. A sharp blade cuts more effectively than a dull one and that translates into less fatigue for the firefighter using the tool, and because fatigue is a leading contributor to fireline accidents, a sharp tool is a safer tool.

Hand Tool Safety

Hand tools such as axes, Pulaskis, McLeods, and brush hooks should have smooth, well-maintained handles and sharp cutting edges, allowing firefighters to use short, sharp cutting strokes. Sharp tools will reduce the need for firefighters to raise these tools above their heads (Figure 8.79). However, the sharp edges of the tool blades should be covered by blade guards when the tools are transported. Tools should be held at the balance point and carried at the side close to the body, not on the shoulder (Figure 8.80). When carrying hand tools, firefighters should maintain a distance of at least 10 feet (3 m) between themselves and other firefighters.

Figure 8.79 Short strokes should be taken with axes and other cutting tools. *Courtesy of NIFC.*

Figure 8.80 Hand tools should be carried properly. *Courtesy of NIFC.*

Chain Saw Safety and Maintenance

One of the most useful tools available to wildland firefighters is the chain saw, but it is also potentially one of the most dangerous. Chain saws are notoriously unforgiving of mistakes. The basic rules for the safe use of chain saws are as follows:

- A blade guard should be in place when a saw is carried more than 10 feet (3 m) in rough country.
- The same distance between firefighters should be maintained when carrying chain saws as any other tool — at least 10 feet (3 m).
- The motor should be stopped whenever a firefighter is carrying, adjusting, repairing, or cleaning a saw.
- The motor should be stopped before refueling a saw.
- If fire conditions allow, chain saws should be allowed to cool before they are refueled.
- Saws should be refueled on bare ground a safe distance from the working area.
- Operators should wear proper safety equipment such as chaps, gloves, hard hat, and eye and ear protection (Figure 8.81).

An important aspect of chain saw safety is proper and timely maintenance of the saw. A regular schedule of daily, weekly, and monthly inspections and maintenance should be followed. Chain saws should also be cleaned and inspected after each use and the chain sharpened according to the manufacturer's specifications. During protracted fire fighting operations, chains may have to be sharpened in the field (Figure 8.82).

Figure 8.81 Chain saw operators should wear full wildland PPE, including chaps. *Courtesy of Monterey County Training Officers.*

Figure 8.82 A chain saw is sharpened in the field. *Courtesy of NIFC.*

ELECTRICAL SAFETY

Power lines on the ground can be dangerous without even being touched. When an energized electrical wire comes in contact with the ground, current flows outward in all directions from the point of contact. As the current flows in all directions away from the point of contact, the voltage drops (Figure 8.83). This is called *ground gradient*. Depending upon the voltage involved and other variables such as ground moisture, this energized field can extend for several feet (meters) from the point of contact. A firefighter walking into this field can be electrocuted because of the differing potentials between each foot (Figure 8.84). To avoid this hazard, firefighters should stay away from downed wires a distance equal to one span between poles until they are certain that the power has been turned off. When it is confirmed that the power is off, this should be communicated to all personnel at scene.

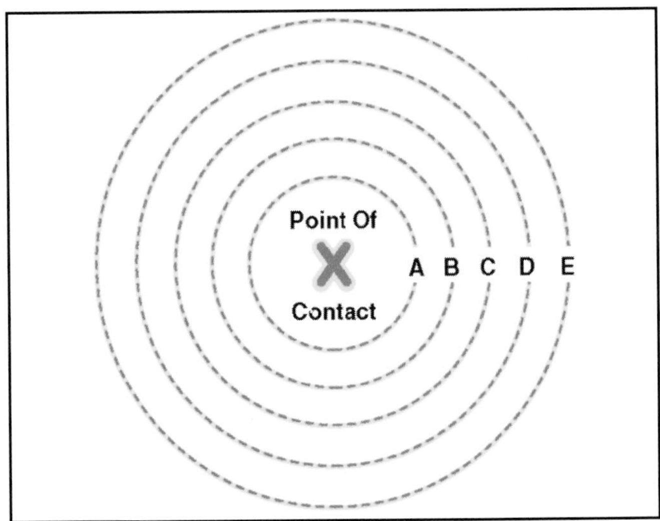

Figure 8.83 Current spreads in all directions from the point of contact.

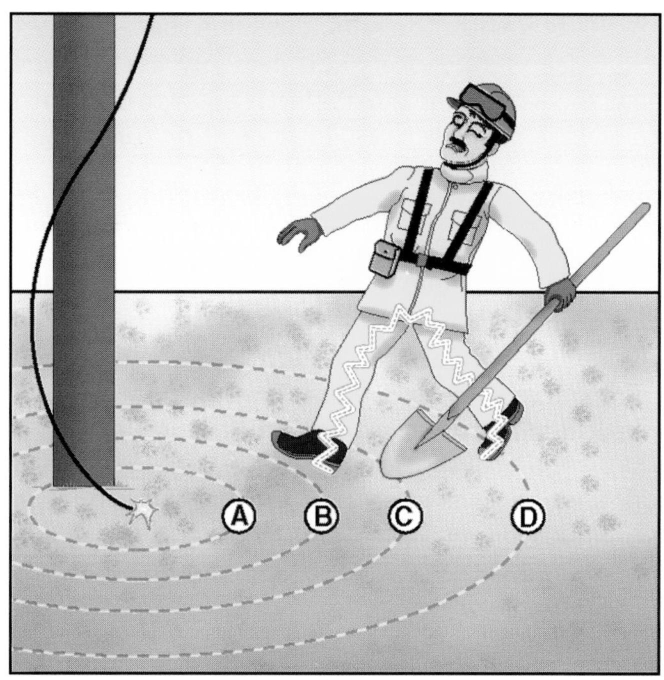

Figure 8.84 Firefighters may be electrocuted by walking near an energized wire.

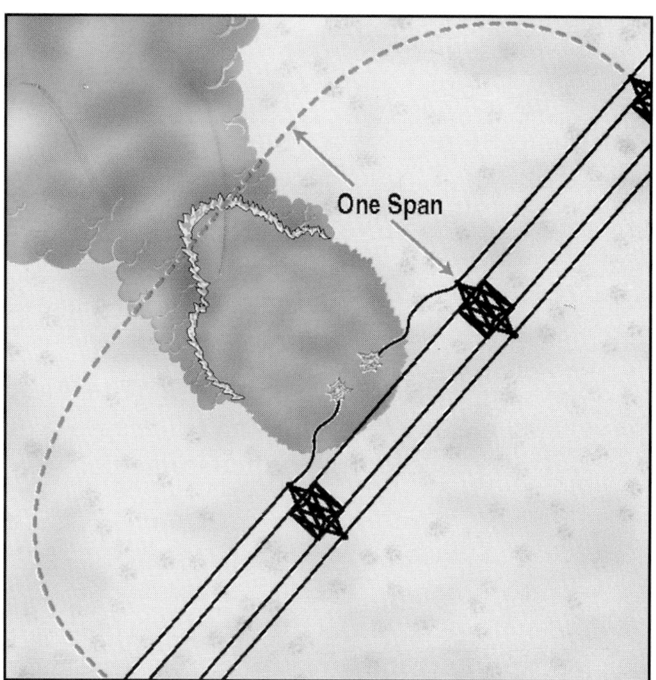

Figure 8.85 A safe distance must be maintained between the downed wire and the fire crews.

Guidelines for Electrical Emergencies

There are three general guidelines for dealing with downed power lines when there is no fire. When firefighters encounter downed power lines, they should always do the following:

- Assume that all lines are energized.
- Call for the power provider to respond.
- Control the scene.

If an energized electrical line falls across a metal fence or guardrail, the entire length (as long as it is continuous) can become charged. This can make controlling the scene difficult because of the length of the fence. Engines and mechanized equipment can be energized in the same way, so they should never be parked under power lines.

Fire Fighting Near Power Lines

It is common for energized electrical wires to start fires when they fall into dry grass. If the fire has spread away from the downed lines a distance equal to one span between poles or towers, the fire can be fought like any other wildland fire. If the fire has not yet burned that far from the downed wire, the attack should be delayed until it has (Figure 8.85). Because solid fire streams are conductive, firefighters should only use fog streams in the vicinity of downed power lines.

Whenever power lines are in or near a wildland fire, all incoming units should be notified of this fact, and all incoming aircraft should be warned of the poles or transmission towers. Even power lines that have not fallen can increase the possibility of firefighters being electrocuted. High brush and small trees under power lines can sometimes cause a phase-to-ground short. Constructing a control line around the fire a distance equal to one span from the power lines is the safest tactic. Once the fire has spread away from the point of contact, fog streams may be used to extinguish the fire.

Firefighters should be able to safely work around the power lines whenever they can operate at least one span distance from the power lines. Fire apparatus may be used to move equipment and firefighters around the scene or to apply water to wet the control line. This is usually safe as long as the one-span clearance is maintained from overhead power lines.

When fire crews must work in the vicinity of overhead power lines, the power company should deactivate these lines if possible. All personnel should be cautioned against directing water streams or aerial retardant onto high-tension lines. They should also be warned against walking into heavy smoke that might obscure electrical wires that are dangling or on the ground. They should be warned that deactivated power lines continue to be hazardous because automatic controls may be programmed to periodically reenergize them in an attempt to restore service. Whenever crews must operate in these areas, the following "dos and don'ts" should be used:

- **DO** identify, map, and discuss at briefings all electrical lines in the incident area.
- **DO** announce *"power line down"* to alert personnel if a line falls, and make sure that all acknowledge the announcement by repeating it back.
- **DON'T** leave a vehicle if a power line falls on it until the power company deactivates the line. If you must exit the vehicle because it is on fire or fire is approaching, jump well clear of the vehicle and hop away with feet together.
- **DON'T** operate heavy equipment under power lines.
- **DON'T** drive vehicles with long antennas under power lines.
- **DON'T** refuel vehicles under power lines.
- **DON'T** stand near power lines during retardant drops.
- **DON'T** go near or move downed power lines.
- **DON'T** direct fire retardant or water onto power lines.
- **DON'T** stand or work in dense smoke near power lines.

Electric Fences

Electric fences are commonly used to confine livestock. Manufactured systems operate at very low amperage and are not a safety hazard to firefighters — unless the system has been altered (Figure 8.86). However, homemade systems may use enough amperage to be harmful to firefighters. Because firefighters can never be sure what type of system they are dealing with, they must assume that any electric fence is dangerous and use only fog streams near the wires. The most effective way to deal with an electric fence is to de-energize it by turning off the power supply.

Figure 8.86 Firefighters should not assume that an electric fence is harmless.

HAZARDOUS MATERIALS SITUATIONS

Firefighters may have to deal with hazardous materials during a wildland fire, especially in the wildland/urban interface. The materials may range from barrels of agricultural pesticides in farm structures to propane in large tanks at fuel distribution points (Figure 8.87). In some cases, hazardous wastes are illegally dumped in the wildland (Figure 8.88). The hazardous material may have even started the fire to which the firefighters were called.

Clandestine drug labs, often located in rural areas to reduce the chances of detection, may contain several different toxic and/or explosive chemicals. These chemicals can start or contribute to the growth of fires in the wildland. An associated

Figure 8.87 Hazardous materials may take many forms in the wildland. *Courtesy of NIFC.*

Figure 8.88 Hazardous materials are often found in illegal dumps.

illegal industry — growing marijuana — also can put firefighters at risk, though not necessarily from hazardous chemicals. These clandestine operations are often protected by armed guards, attack dogs, and/or potentially lethal booby traps.

Other hazardous materials may be found in the wildland/urban interface. Some of the more common ones are residential or commercial propane tanks, aboveground diesel or gasoline tanks, and ordinary gardening chemicals such as fertilizers and pesticides.

Regardless of what the hazardous material is or how it came to be where it is, firefighters must be able to recognize and isolate it. When such materials are discovered, it should be reported through the chain of command and the appropriate resources requested. Firefighters should not approach any unidentified or uncontained hazardous material but should first attempt to identify it by reading any placards or labels through binoculars from a safe distance uphill and upwind of the material (Figure 8.89). They should control the scene and keep everyone out until relieved by those trained and equipped to handle these situations.

Figure 8.89 Firefighters should size up hazardous materials incidents from a safe distance.

If firefighters find what they suspect may be hazardous materials, they should generally do the following:

- Stay upwind, uphill, and avoid breathing smoke.
- Isolate the area.
- Warn others in the immediate vicinity.
- Notify the IC of the potential problem so that trained specialists can be brought in.
- Remember and follow: *"If you don't know, don't go, it may blow."*

WORKING IN BURNED-OVER AREAS

While burned-over areas are relatively safe places to work because of the absence of unburned fuels, there are safety hazards in them. In addition to the smoke from smoldering stumps and other materials, whirlwinds may stir up clouds of particulates, so filter masks and eye protection may need to be used. However, the most significant hazards that firefighters may find in the black are snags, holes, loose rocks, and materials still burning. To protect themselves, firefighters working in the black must remember to "look up, look down, look around," and they *must* continue to wear their protective clothing until the fire is completely extinguished (Figure 8.90).

Figure 8.90 Firefighters should wear full wildland PPE during mop-up. *Courtesy of NIFC.*

Snags

Snags are standing dead trees (Figure 8.91). They present a variety of hazards to firefighters working near them. They often smolder long after the main fire has been extinguished, so they must be cut down (felled) during mop-up. Dead limbs can break off and fall on firefighters attempting to fell a snag. A spotter is usually needed to watch for falling limbs during felling operations (Figure 8.92). All personnel working nearby must be warned of the impending fall of a

Figure 8.91 An especially dangerous type of snag. *Courtesy of NIFC.*

Figure 8.92 A spotter watches as a snag is being felled.

snag. If there is fire within the trunk, as there often is, the trunk must be opened in order to extinguish it.

Felling snags or large trees, normally over 20 inches (508 mm) DBH (diameter at breast height), is a potentially dangerous operation and should be done only by a qualified feller. Smaller trees, less than 20 inches (508 mm) DBH, may be felled by crew personnel who have been trained to do so. When snags or trees must be felled, the following guidelines should be used:

- Select a clear escape route before starting the cut.
  - Note that the quadrant opposite the planned fall of the tree may be the most dangerous. Choose an escape route at right angles to the planned direction of fall, preferably on the same contour, unless there are some peculiarities in the particular situation (Figure 8.93).
  - Stand behind another tree of sufficient size to provide protection if possible.
  - Watch for whiplashed branches and other broken parts.
  - Stay clear of the butt.
  - Watch for falling branches; continue to watch until all broken branches have fallen.
- Be aware of other nearby workers. Notify workers not on the felling crew when tree felling will be occurring in their work area. Felling should not be closer to other personnel than double the average height of the trees or snags being felled.

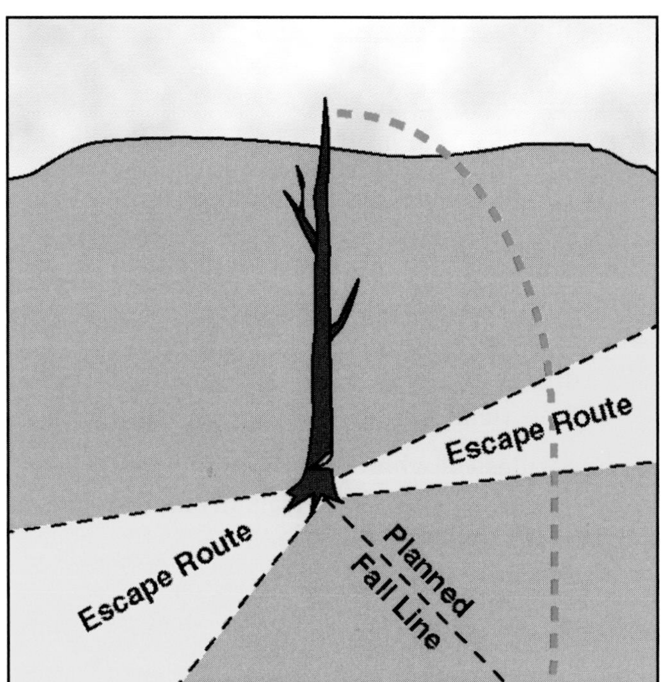

Figure 8.93 Escape routes are at right angles from a snag.

- Do not fell trees upslope from other crews.
- Station a lookout to watch and warn the sawyer of falling limbs and tops when felling trees. Due to chain saw noise, provide the lookout with a system, such as a portable air horn, to signal the sawyer in the event of danger.

Holes

The surface of the ground in burned-over areas is opaque and often covered with dark gray or black ash. This can make holes in the ground difficult to see. Personnel working in the black can be injured by stepping into an unseen hole. Apparatus being driven in the black can be damaged if a wheel drops into a hole.

Loose Rocks

Loose rocks in the black can cause firefighters to slip and fall. On steep hillsides, loose rocks can roll down the hill and strike other firefighters and/or fire apparatus. Since some of these rocks can be quite large, they can do serious damage to anyone or anything they hit (Figure 8.94).

Figure 8.94 Firefighters should watch for boulders rolling downslope.

TRAFFIC HAZARDS

Many wildland fires originate along roads and highways because lighted materials are intentionally or thoughtlessly thrown from vehicles (Figure 8.95). Many fires start in the median strip of divided highways. When firefighters have to fight these fires, they often must position their apparatus on the shoulder of the road and walk along the roadway with hoselines. In many cases, smoke blows across the road and obscures vision (Figure 8.96). Under these circumstances, firefighters on foot are at considerable risk of being hit by passing vehicles.

Prior to the arrival of law enforcement personnel, firefighters may have to direct traffic as well as fight the fire. In addition to a supply of fusees, all fire apparatus should be equipped with traffic cones to place in the roadway to guide motorists around parked apparatus. If the apparatus must be positioned on the roadway or the shoulder, the driver/operator should set the cones or fusees before doing anything else (Figure 8.97). If fusees are used, they must not be allowed to roll into vegetation at the side of the road.

Figure 8.95 Many wildland fires start along roads and highways.

Figure 8.96 Smoke from roadside fires can obscure the roadway.

Figure 8.97 An apparatus driver sets out traffic cones for safety.

Emergency vehicles should be driven with their headlights on, even if emergency lights aren't in use. This is especially important when vehicles are driven in smoke, dust, or anything else that obscures vision (Figure 8.98). Seeing headlights shining through the haze enables firefighters on foot to move out of the way. Unless the apparatus is parked and the engine turned off, all lights should be on for safety.

Figure 8.98 Apparatus should have their lights on whenever they are operating. *Courtesy of New Jersey Forest Fire Service.*

When potential or actual smoke-related traffic safety problems are recognized in the vicinity of a wildland fire, firefighters must act decisively to prevent accidents from occurring. Agency policy must always be followed, but in general, the following actions should be initiated:

- When potential smoke-related problems are recognized, do the following:
  - — Advise the IC or operations chief.

— Implement applicable provisions of the incident traffic plan such as posting warning signs (Figure 8.99).

— Ensure proper equipment is ready and appropriate personnel are briefed on contingency plans and available to control traffic.

— Notify local law enforcement units of potential problem.

— Establish periodic patrols to monitor problem areas.

Figure 8.99 Roads may have to be closed due to smoky conditions. *Courtesy of NIFC.*

- If smoke-related traffic problems do occur, first personnel at scene should do the following:
  - — Take immediate action to prevent accidents (Figure 8.100).
  - — Establish control points on both sides of problem area.
  - — Slow or stop traffic entering problem area.
  - — Advise drivers of alternate routes.
  - — Implement radio/television traffic advisories for the problem area.
  - — Keep a log of all actions taken.

Figure 8.100 Firefighters may have to direct traffic until law enforcement personnel arrive.

- If smoke-related traffic accidents occur, firefighters at scene should do the following:
  - — Make every effort to assist and protect those involved.
  - — Notify, if necessary, appropriate medical units and request assistance.
  - — Notify appropriate law enforcement units.
  - — Provide additional personnel for traffic control if required.
  - — Notify IC so that local safety and tort claims personnel may be assigned.
  - — Record facts of accident and names, addresses, and statements of witnesses.
  - — Record license plate numbers of all vehicles in the vicinity of the accident.
  - — Follow agency policy regarding making statements to other than law enforcement officers.

SMOKE

Wildland fires almost never burn cleanly. In most cases, they produce large amounts of very heavy smoke (Figure 8.101). Wildland fires often produce a smoke column that can be seen for miles (kilometers). A smoke column may obscure the sun to the point that it is quite dark on the ground in the middle of the day.

Figure 8.101 Wildland fires often produce heavy smoke. *Courtesy of John Hawkins.*

Smoke produced by wildland fires is a suspension of small particles of carbon, tar, and dust floating in a combination of heated gases. The particles provide a site for the condensation of some of the gaseous products of combustion, especially aldehydes and organic acids formed from carbon. Depending upon what is burning, the smoke can contain any or all of the following gases: carbon dioxide (CO_2), carbon monoxide (CO), hydrogen cyanide (HCN), and hydrogen chloride (HCl).

Many of the gases produced in fires are toxic; some have a delayed effect, and so the victim doesn't know to seek medical aid. By far the most common fire gas to which wildland firefighters are exposed is CO, which is discussed later in this section.

Visibility

One of the obvious hazards associated with smoke is that it obscures vision. Smoke not only presents an opaque barrier to clear vision, it is irritating to the eyes and often causes heavy tearing. Firefighters must be careful where they step when they cannot clearly see what is in front of them. If they must move in heavy smoke, firefighters should leave both feet on the ground and slide their feet along the surface of the ground. They must also be especially watchful for fire apparatus approaching them through the smoke. Likewise, fire apparatus driver/operators must be very careful where they drive in smoky conditions. They must keep all lights on to make the apparatus more visible to firefighters on foot.

Particulates

A large portion of visible smoke is composed of particulate matter such as carbon, tar, and dust. Some of the particles suspended in smoke are merely irritating if inhaled, but others can be quite harmful. The size of the particle determines how deeply it penetrates into the lungs. If firefighters cannot stay out of the smoke, they should use filter masks or other respiratory protection to prevent particulates from entering their lungs (see Chapter 3, Ground Support Equipment and Personnel Considerations).

Carbon Monoxide

Carbon monoxide is an odorless, tasteless, invisible gaseous by-product of the combustion of organic (carbon-containing) materials. It is produced when there is insufficient oxygen available to form CO_2 as part of the combustion process. Even though there appears to be an unlimited supply of oxygen available to fires in the wildland, this is not always the case. Fires burning in thickly matted grasses, for example, cannot get enough oxygen to all layers of the fuel to form CO_2, so CO is formed instead.

CO is present at some level in all smoke from wildland fires and can even be found in higher concentrations in inversions. When both oxygen and CO are available, the body assimilates CO 200 times more readily than oxygen. Also, the CO absorbed by the body is cumulative, so repeated exposures over a period of time can result in very high levels of CO in the bloodstream. This can have long-term health consequences for firefighters, so they must avoid breathing smoke whenever possible. However, CO exposure can have short-term effects as well. Table 8.1 shows the immediate signs and symptoms of CO exposures.

Research shows that there is little if any chance of firefighters in the wildland being exposed to significant levels of CO. However, crew leaders and supervisors should limit each firefighter's possible exposure to CO by rotating personnel from areas of heavy smoke to areas with little or no smoke as the situation allows. Personnel operating chain saws or fire pumps in confined areas should be monitored for signs of CO exposure. Any firefighter

TABLE 8.1
Signs/Symptoms of Short-Term Carbon Monoxide Exposure

Blood CO Level (Percent)	Symptoms
0–10	No symptoms
10–20	Shortness of breath during physical exertion; tightness across forehead
20–30	Headache, shortness of breath
30–50	Confusion, severe headache, dizziness, fatigue, collapse from exertion
50–70	Unconsciousness, respiratory failure and death if exposure is continued

Source: IFSTA Self-Contained Breathing Apparatus.

exhibiting symptoms of CO exposure should not be allowed to operate a chain saw or any other potentially dangerous piece of equipment. Nor should these firefighters be allowed to drive fire apparatus until they have recovered from the exposure. Agency policy should be followed with regard to giving oxygen to affected personnel.

COMMUNICATIONS

The degree to which firefighters can communicate on the fireground can have a significant effect on their safety and survival. If orders are not clearly communicated and not completely understood, the chances of serious accidents occurring increase considerably.

Face to Face

Direct, face-to-face communication is preferred whenever possible. Because the participants can read each other's facial expressions and body language during the exchange, the chances of miscommunication are greatly reduced. However, the obvious limitation is that it is confined to very short distances between the participants. Even when orders are clearly heard and understood, they may be forgotten later. Therefore, whenever possible, orders should be written down for later reference (Figure 8.102).

Figure 8.102 Firefighters should write down important instructions and information.

Safety Briefing

As described earlier in this chapter, the safety briefing given to crews before they go onto the fireline is one of the most valuable communications they receive. The information passed on during these briefings can have a direct effect on firefighter safety and survival. This information should also be written down for later reference.

Importance of Portable Radios

As mentioned earlier, direct, face-to-face voice communication is preferred whenever the situation allows. However, the variety of environments in which firefighters must attempt to communicate usually make the use of mobile and portable radios a necessity (Figure 8.103). Communications using some type of radio equipment is vital to coordinated fireground activities. They provide instantaneous communication between operating units, between them and the communications center, and between them and the rest of the fireground organization through the chain of command.

Figure 8.103 The use of radios is often necessary because of ambient noise levels. *Courtesy of NIFC.*

Radios allow communication to take place when face-to-face communications are not possible. The advantages of radio communications are the speed with which messages can be sent and received and the ability to communicate over long distances. The main disadvantage is its one-dimensional nature — voice only.

Radios capable of monitoring and transmitting on numerous frequencies selectively are the most flexible; however, adequate procedural controls must be established to prevent units from transmitting on any but their assigned frequency. Personnel should be trained to avoid using agency-specific codes or terminology and to use clear text (standard English) and general terminology to avoid misunderstandings between units and agencies. Especially on large, complex incidents involving mutual aid units, all radio communications must be conducted according to the incident communications plan.

Unless a battery charger is available to them, personnel assigned to large-scale incidents should carry spare batteries for their portable radios. In addition, when firefighters such as pump operators and chain saw operators must work for long periods in areas of high ambient noise, they should wear hearing protection. However, so that they can continue to monitor fire-related radio traffic, they should wear hearing protection with built-in receivers (Figure 8.104).

Figure 8.104 Equipment operators and apparatus driver/operators need built-in communications equipment to hear over the sounds of the apparatus or equipment.

FIRE PREVENTION AND INVESTIGATION 9

LEARNING OBJECTIVES

This chapter provides information that addresses the following objectives of NFPA 1051, *Standard for Wildland Fire Fighter Professional Qualifications* (1995 edition):

Wildland Fire Fighter II

4-5.6* Secure the area of suspected fire origin and associated evidence, given a wildland fire and agency procedures, so that all evidence or potential evidence is protected from damage or destruction and reported to a supervisor.

4-5.6.1 *Prerequisite Knowledge:* Knowledge of types of evidence and the importance of site security and evidence preservation.

4-5.6.2 *Prerequisite Skills:* Evidence preservation techniques and use of marking devices for site security.

Chapter 9
Fire Prevention and Investigation

INTRODUCTION

Both fire prevention and investigation are included in this chapter because these two activities are closely related. Fire prevention programs focus on preventing fires of accidental origin and limiting the spread of any fires that do start, but such programs have little effect on the number of incendiary fires (malicious burning of property) in a given jurisdiction. However, investigating the cause and origin of every unfriendly fire to which a department is called provides information that is essential to both preventing accidental fires and to apprehending and prosecuting those responsible for incendiary fires. Even though arson investigations in many jurisdictions are now conducted by law enforcement personnel, firefighters have a responsibility to protect and preserve any possible evidence they observe on the fireground.

Because the success of both prevention and investigation is heavily dependent upon an accurate determination of the causes of fires within the jurisdiction, this chapter discusses the elements of fire-cause determination before going on to discuss fire prevention. Included in the discussion of fire-cause determination are scene preservation, cause and origin determination, fire investigation tools/equipment, search documentation, and how to preserve the evidence discovered. Included in the discussion of fire prevention are risk and hazard assessment, fire-cause analysis, mapping, components of a fire prevention program, fire prevention program application, and how to conduct inspections.

USES OF CAUSE/ORIGIN DETERMINATION

An accurate determination of the cause and origin of a wildland fire is extremely important for a variety of reasons. As mentioned earlier, this data may become an important piece of evidence in the prosecution of those responsible for starting the fire, whether intentionally or negligently. However, one of the most important uses for this information is its value to the fire prevention program.

Cause/origin data are used as a basis for prioritizing and focusing a department's fire prevention activities. For example, if a high percentage of wildland fires is caused by sparks from weed mowers or similar equipment, this information can form the basis for a public education and/or inspection program on spark arrestors for this type of equipment. Similarly, if a particular person or vehicle has been seen in the vicinity of several recent arson fires, area residents should be informed about what to do if they see this person or vehicle in their neighborhood.

FIRE-SCENE PRESERVATION

Even before they arrive at the scene of a wildland fire, firefighters should observe a number of things that may contribute to a subsequent fire investigation. Among these are the amount and color of the smoke. Is the amount of smoke consistent with other wildland fires in similar fuels and weather conditions? Is the color of the smoke normal for wildland fuels, or does it suggest that an accelerant was used? Make mental notes (written notes, if possible) of people and vehicles leaving the scene. Are any of them fleeing noticeably faster than others? Are any of them carrying safety cans or other unusual items? If it is nighttime, are any of the vehicles being driven without headlights? If the answer to any of these questions is "yes," note as many identifiable features — sex, age, height, weight, hair color, race, clothing, etc. — about these people as possible. The same is true for their vehicles — color, make, model, license plate number, damage, etc.

Firefighters should also observe the status of gates, locks, fences, and bridges at or near the scene. Is there evident of tampering with gates or their locks? Are fences intact, or have they been knocked down or the wires cut? Are there tire tracks through downed fences? Have bridges been damaged or destroyed? Are there any other unusual conditions that might have been created to impede the response of the fire department? These, too, should be reported.

On arrival, firefighters should observe the bystanders, even those who volunteer to help fight the fire. Do any of the people look familiar — have they been seen at other fires? Do any of them seem unduly excited, emotional, or angry? Are any of them overly eager to volunteer information or to help fight the fire? If so, firefighters should report these observations to their immediate supervisor.

Finally, once at scene, firefighters should mark and protect any potential evidence they observe while carrying out their fire-suppression assignments. Especially when working near the heel of a fire, they should remember that evidence can be destroyed while spraying water, dragging hoselines, using hand tools, or driving apparatus in the black (Figure 9.1). When near the heel where a fire tends to be less intense than in other areas, firefighters should operate in the green whenever possible. If they must enter the black near the heel, firefighters should apply their knowledge of basic fire behavior to help them estimate the general area of origin. If possible, they should avoid operating in that area. But if their assignment requires them to work there, they should be especially observant for any possible evidence of the fire cause and be very careful to not disturb it. Any potential evidence should be reported to the immediate supervisor. Depending upon its nature and size, evidence should be protected by covering it if possible and by marking its location with fireline tape, rope, or hand tools (Figure 9.2). Evidence should not be moved unless it is necessary to protect it from being destroyed.

Figure 9.1 Evidence can be destroyed by driving apparatus in the black.

Figure 9.2 When the area of origin has been found, it should be cordoned off. *Courtesy of Bill Lellis.*

DETERMINING AREA OF ORIGIN

The first step in investigating the cause of any wildland fire is to locate the area of origin. By applying an understanding of basic fire behavior, an investigator can estimate where, in general, a fire started. Knowing how wind, topography, and fuels affect fire spread and that the area of origin is probably closer to the heel than to the head, an investigator can narrow the search area considerably. The investigator can further reduce the area to be searched by interviewing the reporting party and the first-arriving firefighters to confirm the location and size of area involved early in the fire and the direction in which the fire spread (Figure 9.3).

Figure 9.3 A fire investigator interviews a first-arriving firefighter.

Principles of Fire Spread

From its point of origin, a wildland fire burns outward in all directions. On flat ground with a consistent fuel bed and no wind, a fire would burn equally in all directions, and the point of origin would be in the center of the circular burn pattern. However, wildland fires rarely occur in these conditions. In reality, the rate and direction of spread of wildland fires are affected by wind, slope, and fuel variations.

The initial area of involvement, as reported by witnesses or the initial attack crew, contains fire-spread indicators that are used to determine the area of origin; others are found at the head of the fire (see Indicators section). The head burns hotter than other parts of a wildland fire; therefore, it consumes more of the fuel and leaves less unburned debris. Wherever little debris remain, the fire probably burned outward from its origin at the greatest rate of speed. Also, the deepest char is found on the surfaces of brush and tree limbs near the head of the fire (Figure 9.4). Some limbs bend in the direction in which the fire was moving, sometimes "freezing" in that position (see "Freezing" of Branches section). However, at the head, windblown sparks and embers may cause spot fires ahead of the main fire, which may create confusion about where the main fire originated.

Figure 9.4 An investigator checks the char on a bush in the black.

On the other hand, the fire near the origin is generally slower and less intense than it is at the head, but the direction and speed of fire spread near the origin can be affected by small changes in wind speed or direction or in the slope of the terrain. These factors make it more difficult to determine the point of origin. More unburned materials are left in the area of origin, and the effects of flame on the fuels are considerably less here than at the head. In a backing fire (one burning downslope and/or against the wind), a great deal of debris are left, and the fire scars upon them are slight compared to the rest of the fire area.

Radiant heat from burning materials affects other fuels. As the fire burns past any given area, the flames scorch or char surfaces exposed to the fire. But the stems of the fuel, whether grass or timber, protect their backsides. This is true even though the fuel is consumed and turned to ash. White ash is a product of more complete combustion, and it appears on the exposed sides of the remaining debris.

The amount and type of fuel and the intensity of a fire have a great effect on the rate of spread. As a fire moves along the ground, it does not burn with an even or equal intensity at all times. A fire burns with more or less intensity because of barriers and changes in wind, slope, or fuel. Barriers, such as rock formations, may cause a fire to slow down or even go out. Almost any barrier reduces the intensity of a fire. Barriers often cause wind eddies that change the direction of a fire. Fire burns faster and more completely in dry fuels than in fuels that are not completely cured. Fire burns slower and less completely when fuel moisture and humidity are high. These changes are visible to an investigator in the examination of a burn.

Fire-Origin Indicators

Indicators are evidence of how a fire burned, and they are used to determine the direction of fire spread. Indicators are apparent on both large and small fuels. An investigator should consider the following when using indicators to locate the area of origin:

- Indicators are smaller in size the closer they are to the point of origin (as a fire burns away from the point of origin, more of the fine fuels are consumed).
- The majority of evidence should be used to determine the direction of fire spread.
- Several indicators should be used to determine the exact point of origin.

GRASS-STEM INDICATORS

As fuel stems burn off at the bottom, they fall in various directions. Those falling into the burned area remain unburned after the fire has passed; however, fuel stems falling into the unburned area are subsequently burned. The heads of grass found within the burned area indicate the direction *from which* the fire came (Figure 9.5).

Figure 9.5 Unburned grass stems often fall toward the point of origin. *Courtesy of Hugh Graham.*

In most cases, a fire front progresses in a straight or slightly curved line, leaving about 180 degrees in which the grass could fall into the burned area. However, over time, wind can alter the position in which grass stems are found. Therefore, investigators should not rely on this single factor to determine the direction of fire spread.

PROTECTED FUEL

A slow-burning, low-intensity fire produces deeper char on the side of the fuel exposed to the approaching fire (Figure 9.6). Sometimes, stems protected by other stems or debris do not show any signs of being burned. A fire burning slowly over a large area appears dark when looking toward the point of origin. When looking away from the point of origin, it appears light because of the ash produced in more complete combustion.

A leaf, limb, or board shows more complete combustion on the side from which a fire approached. This edge shows more burn stains, white

Figure 9.6 Vegetation may show more char on the side from which the fire spread.

ash, and charring, while the other side (lee side) is protected and shows fewer signs of burning.

Anything lying on the fuel protects it from the fire and leaves a definite pattern indicating the direction of fire spread. Unlike the char, white ash, and burn stains found on the exposed side of the fuel, the protected area is very distinct. There is a clean burn line on the exposed side (windward side) and a ragged burn line on the lee side (Figure 9.7).

Figure 9.7 The direction of spread can be determined by the pattern left by objects. *Courtesy of Bill Lellis.*

CUPPING

"Cupping" normally occurs on the windward side of the fuel (Figure 9.8). The windward side is exposed to the most wind and burns the deepest, while the lee side remains more protected and cooler. Cupping can be clearly seen on most standing fuels, even on stems of grass.

Figure 9.8 Fuels tend to "cup" on the side from which the fire spread.

CHAR PATTERNS

A fire burning upslope and/or with the wind produces a char pattern on trees that slopes at an angle greater than the slope of the terrain (Figure 9.9). This pattern is caused by a reduced pressure on the back (lee side) of the tree. The reduced pressure draws heat and flames into an eddy on that side and raises them up the tree by convection. Fuel accumulations on the uphill or windward side of the tree have little effect on this char pattern. This char pattern remains evident on trees for many years after a fire.

A fire burning downslope or against the wind creates a char pattern parallel to the ground slope (Figure 9.10). An accumulation of debris may cause char on the side of a tree directly above the location of the accumulation (Figure 9.11). However, the burning debris have little effect on the char pattern around the rest of the tree trunk.

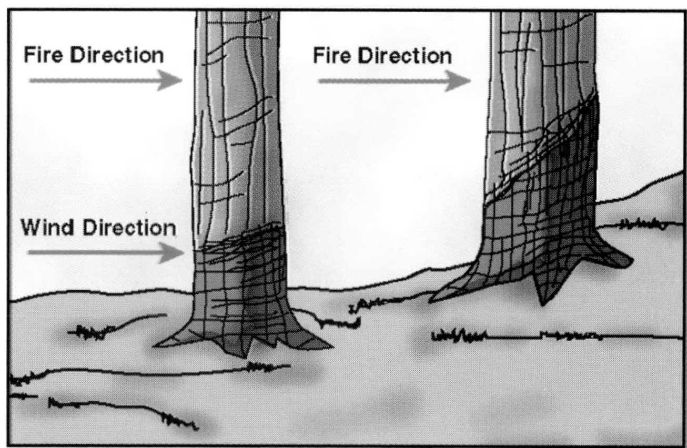

Figure 9.9 Char pattern on trees may indicate fire spread with the wind or upslope.

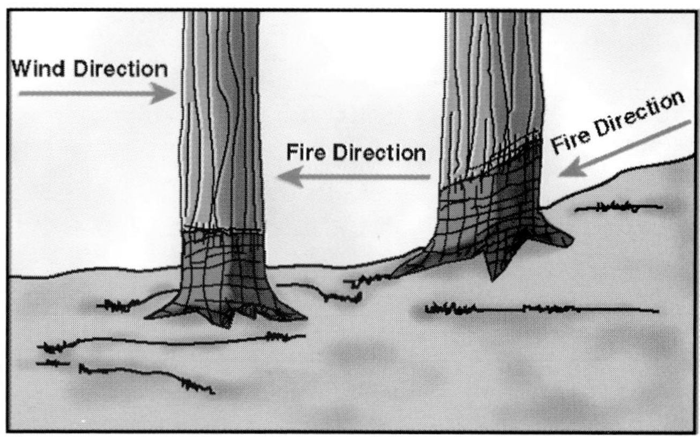

Figure 9.10 Other patterns may indicate fire spread against the wind or downslope.

Figure 9.11 Char on tree trunks from accumulated debris burning creates its own distinct pattern.

In the absence of wind, the sweep of a convection column through trees and brush creates a distinctive burn or char pattern (Figure 9.12). The lighter the wind, the more vertical the burn pattern becomes. In high winds, this pattern may be almost parallel to the terrain (Figure 9.13).

Figure 9.12 Char pattern in trees in the absence of wind.

Figure 9.13 With wind, the pattern may be more parallel to the ground.

Very often, a fire burning near the point of origin does not yet produce very much heat. The surface fuels burn leaving the canopy (crown fuels) intact, except for some possible scorching of the lower branches. As the fire spreads away from the point of origin, it becomes hotter, and more of the canopy becomes involved (Figure 9.14).

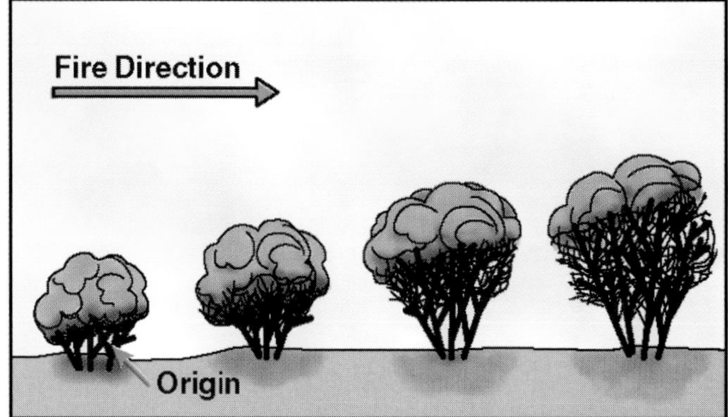

Figure 9.14 More of the canopy is involved as fire spreads away from the point of origin.

ALLIGATOR EFFECT

The *alligator effect* (also called *alligator char*) is a form of charring commonly found on fence posts, boards, structures, and signposts after a fire has burned past. The char is deeper on the side from which the fire approached (Figure 9.15).

Figure 9.15 Alligator effect on a fence post.

"FREEZING" OF BRANCHES

Leaves and small stems soften when they are heated. Prevailing wind or indrafts created by a fire cause foliage to point *away from* the approach of the fire (Figure 9.16). The foliage often remains or "freezes" in this position when it cools. However, as mentioned earlier, no single indicator should be relied upon to indicate the direction of spread. "Frozen" branches should be considered along with other indicators to confirm the fire direction.

Figure 9.16 Small branches may "freeze" pointing toward the head.

STAINING

Rocks and other objects stain on the side exposed to the oncoming fire when adjacent fuels burn and deposit minute ash particles on the object's surface (Figure 9.17). This stain pattern can be found on a variety of items including fence posts, aluminum cans, pieces of scrap metal, dirt clods, and unburned vegetation.

SOOT

Soot is deposited on the side of objects exposed to the approaching fire, particularly when a fire burns in very oily fuels or when burning is incomplete (Figure 9.18). When a fire passes through a wire fence, soot is most likely found on the lower wires. Less soot is deposited on the upper wires.

Figure 9.17 Objects develop a stain on the side from which the fire approached. *Courtesy of Bill Lellis.*

Figure 9.18 Soot is deposited on the side of objects exposed to the approaching fire.

DETERMINING FIRE CAUSE

Once the area of origin is located, the specific cause of the fire should be determined. Initially, an investigator eliminates all natural and accidental causes before starting to look for an intentional cause. Examples of natural causes would be a lightning strike or the friction between dehydrating logs piled on a log deck. Examples of accidental causes would be sparks from an exhaust pipe with a faulty or missing spark arrestor or the backfire from a poorly tuned engine.

An investigator also looks for the most likely cause under the circumstances. If a lightning storm passed through the area shortly before a fire was discovered, the investigator looks for signs of lightning strikes in the area of origin. If the area of origin includes a campsite, a campfire and/or careless use of smoking materials are likely causes. If a homeowner's burn barrel or incinerator is in the area of origin, this should be investigated as a possible cause. If heavy equipment was operating in the area prior to the fire, sparks from an exhaust or those generated by steel striking stone should be considered possibilities. If leaf burning or other "controlled" burning was being done before the fire was discovered, this is a possible cause. If there is a railroad line through the area of origin, the cause may be related to trains passing through shortly before the fire was discovered. If the area of origin is along a road or highway, the likelihood of a fire being caused by something thrown from a vehicle is high. If an automobile equipped with a catalytic converter was parked in dry grass, it is a likely source of ignition. If it is a holiday/celebration, or within a few days before or after, fireworks may be involved. A host of other natural or accidental causes of wildland fires are possible such as power lines falling or airplane crashes. An investigator should consider any and all possible causes that the circumstances surrounding the area of origin indicate. However, if the circumstances do not suggest any of the most likely natural or accidental causes, an investigator must consider the possibility of the fire being set intentionally.

Intentionally set fires range from controlled burns that get out of hand to children playing with matches to arson fires set for profit, revenge, or pyromania. While some accidental fires may result in litigation if negligence was involved, all intentionally set fires that produce an emergency response by the fire department are potentially litigable. Litigation may be necessary to try those accused of crimes or to enable fire protection agencies to recover the cost of fire suppression. The possibility of subsequent litigation makes a thorough investigation, accurate cause determination, and proper documentation extremely important.

Fire Causes and Indicators

Some of the more common fire causes (natural, accidental, and intentional) are discussed in this section. Each cause has certain telltale characteristics that investigators should look for. These causes and characteristics are as follows:

LIGHTNING

Trees, poles, and logs struck by lightning may show strike marks on their surfaces (Figure 9.19). Live trees and deadwood with a high moisture content literally explode when the heat from a lightning bolt instantly turns the internal moisture to steam. The explosion often splinters trees, logs, and roots (Figure 9.20). Soil at the site of a strike may be disturbed, and there may be fused, glassy clumps beneath the surface.

CAMPFIRES

Even a campsite that burns completely leaves evidence of its existence. Metal or glass containers and metal grommets from a tent may be found. The campfire itself may be detected by a circle of rocks enclosing a large concentration of ash, or there may be ash from pieces of wood lying in a definite pattern (Figure 9.21).

Figure 9.19 A tree showing a lightning scar on its surface. *Courtesy of NIFC.*

Figure 9.20 A shattered tree burns after being struck by lightning.

SMOKING MATERIALS

While it is extremely difficult to start a wildland fire with a cigarette, under ideal conditions it may happen. At least 30 percent of the length of a

Figure 9.21 The remains of a campfire. *Courtesy of Hugh Graham.*

smoldering cigarette must be in contact with very fine fuels for it to cause a fire (Figure 9.22). However, a lighted match thrown into dry, fine fuels can easily start a fire. Fires resulting from discarded smoking materials (matches or cigarette or cigar butts) most often occur in grass and other extremely fine fuels when the relative humidity is less than 25 percent.

Figure 9.22 Ash from a cigarette is sometimes recognizable. *Courtesy of Bill Lellis.*

DEBRIS BURNING

Fires at dump sites or other debris-burning operations often spread to surrounding wildland areas. A burn barrel or incinerator near or in the burn, a history of other fires at the location, windy conditions at the time the fire started, or civilians at a debris-burning site may indicate that the debris-burning operation was the cause of the fire (Figure 9.23). Ideally, one or more witnesses will come forward to confirm that debris burning was the cause.

Figure 9.23 Burn barrels are a frequent source of accidental fires.

INCENDIARY FIRES

Unlike debris-burning or other friendly fires that accidentally spread to wildland fuels, *incendiary fires* are those set with malicious intent. They are deliberately set to burn property for illegal purposes. Some of the common indicators of incendiary fires are two or more sets (often along roads or trails) or sets in places frequently used for parties or gatherings. Igniting devices, such as fusees or matchbooks, may be found at the point of origin. Delayed igniting devices are sometimes found. These devices may consist of combinations of items not normally found in the wildland — cigarettes, matches, candles, rope, wire, tape, or rubber bands. People who have grudges against or disputes with property owners should be investigated to determine their whereabouts and actions at the time of a fire. Also, people who may want jobs helping fight fires, desire to be "heroes," or have personal histories of firesetting (perhaps of pyromania) should be investigated.

HEAVY-EQUIPMENT USE

Heavy-equipment operations can start wildland fires in a number of ways. Operating heavy equipment without a spark arrestor (or with a defective spark arrestor) on the exhaust pipe is a frequent cause of wildland fires (Figure 9.24). When the steel blade or tracks strike rocks, a spark can be produced that is capable of igniting very fine, dry fuels.

Figure 9.24 Heavy equipment without spark arrestors start many fires. *Courtesy of NIFC.*

RAILROADS

Railroads also are sometimes sources of wildland fires. These fires often result from railroad crews intentionally burning rights-of-way or old ties. However, they are also started by defective spark arrestors on engines, hot fragments from brake shoes, fusees or flares, or crews discarding lighted smoking materials along the rights-of-way (Figure 9.25).

Figure 9.25 Railroad engines should also have spark arrestors. *Courtesy of NIFC.*

CHILDREN

Children are also responsible for many wildland fires. These fires are usually the result of children's natural curiosity about fire or their carelessness with fireworks, and they most often occur near residential areas, playgrounds, or schools. Most children are not very clever about concealing evidence of their activities, so burned matches or spent fireworks are often found at the point of origin (Figure 9.26).

Figure 9.26 Fireworks are a common cause of fires.

MISCELLANEOUS

Wildland fires that do not fit into any of the categories just discussed are classified as being of *miscellaneous origin.* In this category are fires resulting from blasting operations, electric fences, downed power lines, fireworks, tracer rounds, structure or vehicle fires, and logging operations.

UNDETERMINED

Despite a thorough and painstaking search of the area of origin, there are times when no tangible evidence of a fire cause is found. This most often happens when a fire is intentionally set and the firesetter takes the igniting device when leaving the scene. In these cases, the fire cause is listed as *undetermined.*

Elimination of Causes

As mentioned earlier, after locating the point of origin, the fire cause may be obvious. However, even if it is not, the investigator must first eliminate all natural or accidental causes of the fire. If no lightning strikes were recorded in the area for weeks before the fire started, the investigator can probably eliminate lightning as a cause. Likewise, if there are no railroad lines within several miles (kilometers), railroad operations are not likely to have been the cause. In the middle of an inaccessible wilderness area, heavy equipment can be eliminated as a possible cause. This process helps an investigator concentrate on the most likely cause while analyzing the indicators discussed earlier.

Methods of Search

In many cases, investigating the fire-origin indicators reveals the cause of a fire. If a fire was accidental, the specific source of ignition will still be there. However, as mentioned earlier, if the fire was intentionally set, the igniting device may have been removed. If the cause is not readily visible, the area of origin must be searched to find the cause or to determine that the cause was removed. It is important that the investigation be started while suppression activities are still being conducted so that control of the scene is not relinquished prematurely. Once the fire department releases the scene, investigators may have to obtain a search warrant to reenter the property.

There are many ways of searching the area of origin for the specific cause of a fire. If the area of origin is small, the entire area can be searched as a whole. However, it is often useful to systematically divide the area into segments (Figure 9.27). Beginning with the outermost segments, each segment is thoroughly searched (Figure 9.28). By working from the outside toward the middle, each new segment is searched from one that has already been searched (Figure 9.29). If necessary to locate the cause, each segment is searched twice, preferably by different investigators each time. For example, Investigator A searches Segment 1 from

Figure 9.27 The area of origin should be divided into segments.

Figure 9.28 An investigator checks the first segment.

Figure 9.29 A different investigator checks the first segment.

Figure 9.30 Investigator uses a straightedge and magnifying glass looking for evidence.

Figure 9.31 Investigator uses a magnet to search for metallic fragments.

outside it, and then moves around to Segment 3, while Investigator B searches Segment 1. If neither investigator finds anything suspicious in Segment 1, Investigator A can walk in that segment to search Segment 2, while Investigator B searches Segment 3, etc. This process is repeated until either the cause is found or it is determined that the cause was removed by the firesetter.

In searching each segment of the fire origin, an investigator first conducts a macrosearch by observing the entire area within the segment. Then the investigator conducts a microsearch by concentrating on the surface details within the segment. To do this, the investigator may want to focus the point of vision along a straightedge placed across the width of the segment (Figure 9.30). After each focus area is searched, the straightedge is moved about 1 foot (305 mm) laterally to establish a new focus area. This process is repeated until the entire length of each segment is searched. In addition to the visual search, a strong magnet should be passed over the area to pick up carbon particles or metal fragments that may be associated with the fire cause (Figure 9.31).

If an igniting device is found, it should be photographed at close range from several different angles (Figure 9.32). The device should be packaged and labeled according to departmental general operating guidelines (Figure 9.33). Because of

Figure 9.32 Investigator photographs an igniting device.

the possibility of subsequent litigation, care must be taken to maintain the chain of evidence (record of custody) and to fully document the investigation (see Documenting the Search section).

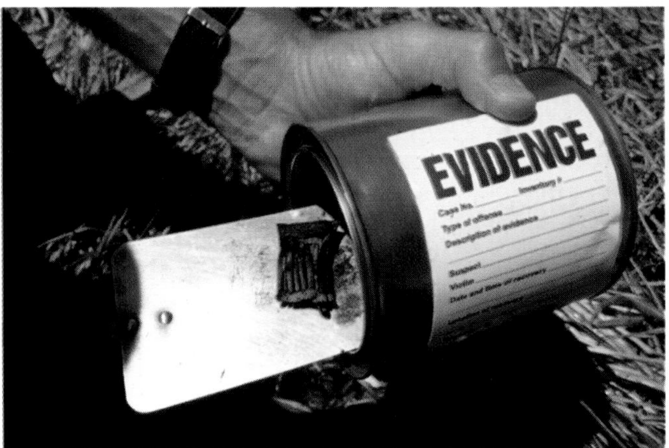

Figure 9.33 The igniting device is placed in a container. *Courtesy of Bill Lellis.*

FIRE INVESTIGATION TOOLS/EQUIPMENT

There are certain basic tools that investigators use in the search for the cause and origin of a wildland fire. While not necessarily complete, the following list of tools should be available to the fire investigator:

- Fireline tape and wooden stakes (to mark the area of origin and to divide it into segments for a detailed search)
- Tape measure (to accurately measure distances between relevant points on the fire scene)
- Straightedge (to help the investigator focus on a small segment)
- Magnet (to locate metal fragments or carbon particles)
- Camera (to photograph evidence)
- Ruler (to show relative size of objects photographed)
- Containers (to store, preserve, and protect evidence)
- Writing materials (to make sketches and notes during the investigation)

Cameras

While a professional-quality still camera is not absolutely necessary for photographing evidence on the fireground, a reliable 35 mm camera with zoom lens, flash, and dating feature should be available to both firefighters and investigators (Figure 9.34). The primary disadvantage of most autofocus electronic cameras is that they usually are equipped with a wide-angle, landscape lens. With that type of lens, the subject appears larger in the viewfinder than it does in the slide or print. Many investigators are also using video cameras (camcorders) in addition to still cameras (Figure 9.35). Firefighters and investigators should practice beforehand with the cameras they will use for photographing evidence. Still cameras should be loaded with color slide film because slides are easier to use in court and/or in training, and it is easier to make prints from slides than slides from prints.

Figure 9.34 A quality camera with flash and zoom lens.

Figure 9.35 A camcorder is used to record the scene.

Evidence Containers

Any clean, sealable container can be used to protect and preserve evidence of a fire cause. Depending upon the size and nature of the evidence, anything from an ordinary envelope to a large packing crate may be used (Figure 9.36). New, unused containers, with no interior coating should be used to avoid any possible contamination of the evidence by traces of the previous contents of the container. New, unused paint cans are available at most paint stores, and these cans make excellent evidence containers (Figure 9.37).

Figure 9.36 Evidence is placed in a container. *Courtesy of Bill Lellis.*

Figure 9.37 A new, unused paint can makes an ideal evidence container.

DOCUMENTING THE SEARCH

As mentioned earlier, fully and properly documenting the search is an extremely important part of an investigation. Regardless of whether litigation results from a fire, having the facts of the incident clearly documented provides the basis for cost recovery, future fire prevention activities, and training.

Chain of Evidence

Because of the possibility of litigation soon after a fire or at some point in the future, documenting the *chain of evidence* (sometimes called *chain of custody* or *chain of possession*) is critically important. A written record of who had custody of any piece of evidence must be maintained. This record must show when and how or from whom someone took custody of the evidence and when and to whom they surrendered it (Figure 9.38). Every person who had custody of the evidence at any time must be prepared to testify as to how and where the evidence was kept while in their custody. If the chain of evidence is broken by anyone failing to maintain full and complete control over the evidence, such as leaving it where others could accidentally or intentionally contaminate it, any subsequent prosecution will probably be jeopardized.

CRIME SCENE SEARCH EVIDENCE REPORT

Name of Subject ....

Offense ....

Date of Incident .... Time .... AM-PM

Search Officer ....

Evidence Description ....

....

Location ....

....

CHAIN OF POSSESSION

Received From ....

By ....

Date .... Time .... AM-PM

Received From ....

By ....

Date .... Time .... AM-PM

Received From ....

By ....

Date .... Time .... AM-PM

Received From ....

By ....

Date .... Time .... AM-PM

Received From ....

By ....

Date .... Time .... AM-PM

Received From ....

By ....

Date .... Time .... AM-PM

Received From ....

By ....

Date .... Time .... AM-PM

Figure 9.38 A typical chain-of-custody form.

Photographing/Videotaping a Fire Scene

Ideally, only trained personnel are assigned to photograph and/or videotape a fire scene and any evidence found there. However, because of the remoteness of some fire scenes and the difficulty of getting a trained photographer/videographer there before the scene is disturbed by wind or contaminated by foot traffic, firefighters or crew leaders may have to perform this duty. As mentioned earlier, they should practice beforehand with whatever cameras are available to them.

PURPOSES

There are several purposes for photographing or videotaping a fire scene. The primary one, of course, is to preserve a visual record of the evidence. But fire-scene photographs/videotapes can do much more than that. Photographs and videotapes of a fire scene can do the following:

- Present a picture of the scene that would take many pages of text to describe.
- Preserve an image of perishable evidence.
- Show geographic relationships and proximities.
- Show evidence too large to be brought to the courtroom.
- Verify oral or written testimony.

METHOD

As mentioned earlier, the fire cause should be photographed or videotaped from several different angles. However, this should be done only after taking a series of shots of the overall scene from the perspective of a witness approaching on foot. These shots should be taken from eye level as a witness would view the scene.

RANGE

The cause of a fire should be photographed or videotaped from close range to capture as much detail as possible. But a series of shots from various distances must also be made to establish geographic relationships of various features and items of evidence. There should be one or more establishing shots at relatively long range to show the overall scene, one or more intermediate-range shots to show the relative positions of various items, and several close-ups to show the actual cause.

RECORDS

Because it is difficult to later recall the exact circumstances under which any particular photograph was taken, it is very important to make a written record of every shot used to document a fire scene, including lens, film type, and *f*-stop (Figure 9.39). Since video camcorders also allow audio recording, verbal notes should be dictated as the tape is being shot. Considering that testimony concerning a given incident may not be required for a considerable length of time, perhaps years, it is difficult to overemphasize the importance of such records.

Figure 9.39 Photographs should be carefully documented. *Courtesy of NIFC.*

Sketching a Fire Scene

Even though a fire scene is thoroughly photographed or videotaped and a detailed report is written, an accurate sketch of the scene can also be an important part of documenting the investigation. The main purpose of a sketch is to help clarify the facts and circumstances relating to a fire. In some cases, a sketch presents information more clearly than any other medium by eliminating everything except the most essential information. In the absence of a camera, a sketch may be the only visual record of the scene.

WHEN TO SKETCH

A sketch should be made of any fire scene in which the fire resulted from human activity, whether accidental or intentional, and may be made of scenes of natural-origin fires. A sketch should be made even if the scene is also photographed or videotaped.

HOW TO SKETCH

Landmarks and other relevant features must be shown, but a fire-scene sketch should be kept as simple as possible. Distances, measurements, and geographic proximities must be shown accurately. Graph paper (sometimes called *quadrille-ruled* paper) should be used to make it easier to draw to scale (Figure 9.40). North is indicated on the sketch to orient the scene in relation to the compass. A fire-scene sketch need not be a work of art, but since it may become a critical piece of evidence in court, it should be neatly drawn and clearly labeled.

AFTER DETERMINING FIRE CAUSE

Once the cause of a fire has been determined and the scene photographed/videotaped and/or sketched, the relevant facts must be documented. Depending upon local laws and protocols, if the fire was the result of negligence or arson, law enforcement personnel may have jurisdiction over the scene and become responsible for any further investigation or other action. If this is the case, firefighters and fire investigators need only remain available to assist law enforcement if asked.

Reporting

If the fire department retains jurisdiction and responsibility for an investigation, investigators report their findings to the department administration through channels. If a fire resulted from arson or negligence, the district attorney's office or other appropriate law enforcement agency is also notified.

Documentation

The most thorough and painstaking investigation is of little value if it is not fully and properly documented. Many fire departments have standard forms on which to document a fire investigation (Figure 9.41). However, many of these forms are designed primarily for investigating structure fires, not wildland fires. Whether standard forms are available or not, the documentation must include the following information:

- ***Incident information*** — General data regarding the incident. Such things as the incident number, date, time of alarm (military time), and the address or location of the point of origin are listed.
- ***Investigation information*** — Name(s) of the investigator(s), date and time the investigation was requested, name and rank/title of the person making the request, and the date and time the investigation was initiated (if different from the time of request). Also listed are the names of those assisting in the investigation such as the photographer/videographer.
- ***Scene information*** — Reported and actual address or location (if not the same) of the fire scene, type of incident, weather conditions at the time of the fire, wind direction and speed, temperature, and the relative humidity.
- ***Persons affected*** — All those who were affected by the fire. The names, addresses, phone numbers (home/business), races, sexes, dates of birth, and social security numbers (optional) of the property owner(s), tenants, visitors, and/or victims are listed.
- ***Insurance information*** — Insurance carrier(s) insuring the property and life of any of the affected persons. The names, addresses, and phone numbers of each company and local agent or broker are listed. In some jurisdictions, policy numbers, coverages and exclusions, beneficiaries or payees, and other details can be obtained from the agent or broker. In others, a subpoena may be required to get these data.
- ***Cause/origin information*** — Known or most likely cause and origin of the fire and a damage estimate. Also listed are the name and personal data of the person who discovered the fire, who reported it, and any witnesses.
- ***Narrative*** — Description of the entire incident. Included in the narrative are the observations of those responding to the fire, statements of witnesses, property

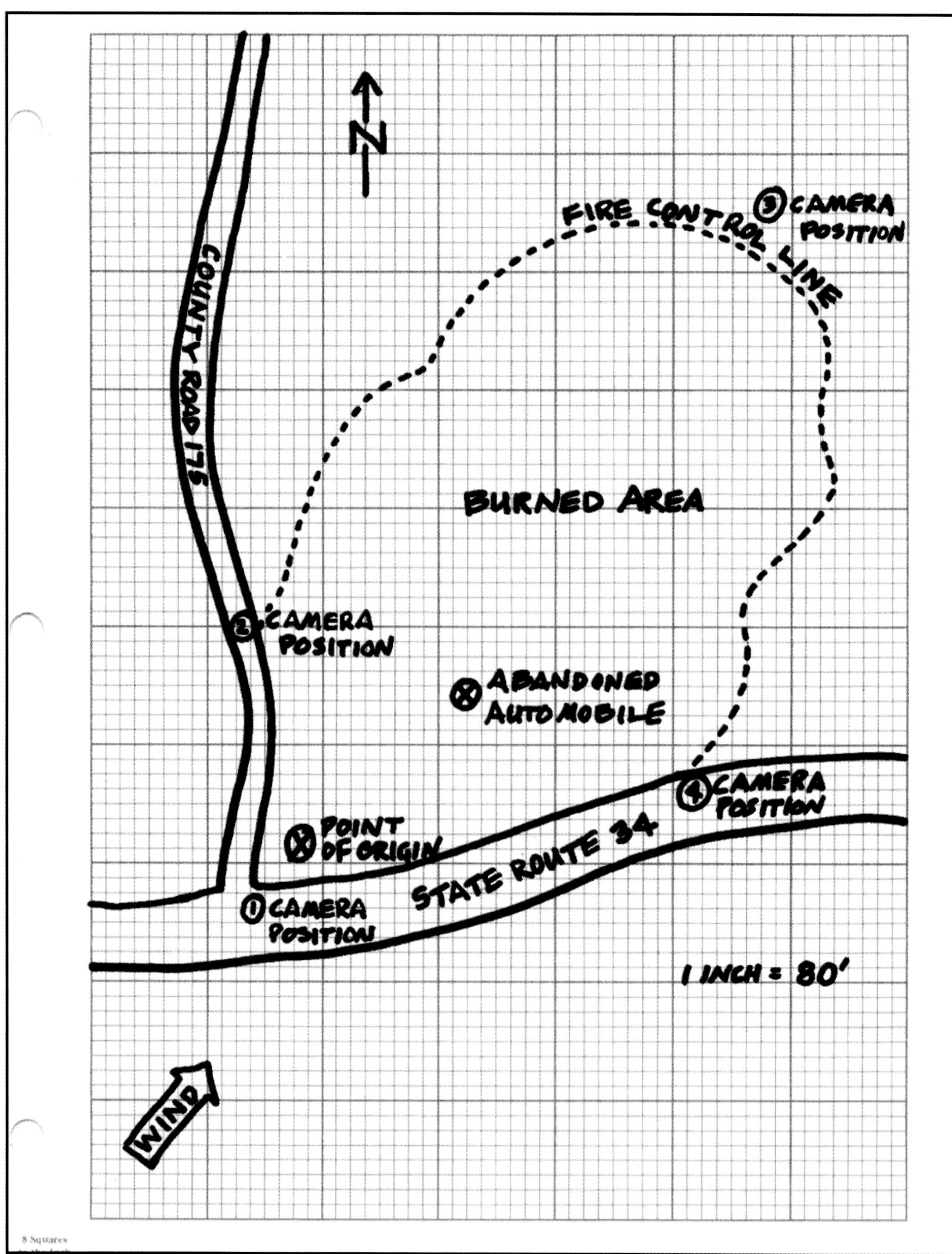

Figure 9.40 A fire-scene sketch.

CITY OF Santa Rosa Fire Department	**WILDLAND FIRE**		
	Incident #:	Incident Date:	Incident Time:
	Incident Location:		A.P.N. #:
	Victim's Name:		
	Address:		Phone:

Property Description: [ ] Vacant [ ] Residential [ ] Commercial [ ] Industrial [ ] Farmland [ ] Other
Brief description of property:
Fire Damage: [ ] Less than 1 acre [ ] 1 to 5 acres [ ] Greater than 5 acres [ ] Damaged structures
Other Properties Involved: [ ] Yes [ ] No If yes, Describe:
Terrain: [ ] Flat [ ] Sloped [ ] Wooded [ ] Grassland [ ] Cultivated [ ] Inaccessible by vehicle
Security: [ ] Open [ ] Fenced [ ] Locked [ ] Gates [ ] Roadside [ ] No Security
Access: [ ] Public Way [ ] Private Road [ ] Overland [ ] Inaccessible Unless Hiked To [ ] Hiking Trail
Fire Travel By: [ ] Ground [ ] Crown [ ] Radiant Heat [ ] Airborne Embers [ ] Multiple Origins
Fuel Load: [ ] Light [ ] Medium [ ] Heavy [ ] Downed Material Increased Fuel Load [ ] Not Contributing
Weather: Temperature:______ Humidity: ______ Wind Direction: ______ Speed: ______
Area of Origin: (describe)

Ignition Sequence:	Heat of Ignition:	
Materials Ignited:		
Ignition Factors:		
If Equipment Was Involved, Type:	Make:	Model:
People In Area: [ ] Yes [ ] No [ ] Undetermined	Vehicle Description:	
Eliminated: [ ] Lightning [ ] Power Lines [ ] Portable Equipment [ ] Roadside [ ] Vehicle Exhaust		
Suspect(s):		
Suspect Travel: [ ] Foot [ ] Bicycle [ ] Motorcycle [ ] Car [ ] Truck [ ] Other [ ] Unknown		
Prepared By:	Date:	
Reviewed By:	Date:	

Figure 9.41 A typical fire-investigation form.

description, information on injuries and/or fatalities, fire-scene description, fire-suppression operations, investigation conducted, evidence collected, and conclusions drawn. Copies of alarm records, run reports, and other forms may be attached.

FIRE PREVENTION PROGRAM

Fire prevention includes those activities intended to reduce the number and/or severity of fires within the jurisdiction. Since the cost of preventing a wildland fire is almost certainly less than the cost of suppressing it, fire prevention is one of the most cost-effective activities of the fire department. An accurate determination of the causes of fires that have occurred within the jurisdiction provides a statistical basis for establishing fire prevention priorities and focusing fire prevention activities. In general, fire prevention programs involve a combination of code enforcement, education, and engineering activities. To be most effective, the department's fire prevention program must use the fire-cause data to identify ways to prevent fires or reduce the severity of fires within the jurisdiction. The first step in this process is to identify the risks and hazards that exist within the jurisdiction.

Risks

According to NFPA 921, *Guide for Fire and Explosion Investigations*, *risk* is defined as "*(a) The degree of peril; the possible harm that might occur. (b) The statistical probability or quantitative estimate of the frequency or severity of injury or loss.*" In more common terms, this definition translates into an assessment of what is at risk within the jurisdiction — who may be injured and what is there to burn.

One example of a risk-rating system for planning fire prevention activities in existing wildland/urban interface areas is the one developed by the Montana Department of State Lands. Using information gathered at the site of a development, such as road access, topography, types of fuels, types of building construction, and available water supplies, the system helps planners assess hazard areas and establish priorities for allocating fire prevention resources and focusing fire prevention activities. A secondary benefit derived from gathering the information needed to apply the system is that those conducting the surveys become more familiar with the hazardous areas and the values at risk within the jurisdiction. This information helps in fire-suppression planning and decision making. See Appendix D for instructions and samples of the forms used in applying this system.

A risk-rating system is applied to wildland or rural areas that have sufficient permanent, seasonal, recreational, or commercial habitation to pose a fire hazard and/or fire risk. Typical applications include rural subdivisions, scattered residential developments, permanent camps, lodges, and resorts. A rating area should consist of one relatively homogeneous development with similar site factors. It may be large or small, or it may even be a part of a large development. A rating area should be large enough to include a representative portion of wildland fuels, topography, etc., in close proximity to the developed area.

There is at least one wildland fire-hazard assessment device on the market. This device helps the property owner assess the degree to which structures are at risk based on slope, aspect, and fuels where the structures are located (Figure 9.42). While the device is based on sound scientific principles, it is unclear how accurate the results are when it is used by someone untrained and inexperienced in assessing wildland fire hazards.

Hazards

NFPA 921 defines a *hazard* as "*any arrangement of materials and heat sources that presents the potential for harm, such as personal injury or*

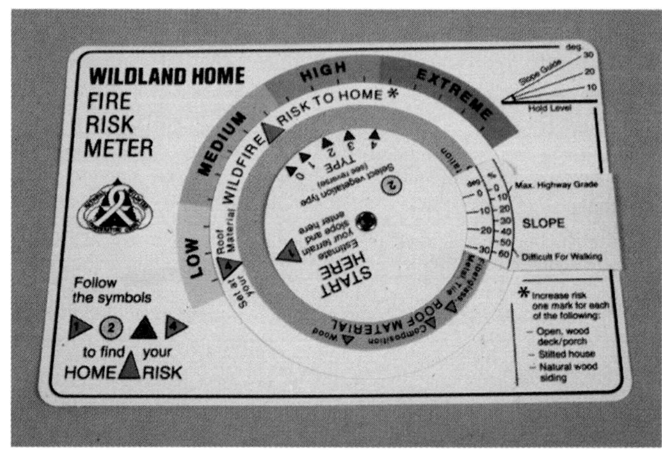

Figure 9.42 A Wildland Home Fire Risk Meter.

ignition of combustibles." In more common terms, hazards are the most likely sources of ignition and fire potential within the jurisdiction. The risk-rating systems just discussed focus primarily on the fuel component of the hazard — what there is to burn. The other major component of the hazard is the likelihood of the available fuels igniting. This component provides ample opportunities for the fire department to have a positive impact on the wildland fire potential within the jurisdiction. A combination of code enforcement and public education can significantly reduce the number and severity of these fires.

Figure 9.43 An improved access way into an interface development.

Figure 9.44 A tractor mows weeds to reduce available fuel.

Cause Analysis

An analysis of the causes of fires occurring within the jurisdiction over a period of time, especially the most recent ones, provides a statistical basis for developing and implementing a fire prevention program. If a statistically significant number of fires have been the result of debris burning, for example, the local fire code can be used to restrict this activity to those times of the year when it is relatively safe. Or the code may be used to limit the types of incinerators allowed. Educating the public about code requirements and safe burning practices can also reduce the number of these fires.

Mapping

Another important result of the process of gathering information about existing fire risks and hazards, and an analysis of fire causes, is the opportunity these data provide for mapping the areas where fire prevention efforts will produce the greatest results. By combining these three elements — risk, hazard, and cause analysis — zones can be identified based on fire potential. The zone(s) with the highest values at risk may justify engineered improvements such as wider roads or a better water supply (Figure 9.43). Zones with high concentrations of susceptible fuels may lend themselves to an enforced weed-abatement or fuel-modification program (Figure 9.44).

Prevention Program Components

As mentioned earlier, a fire prevention program has three major components — education, code enforcement, and engineering. Each component represents a different approach to the common goal of reducing the number and severity of fires within the jurisdiction. It is important for fire department personnel to understand how each of these components acts to protect citizens and preserve their properties.

EDUCATION

Many wildland fires are the result of human activity in areas where conditions are susceptible to ignition. Especially during periods of high fire danger, public service announcements over local radio and television warning of the potential danger may be worthwhile. These announcements are used to remind the public of the fire danger and what they can do to reduce the likelihood of unfriendly fires starting. Of particular importance is public education for people who live in extraordinarily susceptible areas such as the wildland/urban interface. This may be done on a one-to-one basis with area property owners or through public meetings at neighborhood or community centers as well as media campaigns (Figure 9.45).

Figure 9.45 A firefighter addresses a community group about fire prevention. *Courtesy of NIFC.*

Public education also involves informing people of local fire regulations, such as those governing outdoor debris burning, or explaining the engineering services available to them. If the jurisdiction has adopted all or part of NFPA 299 (*Standard for Protection of Life and Property from Wildfire*), NFPA 1141 (*Standard for Fire Protection in Planned Building Groups*), or the International Fire Code Institute's *Urban-Wildland Interface Code*, the public should be made aware of the requirements of these documents. Education may also be aimed at correcting a problem in a specific segment of the population. If fire-cause analysis indicates that careless smokers or careless campers have been responsible for numerous recent fires, these groups may be targeted for an intense informational/motivational fire prevention campaign. For more information on public education techniques, see the IFSTA **Fire and Life Safety Educator** manual.

CODE ENFORCEMENT

Code enforcement is a very important component of a fire prevention program. When needed, the code provides the statutory basis for compelling people to behave in a manner that reduces the risk to themselves and others. Even though it is only one of the components of the program, people sometimes think of it as the entire program because it is such a visible (and sometimes annoying) component. This perception presents an opportunity for firefighters to use the code as an educational tool.

Every requirement in the fire code is a solution to a potential fire problem. Each one is based on a history of fires starting from a specific cause, a likelihood of one starting from that cause, or a way of reducing the effect of one that does start. In other words, the code is intended to help citizens reduce the likelihood of a fire starting in their properties and/or to limit the amount of damage a fire will do if it does start. If property owners are aware of this connection between the code requirements and avoiding injury or loss, they are more likely to willingly comply.

There is also another way in which code enforcement is used to reduce fire loss. Enforcement of local codes helps reduce the number and severity of fires within the jurisdiction by aggressively prosecuting the perpetrators of unfriendly fires, whether negligent or incendiary. If the successful prosecution of these cases is fully and prominently reported in the local news media, it provides incentive for law-abiding citizens to be more careful with fire and discourages would-be firesetters (Figure 9.46).

Figure 9.46 An arson suspect is arrested.

Another way that code enforcement positively impacts the local fire problem is through fuel reduction. As mentioned earlier, jurisdictions may enact a mandatory weed-abatement or fuel-modification program that involves mowing, plowing, or burning fields not under cultivation (Figure 9.47). The owners of these properties are notified of the code requirement that wildland fuels be removed at the beginning of fire season. They are given an opportunity to remove the fuels or to appeal the requirement to the local governing body. If their appeal is unsuccessful or if they simply do nothing, the fuels are removed by an agent of the jurisdiction after a specified period, and the costs are added to the property tax liability for that piece of property.

Figure 9.47 Controlled burning can eliminate potential fuel. *Courtesy of NIFC.*

Local codes can also require that fire-resistant vegetation be planted around structures surrounded by wildland. Codes may require a "greenbelt" around developments in the wildland/urban interface. These greenbelts can be playing fields, parks, golf courses, or landscaped roadways. Codes also are used to protect structures or developments in the wildland by requiring fuel breaks or firebreaks. *Fuel breaks* are wide strips around developments from which most, but not all, fuel is removed (Figure 9.48). *Firebreaks* are strips in which all fuel is removed down to mineral soil (Figure 9.49).

Other typical code requirements relate to building features and materials. Spark-arresting screens on chimneys are almost universally required. If allowed at all, burn barrels or incinerators must be properly designed and located. Adequate road access to structures must be developed and maintained. Noncombustible roof coverings or roof sprinklers may be required, along with an adequate water supply.

ENGINEERING

Engineering involves the alteration of a structural design or a physical situation to reduce the likelihood of a fire starting or to reduce its effects if

Figure 9.48 A typical greenbelt (fuel break). *Courtesy of NIFC.*

Figure 9.49 A typical firebreak. *Courtesy of NIFC.*

one does start. These objectives can be met by eliminating fire causes, reducing the available fuel, improving access for fire apparatus, increasing the amount of water available on site, etc. Engineered solutions to specific fire problems take several forms, most of which are intended to satisfy local fire-code requirements. In most jurisdictions, builders must submit plans that satisfy the code requirements to obtain a building permit for new construction or major remodels. In order to be reasonable, codes are written to allow some flexibility in the way that the intent of the code is met in a particular situation. This flexibility often involves some engineering decisions. For instance, the code may require noncombustible roof coverings, *or equivalent*, in a particular fire zone. Possible *equivalent* alternatives might include installing a roof sprinkler system or using combustible roofing materials that have been pressure-treated with a long-term fire retardant. The builder may need the help of a fire protection engineer to evaluate the adequacy, practicality, and cost of each alternative (Figure 9.50). The fire chief (or a designated agent) is usually empowered to decide whether the proposed solution is equivalent and meets the intent of the code.

Engineering decisions may also be required to evaluate the design of the means of access to and/or egress from a particular property or the adequacy of its private water supply. The fire department may require the builder to submit an engineering study to verify that the proposed development meets these code requirements.

Figure 9.50 A builder consults an engineer about code compliance.

Implementing a Fire Prevention Program

Because every fire department's resources are finite, priorities must be established, goals and objectives identified, and decisions made about how and where these limited resources can be invested to do the most good. To do this, fire prevention planners must follow a very structured process when implementing a program to reduce the number and severity of fires within the jurisdiction. The steps in this process are to identify the problem, identify solutions, apply the solutions, and evaluate the results.

IDENTIFY THE PROBLEM

The first step in the process is to identify the problem. More precisely, planners usually must identify the most significant or most pressing problem among several problems. Sometimes, when more than one problem of equal importance exists, the decision about which problem has the highest priority is based on which problem is the most manageable or which will respond better to the available solutions.

As mentioned earlier, an analysis of fire causes within the jurisdiction, especially the most recent and most destructive ones, may reveal the specific *problem* on which to focus fire prevention efforts. For instance, if the majority of fires resulted from debris-burning fires that got out of control, homeowners would be the most likely target audience. If a disproportionate number of fires have occurred in one particular zone, this may tell *where* efforts should be focused. Obviously, to assist with this planning process, complete and accurate records of all fire calls within the jurisdiction should be kept, including the specific cause of each fire (Figure 9.51). Regardless of how basic or how sophisticated the planning tools available to the department are, they all depend upon accurate and up-to-date fire-cause data.

Figure 9.51 A firefighter checks fire activity records to identify problem areas.

IDENTIFY SOLUTIONS

In the process of identifying possible solutions, *brainstorming* and other idea-generating techniques may be useful. However, once a target fire cause has been identified, there may only be a limited number of workable solutions available.

Some solutions, such as prohibiting smoking throughout the district, may be impractical. Others, such as providing a 24-hour fire patrol in all areas of the district, may be too expensive. In most cases, a careful analysis of the department's fire-cause data will suggest not only a specific fire cause on which to focus fire prevention efforts but also the solution most likely to produce positive results. As mentioned earlier, this analysis may also reveal to whom the campaign should be directed (target audience) as well as where within the district that effort should be concentrated (target area).

APPLY SOLUTIONS

Once the problem and the solution(s) are identified, as well as to whom and where the efforts are to be directed, the next step is to implement the plan or apply the solutions. For any solution to produce the desired results, it must be applied properly. In other words, how the solution is applied must match the nature of the problem. If local laws (statutory basis for certain corrective actions) are applied in a heavy-handed manner based only on the coercive power of the law, the results are likely to be temporary, and a lot of goodwill and public support can be destroyed in the process. A more positive approach might be to combine enforcement with education so that even if citizens do not like complying with a law, at least they understand why it is important for them to do so.

EVALUATE RESULTS

Once a solution has been implemented, it must be given a reasonable amount of time to show results. For example, if fires resulting from outdoor debris burning were identified as the target fire cause and homeowners as the target audience, a public education program on safe-burning practices might be initiated in combination with increased enforcement of burning permits. If after a reasonable period of time (perhaps two to four weeks), there is little or no impact on the number of fires spreading from outdoor burning, an outright ban on debris burning during fire season might have to be implemented.

Inspections

Fire inspections serve a variety of purposes. They help firefighters become familiar with the risks and hazards in their jurisdictions, provide an opportunity to educate the public about fire safety, and facilitate enforcement of and compliance with local fire codes. Compliance with the codes translates into fewer fires starting and limiting the effects of those fires that do start.

In most cases, fire inspections should be conducted by the firefighters who will respond if a fire is reported at the particular location or address. In those cases where the occupancy to be inspected contains materials, processes, or other items that require the expertise of a fire inspector or engineer, the firefighters should accompany the inspector during the inspection.

Departmental guidelines and local protocols must always be followed; however, it is generally good practice to notify property owners in advance of any planned inspections. Notification can be done on an individual basis by letter or on a collective basis by publishing a general notice of the impending program in a local newspaper. If done on an individual basis, the notification letter should include the following:

- Statutory basis for the inspection
- Need for hazard reduction/abatement
- Approximate or specific date of the proposed inspection
- Scope of the inspection
- Name and telephone number of the fire department contact person

With the exception of the specific date for an inspection, the same information should be included in the public notice if notification is done collectively. Unless the statement regarding the scope of the inspection is expressed in very general language, the scope may vary from one occupancy to another depending upon the nature, size, and complexity of the specific occupancy.

While the purpose of all fire prevention inspections remains the same, the scope or focus of the inspection varies with the occupancy. If the occupancy is an industrial site in a rural or wildland setting, the inspection will be conducted with a different focus than one done on a single-family residence in the wildland/urban interface. An inspection of an industrial site would most likely focus on preventing fires from starting on site and spreading to the surrounding wildland (Figure 9.52). One conducted on a single-family residence may focus on both preventing a fire in the house from spreading to the wildland and preventing a wildland fire from involving the house. In the case of the residence, the inspection would include items identified in previous chapters of this manual — combustible vegetation or other fuels within 30 feet (9 m) of the structure, accumulations of dry leaves or needles in rain gutters or on rooftops, and the adequacy of water supplies and access roads (Figure 9.53).

Figure 9.52 An industrial complex surrounded by wildland area.

Figure 9.53 A firefighter discusses fire prevention measures with a homeowner. *Courtesy of NIFC.*

The results of any inspection should be reviewed and discussed with the occupant. If no code violations or other hazards are found, the occupant should be congratulated. If anything was found that needs to be corrected, this should be explained to the occupant in a matter-of-fact but professional manner. The specific hazard or violation should be pointed out, why it is a hazard should be explained, and what steps the occupant must take to correct the situation should be outlined. Any code violations require a follow-up inspection. The occupant should be allowed a reasonable period of time to correct the violation, depending upon its nature and the difficulty of compliance. Violations that involve immediate threats to life safety, such as obstructed or locked exits, must be corrected before the inspector leaves the premises.

RIGHTS-OF-WAY

Many wildland fire prevention inspections do not involve structures at all. They instead focus on open areas such as rights-of-way. These strips of land are used for roads, highways, railroads, and power lines (Figure 9.54). The primary concern is that these areas be maintained relatively free of vegetation that would serve as fuel for a wildland fire. One possible difficulty with this type of inspection is identifying and locating the entity or person responsible for maintaining the right-of-way.

Figure 9.54 Rights-of-way must be maintained relatively free of vegetation.

RECREATION AREAS

Every summer throngs of city dwellers try to escape the heat and other conditions in the cities by vacationing in parks and wilderness areas. Many of these urbanites lack knowledge of wildland fire safety. While park service personnel are well trained and very conscientious, fire safety is but one of their many jobs when the parks are open. Depending upon the scope of their authority under local codes and ordinances, firefighters can help maintain the fire safety of campsites within and outside state, provincial, or federal parks (Figure 9.55). Under extreme fire-danger conditions, maintaining fire safety may involve closing the parks until conditions moderate (Figure 9.56).

Figure 9.55 A firefighter cautions campers about fire safety in a park.

Figure 9.57 A firefighter inspects a barbeque pit in a park.

Figure 9.56 Parks may have to be closed during periods of extreme fire danger.

Figure 9.58 Firefighters should inspect standpipes and other water sources.

Figure 9.59 A firefighter posts a fire-danger warning sign.

When inspecting campgrounds, firefighters should see that fires are confined to relatively safe areas, such as built-in barbeques and burn pits, and that sufficient clearance is maintained between open fires and surrounding vegetation (Figure 9.57). The availability of water in case a fire threatens to spread to the vegetation should also be verified (Figure 9.58). Firefighters may also inform campers of the specific requirements of local codes and may post signs or posters warning of the fire danger (Figure 9.59).

INDUSTRIAL OPERATIONS

Industrial operations and construction projects located in or adjacent to wildland areas require thorough and conscientious inspections. Industrial sites must be inspected to monitor their hazardous materials and processes and to check their emergency exits, fire extinguishers, alarms, suppression systems, and other fire fighting equipment. In addition, these occupancies tend to maintain relatively heavy fire loads, so a major fire on the site is very likely to threaten the surrounding wildland. This threat means that every effort must be made to identify potential sources of ignition, and those that cannot be eliminated must be properly contained (Figure 9.60).

Figure 9.60 A tank battery is diked to contain the oil should the tanks leak.

CONSTRUCTION PROJECTS

Construction projects present very different challenges for fire prevention inspectors. The primary ignition sources — heavy equipment and welding operations — are not in fixed locations as they often are on industrial sites. Heavy equipment, primarily bulldozers and other earth-moving machines, move around on the site and can start fires anytime they are near dry fuel. These machines must be inspected for functional spark arrestors on their exhaust pipes (unless they are turbocharged) and to ensure that each one has a functional fire extinguisher. Welding operations also tend to move from place to place on the construction site. This means that firefighters should inspect these sites at relatively frequent intervals.

FIRE PROTECTION PLANNING 10

Chapter 10
Fire Protection Planning

INTRODUCTION

Fire protection planning is different from the fire prevention activities and prefire planning discussed in earlier chapters. Fire prevention activities focus on preventing fires from starting and/or reducing the *impact* of those that do start. Prefire planning focuses on identifying and acquiring the resources needed to suppress anticipated fires. Fire protection planning focuses on reducing the impact of fires in or around subdivisions and other developments in the wildland environment. Reducing the impact of fires in these areas requires a combination of fire department programs and activities along with the cooperation and involvement of local government, builders, property owners, and the community as a whole.

This chapter discusses fire-defense improvements such as fuel management, land-use restrictions, and code requirements. Also discussed are building construction standards, spacing, and density. Pre-incident planning is discussed from a different standpoint than it was earlier in this manual. Planning for mass evacuations is covered, as well as community involvement. Also covered are communications plans and other planning activities.

FIRE-DEFENSE IMPROVEMENTS

Fire-defense improvement programs can contribute significantly to protecting property from wildland fires and to protecting the wildland from fires starting in developed areas. These programs generally include one or more of the following:

- Separating communities or groups of structures from the wildland
- Providing access for fire apparatus and equipment
- Providing adequate water supplies

These general fire-defense improvement goals can be realized by the implementation of specific programs. Each program provides a limited but important improvement in the overall fire defenses of a target area or development. To be most effective, several programs should be used in concert to produce a synergistic effect.

Fuel Management

The first of these fire-defense improvements, separating the developments from the wildland, most often involves one or more fuel-management programs. *Fuel management,* sometimes called *fuel modification,* is a general term for a number of specific techniques for reducing the amount of fuel available to any fire that starts. The purpose of fuel management is to create defensible space between developed areas and the wildland and vice versa. In most cases, these spaces are initially created by the developer and should be subsequently maintained by the property owner. In developed areas in or adjacent to the wildland, the most common fuel-management techniques involve the creation of *fuel breaks, firebreaks*, and/or *greenbelts.*

FUEL BREAKS

Fuel breaks surround or separate developments from the adjacent wildlands. They are usually strips of land where the grass is mowed and brush is thinned or removed to reduce the amount of standing fuel (Figure 10.1). Within fuel breaks, trees are trimmed to reduce interlocking crowns, reduce brush, and break the continuity of these fuels.

Figure 10.1 A tractor mows a fuel break.

FIREBREAKS

Firebreaks differ from fuel breaks in that the vegetation in a firebreak is not just mowed but is completely removed (Figure 10.2). The fuel is either burned off during relatively wet periods or it is plowed under or scraped away down to mineral soil with heavy equipment. This technique can be used around an entire development or around individual structures. As mentioned in an earlier chapter, homeowners are advised to clear combustible vegetation from around their homes for a distance of 30 to 200 feet (9 m to 60 m) (Figure 10.3).

Figure 10.2 A dozer starts a firebreak.

Figure 10.3 A home on a hillside may need more than a 30-foot (9 m) clearance.

Firebreaks are also created by providing all-weather access roads into and around developed areas. For more information on access roads, see the Access for Fire Apparatus/Equipment section.

GREENBELTS

A more permanent way of separating developments from the wildland is by creating a greenbelt between structures and the adjacent wildland. This is done by planting wide strips of permanently green vegetation, perhaps with irrigation sprinklers such as on golf courses, or by planting drought-resistant shrubs (Figure 10.4). For a list of fire-resistive landscaping, see *Fire Safe, California* (discussed in the Fire Safety Education/Information section later in this chapter) or consult the local agricultural extension agent.

Figure 10.4 Homes can be protected by permanently green vegetation. *Courtesy of John Hawkins.*

Access for Fire Apparatus/Equipment

The second major category of fire-defense improvements is providing access for fire apparatus and equipment. Access requirements relate to wheeled vehicular traffic such as fire engines, mobile water supply apparatus, and other motor vehicles. The specific requirements for roads, streets, and ways are delineated in NFPA 299, *Standard for Protection of Life and Property from Wildfire*, in NFPA 1141, *Standard for Fire Protection in Planned Building Groups*, and to some extent in the *Urban-Wildland Interface Code* published by the International Fire Code Institute.

The phrase *access ways* includes *roads, streets, fire lanes, private streets, parking lots, driveways,* and *ways* — any access way to an area by land-based vehicles. The primary function of an access way is to provide a means for emergency vehicles to enter the area and for civilians to evacuate the area *simultaneously*. In other words, for an access way to be considered adequate, it must allow for vehicular traffic in both directions at the same time. An adequate access way would allow large vehicles such as water tenders to pass each other.

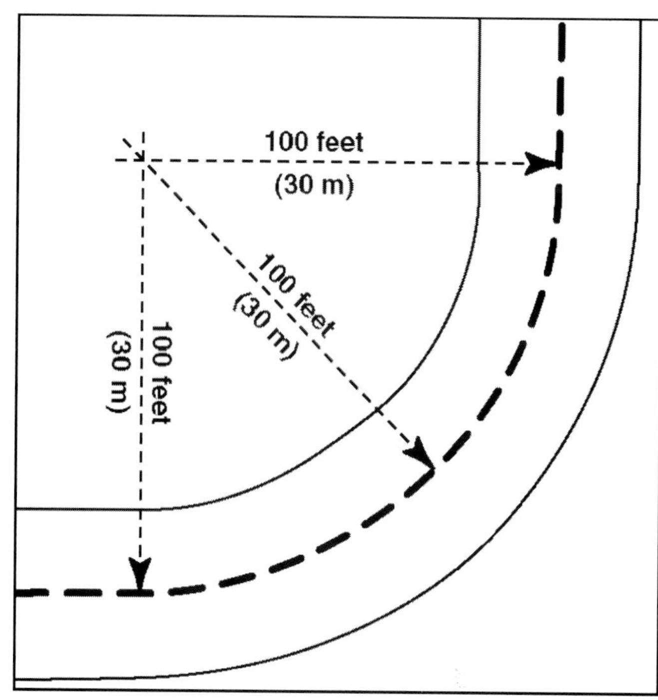

Figure 10.6 Major feed roads must have at least 100-foot (30 m) radius curves.

ACCESS SPECIFICATIONS

While the specifications (minimum requirements) for the design and construction of access ways into any particular development are always subject to local codes and standards, NFPA 299, NFPA 1141, and the *Urban-Wildland Interface Code* set forth recommended specifications. The balance of this section represents recommended specifications drawn from those sources.

According to these codes and standards, developments must be accessible from more than one direction, and the access ways must be looped wherever possible (Figure 10.5). These documents contain specifications for the widths, surfaces, grades, turns, and curves of streets and roads used as access ways (Figure 10.6). The design and maximum length of dead-end roads are also specified, along with the design and location of gates.

Figure 10.5 Developments should have a looped access road. *Courtesy of John Hawkins.*

SIGNAGE

In addition to recommended specifications for access ways, NFPA 299, NFPA 1141, and the *Urban-Wildland Interface Code* contain specifications for the appropriate signage associated with these streets and roads. For example, all streets, roads, and buildings are required to have names or numbers clearly visible from the roadway on which they are addressed (Figure 10.7). Signs indicating special conditions, such as "Dead End," "Not a Through Road" or "No Outlet," must be posted at the entrance to any such roads (Figure 10.8). Signs must also be posted indicating the maximum safe load on bridges

Figure 10.7 Address numbers must be clearly visible.

Figure 10.8 Dead-end roads must be clearly marked.

(Figure 10.9). Street signs should be mounted 6 to 8 feet (1.8 m to 2.4 m) above the ground, have a reflective background, and have numbers and letters at least 4 inches (100 mm) high with a stroke at least ½ inch (13 mm) wide. The numbers and letters must also contrast with the background (Figure 10.10).

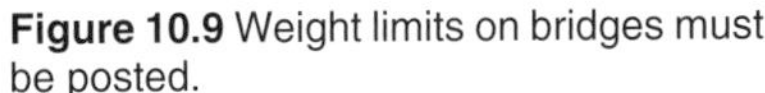

Figure 10.9 Weight limits on bridges must be posted.

Figure 10.10 Street signs must conform to standards.

Water Supplies

The third major category of fire-defense improvements involves the water supplies available for fighting fires that may occur in or near a development. While all three of the sources cited address emergency water supplies to some extent, most of the applicable water-supply requirements are contained in NFPA 1231, *Standard on Water Supplies for Suburban and Rural Fire Fighting.* NFPA 1231 provides detailed instructions for determining minimum water-supply requirements for various construction situations such as single and multiple structures, both with and without exposures. Fire department officials should work with planners, builders, and property owners to ensure that adequate supplies of water are available for fire fighting during and after construction of any developments within the jurisdiction.

In general, the applicable NFPA standards require a minimum water-supply storage of at least 2,000 gallons (7 570 L) and a minimum fire flow of 250 gpm (946 L/min) for 2 hours. In other cases, a minimum of 3,000 gallons (11 356 L) of water must be available at flows in excess of 1,500 gpm (5 678 L/min) for 2 hours. These flows are to be delivered at a minimum of 20 psi (140 kPa). However, water storage of as much as 500,000 gallons (1 892 700 L) may be required depending upon the specific situation (Figure 10.11). The specific water-storage and fire-flow requirements for any particular development must be determined by local authorities using the codes and standards adopted locally.

Figure 10.11 Interface developments may require an elevated water source.

Both NFPA 299 and NFPA 1141 specify minimum requirements for fire hydrant spacing and marking where municipal-type water supplies are provided. Except in dedicated open spaces, such as permanent greenbelts, fire hydrants of a type appropriate for the climate should be provided and spaced as specified in local codes. Hydrants should be protected by appropriate physical barriers where needed, such as in parking areas, and marked in a uniform manner such as specified in NFPA 291, *Recommended Practice for Fire Flow Testing and Marking of Hydrants* (Figure 10.12). Hydrant markers may be needed at the entrance to access ways and/or within 3 feet (1 m) of individual hydrants.

Signage for other sources of water for fire fighting may be needed. Signs meeting the same

Figure 10.12 Some hydrants may need to be protected.

minimum standards as the street signs should be located near the access to these water sources. The signs should identify the type of water source such as *DRAFT WATER* or *PRESSURE WATER*.

Land-Use Planning

Land-use planning is necessary to balance the needs of the community for proposed developments with the more general but equally important need to protect the adjacent wildlands. To do this, both the land itself and the proposed use must be analyzed. One of the most important elements of the planning process is to determine whether a proposed development is an appropriate use for the land in question. While the fire department representative does not make this decision, fire protection planners should contribute to the body of data on which the decision is based. As mentioned earlier, fire department representatives must work with community planners, developers, and builders to ensure that the planned fire protection is adequate and in compliance with local laws, codes, and ordinances.

When planning for the construction of developments in or adjacent to wildland areas, fire protection must be addressed along with other elements such as drainage, soil erosion, flood control, and sanitation. To ensure that all of these elements are adequately addressed, many jurisdictions require that a representative of the department or agency responsible for each element review and approve the plans before a building permit is issued. This review and approval process allows fire department representatives to see whether adequate fire-defense improvements are provided and to require specific changes that include the needed improvements if they are not provided.

FIRE PROTECTION ANALYSIS

While both the *Urban-Wildland Interface Code* and NFPA 299 provide means for analyzing both the land and the proposed use or occupancy, only NFPA 299 differentiates between the wildland/urban interface and the wildland/urban intermix. For purposes of this manual, both interface and intermix are included when the term *development* is used. The fire protection analysis specified in NFPA 299 has six components:

- Development boundaries
- Fuel hazard
- Slope hazard
- Structure hazard
- Additional factors
- Development hazard

The analysis requires that the development in question be mapped, subdivided into logical units or areas, and given a name or number. This defines the development and limits the analysis to the areas within it. The predominant fuel types, slopes, planned structures, and other factors in each area of the development are assigned a hazard rating according to tables provided. However, the ratings assigned to the structures planned in the development are based only on the proposed construction materials and not on the intended use or occupancy of the buildings.

Both the *Urban-Wildland Interface Code* and NFPA 1231 classify planned structures according to occupancy or use. While the code uses 14 specific classifications, the standard assigns structures to one of five occupancy hazard classifications. The hazard classifications used in NFPA 1231 are as follows:

- ***Severe hazard*** — Aircraft hangars, cereal or flour mills, explosives manufacturing
- ***High hazard*** — Auditoriums, theaters, freight terminals, mercantiles, warehouses
- ***Moderate hazard*** — Amusement parks, libraries, restaurants, textile manufacturing plants
- ***Low hazard*** — Parking garages, bakeries, breweries, churches, service stations

- ***Light hazard*** — Apartments, dwellings, fire stations, hospitals, offices, prisons, schools

These examples are but a few of the many occupancies classified in the standard, and even the list in the standard is not all-inclusive. However, by using NFPA 299, NFPA 1231, the *Urban-Wildland Interface Code,* and other local codes and standards, planners can determine whether the planned development is an appropriate use of the property in question. Using these planning tools also helps to further define the fire protection needs that would be created by the planned development.

PROTECTION DURING CONSTRUCTION

Another aspect of planning for development in or adjacent to wildland areas involves the fire protection provided during construction. While fire protection requirements during construction are always the jurisdiction of the local fire department, NFPA 1141 lists a number of basic requirements that should be enforced as a minimum. The standard specifies that all-weather access ways must be provided (Figure 10.13). Of at least equal importance is a requirement that fire protection water supplies, including hydrants, be installed and operable before any combustible building materials are brought onto the site (Figure 10.14). As construction progresses, additional forms and levels of fire protection may be required. For instance, at least one fire extinguisher (of a specified minimum capacity) must be provided. Extinguisher(s) must be located so that workers need not travel more than 75 feet (23 m) to reach one. In addition, the standard requires that workers normally on the site be trained in the use of the fire extinguishers provided. Flammable and/or combustible liquids must be stored, handled, and used in a safe manner. Combustible debris must not be allowed to accumulate. Finally, because unoccupied construction sites are often attractive to children and others, some form of security must also be provided after hours.

Figure 10.13 All-weather roads into developments may be required.

Figure 10.14 Hydrants should be functional before construction is begun.

Codes and Standards

Compliance with applicable national and local codes and standards is critically important for preventing fires from occurring in or near the wildland, reducing the impact of those that do start, and securing firefighter safety. A fire that does not start cannot threaten property or place firefighters in jeopardy. Reviewing the engineering aspects of plans for developments in or adjacent to wildland areas may involve a number of different codes and standards. However, unless the local governing body (authority having jurisdiction) adopts a specific edition of a code by reference, national codes and standards do not have the force of law and are advisory only. It is very important for local fire officials to know which, if any, of the national codes and standards have been adopted locally. Also, even if a particular code or standard has been adopted, only a specific edition of that code is enforceable. Changes in the requirements in subsequent editions of that code are not applicable unless the new edition is adopted locally.

Of the sources already mentioned, the *Urban-Wildland Interface Code* and NFPA 299 include specifications for the design and construction of buildings in or adjacent to wildland areas. In addition, NFPA 1141 and NFPA 1231 are very important to the plans-review process. Other NFPA codes and standards, local codes, and certain regional or national codes may also apply. Prominent among these in the United States are the *National Electrical Code®* (NEC); NFPA 101®, *Code for Safety to Life From Fire in Buildings and Structures* (commonly referred to as the *Life Safety Code®*); and the particular building code that applies to that region (see Building Codes section).

In Canada, many of the same NFPA codes and standards are also used. As in the U.S., these codes and standards are only advisory unless adopted by the authority having jurisdiction.

NATIONAL ELECTRICAL CODE

The NEC is one of the few codes that applies across the entire United States (Figure 10.15). This code covers the design, installation, identification, and inspection of electrical applications in virtually every environment. Its companion publications — the *NEC Handbook* and the *National Electrical Safety Code*—help workers interpret and apply the NEC requirements in a safe manner. Development plans should comply with these codes to reduce the likelihood of fires of electrical origin and to increase the safety of those who install and use electricity.

Figure 10.15 The *National Electrical Code®* is used coast to coast.

LIFE SAFETY CODE (NFPA 101)

This code, along with its companion publication, *Life Safety Code Handbook,* cover the construction, protection, and occupancy features needed to minimize threats to life from fire, smoke, and the panic that may ensue (Figure 10.16). Compliance with this code is essential to protect those who occupy any planned developments.

Figure 10.16 The *Life Safety Code®* is used internationally.

BUILDING CODES

While there is growing interest in the creation of a single building code that would apply to the entire United States, at the time of this manual's publication no such code had been adopted. Across the United States and Canada, there are a number of different building codes in effect. In the western U.S., the *Uniform Building Code®* (UBC) applies. In the Midwest, the *Building Officials and Code Administrators* (BOCA) code is used. In the southern U.S., the code written by the Southern Building Code Congress International (SBCCI) applies. In most other areas, the *National Building Code* is used.

In Canada, each province adopts the *National Building Code* by reference. However, they may not necessarily adopt the code in its entirety. At the time of adoption, each province is free to amend, add to, or delete from the model code. Once adopted, that edition of the code, as amended, remains in effect until a later edition is adopted.

OTHER CODES

Plans for developments in or near the wildland may need to be checked for compliance with other national codes and standards also. For instance, if the development includes buildings that are required to have automatic sprinkler systems, the plans must comply with NFPA 13, *Standard for the Installation of Sprinkler Systems.* Depending upon the situation and local requirements, NFPA 13D, *Standard for the Installation of Sprinkler Systems in One- and Two-Family Dwellings and Manufactured Homes,* or NFPA 13R, *Standard for the Installation of Sprinkler Systems in Residential Occupancies Up to and Including Four Stories in Height,* may also apply. The provisions of NFPA 256, *Standard Methods of Fire Tests of Roof Coverings,* may be applicable. In some areas, the *Uniform Fire Code®* has been adopted, and it must be checked as well.

Maps

As mentioned earlier, NFPA 299 includes a system for analyzing the fire protection planned for a development. One component of this system involves mapping the development under analysis. The mapping process involves subdividing the area into logical units and designating each unit by a number or name. Subdividing the area provides the flexibility needed to specify different forms and levels of fire protection in different units based on conditions within each unit.

Addressing

One of the most important provisions of NFPA 299 and other standards is requiring address numbers on buildings (Figure 10.17). Having prominent and easily read address numbers in place makes it easier for emergency responders to find the correct location and thereby reduce response time. As a minimum, requiring address numbers that meet or exceed NFPA 299 should be a condition of the building-permit process.

Figure 10.17 Homes in the interface should have easily read address numbers.

Fire-Resistive Landscaping

Using the review and approval process to require greenbelts in and around planned developments is an effective way to reduce the potential impact of fires in those areas. As mentioned earlier, the publication *Fire Safe, California* includes a list of fire-resistive plants, shrubs, and trees that property owners can use to landscape around their homes and other structures (see Fire Safety Education/Information section later in this chapter). Local agricultural extension agents can also be helpful in identifying types of fire-resistive plants that are appropriate for the local climate and soil conditions.

BUILDING CONSTRUCTION STANDARDS

Fire department representatives should work closely with local building officials to ensure that planned developments conform to the fire and life safety requirements of the applicable building code and any other applicable codes and standards. Which specific codes and standards apply is entirely a matter of which editions of which codes and standards have been adopted by reference by the local governing body. However, from a purely fire protection standpoint, the requirements in NFPA 299 are the most practical.

Protection of Life and Property From Wildfire

Chapter 7 of the 1991 edition of NFPA 299 requires that structures and developments constructed in or adjacent to wildland areas conform to the applicable building code. It also requires that structures and developments be located, designed, and constructed in ways that minimize the possibility of a wildland fire spreading to the developments or vice versa. Specifically, Chapter 7 requires that such structures have the following:

- Class C roofing (minimum) (Figure 10.18)

  NOTE: While this class of roofing is specified in the standard, it is likely to provide inadequate fire resistance to be acceptable in many wildland environments.

- Building vents covered with corrosion-resistant screen with openings no larger than ¼ inch (6.35 mm) (Figure 10.19)
- Overhangs (eaves, cantilever balconies, stilt construction, etc.) enclosed with ½-inch (12.7 mm) sheathing (Figure 10.20)

  NOTE: Although not specified in the standard, noncombustible sheathing should be used.

- At least ½-inch (12.7 mm) sheathing on exterior vertical walls (Figure 10.21)
- Windows facing vegetative fuels within 30 feet (9.1 m) with closable, solid exterior shutters

Figure 10.18 A typical Class C roof.

Figure 10.19 Attic vents should be covered by wire screens.

Figure 10.20 Supports for balconies and decks should be enclosed.

Figure 10.21 Walls should be covered with noncombustible material or plywood sheathing.

- Chimney outlets with spark arresters of at least 12-gauge welded or woven-wire mesh with openings no larger than ½ inch (12.7 mm) (Figure 10.22)
- Chimney outlets at least 10 feet (3 m) from any vegetation (Figure 10.23)
- Full skirting of at least ½-inch (12.7 mm) sheathing on mobile homes or manufactured homes (Figure 10.24)

Chapter 8 of this standard suggests that fire department representatives prepare a year-round fire prevention and public fire safety education/information plan. See the Fire Safety Education/Information section later in this chapter.

Figure 10.22 Chimney outlets should be covered with wire screen or an approved spark arrestor.

Figure 10.23 Chimney outlets should be at least 10 feet (3 m) from the nearest vegetation.

Figure 10.24 Mobile/manufactured homes should have full skirting.

Building Spacing and Densities

The number of buildings within a development and the spacing between them both affect the likelihood of a structure fire spreading from one building to another. If a fire can be limited to the building of origin, the likelihood of it spreading to the surrounding wildlands is reduced. Also, if buildings are widely spaced, the likelihood of more than one building becoming involved from a spreading wildland fire is likewise reduced.

NFPA 1141 specifies that any building with more than 1,000 square feet (93 m^2) of floor area be set back *not more than* 50 feet (15.2 m) from an access way. This setback allows firefighters to leave their apparatus on the access way and still reach the building with preconnected attack lines that are typically at least 150 feet (46 m) long. The standard allows buildings protected by automatic sprinkler systems conforming to NFPA 13 to be set back as much as 200 feet (61 m) from an access way.

NFPA 1141 also specifies that any building more than three stories high with a ground floor area of more than 3,000 square feet (279 m^2) and with nonrated exterior openings facing other buildings shall be separated from those other buildings by a clear space of at least 50 feet (15.2 m). Such buildings must also be at least 25 feet (7.5 m) from the nearest property line.

PRE-INCIDENT PLANNING

Pre-incident planning is necessary to reduce the number of decisions required when a major fire starts and to reduce the amount of time needed to get the required resources to the scene. The planning process includes surveying fire-defense improvements, obtaining or developing maps, obtaining or taking aerial photographs of the area, and developing a written plan for fire suppression. A typical pre-incident plan for a development in or adjacent to wildlands includes the following:

- Block or unit designation system
- Description of block or unit boundaries and area
- Block or unit map and instructions for its use
- Special instructions for block or unit
- Oblique aerial photo of block or unit
- Locations of water supplies

- Locations of support facilities
- Locations for unloading heavy equipment
- Locations for staging areas
- Locations of special problem areas
- Preferred routes of travel
- Possible helispot locations

The pre-incident plan includes a narrative section that describes access ways and their width, along with trails, fuel breaks, and firebreaks in sufficient detail for firefighters to know where they can gain access to a fire. A description of staging-area and water-source locations and how they are marked is included. In addition, any special fuel, topography, or possible weather problems are noted.

Area Knowledge

As discussed in earlier chapters, firefighters should become as familiar as possible with the area to which they are assigned. This process may begin with a study of maps of the area and any pre-incident plans that have been developed, but it certainly does not end there. Firefighters should tour the area until they become familiar with its topography, features, and landmarks. Of particular importance is knowledge of the roads, bridges, and locked gates in the area.

ROADS

The first step in making the transition from knowing the map to knowing the area is to drive the streets, roads, and fire lanes within the area. However, this process involves more than just randomly driving around the area — although that helps as well. To be most effective, the roads should be driven systematically following some pattern. Over time, firefighters develop methods or systems that work best for them individually. The various systems used by most firefighters have the following things in common:

- The area is subdivided into smaller, more manageable zones — sometimes topographically by major terrain features and sometimes geographically by major arterial routes and secondary roads.
- The fire station where firefighters are assigned serves as the starting point for all tours of the area.
- Every street, road, and fire lane in the area is driven from beginning to end — or to the area boundary, whichever comes first.
- As each road is driven, road conditions, street signs, and landmarks are observed; notes are made of features or potential problems — such as water sources, fuels, topography, and values at risk.

BRIDGES

Some of the most important things for firefighters to learn about an area are the locations of all bridges and their conditions (Figure 10.25). Any bridge that may not be strong enough to support the weight of a fully loaded fire engine or mobile water supply apparatus must be identified and inspected by a structural engineer. One advantage to identifying and inspecting these bridges during pre-incident planning is that there may be time to have them strengthened or to have alternatives constructed before fire apparatus need to cross them to reach a fire. One alternative to a substandard bridge is to identify or develop areas where dry creek beds may be crossed during fire season (Figure 10.26). If substandard bridges can-

Figure 10.25 All bridges in the district should be inspected.

Figure 10.26 Where bridges are unsafe, fire apparatus may have to cross the stream bed. *Courtesy of NIFC.*

not be improved and alternatives cannot be developed, these bridges must be marked on response maps and alternative response routes identified. It is also important that signs be posted specifying the maximum weight limits allowed on such bridges.

LOCKED GATES

Even though most fire apparatus are equipped with bolt cutters or other *master keys,* taking the time to cut a lock or chain increases the response time. If the lock is cut, instead of the chain, there is also the additional expense for the property owner to replace it. If it is necessary to cut a chain to open a gate, the link nearest the lock should be cut (Figure 10.27). A better alternative is for the property owner to install a lock box near the gate with a gate key inside (Figure 10.28).

A growing number of rural gates are electrically operated and controlled (Figure 10.29). Where these gates are installed, the fire department should be provided a universal-access device so these gates can be opened without firefighters having to get out of their vehicles (Figure 10.30).

Figure 10.27 When gates are chained and locked, the chain should be cut, not the lock.

Figure 10.28 Homeowners should provide *lock boxes* if they keep their gates locked.

Figure 10.29 Many gates in the wildland/urban interface operate electrically.

Figure 10.30 Firefighters should be provided with universal-access devices.

Vehicular Traffic Plans

Smoke and other hazards can create severe safety problems for both emergency and nonemergency vehicular traffic in the vicinity of active fires. The following traffic-related items should be addressed in pre-incident plans:

- Identification of public roads that may be impacted by smoke and increased traffic volume
- Identification of equipment and personnel needed to control traffic; may include warning signs, barricades, radios (not on the active fire frequency), and vehicles equipped with flashing lights
- Identification, location, and phone numbers of local hospitals and ambulances that may be needed for emergency medical evacuation
- Identification, location, and local phone numbers of law enforcement agencies with traffic control responsibilities in the area and identification of any existing agreements with these agencies
- Identification of traffic routes subject to temperature inversions as well as other possible hazards such as fog and ice
- Identification of alternate traffic routes
- Identification and phone numbers of radio and television stations that can issue traffic advisories for the area

Predesignated Staging Areas

The pre-incident planning process allows firefighters to identify parking lots and other areas that may be used as staging areas during major fires (Figure 10.31). Because it is impossible to predict with any degree of certainty exactly where a major fire will develop and where it will spread, potential staging areas need to be identified in all parts of the area. Some of these potential staging areas serve that purpose better than others, so they should be evaluated according to their suitability and identified on response maps. When special arrangements with the owners of these areas are required, they should be made during the pre-incident planning process.

Figure 10.31 Potential staging areas should be identified during pre-incident planning.

Seasonal Weather

As discussed in Chapter 1, Wildland Fire Behavior: Fuel, Weather, Topography, and in other parts of this manual, weather is perhaps the single most important influence on a wildland fire. Therefore, another critically important aspect of local area knowledge is being familiar with the seasonal weather patterns in the area. Many parts of a response area can have microclimates that are quite different from those in other parts of the same area. Knowing what these differences are and when to expect them are essential parts of the pre-incident planning process. Knowing how the weather in the area behaves in the daytime, at night, and at various times of the year can mean the difference between the timely control of a fire and a long and costly campaign.

Fire-Behavior Modeling

Another aspect of pre-incident planning is the use of fire-behavior models to plot various possible fire scenarios. These models range from relatively simple checklists and flowcharts that lead the user through an analysis of local fire conditions to more sophisticated computer-based programs. Some examples of computer-based fire-behavior modeling programs are the National Fire Danger Rating System (NFDRS), BEHAVE, and FARSITE. These programs are available from the following sources:

- ***NFDRS and BEHAVE*** — United States Department of Interior (USDI) Bureau of Land Management, National Interagency Fire Center, Boise, ID 83705-5354
- ***FARSITE*** — Dr. Mark Finney, Systems for Environmental Management, P. O. Box 8868, Missoula, MT 59807

EVACUATION PLANNING

Because life safety is always the first and highest priority in fighting a fire, it is also the highest planning priority. While life safety is usually achieved by eliminating the threat to life — putting out the fire — this is not always possible before some property owners have to evacuate their homes. Planning for this contingency results in fewer problems and a more efficient evacuation if one becomes necessary. When planning for the evacuation of an entire development, planners must first decide which agency will be primarily responsible for conducting the evacuation. In most cases, this responsibility is given to law enforcement officials because they are usually better equipped to conduct this type of operation. Planners must then decide how to provide the three major components of a mass evacuation — time, transportation, and shelter.

Time

Planners must anticipate that once a decision to evacuate an area is made, it takes time to notify, organize, and transport evacuees to shelters or areas of safe refuge. To provide the time needed to initiate a mass evacuation, planners must identify certain geographic benchmarks for the IC to use when a wildland fire threatens to overrun a development. If an approaching fire reaches these predesignated points and the resources needed to arrest the fire's progress are not at scene, the IC then knows to order an evacuation. The time needed for evacuation can be reduced if notification is done through a computerized or other automatic telephone alerting system (Figure 10.32). The amount of time needed can also be reduced if those being

Figure 10.32 Automated alerting systems can speed evacuation notification. *Courtesy of NIFC.*

displaced can transport themselves, but this also has certain disadvantages. These disadvantages are discussed in the next section.

Transportation

If all residents drive themselves out of the area, the streets and roads can become seriously congested. This is especially true if, in an attempt to save all of their vehicles, individual members of each household drive separate vehicles. Because some of these vehicles are likely to run out of fuel or experience some mechanical breakdown, planners should see that tow trucks are available at strategic points during an evacuation. If the access ways into and out of a development have been properly planned and the codes diligently enforced, the roads should be able to handle the increased traffic flow. If not, plans for buses and other mass-transport vehicles must be included.

The major transportation problem for those conducting an evacuation may be removing nonambulatory residents. Some people may be bedridden and/or dependent upon life-support machines or devices. To deal with these situations, planners must anticipate these needs and have one or more ambulances on the evacuation resource list.

Shelter

Planners should provide for three possible evacuation options: shelter in place, safe refuge, and/or relocation centers. While technically not a means of evacuation, *shelter in place* is an accepted means of protecting building occupants from a fast-moving wildland fire. Shelter in place is especially useful with certain groups, such as the occupants of nursing homes and similar care facilities, that would require a significant amount of time and resources to evacuate. In this mode, firefighters concentrate their efforts on protecting the structure and keeping it from becoming involved while allowing the fire to burn past.

Depending upon the situation at hand, the IC may decide that moving residents to an area of *safe refuge* is more appropriate than sheltering in place or conducting a full-scale evacuation. As discussed in earlier chapters, this concept involves moving people from their homes to the nearest open space that is free of vegetative fuels. Planners should anticipate this possibility and identify such areas near developments.

Planners should also anticipate that full-scale evacuations may be necessary and that one or more *relocation centers* should be identified. While some evacuees in a mass evacuation will be able to provide their own shelter with family or friends, most of those displaced will need to have shelter provided. Depending upon what is available in the local area, churches, schools, and gymnasiums are appropriate sites (Figure 10.33). It may be necessary to make formal arrangements with those responsible for potential relocation centers in advance. Plans should also be developed in cooperation with the Federal Emergency Management Administration (FEMA), the Red Cross, and similar organizations to provide long-term shelter should that become necessary.

Figure 10.33 Large public buildings may be used as relocation centers.

Planning for the Media

Large wildland fires, especially those that threaten homes and businesses, are news stories in which the public is interested. To accommodate this legitimate public interest and to reduce the likelihood of aggressive reporters interfering with fire suppression or evacuation operations, planners must devise effective ways of dealing with the media. One of the most effective ways of providing the media with the information they seek is for the IC to appoint an Information Officer as part of the command staff (Figure 10.34). As described in Chapter 5, Initial Fireground Command, the Information Officer acts as the contact point for all news releases so that reporters have no need to bother those directly involved in mitigating the incident. If planners can ensure that the Information Officer is able to distribute photos and/or videotapes of fireline action, evacuations in progress, and other newsworthy items, the likelihood of reporters and photographers interfering or endangering themselves is greatly reduced. Planners should also recognize that the broadcast media can also be an asset during emergencies by broadcasting emergency instructions. Arrangements should be made in advance of the need for this sort of cooperation.

Figure 10.34 An Information Officer keeps the media informed about a major fire. *Courtesy of NIFC.*

Another aspect of planning for the media is training. Most firefighters are highly trained in fire suppression and other emergency skills, but few are trained in how to interface effectively with the media. Preparing a firefighter to function successfully as an Information Officer requires extensive and specific training for this highly specialized assignment. This training is available from a number of different state, provincial, and federal agencies.

COMMUNITY INVOLVEMENT

In fire protection planning, community involvement is imperative. Regardless of how well designed a plan may be, without community involvement it lacks the political component that is essential to making it work. Community involvement in the development of the plan is absolutely necessary if there is to be the sense of ownership and cooperation needed to implement the plan. However, in order to generate interest and participation by members of the public, it is often necessary to inform and educate them about the nature and magnitude of the community's fire problem.

Fire Safety Education/Information

Fire protection planning should include programs to promote fire safety among those who live, work, or recreate in the wildlands. These programs may include inspection services and educational/informational activities. As mentioned earlier, Chapter 8 of NFPA 299 requires the fire department representatives to prepare a year-round fire prevention and fire safety public information/education plan. The standard specifies that the plan, at a minimum, identify and analyze the following:

- Specific hazards
- Risks
- Fire causes
- Applicable prevention and safety programs
- Target audiences
- Activities

The standard also specifies that various communication techniques be used to achieve the desired results. Some of the techniques recommended in the standard are exhibits and displays, roadside signs, group presentations, and mass media campaigns (Figure 10.35).

Figure 10.35 Roadside fire-safety signs can be effective. *Courtesy of NIFC.*

There are many ways in which firefighters and fire protection planners can inform and educate the public, and no opportunity should be overlooked. For more information on this topic, refer to the IFSTA **Fire and Life Safety Educator** manual.

WILDFIRE STRIKES HOME!

One of the resources that can be used, especially when preparing for presentations to service clubs and other groups, is a publication entitled *Wildfire Strikes Home!* (Figure 10.36). Prepared by

NFPA in cooperation with The National Wildland/Urban Interface Fire Protection Initiative, this booklet focuses attention on the growing trend toward people moving into wildland areas and the resulting effects on fire safety. The publication has four goals:

Figure 10.36 *Wildfire Strikes Home!* contains useful information for homeowners and local officials.

1. To provide basic information on the nature of the wildland/urban interface fire problem
2. To motivate homeowners, local officials, and fire protection agencies to become involved with this important issue
3. To recognize the hard and creative work already being done by leadership-minded people
4. To enhance the networking of individuals already working on the interface problem who want to share their information and learn from others

FIRE SAFE, CALIFORNIA

When preparing mass mailings to homeowners, an informational flyer published by the California Department of Forestry and Fire Protection entitled *Fire Safe, California* makes an excellent mailer (Figure 10.37). It may also be used as the basis for a newspaper article or public service announcement. This pamphlet contains a three-page checklist of things that homeowners can do to protect themselves and their homes from fires. It also includes a list of fire-resistive ground covers, shrubs, and trees that homeowners can use to landscape around their homes.

Figure 10.37 *Fire Safe, California* contains information homeowners need to help them protect their homes.

Emergency Alert System

Another aspect of community involvement in fire protection planning is the Emergency Alert System (EAS). This system is a public service required of commercial broadcasters as a condition of their operating licenses. This means that local radio and television stations *must* cooperate with local officials during emergencies, although most are eager to help without being required. The EAS can be of tremendous value in giving instructions to the public during widespread or catastrophic emergencies. Fire protection planners should include representatives of the various local broadcast media in the planning process so that maximum use is made of these services when needed.

Automatic Aid

Yet another aspect of fire protection planning involves other fire departments. As discussed in Chapter 5, Initial Fireground Command, fire departments that share a common boundary sometimes enter into automatic aid agreements. Under these agreements, both departments automatically dispatch resources to any incident within the area covered by the agreement. These areas are typically along a common boundary, especially where the actual location of the boundary is unclear. In the planning process, any such area on the boundary between two districts needs to be identified so that agreements can be negotiated *before* a fire starts in that area.

Mutual Aid

As also discussed in Chapter 5, preparing mutual aid agreements is another function of fire protection planning. Under these agreements, two or more not necessarily contiguous fire departments commit to sending resources (if available) to another department in the agreement when that department's resources have been depleted by an incident of unusual proportions. With certain negotiated exceptions, the usual arrangement is that the party rendering aid absorbs the cost of sending their resources.

Another aspect of planning related to mutual aid agreements involves training. For these agreements to be effective, any differences in equipment and procedures must be identified before crews have to function together during a fire. The most effective way of identifying and rectifying these differences is for personnel from all departments participating in the agreement to cross-train using the same apparatus and equipment they will use during an actual incident.

Planners must also consider the need for mutual aid from outside the local area. Such aid is available from various state, provincial, and federal emergency services agencies. Unlike most local agreements where resources are often requested and supplied one unit at a time, mutual aid from outside the area is almost always supplied in the form of strike teams or task forces.

COMMUNICATIONS PLANS

Without question, the single most important aspect of planning for major wildland incidents is communications. Communications is the most common problem identified in post-incident critiques of large-scale emergencies. As discussed in earlier chapters, it is impossible for a complex fireground organization to function efficiently without effective communications. The chaos that would result from too many units attempting to transmit and receive on the same frequency or from diverse units not having common frequencies is simply not acceptable. Therefore, before a major incident occurs, planners must anticipate the need for communication capabilities between units from various agencies with a variety of communications equipment. The product of such planning should be put in writing and distributed to all agencies within the affected area. Caches of portable radios with scanners and multichannel capability are available from some state and federal emergency services agencies (Figure 10.38). For example, the National Interagency Fire Center (NIFC) in Boise, Idaho, controls a large block of dedicated frequencies and maintains a large cache of radios for use in major wildland fires. Planners must identify where other communications resources are located and how to access them. In addition, new radio technology must be considered in communications planning. Also to be considered is how to coordinate the transition from existing very high frequency (VHF) systems to those incorporating ultrahigh frequency (UHF), 800 megahertz (MHz), or other technologies. All of this information should then be included in a master communications plan for the jurisdiction.

Figure 10.38 The Great Basin Fire Cache located at the National Interagency Fire Center in Boise, Idaho.

Agencies planning to upgrade their communications hardware should investigate all of the various options currently available. When specifications for new radios are being written, multichannel capability should be included. Also considered is the compatibility of new hardware with that used by surrounding agencies. Compatibility is important for a variety of reasons, but one that is sometimes overlooked is *frequency sharing*. Any entity licensed to a radio frequency can agree to share it with another agency. A frequency-sharing agreement specifies the conditions of use, and this allows any agency to have emergency access to many more frequencies than those to which they are individually licensed. This greatly improves mutual aid capabilities.

Some departments have converted to 800 MHz equipment. These systems have advantages and disadvantages, so it would be prudent to ask those already using this equipment about their experiences with it in both routine and emergency use. Other agencies are acquiring digital communications equipment. For more information on this technology, contact the world headquarters of APCO International at 2040 S. Ridgewood Avenue in South Daytona, FL 32119-8437.

OTHER PLANNING ACTIVITIES

There are other aspects of fire protection planning that do not necessarily fit into any of the previously discussed categories. Among these are identifying other available resources and training needs.

USFA Resources

The U.S. Fire Administration (USFA) can provide planners with lists of fire protection resources available from a variety of federal and state sources. Colocated with the USFA at the Emergency Management Institute in Emmitsburg, Maryland, the National Fire Academy is the repository for numerous research documents and other publications related to all facets of fire protection. These agencies can assist fire protection planners in numerous ways, and they can be contacted by mail at 16825 S. Seton Avenue, Emmitsburg, MD 21727, or by telephone at 301-447-1080.

Training

Finally, one of the most important fire protection planning activities is in the area of training. The value of training to life safety, property protection, and the preservation of natural resources cannot be overemphasized. Therefore, the time spent identifying training needs and arranging to meet those needs is some of the most productive for fire protection planners. Training plans should address all levels within the organization from basic firefighter through administration. The table of contents of this manual can serve as an outline of training needs for fire departments that may have to fight fires in wildlands or in the wildland/urban interface.

NFPA 101®
Life Safety Code®
ANSI/NFPA 101
1994 Edition
An American National Standard
February 11, 1994
NFPA
National Fire Protection Association

APPENDICES

Appendix A
Clear Text Terminology

WORDS AND PHRASES	APPLICATION
Unreadable	Used when signal received is not clear. In most cases, try to add the specific trouble. Example: "Unreadable, background noise."
Loud and Clear	Self-explanatory
Stop Transmitting	Self-explanatory
Copy, Copies	Used to acknowledge message received. Unit radio identifier must also be used. Example: "Engine 2675, copies."
Affirmative	Yes
Negative	No
Respond, Responding	Used during dispatch - proceed to or proceeding to an incident. Example: "Engine 5176, respond . . ." or "St. Helena, Engine 1375 responding."
Enroute	Normally used by administrative or staff personnel to designate destinations. Enroute is *NOT* a substitute for responding. Example: "Redding, Chief 2400 enroute RO II."
In-quarters, with Station Name or Number	Used to indicate that a unit is in a station. Example: "Morgan Hill, Engine 4577 in-quarters, Sunol."
Uncovered	Indicates a unit is not in-service, because there are no personnel to operate it.
Out-Of-Service	Indicates a unit is mechanically out of service. Example: "Auburn, transport 2341, out-of-service." Note, when repairs have been completed the following phrase should be used: "Auburn transport 2341, back in-service, available."
In-Service	This means that the unit is operating, not in response to a dispatch. Example: "Fortuna, Engine 1283, in-service, fire prevention inspections."
Repeat	Self-explanatory
Weather	Self-explanatory
Return to	Normally used by ECC to direct units that are available to a station or other location.

What is your Location?	Self-explanatory
Call_____by Phone	Self-explanatory
Disregard Last Message	Self-explanatory
Stand-By	Self-explanatory
Vehicle Registration Check	Self-explanatory
Is_____Available for a Phone Call?	Self-explanatory
At Scene	Used when units arrive at the scene of an incident. Example: "Perris, Engine 6183, at scene."
Available at Residence	Used by administrative or staff personnel to indicate they are available and on-call at their residence.
Can Handle	Used with the amount of equipment needed to handle the incident. Example: "Susanville Battalion 2212, can handle with units now at scene."
Burning Operation	Self-explanatory
Report on Conditions	Self-explanatory
Fire under Control	Self-explanatory
Emergency Traffic Only	Radio users will confine all radio transmissions to an emergency in progress or a new incident. Radio traffic which includes status information such as responding, reports on conditions, at scene and available will be authorized during this period.
Emergency Traffic	Term used to gain control of radio frequency to report an emergency. All other radio users will refrain from using that frequency until cleared for use by ECC.
Rescue Normal Traffic	Self-explanatory.

Appendix B
Sample Use and Rental Criteria

CHAPTER 7760 — NON-CDF PERSONNEL AND EQUIPMENT 7760

Managing major emergencies will usually require hiring non-CDF personnel and equipment and requesting assistance from various governmental and other organizations to supplement the CDF's regular fire control forces.

Although wide latitude is granted to field personnel for acquiring these resources, established rules and procedures must be followed.

This chapter presents rules and procedures for using hired personnel and equipment, as well as emergency firefighters (EFF), OES, military, National Guard, and organized fire departments on CDF fires. (Also see the following section of the Incident Fiscal Management Handbook: E-Fund Policy 3821, OES 3827, Local Government 3836.15 and 3847.7 and National Guard 3836.16.)

HIRING GUIDES 7761

PERSONNEL 7761.1

The following applies to employment for fire suppression on a temporary or emergency basis:

- No one under 18 years of age shall be hired. (See Labor Code, §1285, et seq.) See also Section 3836, Incident Fiscal Management Handbook.
- Agricultural workers should not be hired during harvesting season except in extreme emergency.
- No one convicted of arson shall knowingly be hired.
- No person should be hired to protect their own property or property bearing their interests.
- Volunteers requested by or receiving specific instructions from CDF officials may be hired.
- Volunteers must be properly clothed, physically qualified, properly trained and experienced.

EMPLOYMENT PROCEDURE 7761.2

A person who is hired as paid pickup labor should sign an FC-42 at the time of employment, or as soon after as is reasonable. Refer to the current Emergency Worker rates published annually by temporary directive. See section 3836 of the Incident Fiscal Management Handbook.

SUPERVISION AND DIRECTION 7761.3

- Paid pickup laborers hired by CDF must be supervised commensurate with their position and qualifications the same as regular CDF employees.
- In order to operate CDF equipment, pickup laborers must possess the same type license as required for regular CDF employees. Normally, they should be certified by the local ranger-in-charge as being qualified to operate the equipment.
- In case of injury, the same reports and treatment are required as for regular CDF employees. (Refer to Incident Fiscal Management Handbook 3800 and refer to temporary directives on SCIF 3301.)

GROUND EQUIPMENT 7761.4

Equipment may be rented from any owner (except one protecting his own interests) who is willing to rent. There is no practical provision for forcing an owner to rent equipment for emergency use.

Public entities such as cities, counties, and state agencies may be willing to make equipment available to CDF in emergencies. In general, the use of publicly owned equipment (3800 Handbook, Section 3826) is subject to the same conditions as private equipment; in addition, advance approval of the public official responsible for the equipment must be secured (refer to 3800 Handbook Section 3836.15 regarding local government personnel on FC-42).

Hiring privately-owned equipment (3800 Section 3833.2.7) to perform work for the CDF in emergencies is permitted under GC § 14784 authorizing exception from contracting. Rental of "nonstate" equipment for fire control use will be reported to Department of General Services after its use. The reporting process will be handled directly through the director's mobile equipment section in conjunction with the departmental accounting office and will not require field input.

The problems inherent in hiring motorized equipment on a large fire are many and varied and will usually require the full-time attention of competent individuals. It is, therefore, quite important that the position of equipment manager be filled whenever a large amount of equipment is to be hired.

HIRING GUIDES - GENERAL 7761.5

The following items should be considered and appropriate actions taken whenever equipment is hired, regardless of the number of pieces (see Handbook 3800 on Hired Equipment, Section 3833.2):

- There must be a genuine need for the equipment.
- The contractor must enter into a pre-hire contract on an FC-100 establishing terms of employment, rate and method of payment, and equipment conditions.
- An inspection of each piece of equipment should be made by a qualified CDF employee from the Ground Support Unit before the unit is hired using form OF-296 (see 3800, Section 3833.3 Vehicle/Heavy Equipment Inspection Checklist).
- Only equipment that is judged by a qualified CDF employee to be mechanically sound and designed or equipped to do the assigned job should be hired (see Section 3833.2 of 3800 Handbook).
- All visual damage will be noted on the OF-296 Vehicle/Heavy Equipment Inspection Checklist.
- Regular qualified operators and their relief should be obtained with each piece of rented equipment.

It is often desirable to assign and paint number on equipment (particularly dozers) for ease of identification later.

- Conditions-of-hire statements on Equipment Rental Agreement form FC-100 should be completed as soon as possible, preferably before the equipment is used.
- Before being released from the fire, each piece of hired equipment should be inspected by a qualified CDF employee on form OF-296 (see 3800 Handbook, section 3833.3).
- After a qualified CDF employee has evaluated the extent and cause of the damage, the state will consider making payment for repairs of equipment actually damaged because of the unusual rigors of fire work. The determination will be made by the Davis Mobile Equipment Unit.
- The state will not consider making payment for repairs which are deemed due to normal wear and tear or which are due to negligent operation or inefficient operation by the owner or operator.

SPECIFIC HIRING GUIDES - BULLDOZER AND WATER TENDERS 7761.6

SYSTEM DESCRIPTIONS AND DEFINITIONS 7761.6.1

HIRING SYSTEM 7761.6.1.1

The intent of the Department is to maximize the utilization of planned need resources as defined in this policy to meet the goals as stated herein.

The hiring system will consist of a Fire Ready Group and a Support Group. Utilization of equipment from these groups will be based on incident need — either immediate or planned. All equipment must be hired in accordance with the Rates Procedure Handbook (3900), Section 3934 (California Interagency Emergency Hired Equipment Rates).

The selection of the most appropriate resources to assign (fill) a request must be based on time frame as well as specific mission requirements and equipment capabilities. It is the responsibility of the incident commander (IC) to determine the appropriate resource and time of need to be utilized.

DEFINITIONS: EQUIPMENT 7761.6.1.2

Fire ready refers to equipment that meets minimum response time and performance requirements detailed in Section 7761.6.9 of this policy.

Support refers to equipment that is unable to meet the requirements for the Fire Ready Group, but meets all of the requirements detailed in Section 7761.6.10 of this policy.

Planned need equipment consists of Disabled Veteran Business Enterprise (DVBE) vendors on the Unit Fire Ready and Support Equipment lists. A vendor's designation as either "fire ready" or "support" has nothing to do with when the vendor was hired for a particular incident. The designation is based solely on which set of criteria a vendor is able to meet the time the vendor signs up.

DEFINITIONS: TIME FRAMES 7761.6.1.3

Immediate need is defined as those times when, due to the threat of life and/or property, there is a need for a resource(s) to be available without delay. The minimal acceptable response criteria is detailed in Section 7761.6.6.3.

Planned need is defined as the time from the **BEGINNING** of the second operational period following the initial attack period of an incident. This allows for an initial attack period of approximately 18 hours and one 12-hour operational period. For further direction in the application of planned need, see Section 7761.6.6.4.

SIGN-UP AND LISTS — GENERAL 7761.6.2

SIGN-UP 7761.6.2.1

Annually, each unit is responsible for recruiting and signing up resources in order to meet the time frames and all other requirements set forth in this and other departmental policies and procedures governing hired equipment.

Each piece of equipment can be under the control of **ONE** vendor and can only be signed up in **ONE** unit.

DOCUMENTATION 7761.6.2.3

All vendors will provide the following information:

- All information as required in the 3900 Rates Procedures Handbook.
- A single contact number. Only one method of contact (phone, phone pager, answering service, answering machine, etc.) at the vendor's choosing will be accepted. This will be the sole method used by the Department to contact the vendor to fill incident requests.

Support/Planned Need
Each vendor will be contacted for all requests regardless of time frame. If it is the vendor's desire to reduce unnecessary contact, he/she may request that their name be flagged. This flag would give the Command Center (CC) duty officer the authority to determine if a vendor could reasonably meet the report time to the incident. This time will be considered to be from the time the vendor would be contacted to the report time at the incident. Vendors desiring to do this will be bypassed and left in position in the rotation when it is determined that the response time cannot be met.

By requesting this flag and the associated conditions, the vendor agrees to accept any decision made by the duty officer as it pertains to the implementation of this section.

HIRING GOALS 7761.6.2.3

While California law does not provide any actual preference for vendors meeting registration criteria, each agency of state government has goals of conducting a minimum of 15% of its total contracted work with minority-owned business enterprises, 5% of its contracted work with women-owned businesses, and 3% of its contracted work with disabled veteran-owned businesses (ref.: Public Contract Code Sec. 10115[c]).

To assist in meeting these goals, each unit will implement an "outreach" program to encourage participation by the business enterprises identified above.

All participating vendors should be provided with the following information that is necessary to qualify at the time of initial contact:

MBE/WBE certification: Any business claiming minority or women-owned business status must complete the "Minority/Women Business Enterprise Self-Certification" form. The self-certification will be provided to CDF at the time the vendor is listed in the Emergency Resource Directory or when the hiring package is prepared.

DVBE certification: Disabled veteran business enterprises must be certified by the Office of Small and Minority Business (OSMB) in accordance with Title 2, California Code of Regulations, Section 1896.94. A copy of the certification will be provided to CDF at the time the vendor is listed in the Emergency Resource Directory or when the hiring package is prepared. Interested vendors can contact OSMB at 1808 14th Street, Suite 100, Sacramento, CA 95814, (916) 322-5060.

SINGLE-UNIT SIGN-UP AND LISTING 7761.6.2.4

Individual dozers and water tenders may only be signed up and listed in the Emergency Resource Directory (ERD) of **ONE** ranger unit. However, individual vendors that assign and operate multiple pieces of equipment in different ranger units may sign up their individual pieces of equipment to the extent that each piece of equipment may only be listed in one unit. A vendor that is signed up in multiple units may **NOT** make available any equipment that is signed up in any unit other than the one placing the request. Access by a unit to any fire ready or support equipment signed up in another unit will be governed by the procedures detailed in Sections 7761.6.4 and 7761.6.6 of this policy and the Command and Control Procedures Handbook (8100), Procedure 006G. Further, any brokering or sharing of equipment or equipment cooperatives of any type or manner that in the determination of the Department circumvents the intent of any part of this policy is expressly prohibited.

MID-SEASON SIGN-UP 7761.6.2.5

Vendors may sign up their equipment any time after the initial sign-up period (this will be referred to as "mid-season" sign-ups). All vendors making mid-season sign-ups must comply with all applicable sections of this policy. Qualifying resources making mid-season sign-ups will be incorporated into the appropriate hired equipment list as detailed in this policy. Any resources, signing up as provided for in this section, will be placed last in the rotational sequence on any list for which they qualify.

LISTS 7761.6.3

GENERAL 7761.6.3.1

The system will be comprised of three types of lists: a Unit Fire Ready Equipment List, a Unit Support Equipment List; and a Region Planned Need List (one list each for bulldozers and water tenders), as follows:

1. **Unit Fire Ready Equipment Lists:** Each unit emergency command center (ECC) will maintain a Fire Ready Equipment List for resources signed up in that unit. These lists will be utilized to fill **ALL** immediate need requests for fire ready resources and may be used to fill other fire ready, support or planned need resource requests as permitted in this policy. Also, each unit list will include the fire ready DVBE vendors from the adjacent unit(s) as per Exhibit 7761.6.3.1, Adjacent Units. This information will be provided for inclusion in each unit's list by their region CC as detailed in Section 7761.6.4.3, Region Planned Need Lists. These lists will be established and administered following the guidelines for fire ready equipment outlined in Section 7761.6.4.1 and other applicable sections of this policy.
2. **Unit Support Equipment Lists:** Each Unit ECC will maintain a Support Equipment List for resources signed up in that unit. These lists will be used to fill **ALL** requests for immediate need support equipment other than as provided for in the unit fire ready and region planned need guidelines outlined in this section. These lists will be established and administered following guidelines for support equipment outlined in Section 7761.6.4.2 and other applicable sections of this policy.

   NOTE: The Unit Support Equipment Lists may include equipment that meet all fire ready criteria except the availability/response criteria.
3. **Region Planned Need Lists:** Each region CC will maintain a list of resources signed up in that region as detailed in Section 7761.6.3.1. Utilization of the Region Planned Need Lists will be requested by the units (following 8100 Handbook Procedure 006G) to fill incident requests for fire ready and support resources where the criteria set forth in the definition of planned need is met. It is intended that, within the time frames established, these resources will be on the incident by the beginning of the second operational period following initial attack (this is normally after 30 hours from the start of the incident).

   NOTE: The Region Planned Need Lists will include equipment that meets fire ready, and support equipment criteria.

MBE/WBE/DVBE DOCUMENTATION 7761.6.3.2

All hiring lists will identify each vendor that has provided proof of certification as detailed in Section 7761.6.2.

FIRE READY DOCUMENTATION 7761.6.3.3

All Support and Planned Need Lists will also identify each vendor that meets **ALL** fire ready criteria except the response criteria as detailed in Section 7761.6.9. This information will be gathered by the units from each vendor at the time of sign up and forwarded as detailed in Section 7761.6.4.

PLANNED NEED DOCUMENTATION 7761.6.3.4

The Region Planned Need Lists will identify each vendor that desires to have their name flagged as detailed in Section 7761.6.2.2. This information will be gathered by the units from each vendor at the time of sign up and forwarded as detailed in Section 7761.6.4.

ESTABLISHING LISTS 7761.6.4

UNIT FIRE READY LISTS 7761.6.4.1

By May 1 of each year, each unit will compile two Unit Fire Ready Lists (one for bulldozers and one for water tenders). All vendors signing up in a unit and meeting the fire ready equipment criteria as detailed in Section 7761.6.9 will be placed on the appropriate Unit Fire Ready Hiring List. These lists will be arranged by Incident Command System (ICS) equipment kind and type, as per ICS 420-1 Field Ops Guide, Chapter 11. Equipment not meeting all of the “fire ready” criteria, but meeting the “support equipment” criteria will be placed on the Unit “Support Equipment” Lists.

UNIT SUPPORT EQUIPMENT LISTS 7761.6.4.2

By May 1 of each year, each unit will compile two Unit Support Equipment Lists (one for bulldozers and one for water tenders). All vendors signing up in a unit and meeting the support equipment criteria will be placed on the appropriate Unit Support Equipment List. Also included on these lists will be all fire ready equipment signed up in that unit.

REGION PLANNED NEED LISTS 7761.6.4.3

A. Information to be forwarded by units: By May 1 of each year, each ranger unit will forward a list of all DVBE vendors and their equipment from their Fire Ready and Support Equipment Lists to their region CC.

The region CC will compile the Region Planned Need List from this information. The information will include: the appropriate list — Fire Ready or Support — for each piece of equipment, flagging requested by the vendor (as detailed in Section 7761.6.2.2), a clean copy of the completed OF-294 and a clean copy of the DVBE certification form.

B. Compiling lists: By May 15 of each year, each region CC will compile two Regional Planned Need Equipment Rotational Lists (one for water tenders and one for bulldozers). These lists will be compiled from vendor information provided by the ranger units in that region and will include **ALL** ranger unit fire ready and support water tenders and bulldozers that have provided proof of registration as detailed in Section 7761.6.2.

C. Rotation of lists: Each list will be rotated until a vendor is hired or until the list is exhausted.

For example, if 4 dozers are requested, the list would be used until a vendor is hired. If a vendor could not be secured from the list, the request would be returned to the ranger unit ECC for hiring off of their lists.

D. Distribution of lists: Once these lists are compiled, the region command center chief will forward them to the region command center chief in the other region(s).

LIST UTILIZATION 7761.6.4.4

Generally, the need/time frame of a request will determine the type of resource/list that is used. This will **NOT** be considered the sole criteria. The specific needs of a mission will be considered in selecting the resource to fill a given request. It may be necessary or beneficial to fill a planned need request with a fire ready resource or an immediate need request with a support resource due to: time frame, capability, equipment or training requirements of the mission. The provisions allowed by this paragraph shall not be used to circumvent the use of planned need equipment to assist the department in meeting the hiring goals for state government (see Section 7761.6.2.3).

LISTS ESTABLISHED RANDOMLY 7761.6.4.5

All equipment lists (unit and region) will be established annually on a random basis either by drawing or other means that guarantees no biased placement on the list.

LIST ADMINISTRATION AND COORDINATION 7761.6.5

UNIT FIRE READY EQUIPMENT LISTS 7761.6.5.1

The administration and coordination of the Unit Fire Ready Equipment List will be the responsibility of the unit ECC chief.

UNIT SUPPORT EQUIPMENT LISTS 7761.6.5.2

The administration and coordination of the Unit Support Ready Equipment List will be the responsibility of the unit ECC chief.

REGIONAL PLANNED NEED LISTS 7761.6.5.3

The administration and coordination of the Regional “Planned Need” Lists will be the responsibility of the region CC chief.

REQUEST AND DISPATCH PROCEDURES 7761.6.6

DEPARTMENTAL CONTACT NUMBER 7761.6.6.1

At the time of dispatch, the vendor will be provided with a contact number to utilize when contacting the Department. This number will be used to make **ALL** contacts and notifications concerning changes in resource status or for any other communication concerning the response.

REQUEST NUMBERS 7761.6.6.2

The vendor will be provided with one request number for each resource dispatched from one of the lists. There will be no request numbers issued for any other equipment (i.e. pilot vehicles, transports, service units etc.), per the CDF Operation Option in the 3900 Handbook.

DISPATCH OF IMMEDIATE NEED RESOURCES — “CLOSEST RESOURCE” CONCEPT 61.6.6.3

In hiring fire ready or support equipment to fill immediate need requests for work on emergency incidents, it is CDF’s intent to utilize the closest available resources which meet the mission requirements, without regard for administrative boundaries. When the hired equipment resources in the unit ERD have been exhausted, the unit ECC will place the request for immediate need hired equipment with the region command center.

In filling immediate need requests, if fire ready or support resources are **KNOWN** to be available in an adjoining ranger unit and are within a one-hour activation time and a one-hour travel time of the incident (maximum total of two hours from contact to arrival at scene), the ECC may place the order for the resource directly with the adjoining unit ECC. This is for initial attack requests **ONLY**. Units will place **ALL** subsequent resource requests for hired equipment with their region CC following 8100 Command and Control Procedures Handbook, Procedure 006G. The unit with the list on which a piece of equipment is signed up (the sending unit) will be responsible for dispatching all hired equipment requested from that list. It is the sole purpose and intent of this section to clearly state that it is the Department's goal to place resources appropriate for the assignment available at the scene of the incident in the shortest possible time frame. To meet this goal, **ALL** requests for resources to fill immediate need requests (as defined in Section 7761.6.1.3) will be placed with vendors that can best meet these criteria. To this end, in filling immediate need request (**ONLY**), each vendor will be assigned as many requests as they can fill and will then be rotated to the bottom of the list. When filling these requests, list rotation will be followed, but only to the extent possible to meet the expressed intent of this section. Vendors will be bypassed to access the vendor(s) that can be on scene and available in the shortest time frame possible. The minimal acceptable response criteria are detailed in Section 7761.6.8.

ROTATIONAL DISPATCH OF RESOURCES FOR PLANNED NEED REQUESTS — 7761.6.6.4

In filling **ALL planned need** requests with hired equipment governed under this policy, **ALL** contact and hiring will be conducted as follows. Resources listed on the Region Planned Need Lists will be dispatched on a region-wide rotational basis. Resources on the unit lists (fire ready and support) will be dispatched on a unit rotational basis. However, if after contacted, the vendor advises that they cannot meet the time, performance or equipment requirements of the particular mission, that vendor may be bypassed for that request.

UTILIZATION OF PLANNED NEED RESOURCES — 7761.6.6.5

In some cases it may be in the best interest of the State to release immediate need resources already on the scene once their initial assignment has been completed, and replace them with resources from Region Planned Need List to perform other types of assignments, such as mop-up and rehabilitation.

DISPATCH OF PLANNED NEED RESOURCES — 7761.6.6.6

A. Method of dispatch: Once a need for planned need resources is determined by a unit, a request will be placed to their region CC per 8100 Command and Control Procedures Handbook, Procedure 006G. There will only be **ONE** resource requested from a vendor for each call to that vendor.

If the vendors on the Region Planned Need Rotational List are exhausted, or if the vendors cannot fill the requests, the region duty officer may authorize either of the following:

1. The requesting unit ECC to contact vendors listed on that unit's appropriate equipment list to fill any open requests, or
2. Place the request(s) to the other region to be hired from that Region's Planned Need Equipment List.

B. Area of response: All vendors on the Regional Planned Need Lists will be considered to be available to respond region-wide and may in no way to seek restrict or define the geographic area to which they will respond.

VENDOR CONTACT — 7761.6.6.7

Vendors will be contacted by phone as provided for in Section 7761.6.2.2. The vendor will be given 10 minutes from the time of the attempted contact to respond and either accept or decline the request. A vendor not returning a call within the 10-minute time frame will be moved to the bottom of that rotational list.

RECORD OF CONTACT 7761.6.6.8

A record will be kept for **ALL** transactions made in furtherance of this policy. This record will include as a minimum:

1. Each attempted contact, whether or not actual contact was made, and the vendor's name and I.D. number.
2. What type of contact it was; personal contact, pager, answering machine, answering service, voice mail, etc.
3. Whether or not there was a response.
4. If a contact was made or there was a response to the attempt, the name of the individual contacted or returning the call.
5. If there was no response to the attempted contact.
6. The number called.
7. Date and time of every transaction.
8. Name of the Department employee making the contact.
9. Disposition of the contact; accepted, declined, unable to fill, etc.
10. Order and request number.

LIST ROTATION 7761.6.6.9

Vendors will be rotated to the bottom of a list for the following reasons:

- **Unanswered calls:** An unanswered call is considered a call. The vendor was given the opportunity to be hired and for whatever reason did not answer or did not return the call within the allotted time frame.
- **Equipment unavailable:** A vendor's equipment is unavailable at the time of dispatch.
- **Vendor declines:** If the vendor declines the opportunity to be hired for any reason.
- **Vendor accepts:** When a vendor accepts a request.
- **Failure to comply:** Violation(s) of any part of Section 7761.6.8

  NOTE: See the 8100 Command and Control Handbook, Procedure 006G, for details on this process.

YEAR-END SUMMARY 7761.6.6.10

By January 15th of each year, this contact and use information will be summarized into a unit report for the preceding year. The report will show the total use of all hired equipment (fire ready, support and planned need) and will break out the contact and use of BE, WBE and DVBE program contractors.

EMERGENCY PICKUP/HIRES 7761.67

RENTAL AT SCENE 7761.6.7.1

On occasion, incident commanders (ICs) may have to hire equipment which has not been inspected and signed up by a CDF or a cooperating agency. When that occurs, CDF will only hire equipment and operators which meet all criteria as established in this chapter. The IC will be responsible for notifying the ECC and acquiring a request number for the hire. The IC must make certain that all inspections, fiscal procedures and documents are properly completed.

NO HIRING UNLESS REQUESTED 7761.6.7.2

It is the policy of the Department to discourage equipment owner/operators from showing up at incidents hoping to be hired. ICs will not hire rental dozers (or water tenders) who show up un-requested unless in the judgment of the IC (or other Department representative) there is clearly an imminent threat to life and or property and other appropriate resources are not available to meet the immediate need.

PROMPT RELEASE OF SUB-STANDARD OR UN-REQUESTED EQUIPMENT 7761.6.7.3

Certain other actions are required of the IC when the equipment hired to meet these imminent threat situations arrives without being requested or when the equipment does not meet the basic safety criteria established in this chapter. The IC must immediately order appropriate replacement equipment and then release the "pick-up rental" when the pressing emergency necessitating the hire has passed, or when the replacement equipment arrives. **IN NO CASE WILL THIS TYPE OF HIRE EXTEND BEYOND 12 HOURS DURATION TIME WITHOUT SPECIFIC WRITTEN AUTHORIZATION OF THE INCIDENT COMMANDER.**

VENDOR PERFORMANCE 7761.68

GENERAL 7761.6.8.1

Once a vendor accepts an assignment, they must comply with all subsections of this policy. The vendor is required to notify the Department **IMMEDIATELY** in any instance in which they cannot comply with this policy or any requirements of the assignment.

RESPONSE AND ARRIVAL AT SCENE 7761.6.8.2

A vendor must make notification of any change in their status or ability to meet the assigned report time to the incident or other reporting location. Notification will be to the Department at the contact number provided at the time of dispatch. See Section 7761.6.6 for further details.

The incident commander or his/her designee shall have the authority to release any resource not arriving at the incident by the required report time. This release may occur prior to arrival (upon phone contact) or upon arrival at the incident. If it is determined that there is still a need for the resource, the resource may be given another assignment on the incident.

Any open assignments on an incident caused by resources not reporting on time may be filled with any appropriate resource to meet operational needs as determined by the incident commander.

OPERATOR ABILITY TO PERFORM UPON ARRIVAL 7761.6.8.3

The resource operator must arrive at the incident by the required time in a condition in which they can safely and legally operate their equipment and perform their assigned mission. Notification will be to the Department at the contact number provided at the time of dispatch. See Section 7761.6.6 for further details.

EQUIPMENT LOCATION 7761.6.8.4

At the time of a request, it is the vendor's responsibility to notify the Department of any changes in a resource's location or any condition that may affect its availability or response.

QUALIFIED OPERATOR 7761.6.8.5

Operators must meet all training/safety requirements for the type of response (fire ready or support) or assignment. See Section 7761.6.9 and 7761.6.10 for further details.

FAILURE TO COMPLY 7761.6.8.6

Failure to comply with any of the requirements in this policy may result in any of the following actions based on the judgment of the Department. Non-compliance will include, but is not limited to the following:

- Contacting any unit, incident or other location soliciting hiring by the Department.
- Responding to the incident without being requested.
- Misrepresentation of response time.
- Failure to arrive at requested time without making contact.
- Misrepresentation of condition or location of equipment.
- Arrival or operation at the incident without protective gear or any other required equipment.
- Unsafe or negligent equipment operation.
- Failure to follow directions or complete assignments or any other failure in performance.
- Any brokering of equipment or equipment cooperatives of any type or manner that circumvents the intent of any part of this policy.
- Misrepresentation of minority, women or disabled veteran-owned business status.
- Operating equipment on fire ready assignments without the required training.
- Willful violation of fair hiring practices.
- Any other action that violates the intent of this policy.
- Misrepresentation of time worked on incident.

As appropriate, any of the following actions may be taken by the Department. Action on the part of the Department may result from a single action or a combination of actions on the part of a vendor.

- Placement at the end of rotation.
- Removal from list(s) for a specified period of time.
- Removal from list(s) for the remainder of the year.
- Civil and/or criminal action.

The region chief or his/her designee will be the final appeal point. Notice of any action will be forwarded to the vendor in writing.

VENDOR DOCUMENTATION 7761.6.8.7

The operator of each piece of equipment will complete a Unit Log (ICS 214) for each shift worked. This will be in addition to any other required documentation. The Unit Log will be turned in to the incident commander or his/her designee at the end of each shift.

DOCUMENTATION OF SUBSTANDARD PERFORMANCE 7761.6.8.8

Employees witnessing or having knowledge of vendor performance not in compliance with this or any applicable policy will thoroughly document any and all facts. This information will be forwarded without delay to the Ground Support Unit leader or incident commander.

Immediate action will be taken to resolve any issues that involve safety, productivity, operational effectiveness or any other issue that in any way compromises the incident operations. All violations will be referred with documentation to the unit chief of the ranger unit, or his/her designee where the vendor is signed up, for review and possible action.

PERFORMANCE EVALUATIONS 7761.6.8.9

A Performance Evaluation Form (ICS 225) will be completed on all vendors when they are demobilized from each incident. Additional evaluations will be completed by supervisors during the course of an incident to document exemplary or substandard performance. A copy of the evaluation will be forwarded to the hired equipment coordinator of the unit in which the resource was originally signed up to be included in that vendor's file.

REQUIRED EQUIPMENT 7761.6.8.10

All personnel and resources must arrive with all required safety equipment and other equipment in good repair and operating condition. See Sections 7761.6.9 and 7761.6.10 for further details.

WATER TENDER INSPECTION 7761.6.8.11

All SE plated water tenders and all 2-axle commercial vehicle water tenders not currently required to be inspected, or not subject to the DOT inspection standard **will** be inspected annually prior to use. Each inspection will be valid for one year only. In no event will an inspection date of greater than 12 months be accepted. **This inspection must be performed by a facility licensed to operate as a commercial motor vehicle repair and maintenance shop**. Forest Service and CDF personnel will not perform this inspection. All costs will be borne by the contractor. The inspection level will meet the inspection criteria of Federal Motor Carrier Safety Regulations Part 393, Parts and Accessories Necessary for the Safe Operation, 393.1 through 393.209 and Title 49, Parts 40, 325, and 350 through 399, and the California Vehicle Code Division 14.8, 34500, Safety Regulations and California Code of Regulations Title 13. **Proof of inspection for the current fire season must be on file with the Hired Equipment Coordinator prior to listing the vendor in the Emergency Resource Directory**. All water tenders, regardless of type of vehicle registration, must:

1. Carry a copy of the inspection with them, and
2. Provide a copy to the Forest Service or CDF unit that originally signed up their equipment.

The commercial motor vehicle shop's DOT inspection form will be acceptable as long as the following as shown:

1. The shop's address and telephone number.
2. The vehicle's identifying numbers (license and serial numbers).
3. The inspection form must have all critical items identified in the regulations listed above.
4. Repair documentation must show, in addition to 1 and 2, itemization of completed repairs.
5. All forms and job orders must be readable.

As an alternative, the R5-5100-2T Inspection Form may be used by the vendor. If this form is used, items 1, 2, and 4, above (when applicable), must be shown and must be accompanied by a copy of the inspector's work order(s).

Finally, either of the following California Highway Patrol Inspection Forms is also acceptable:

1. Form CHP 407F, Safetynet Driver/Vehicle Inspection Report, or
2. Form CHP 343A, Vehicle/Equipment Inspection Report, Motor Carrier Safety Operations

   NOTE: CHP 108, Truck and/or Tractor Maintenance & Safety Inspection **is not accepted as the safety inspection**.

Any vehicle arriving at an incident without proof of successful completion of this inspection, along with proof that noted defects have been corrected will be rejected and will not be inspected until such documentation is produced. No payment will be made for any time incurred by the contractor. An exception is made for imminent threat situations as described in Sections 7761.6.7.2 and 7761.6.7.3. This type of hiring does not exempt the vehicle from the required inspection process necessary for listing in the unit Emergency Resource Directory.

Any vehicle that does not pass the initial inspection at the incident will be afforded the opportunity for a second inspection. The contractor will be charged $75.00 for the second inspection. If the vehicle does not pass the second inspection, it is rejected. No payment will be made for any time incurred by the contractor and the equipment will not be hired on that incident.

If the vehicle passes the second inspection, the total cost of the second inspection will be deducted from his/her final payment. Travel time to the incident will be paid. Time that the equipment was unavailable will not be compensated.

FIRE READY CRITERIA 7761.6.9

RESPONSE CRITERIA FOR FIRE READY EQUIPMENT 7761.6.9.1

The Fire Ready Dozer Group and Water Tender Group will be required to initiate a response within one hour of notification. **However, if the vendor cannot meet the response criteria for the particular incident, e.g., the required ETA to the incident, or if the vendor does not have the equipment of the size and type needed for the mission, CDF may bypass the contractor for that incident and proceed to contact the next vendor on the rotational list.**

The hiring of vendors governed under this section will be consistent with the criteria set forth in Section 7761.6.6 and specifically Section 7761.6.6.3.

It is the responsibility of the vendor to provide a SINGLE contact number that is available 24 hours per day, 7 days a week, to be used by CDF to make contact for the dispatch of resources (e.g. , telephone, pager, radio paging service, etc.).

EQUIPMENT CRITERIA FOR FIRE READY DOZERS 7761.6.9.2

Each bulldozer listed in the Fire Ready Dozer Group will be supported by a vendor-supplied transport truck of appropriate size and a bulldozer service unit equipped to service the bulldozer after 12 hours of continuous work. Dozers assigned to the Fire Ready Group will also meet or exceed the following requirements:

- backup alarm
- safety belt
- Roll Over Protection System (R.O.P.S.) with heavy mesh screen sides and rear
- lights, 4 front and 2 rear
- belly pans and rock guards
- radiator protection
- radio programmable to CDF frequencies car/car only (see Section 7761.6.12.1)
- protective fire curtains inside of R.O.P.S. on sides and rear
- (1) 10# ABC fire extinguisher
- 1 shovel
- 1 axe/Pulaski

- Operator personal protective equipment will comply with all 3900 Handbook and applicable Title 8 requirements and as a minimum will include:
  — Nomex shirt and trousers — CAL OSHA specification — Nomex shroud is recommended.
  — hard hat
  — goggles
  — gloves
  — fire shelter — USFS Specification
  — lace-up leather boots
- canteen with water, 1 gallon minimum

EQUIPMENT CRITERIA FOR FIRE READY WATER TENDER GROUP 7761.6.9.3

The Fire Ready Water Tender Group will meet or exceed the criteria for the Support Water Tender Group (7761.6.10.3), plus:

- 2000-gallon minimum capacity
- 250 GPM pumping capability
- 1 each 1" NPSH thread discharge and 2 each 1-½" NH thread discharge
- radio, programmable to CDF frequencies car/car only (see Section 7761.6.12.1)

IDENTIFICATION NUMBER 7761.6.9.4

Fire ready resources will prominently display a pre-designated identification number on both sides of their equipment. The contractor will provide number/letters at least 6" high and ¾" wide stroke that must be a reflective material. The number to be displayed consists of two numbers and the letter E. The numbers will denote the ranger unit where the equipment was signed up. For example, a dozer signed up in Shasta would display "24E." This not only denotes the ranger unit but also provides a place for the procurement personnel to add the request number of that machine.

Example: a bulldozer signed-up in SHU would have a contractor-provided number of "24E." At the time of hire on an incident, if the dozer was being used to fill request #E205, the Equipment Manager would add "205" behind the "24E," (using water-color paint or similar removable marking) and the equipment's identifying number for that incident would be "24E205."

ANNUAL TRAINING REQUIRED 7761.6.9.5

Eight hours of specific annual training is required for the Immediate Need Groups. Training details are as shown in Exhibit 7761.6.9.5. Each operator who attends the required training will be issued a document by the Department certifying that they have successfully completed the training. While operating for the Department, all operators will have this certification documentation in their possession and present it upon request by a representative of the Department.

SUPPORT EQUIPMENT CRITERIA 7761.6.10

RESPONSE CRITERIA FOR SUPPORT EQUIPMENT GROUPS 7761.6.10.1

Planned need resources need not meet the one-hour response time criteria established for immediate need resources, but they must be able to arrive at the incident location within the time prescribed by the agency dispatcher at the time of contact. However, if after contact the vendor cannot meet the response criteria for the particular incident, e.g., the required ETA to the incident, or if the vendor does not have the equipment of the size and type needed for the mission, CDF may bypass the contractor for that incident and proceed to contact the next vendor on the rotational list.

EQUIPMENT CRITERIA FOR SUPPORT DOZER GROUP 7761.6.10.2

The Planned Need Dozer Group will meet or exceed the following equipment requirements:

- R.O.P.S. with heavy mesh screens on sides and rear
- lights, 2 front and 2 rear
- belly pan and rock guards
- radiator protection
- 1 shovel
- 1 axe/Pulaski
- backup alarm
- seat belts
- Operator personal protective equipment will comply with all 3900 Handbook and applicable Title 8 requirements and as a minimum will include:
  — Nomex shirt and trousers — CAL OSHA specification — Nomex shroud is recommended.
  — hard hat
  — goggles
  — gloves
  — fire shelter — USFS Specification
  — lace-up leather boots

EQUIPMENT CRITERIA FOR SUPPORT WATER TENDER GROUP 7761.6.10.3

The Support Water Tender Group will meet or exceed the following requirements:

1. When loaded (including operators and accessory equipment), water tenders will conform to the manufacturer's gross vehicle weight rating (GVWR). This includes balancing the load in a manner **so that all axle weights comply with the manufacturer's gross axle weight ratings.** Preseason sign up will require the unit to be loaded with the contractor providing weight tickets for the load from a certified scale. The weight tickets will be by individual axle weight. Loaded/empty weights may be necessary to certify gallons.
2. Water tanks are to be baffled, meeting the standards of NFPA, the American Society of Mechanical Engineers, or other industry-accepted engineering standards.
3. Tender shall be configured in a manner that the vehicle center of gravity is within the design limits of the equipment.
4. Tender shall be equipped with a back flow protection device for hydrant filling.
5. Pump assembly is to be driven either by power-take-off or auxiliary engine drive. The pump shall be plumbed with a suction outlet so water may be drafted from a water supply such as a pond, river or creek to refill the tank or pump directly to the fire. An auxiliary engine-driven pump assembly is acceptable if the auxiliary unit is permanently mounted and plumbed to the tank, discharge and suction outlets, and if the pump meets the minimum GPM for the group.
6. All discharge outlets shall be plumbed to the pump. Gravity discharge systems are not acceptable.
7. Tender shall be equipped with a minimum 24 feet of appropriate diameter suction hose equipped with a screened foot valve or strainer.

8. Tenders shall have a 2-½" valve with National Hose threads (NH) installed so that pressure or suction lines (hose) can allow filling or drafting by other engines. Adaptation of dump valve with fittings is acceptable.
9. A dump valve of a minimum 4" diameter or equivalent is desirable so that water may be discharged into portable tanks. This valve is to be at the bottom rear of the tank to allow complete water discharge and should have a minimum clearance of 34" from the ground to the bottom of the outlet.
10. It shall be the contractor's responsibility to ensure that the equipment meets the standards or specifications. Should the equipment be designed to any industrial standard other than NFPA or ASME, the contractor will be required to provide a copy of the industrial standard met. It is the contractor's responsibility to demonstrate that the equipment meets the standard. Beyond NFPA standards, agency personnel at the federal, state, and local levels are not knowledgeable in design standards; therefore, the contractor must determine the requirements of the standard and bring his/her equipment up to that standard. The knowledge and expertise of the standards rest in the private sector with mechanical engineers, automotive engineers, manufactures or other experts. Contractors should consult these sources to bring equipment into compliance.
11. Tender must be equipped with the following minimum safety and equipment items:
    - reflectors, 1 set of 3
    - fire extinguisher (10BC or better)
    - chock blocks of appropriate size for tire diameter
    - flashlight
    - electronic backup alarm, minimum 87 dbls
    - one axe/Pulaski and 1 shovel
    - operator personal protective equipment
    - Nomex shirt and trousers — CAL OSHA specification. Nomex shroud is recommended.
    - hard hat
    - goggles
    - gloves
    - fire shelter — USFS Specification
    - lace-up leather boots
    - canteen with water, 1 gallon minimum
    - pump, minimum 150 GPM
    - 100 ft. 1-½" serviceable cotton jacketed fire hose with nozzle
    - one 1" combination fog/straight stream nozzle with 1-½" NH to 1" NPSH reducer
    - tender shall have the following discharge outlets as a minimum:

Group 1	2 each 1-½" NH thread
	1 each 2-½" NH thread
Groups 2 & 3	1 each 1" NPSH thread
	1 each 1-½" NH thread

- one fire hose clamp

NOTE: If the water tender arrives at the incident without the required minimum safety items, required equipment or appliances, and does not meet the specifications for pump, plumbing, buildup, etc., **the water tender will be rejected**. No payment will be made for a rejected water tender for any time incurred by the contractor and the equipment will not be hired on the incident.

IDENTIFICATION NUMBER 7761.6.10.4

Same as detailed in Section 7761.6.9.4.

ANNUAL TRAINING 7761.6.10.5

The Support Equipment Groups are encouraged to attend the Fire Ready Equipment Group training, but this training is not required. (See Section 7761.6.9.5.)

COORDINATION/SUPERVISION 7761.6.11

RENTAL EQUIPMENT COORDINATOR 7761.6.11.1

To assist in the routine process of pre-planning the emergency use of hired equipment, all ranger unit chiefs will designate a rental equipment coordinator for their unit. The name of this designee shall be forwarded to the region command center chief for inclusion on the region-wide coordinator list. A suggested list of duties for the coordinator is shown in Exhibit 7761.6.11.1.

TECHNICAL SPECIALIST, HIRED EQUIPMENT 7761.6.11.2

On incidents of significant magnitude with multiple hired resources, the incident commander is encouraged to utilize the position of technical specialist - hired equipment. This position would work for the equipment manager within the Ground Support Unit. A suggested duty statement for this position is shown in Exhibit 7761.6.11.2.

DIRECT SUPERVISION 761.6.11.3

All incident commanders must provide direct supervision for all hired resources. Line supervisors are responsible for the safety, proper deployment, and time keeping of their assigned resources. When forming a functional group, strike team or task force, a functional group supervisor or strike team/task force leader should be assigned to perform the following duties:

1. Initiate shift ticket process.
2. Determine equipment readiness and conformance to requirement standards.
3. Respond to incident with the group or task force.
4. Act as group supervisor or task force leader under direction of Operations Section personnel.

The ratio of supervisors to resources should be as follows:

Dozer Strike Team: 2 dozers, 1 dozer tender, 1 STL
Water Tender Strike Team: 5 tenders, 1 STL

Task Force or Functional Group:
1–5 resources: 1 leader/supervisor
5–10 resources: 2 leader/supervisor + 1 tech. spec. - hired equipment
10–15 resources: 3 leader/supervisor + 1 tech. spec. - hired equipment

ANNUAL SAFETY AND OPERATIONAL TRAINING REQUIREMENTS

The following subjects will be presented to rental equipment operators each year prior to fire season during an eight (8) hour safety and operational procedure training program. These subject areas will be taught in accordance with the lesson plans included in CDF Fire Protection Training Handbook 4300. Minimum instructional time frames are as follows:

Section	Subject	Time
(none)	Hiring, Inspection, Timekeeping, and Payment Procedures	25 Min.
4305.1	Wildland Safety Uniform	60 Min.
4306.2	10 Standard Firefighting Orders	30 Min.
4306.3	13 Situations that Shout "Watch Out"	30 Min.
4306.5	Fireline Hazards	30 Min.
4306.13	Bulldozer as a Refuge	20 Min.
4306.15-16	Fire Shelters	60 Min.
4313.14-17	Dozer Safety	75 Min.
4320.2	Parts of a Vegetation Fire	30 Min.
4320.7	Size up & Report on Conditions	60 Min.
4320.8	Vegetation Fire Strategy	30 Min.
(none)	Radio Procedures	30 Min.
Total		**8 Hours**

Exhibit 7761.6.9.5 NO. 14 SEPT. 1996

TECHNICAL SPECIALIST - HIRED EQUIPMENT

The technical specialist - hired equipment, working under the direction of the equipment manager within the Ground Support Unit, is responsible for the tracking and inventory of all incident hired equipment.

The technical specialist facilitates and coordinates the interactive functions of Operations, Finance and Logistics.

The technical specialist maintains an inventory of all hired equipment, reviews shift tickets for accuracy and completeness, and assures that all required documentation necessary for prompt payment is delivered to Finance. Additionally, the technical specialist - hired equipment:

- Obtains briefing from Ground Support Unit Leader or equipment manager.
- Attends incident briefing and obtains Incident Action Plan (for inventory log).
- Obtains briefing from Operations Section chief, Finance Section chief, and Logistics Section chief.
- Develops inventory listing of all hired equipment on the incident.
- Reviews hired equipment status as shown by current check-in and Re-Stat displays; provides corrected information to Re-Stat.
- Assures that all hired equipment is properly signed up, inspected and documented on the OF-296 "Vehicle/Heavy Equipment Inspection Checklist," and that all hired equipment is visibly numbered.
- Assures that line supervisors accurately prepare the OF-297 "Rental Equipment Use Record"("shift ticket") for each piece of hired equipment and National Guard equipment under their supervision, and that the shift ticket is turned in to the equipment time recorder at the end of each operational period.
- Provides input to the Operations Section chief for coordination of demobilization activities.
- Identifies potential problems and or unassigned equipment not shown on the Incident Action Plan.
- Reports to the Ground Support Unit leader/equipment manager on a regular basis.

Exhibit 7761.6.11.2 NO. 14 SEPT. 1996

RENTAL EQUIPMENT COORDINATOR RESPONSIBILITIES

(A Required Position)

1. The ranger unit chief's staff shall approve appointment to this position.
2. The coordinator will, by May 1 of each year, provide the ECC with a current updated list of all rental equipment available for use and will provide an updated list of immediate need and planned need equipment.
3. Conduct yearly pre-season inspections of rental equipment prior to submittal of the list required in Item #2 above.
4. Make certain that the ECC has current FC-100s on file.
5. Conduct a yearly safety and operational procedure training program prior to fire season. (See Exhibit 7761.6.9.5.)
6. Be assured that enough people from each owner are trained so that an immediate need dozer will not arrive without a qualified operator.
7. Provide input to the unit chief's staff on items pertaining to all aspects of all types of rental equipment.
8. Represent the ranger unit at meetings pertaining to rental equipment.
9. Be the first level of review for complaints for owner/operator rental equipment.
10. Verify status as minority, women or disabled veteran-owned business enterprise.

Appendix C

Incident Command/Management Forms

ORGANIZATION ASSIGNMENT LIST ICS-203 1/82

1. INCIDENT NAME	2. DATE PREPARED	3. TIME PREPARED

4. OPERATIONAL PERIOD (DATE/TIME)

POSITION	NAME
5. INCIDENT COMMANDER AND STAFF	
INCIDENT COMMANDER	
DEPUTY	
SAFETY OFFICER	
INFORMATION OFFICER	
LIAISON OFFICER	

6. AGENCY REPRESENTATIVES

AGENCY	NAME

7. PLANNING SECTION

CHIEF	
DEPUTY	
RESOURCES UNIT	
SITUATION UNIT	
DOCUMENTATION UNIT	
DEMOBILIZATION UNIT	
TECHNICAL SPECIALISTS	

8. LOGISTICS SECTION

CHIEF	
DEPUTY	
a. SUPPORT BRANCH	
DIRECTOR	
SUPPLY UNIT	
FACILITIES UNIT	
GROUND SUPPORT UNIT	
b. SERVICE BRANCH	
DIRECTOR	
COMMUNICATIONS UNIT	
MEDICAL UNIT	
FOOD UNIT	

9. OPERATIONS SECTION

CHIEF		
DEPUTY		
a. BRANCH I — DIVISIONS/GROUPS		
BRANCH DIRECTOR		
DEPUTY		
DIVISION/GROUP		
DIVISION/GROUP		
DIVISION/GROUP		
DIVISION/GROUP		
DIVISION/GROUP		
b. BRANCH II — DIVISIONS/GROUPS		
BRANCH DIRECTOR		
DEPUTY		
DIVISION/GROUP		
DIVISION/GROUP		
DIVISION/GROUP		
DIVISION/GROUP		
DIVISION/GROUP		
c. BRANCH III — DIVISIONS/GROUPS		
BRANCH DIRECTOR		
DEPUTY		
DIVISION/GROUP		
DIVISION/GROUP		
DIVISION/GROUP		
DIVISION/GROUP		
DIVISION/GROUP		
d. AIR OPERATIONS BRANCH		
AIR OPERATIONS BR. DIR.		
AIR ATTACK SUPERVISOR		
AIR SUPPORT SUPERVISOR		
HELICOPTER COORDINATOR		
AIR TANKER COORDINATOR		

10. FINANCE SECTION

CHIEF	
DEPUTY	
TIME UNIT	
PROCUREMENT UNIT	
COMPENSATION/CLAIMS UNIT	
COST UNIT	

203 ICS 1/82	PREPARED BY (RESOURCES UNIT)	7540-130-0284

1. BRANCH	2. DIVISION/GROUP	DIVISION ASSIGNMENT LIST	ICS 204 (1-82)
3. INCIDENT NAME		4. OPERATIONAL PERIOD DATE ____ TIME ____	

5. OPERATIONS PERSONNEL

OPERATIONS CHIEF ____ DIVISION/GROUP SUPERVISOR ____

BRANCH DIRECTOR ____ AIR ATTACK SUPERVISOR ____

6. RESOURCES ASSIGNED THIS PERIOD

STRIKE TEAM/TASK FORCE/ RESOURCE DESIGNATOR	LEADER	NUMBER PERSONS	TRANS. NEEDED	DROP OFF PT./TIME	PICK UP PT./TIME

7. CONTROL OPERATIONS

8. SPECIAL INSTRUCTIONS

9. DIVISION/GROUP COMMUNICATION SUMMARY

FUNCTION		FREQ.	SYSTEM	CHAN.	FUNCTION		FREQ.	SYSTEM	CHAN.
COMMAND	LOCAL				SUPPORT	LOCAL			
	REPEAT					REPEAT			
DIV./GROUP TACTICAL					GROUND TO AIR				

PREPARED BY (RESOURCE UNIT LDR.)	APPROVED BY (PLANNING SECT. CH.)	DATE	TIME

7540-130-0285

INCIDENT RADIO COMMUNICATIONS PLAN

1. INCIDENT NAME	2. DATE/TIME PREPARED	3. OPERATIONAL PERIOD DATE/TIME

4. BASIC RADIO CHANNEL UTILIZATION

SYSTEM/CACHE	CHANNEL	FUNCTION	FREQUENCY	ASSIGNMENT	REMARKS

205 ICS 8-78	5. PREPARED BY (COMMUNICATIONS UNIT)

MEDICAL PLAN	1. INCIDENT NAME	2. DATE PREPARED	3. TIME PREPARED	4. OPERATIONAL PERIOD

5. INCIDENT MEDICAL AID STATIONS

MEDICAL AID STATIONS	LOCATION	PARAMEDICS	
		YES	NO

6. TRANSPORTATION

A. AMBULANCE SERVICES

NAME	ADDRESS	PHONE	PARAMEDICS	
			YES	NO

B. INCIDENT AMBULANCES

NAME	LOCATION	PARAMEDICS	
		YES	NO

7. HOSPITALS

NAME	ADDRESS	TRAVEL TIME		PHONE	HELIPAD		BURN CENTER	
		AIR	GRND		YES	NO	YES	NO

8. MEDICAL EMERGENCY PROCEDURES

206 ICS 8-78	9. PREPARED BY (MEDICAL UNIT LEADER)	10. REVIEWED BY (SAFETY OFFICER)

Appendix D

Sample Fire Risk-Rating System

FIRE RISK RATING

FOR EXISTING AND PLANNED WILDLAND RESIDENTIAL INTERFACE DEVELOPMENTS IN MONTANA

MONTANA DEPARTMENT OF STATE LANDS

MARCH 1993

TABLE OF CONTENTS

CHAPTER I - FIRE RISK RATING FOR EXISTING DEVELOPMENTS

INTRODUCTION

The risk rating system for existing developments is a planning tool for fire prevention. It assesses the potential wildfire hazards faced by wildland residential developments.

The system allows prevention planners to assess areas for risk, rank them according to their risk score, and then set priorities for prevention resources and actions. It organizes physical site information such as road access, topography, fuels, construction and water sources so that the planner can easily review all the information at once.

The Montana Department of State Lands (DSL) developed this rating system for its staff and cooperators to use. The risk rating system can help anyone conducting or planning fire prevention activities in existing wildland/residential interface areas.

Using the risk rating system also logically benefits all prevention planners by increasing their familiarity with hazardous locations and the values at risk within their fire protection areas.

DSL continues to develop ways to minimize destruction from wildfire while using public funds most efficiently. The risk rating system adapts existing knowledge, from agencies well experienced with the interface fire problem, to this end.

CHAPTER I - FIRE RISK RATING FOR EXISTING DEVELOPMENTS

HOW TO USE THE RISK RATING SYSTEM FOR EXISTING DEVELOPMENTS

Prevention planners can evaluate wildland and rural areas that have enough permanent, seasonal, recreational or commercial habitation to pose a fire hazard or risk. Logical applications might include rural subdivisions, scattered residential developments, camps, lodges and resorts.

The rating area can be as large or small as deemed necessary. However, a rating area should meet two criteria:

- It should be one relatively homogenous development that has distinct site factors. For example, the Many Lakes subdivision may make a logical rating area, but the entire Flathead Valley would not. Part of a subdivision may constitute a logical rating area if it is somehow different from the rest of the subdivision.

- It should be large enough to take into account the surrounding fuels, topography, nearby risks, etc., that will affect fire occurrence.

Always access separate subdivisions individually. If you have any doubts about what to make a rating area, err on the small side. In the risk rating system, if you assess smaller areas, you will still have useful information; if you rate larger areas, your rating may be inaccurate, meaningless or, worse, misleading.

CHAPTER I - FIRE RISK RATING FOR EXISTING DEVELOPMENTS

PART I INSTRUCTIONS FOR COLLECTING AREA INFORMATION

To collect the necessary information, you will need Form A - Data Collection Form for Existing Developments, (Appendix A), Form B - Residential Tally Sheet, (Appendix B), a 100-foot tape measure, a clinometer or Abney level, a compass, area maps, a vehicle and these instructions.

The item numbers in these instructions correspond with the items on Form A - Data Collection Form.

1. Enter the number of primary access roads in the rating area. These must:
   - lead into or out of the rating area;
   - have two, twelve-foot traffic lanes, paved or gravel;
   - be able to be negotiated by structural fire equipment;
   - be maintained; and
   - be open year-round, not controlled access.

   If there are no roads that meet all of these conditions, enter zero.

2. Enter the number of alternative access roads in the rating area. These are roads which enter or leave the rating area, but do not meet the conditions for primary access roads.

   They must:
   - lead into or out of the rating area;
   - be able to be used by two-wheel drive vehicles as a substitute entry or exit road for the rating area.

3. Enter the width of the road surface, including any serviceable shoulders, on the primary access roads. Do not include turn-outs.

4. Enter the grade (%) of the steepest part of road within the rating area. Include secondary roads as well as primary and alternative access roads. Secondary roads are roads which leave a primary access road to reach homes, buildings, recreational sites, etc. that lie away from the primary road. Driveways over 600 feet are also considered secondary roads.

CHAPTER I - FIRE RISK RATING FOR EXISTING DEVELOPMENTS

5. Check the blank that matches the narrowest secondary road endings in the area. Include driveways longer than 600 feet as secondary roads. If the rating area has only one road, check the blank that describes how that road ends. Loop roads are those which return to a primary access road.

6. Check the blank which applies to the lightest capacity bridge on a primary access road within the rating area. Thirty-eight ton (statutory limit) bridges should be considered 40 ton bridges.

7. Check the blank which applies to the lightest capacity bridge on a secondary road in the rating area. Thirty-eight ton (statutory limit) bridges should be considered 40 ton bridges.

8. Check the blank which indicates the predominant slope (%) within and directly adjacent to the part of the rating area where the homes are.

9. Check the blank which indicates the predominant aspect of the rating area. This can be determined from a topographical map and/or site visit. To facilitate accurate aspect representation, the form asks for ranges by compass azimuth.

10. Check one or more blanks to indicate topographic features in and around the area that would contribute to erratic or extreme fire behavior. The features are listed from least dangerous to most dangerous. If you check more than one, circle the one that is most dangerous/farthest down the list.

11. Check the blank which most closely indicates the predominant fuels conditions in the rating area.

12. Check the blanks which indicate sources of risk in or immediately adjacent to the rating area. Check all present.

13. Check the blank which indicates the worst maintained electrical utilities R.O.W. in the rating area.

14. After actually counting, enter the number of homes in the rating area. An efficient method would be to conduct this count and the spacing and landscaping information (items 14-17) at the same time. Use Form B - Residential Tally Sheet (Appendix B) for tallying this information, then transfer the total to item 14, Form A.

CHAPTER I - FIRE RISK RATING FOR EXISTING DEVELOPMENTS

15. Enter the number of homes that have composition, metal, or tile roofs or other fire resistant roofing (as defined on page 17). Do not include homes with wood shake or shingle roofs. Tally on Form B, then transfer the total to item 15, Form A.

16. Enter the number of homes which have one or more of: unenclosed balconies, decks, eaves, stilts or cantilevered construction. Do not count any of these that have been thoroughly enclosed. Tally on Form B, then transfer the total to item 16, Form A.

17. Visually estimate how far apart the homes are spaced. Check the blank which most nearly describes the distance between the majority of homes in the rating area. Tally on Form B, then transfer the total to item 17, Form A.

18. Tally the number of homes which meet or exceed the fire resistant landscaping guidelines in Appendix F. Tally on Form B, then transfer the total to item 18, Form A.

19. If there are hydrants in the rating area, check "yes."

20. If there are hydrants in the rating area, measure the distance between them. Enter that spacing. If no standard spacing exists, enter an average spacing.

21. If there are hydrants in the rating area, indicate whether they are of 500 gallon per minute or greater capacity by checking either the "yes" or "no." To get this information, you can test the flow or check with the rural fire district, the water company or the county.

22. Check the blank which most closely indicates what draft sources exist and are accessible in the rating area.

23. Check the blank which most closely indicates how close a reliable helicopter dip spot is to the rating area.

24. Indicate whether the rating area is covered by a Rural Fire District (RFD), Fire Service Area (FSA), or municipal fire department by checking either "yes" or "no." Do not include a Volunteer Fire Company unless it has been formed as part of one of the groups above, or by a county governing body or an incorporated town.

CHAPTER I - FIRE RISK RATING FOR EXISTING DEVELOPMENTS

25. Check the blank which indicates the response time from the nearest fire organization to the rating area. Get the response time from the chief officer. If that fire organization is a Volunteer Fire Company, that is not part of a Rural Fire District, Fee Service Area or municipality, write "VFC" in the blank.

26. Indicate whether the rating area has a way to contact homeowners, such as a homeowners association, civic club, development office, etc., by checking either "yes" or "no."

27. If you checked "yes" for item 26, check the blank which most closely describes the group.

28. Using statistics for the most recent 10 year period, indicate the average number of fires per thousand acres in and around the rating area. Follow either method of calculation below.

Example: Size of ABC Gulch Rating Area = 1280 acres
Number of fires, 1977-1987 = 2

$$\frac{2 \text{ Fires}}{1280 \text{ Acres}} = .00156 \qquad\qquad \frac{2 \text{ Fires}}{1280 \text{ Acres}} = \frac{x \text{ Fires}}{1000 \text{ Acres}}$$

x 1000 Conversion Factor

$$2000 = 1280x$$
$$200 = 128x$$
$$1.56 = x$$

=1.56 Fires/1000 Ac./10 Yrs. = 1.56 fires/1000 Ac./10 Yrs.

CHAPTER I - FIRE RISK RATING FOR EXISTING DEVELOPMENTS

PART II INSTRUCTIONS FOR RATING

To actually rate the area, you will need the filled-in Form A - Data Collection Form (Appendix A), Form C - Rating Form (Appendix C), and these instructions.

Using the information on Form A, score each item on Form C by circling the number on Form C that corresponds to the answer you gave on Form A.

For Item 10, circle the number for the most dangerous feature in the area.

For Item 12, count the number of risks in the area and circle that number.

For Item 15, divide the number of homes with fire resistant roofing by the total number of homes in the area.

For Item 16, divide the number of homes with unenclosed balconies, decks, eaves, stilts, etc. by the total number of homes in the area.

For Item 18, divide the number of homes that have fire resistant landscaping by the total number of homes in the area.

Total the numbers circled on Form C - Rating Form and enter that total in the "total score" space.

Apply the following classifications to the totaled score.

Fire Risk Ratings

Score	Classification
<= 110	low risk - low priority
111-135	moderate risk - moderate priority
136-150	high risk - high priority
151-170	very high risk - very high priority
>= 171	extreme risk - extreme priority

Rating areas should also be ranked within these classifications. For example, two separate rating areas with scores of 136 and 150 would both be classified as "high" risk and priority. However, the area with the 150 score would logically pose the higher risk and the higher priority.

The system provides a basic ranking system, which may be expanded using locally significant criteria, such as relative fire costs or cost per acre protected. We encourage all users to consider local factors which will help them clarify priorities.

CHAPTER II - FIRE RISK RATING FOR PLANNED DEVELOPMENTS

INTRODUCTION

As with the risk rating system for existing developments the risk rating system for planned developments is a planning tool for fire prevention. It assesses the potential wildfire hazards faced by developments planned in the Wildland Residential Interface.

The risk rating for planned developments allows prevention planners to assess areas for risk, so that they may communicate the potential hazards to community planners, local government officials and developers prior to final platting or construction. Like the risk rating system for existing developments; it organizes physical site information such as planned road access, topography, fuels, planned construction, and water sources so that the planner can easily review all the information at once.

Both risk rating systems logically benefits the prevention planner by increasing their familiarity with hazardous locations and the values at risk within their fire protection areas.

CHAPTER II - FIRE RISK RATING FOR PLANNED DEVELOPMENTS

HOW TO USE THE RISK RATING SYSTEM FOR PLANNED DEVELOPMENTS

Prevention planners can assess proposed developments in wildland and rural areas which may pose a fire hazard or risk. Logical applications might include planned rural subdivisions, camps, lodges and resorts.

The rating area can be as large or small as deemed necessary. However, a rating area should meet these criteria:

1. It should be one development represented by a single plat, that has distinct site factors. For example: separate subdivisions, phases or plats covered under one Planned Unit Development, Neighborhood Plan, Overall Development Plan or similar document should be evaluated individually. A single development may be rated in separate parts if those portions are somehow unique.

2. It should be large enough to take into account the surrounding fuels, topography, nearby risks, etc., that will affect fire occurrence or impact.

Always rate separate projects individually. If you have doubts about what to make a separate rating area, err on the small side. In the risk rating system, if you choose to break a single development into smaller areas, you will still have useful information. However, excessively large areas may produce inaccurate, meaningless or misleading information.

CHAPTER II - FIRE RISK RATING FOR PLANNED DEVELOPMENTS

PART I INSTRUCTIONS FOR COLLECTING AREA INFORMATION

To collect the necessary information you will need a copy of the project's preliminary plat, Form A - Data Collection Form for Planned Developments, (Appendix D), a topographic map of the project area and these instructions. Though some projects can be evaluated from the office, your evaluation will likely include a site visit. During that site visit you will need a 100-foot tape measure, a Clinometer or Abney level, and a compass.

The items in these instructions correspond with the items on Form A - Data Collection Form. All required information can be obtained by examining the preliminary plat, discussing the project with a community planner or the developer and by visiting the site.

1. Enter the number of primary access roads which will serve the proposed development. Primary access roads usually lead into the development from a highway, county road or arterial. These must:

   - lead into or out of the proposed development (see plat);
   - have two, twelve-foot traffic lanes, paved or gravel (see plat);
   - be able to be negotiated by structural fire equipment (consider maximum grade and curve radius);
   - be maintained; and
   - be open year-round, not controlled access.

   If there are no roads that meet all of these conditions, enter zero.

2. Enter the number of alternative access roads which will serve the development. These are roads which enter or leave the rating area, but do not meet the conditions for primary access roads (may be platted or existing). They must:

   - lead into or out of the proposed development (see plat);
   - be able to be used by two-wheel drive vehicles as a substitute entry or exit road for the rating area.

3. Enter the width of the planned road surface, including any serviceable shoulders on the primary access roads. Do not include turn-outs.

CHAPTER II - FIRE RISK RATING FOR PLANNED DEVELOPMENTS

4. Enter the grade (%) of the steepest part of any road planned within the rating area. Include secondary roads as well as primary and alternative access roads. Secondary roads are those which leave an access road to reach homes, buildings, recreational sites, etc. that lie away from the access road. Driveways over 600 feet are also considered secondary roads.

5. Check the blank that matches the most limiting secondary road endings planned for the area. Include driveways longer than 600 feet as secondary roads. If the project planned only has one road, check the blank that describes how that road ends. Loop roads are those which return to a primary access road.

6. Check the blank which applies to the lightest capacity bridge on a primary access road serving the planned development. Thirty-eight ton (statutory limit) bridges should be considered 40 ton bridges.

7. Check the blank which applies to the lightest capacity bridge on a secondary road within the planned development. Thirty-eight ton (statutory limit) bridges should be considered 40 ton bridges.

8. Check the blank which indicates the predominant slope (%) within and directly adjacent to the part of the rating area where dwellings will be constructed. This can be determined from a topographical map and/or site visit.

9. Check the blank which indicates the predominant aspect of the rating area. This can be determined from a topographical map and/or site visit. To facilitate accurate aspect representation, the form asks for ranges by compass azimuth.

10. Check one or more blanks to indicate topographic features in and around the area that would contribute to fire behavior which would threaten the proposed development or contribute to erratic or extreme fire behavior.

11. Check the blank which most closely indicates the predominant fuels conditions in and around the rating area.

12. Check the blanks which indicate sources of risk which exist or are likely to exist in or immediately adjacent to the rating area. Check all present.

13. Check the blank which indicates how electrical utilities will be installed and maintained in the rating area. For small developments, you may want to consider the utility installations offsite instead.

CHAPTER II - FIRE RISK RATING FOR PLANNED DEVELOPMENTS

14. Enter the number of homes planned for the development at full build-out. If the number of homes are not indicated on the plat, use the number of lots platted.

15. Enter the number of homes that will have composition, metal, tile roofs or other fire resistant roofing. Do not include homes with wood shake or shingle roofs. This information may require consultation with the developer or planner. If this information can not be obtained, make an assumption or enter multiple options.

16. Measure or estimate how far apart the homes will be spaced. Check the blank which most nearly describes the distance between the homes in the rating area. If the preliminary plat does not include proposed building sites, place the building site near the center of the lot.

17. Indicate whether the dwellings will meet or exceed the fire resistant landscaping guidelines in Appendix F. This information will likely require consultation with the developer or planner. The fire prevention planner should not assume adequate defensible space unless it is clearly addressed in developer's covenants or attached as a condition of approval to the preliminary plat. The rating can be completed assuming adequate defensible space if the prevention planner feels confident that the developer has taken measures, or will be required, to take measures to incorporate adequate defensible space.

18. Indicate whether the project design has incorporated greenbelts and/or fuelbreaks to protect the planned development, existing neighboring developments and/or the adjacent wildlands.

19. If fire hydrants are planned for the project, check "yes".

20. If hydrants are planned for the rating area, measure or otherwise ascertain the spacing between them. Enter that spacing. If no standard spacing exists, enter an average spacing. You may have to consult the planner, developer or local fire department to collect this information.

21. If hydrants are planned for the rating area, determine whether their placement meets with the approval or desire of the local fire department. You will have to consult with the local fire department to collect this information.

22. If hydrants are planned for the project, indicate whether they are of 500 gallon-per-minute or greater capacity by checking either "yes" or "no". To get this information, check with the project engineer or the water company.

CHAPTER II - FIRE RISK RATING FOR PLANNED DEVELOPMENTS

23. Check the blank which most closely indicates what draft sources exist and are accessible in the rating area or are planned for the development.

24. Check the blank which most closely indicates how close a reliable helicopter dip spot will be to the planned development.

25. Indicate whether the planned development is or will be protected by a Rural Fire District (RFD), Fire Service Area (FSA), or municipal fire department by checking "yes" or "no". Do not include a volunteer fire company unless it has been formed as part of the groups above, or by a county governing body or an incorporated town. Obtain this information from the county or city governing body and fire district officers.

26. If the planned development is or will be protected by a fire department, check the blank which indicates the response time from the nearest fire station. Get the response time from the chief officer. If that fire organization is a Volunteer Fire Company, that is not part of a Rural Fire District, Fire Service Area or municipality write "VFC" in the blank.

27. Indicate whether the planned development will have a way to contact homeowners, such as a homeowner's association, civic club, developer's office, etc., by checking either "yes" or "no". This information can be obtained by examining the developer's covenants or consulting the developer.

28. If you checked yes" for item 25, check the blank which most closely describes the group.

29. Using statistics for the most recent 10 year period, indicate the average number of fires per thousand acres in and around the rating area. Follow either method of calculation below.

Example: Size of Piney Woods Condos Rating area = 1280 acres
Number of Fires, 1982-1992 = 2

$\frac{\text{2 Fires}}{\text{1280 Acres}} = .00156$	$\frac{\text{2 Fires}}{\text{1280 Acres}} = \frac{\text{x Fires}}{\text{1000 Acres}}$
x 1000 Conversion Factor	2000 = 1280x 200 = 128x 1.56 = x
=1.56 Fires/1000 Ac./10 Yrs.	=1.56 fires/1000 Ac./10 Yrs.

CHAPTER II - FIRE RISK RATING FOR PLANNED DEVELOPMENTS

PART II INSTRUCTIONS FOR RATING

To actually rate the area, you will need the filled-in Form A - Data Collection Form (Appendix D), Form B - Rating Form (Appendix E), and these instructions.

Using the information on Form A, score each item on Form B by circling the number on Form B that corresponds to the answer you gave on Form A.

For Item 10, circle the number for the most dangerous feature in the area.

For Item 12, count the number of risk sources in the area and circle that number.

For Item 15, divide the number of dwellings planned with fire resistant roofing by the total number of dwellings planned.

For Item 17, divide the number of dwellings planned to meet the fire resistant landscaping guidelines by the total number of dwellings planned.

Total the numbers circled on Form B - Rating form and enter that total in the "Total Score" space.

Apply the following classifications to the totaled score.

Fire Risk Ratings for Planned Developments

Score	Classification
<= 101	Low Risk
102-124	Moderate Risk
125-139	High Risk
140-158	Very High Risk
>= 159	Extreme Risk

The system provides a basic ranking system, which may be expanded using locally significant criteria, such as relative fire costs or cost per acre protected. We encourage all users to consider local factors which will help them clarify their evaluation.

DEFINITION OF TERMS

Alternative Access Route: A road that two-wheel drive vehicles can use as a substitute exit or entry road for the rating area.

Draft Source: A readily available source of water from which a person can draw water into a pump through a non-collapsible suction hose.

Fire Hazard: A fuel complex defined by kind, arrangement, volume, condition and location that forms a special threat of ignition or presents a suppression difficulty.

Fire Prevention: Activities to reduce the number of fires that start, including public education, law enforcement, and methods of engineering.

Fire Resistant Landscaping: Reducing or replacing flammable vegetation from around a building, thereby reducing the building's possible exposure to radiant heat. Flammable vegetation may be replaced with:

- ivy;
- green lawn;
- decorative stone;
- gardens that are kept damp;
- individually spaced green shrubs, of species that burn poorly;
- individually spaced and pruned trees, of species that burn poorly;
- other non-flammable or fire resistant materials.

Fire Resistant Roofing: Composition, metal, tile, concrete, slate rock, asphalt or fiberglass roofing that is classed A, B or C in the Uniform Building Code (UBC) Standard 32.7.

Fire Risk: The chance of a fire starting because there is a causative agent; the causative agent itself.

Fire Service Area (FSA): An area with legally defined boundaries, in which money is raised for fire protection by fees levied annually on structures rather than by a tax on land. To create a fire service area: at least 30 homeowners (or 51% if there are less than 30) in the proposed area must sign a petition for it; the county commissioners must hold a public hearing; and the county commissioners must pass a resolution creating the fire service area unless 50% of the homeowners protest.

Flow Testing: Using a gauge or visual inspection to measure a hydrant's capacity in gallons of water per minute.

DEFINITION OF TERMS

Fuel Type: An association of fuels that have common, similar or equivalent fire behavior or resistance to control.

Helicopter Dip Spot: An accessible water source large enough for helicopters to be able to fill external water buckets.

Hydrant: A discharge pipe with a valve and fittings at which water can be drawn for the purpose of fighting fires.

Loop Road: A secondary road which leaves a primary access road, circumscribes a given area, and then returns to the primary access road.

Pre-suppression: Activities that fire organizations conduct before a fire to make it easier to suppress the fire later on. Includes recruiting, training, planning, and getting and organizing equipment and supplies.

Primary Access Road: A main entry and exit road serving a rating area. Usually the road(s) that leads into the rating area from a highway, county road or major arterial. Must have an all-weather road surface (paved or gravel), have two twelve-foot traffic lanes, be maintained, and open year-around.

Rating Area: A rural or wildland area with enough permanent, seasonal, recreational or commercial development to pose a fire hazard or risk. Development should be relatively homogenous within the area and different enough from neighboring areas to be distinguished from them. The area should be large enough to take into account the surrounding fuels, topography, etc. that will affect fire occurrence.

Road Surface: The part of the road designed to carry vehicles, including the driving lanes, parking lanes, and any shoulders that can safely support vehicles. The road surface does not include turnouts, turnarounds, cleared but unsurfaced right-of-way, etc.

Rural Fire District (RFD): A district with legally defined boundaries, in which money is raised for fire protection by a tax on land and any improvements on it. To create a rural fire district: 50% or more of the landowners in the proposed district must sign a petition for it; the county commissioners must hold a public hearing; the county commissioners must approve the district. Once an RFD exists, the county commissioners appoint a board of trustees who either form a district fire company or contract with others to protect the RFD.

Secondary Road: A road which leaves a primary access road to reach homes, buildings, recreational sites, etc. that lie away from the primary road. Treat driveways over 600 feet as secondary roads.

DEFINITION OF TERMS

Statutory Limit: Load limits on bridges, enacted and regulated by statute. Usually 38 tons.

Volunteer Fire Company (VFC): A firefighting organization of up to 28 members that trains firefighters, acquires and houses firefighting apparatus, and fights fires. Usually, VFCs protect an area that is unincorporated and not legally defined, and are responsible for raising their own money. However, incorporated towns, rural fire districts, and county governing bodies can also form fire organizations which they call volunteer fire companies.

Wildland/Residential Interface: The area where homes, other buildings, or other human development meet or are scattered among wildland vegetation.

SOURCE LIST

The concepts and criteria that the risk rating system uses were drawn from these sources, and from DSL Fire Bureau experience. The list also provides further reading about fire risks in the wildland/residential interface.

California Department of Forestry, 1980. Fire Safe Guides for Residential Development in California. Sacramento, CA.

Colorado State Forest Service, 1974. Model Wildfire Hazard Area Control Regulations. Ft. Collins, CO.

Colorado State Forest Service. Wildfire Safety Guidelines for Subdivisions and Developments. Ft. Collins, CO.

Fisher, William C. and Brooks, David J., Safeguarding Montana's Homes: Lessons from the Pattee Canyon Fire. Western Wildlands, Missoula, MT., Summer 1977.

Montana Department of State Lands, 1992. Fire Protection Standards for Wildland Development in Montana. Missoula, MT.

National Fire Protection Association, 1985. Homes and Camps in Forest Areas 1985. NFPA Publication 224.

New Jersey Department of Environmental Protection, Division of Parks and Forestry, Bureau of Forest Fire Management. Miscellaneous correspondence during 1988. Trenton, NJ.

USDA Forest Service, Fire Safety Considerations for Developments in Forested Areas.

EXISTING DEVELOPMENT
FORM A - FIELD DATA COLLECTION FORM
(Rev. 3/93)

RATING AREA:______________ DATE:________ RATED BY:______________

1) NUMBER OF PRIMARY ACCESS ROADS ______

2) NUMBER OF ALTERNATIVE ACCESS ROUTES ______

3) WIDTH OF ROAD SURFACE + SHOULDER ON PRIMARY ACCESS ROADS ______

4) MAXIMUM ROAD GRADE IN THE AREA (PRIMARY, ALT., SECONDARY) ______

5) SECONDARY ROADS END AS:

Loops or > 90' Diameter Cul de Sacs ______

70-90' Diameter Cul de Sacs ______

< 70' Diameter Cul de Sacs ______

Dead Ends - No Cul de Sac ______

6) BRIDGES ON PRIMARY ACCESS ROADS ARE:

> 40 Ton Capacity ______

20-40 Ton Capacity ______

< 20 Ton Capacity ______

No Bridges ______

7) BRIDGES ON SECONDARY ROADS ARE:

20-40 Ton Capacity ______

< 20 Ton Capacity ______

No Bridges ______

8) PREDOMINANT SLOPE IN AND AROUND THE INHABITED AREA IS:

0 - 10% ______

11 - 20% ______

21 - 30% ______

> 30% ______

9) PREDOMINANT ASPECT IS:

North (316 degrees through 45 degrees) ______

East (46 degrees through 135 degrees) ______

Level ______

West (226 degrees through 315 degrees) ______

South (136 degrees through 225 degrees) ______

10) DANGEROUS TOPOGRAPHIC FEATURES PRESENT ARE:

None ______

Adjacent Steep Slopes ______

Draws/Ravines ______

Chimneys, Canyons, Saddles ______

11) PREDOMINANT FUEL TYPE IS:

Grass will be the main fuel type in the rating area around more than 90% of existing structures. ______

Low brush fields, or open timber stands will exist in the rating area around more than 10% of existing structures. ______

Dense timber stands or high brush fields will exist in the rating area around more than 10% of existing structures. ______

Slash and\or bugkilled timber stands will exist in rating area and won't be removed by development or dense stands of lodgepole pine trees will remain around more than 10% of existing structures. ______

12) **RISKS PRESENT ARE:**

Campgrounds/Campsites/Picnic Grounds ______

Children (playgrounds, schools, etc.) ______

Commercial Businesses ______

Debris Burning ______

Domestic Wood Heat ______

Farming/Ranching ______

Mills ______

Mines ______

Powerlines ______

Railroads ______

Recreation Sites (gun clubs, 4x4/motorbike areas, kegger sites, etc.) ______

Travel Routes (highways, etc.) ______

Other(s) - Describe each ______

______

______

13) **WORST-CASE ELECTRICAL SERVICE IS:**

All utilities in the existing development rating area are underground. ______

Rating area utilities will include underground and/or well maintained above ground powerlines with cleared rights-of-way. Trees or improvements which could blow over into powerlines do not exist or are properly maintained. ______

Rating area utilities include above ground powerlines. Fuel build-up is present in existing rights-of-way, or improvements exist which could blow over onto powerlines. ______

14) **HOW MANY HOMES ARE IN THE RATING AREA?** ______

15) **HOW MANY HOMES HAVE FIRE RESISTANT ROOFING?** ______

16) HOW MANY HOMES HAVE UNENCLOSED BALCONIES, DECKS, EAVES, STILTS, CANTILEVERED CONSTRUCTION, ETC.? ______

17) HOMES ARE SPACED:

> 100' Apart ______

60-100' ______

< 60' Apart ______

18) HOW MANY HOMES MEET THE FIRE-RESISTANT LANDSCAPING GUIDELINES (See Appendix F) ______

19) ARE HYDRANTS AVAILABLE? Yes______

No______

20) IF YES, AT WHAT SPACING? ______

21) IF YES, ARE THEY 500(+) GPM? Yes______

No______

22) DRAFT SOURCES ARE:

Accessible By Hoselay ______

Within 5 Miles Via Primary Access Roads ______

Available, But Need To Be Developed ______

Distant or Unavailable ______

23) HELICOPTER DIP SPOTS ARE:

Under 2 minute turnaround (< 1 mi.) ______

Within 2-5 minute turnaround (1-2 mi.) ______

Within 6 minute turnaround (3 mi.) ______

Distant or Unavailable ______

24) IS RATING AREA IN A RURAL FIRE DISTRICT, FIRE SERVICE AREA OR MUNICIPAL FIRE DEPARTMENT? Yes______

No______

25) FIRE DEPARTMENT RESPONSE:

Fire dept. can respond w/in 5 minutes - VFC? ______

Fire dept. can respond in 6-15 minutes - VFC? ______

Fire dept. can respond in 16-30 minutes - VFC? ______

26) IS THERE A WAY TO CONTACT HOMEOWNERS? Yes______

No______

27) IF YES, WHAT TYPE OF GROUP(S)?

Formal, Well Organized Group ______

Informal, Loosely Organized Group ______

Multiple Groups ______

28) AVERAGE NUMBER OF FIRES/1000 AC./10 YEARS ______

FORM B - RESIDENTIAL TALLY SHEET

RATING AREA______________________________

1 Total No. Residences	2 No. with Fire Resistant Roof	3 No. with Unenclosed Features	4 60' to Next Residence	5 60'-100' to next Residence	6 100' to Next Residence	7 Meets Landscaping Req. (Appendix F)

DOT OR LINE TALLY EACH ITEM.

1) TALLY TOTAL NUMBER OF RESIDENCES IN RATING AREA.
2) TALLY NUMBER OF RESIDENCES WITH FIRE RESISTANT ROOFING (COMPOSITE, METAL, TILE) NO WOOD SHAKES OR SHINGLES.
3) TALLY NUMBER OF RESIDENCES WITH OVERHANGING FEATURES WHICH ARE NOT ENCLOSED UNDERNEATH DECK OR FLOOR LEVEL (BALCONIES, DECKS, STILTS, ETC.)
4) TALLY NUMBER OF RESIDENCES WHICH HAVE LESS THAN 60 FEET BETWEEN THEM AND THE NEAREST ADJACENT RESIDENCE.
5) TALLY THE NUMBER OF RESIDENCES WHICH HAVE 60' - 100' BETWEEN THEM AND THE NEAREST ADJACENT RESIDENCE.
6) TALLY THE NUMBER OF RESIDENCES WITH 100' BETWEEN THEM AND THE NEAREST ADJACENT RESIDENCE.
7) TALLY THE NUMBER OF RESIDENCES THAT MEET THE FIRE RESISTANT LANDSCAPING STANDARDS FOR THEIR LOCATION.

APPENDIX B

EXISTING DEVELOPMENT
FORM C - RATING FORM
(Rev. 3/93)

RATING AREA:______________________ DATE:_________ RATED BY:________________________

ROADS

ROAD ACCESS - Items 1 and 2

- Multiple primary access roads = 0
- Two primary access roads = 1
- One primary + one alternative access road = 2
- One-way in/out = 3
- No primary access roads = 4

ROAD SURFACE WIDTH, PRIMARY ACCESS ROUTES - Item 3

- > 28' Road Surface + Shoulder = 1
- 28' Road Surface + Shoulder = 2
- 16 - < 28' Road Surface + Shoulder = 3
- < 16' Road Surface + Shoulder = 4

MAXIMUM ROAD GRADE - Item 4

- 0-5% = 1
- 6-8% = 2
- > 8 - 10% = 3
- > 10% = 4

SECONDARY ROAD ENDINGS - Item 5

- Loops or > 90' Diameter Cul de Sacs = 1
- Cul de Sac Diameter 70-90' = 2
- Cul de Sac Diameter <70' = 3
- Dead Ends - No Cul de Sac = 4

BRIDGES - Items 6 and 7

- No Bridges = 1
- 40 Ton(+) limit on access bridges = 2
- 20-39 Ton limit on all access bridges = 3
- < 20 Ton limit any access bridge = 4

TOPOGRAPHY

SLOPE - Item 8

- 0-10% = 2
- 11-20% = 4
- 21-30% = 6
- > 30% = 8

ASPECT - Item 9

- North (315 degrees through 45 degrees) = 0
- East (46 degrees through 135 degrees) = 1
- Level = 2
- West (226 degrees through 315 degrees) = 3
- South (136 degrees through 225 degrees) = 4

MOST DANGEROUS FEATURE - Item 10

- None = 2
- Adjacent Steep Slopes = 4
- Draws/Ravines = 6
- Chimneys, Canyons, Saddles = 8

FUELS

FUEL TYPE - Item 11

- Grass around >90% of structures = 5
- Low brush field, or open timber around >10% of structures = 10
- Dense conifer or brush field exist around >10% of structures = 15
- Slash, bugkill, dense lodgepole pine exist around >10% of sructures = 20

RISK SOURCES - total from Item 12

- 0-4 Risk Sources Present = 5
- 5-8 Risk Sources Present = 10
- 9-12 Risk Sources Present = 15
- 13+ Risk Sources Present = 20

ELECTRICAL UTILITIES - Item 13

- All Underground = 0
- Above Ground/Underground Combination (Well Maintained) = 10
- Above Ground (Poorly Maintained) = 20

HOMES

ROOF MATERIAL - Item 15

- 90-100% of homes have metal, composition, tile or other fire resistant roofing = 5
- 80-89% of homes have metal, composition, tile or other fire resistant roofing = 10
- 75-79% of homes have metal, composition, tile or other fire resistant roofing = 15
- < 75% of homes have metal, composition, tile or other fire resistant roofing = 20

UNENCLOSED BALCONIES, DECKS, EAVES, STILTS, ETC. - Item 16

- < 10% of homes have unenclosed balconies, decks, eaves, stilts, etc. = 1
- 10-20% of homes have unenclosed balconies, decks, eaves, stilts, etc. = 2
- 21-25% of homes have unenclosed balconies, decks, eaves, stilts, etc. = 3
- > 25% of homes have unenclosed balconies, decks, eaves, stilts, etc. = 5

DENSITY OF HOMES - Item 17

- (For 0-30% slope)
  - > 100' between homes = 1
  - 60-100' between homes = 3
  - < 60' between homes = 5

- (For 31-50% slope)
  - > 100' between homes = 2
  - 60'100' between homes = 4
  - < 60' between homes = 6

LANDSCAPING - Item 18

- 76-100% homes meet the fire-resistant landscaping guidelines in the Appendix F = 2
- 51-75% homes meet the fire-resistant landscaping guidelines in the Appendix F = 4
- 26-50% homes meet the fire-resistant landscaping guidelines in the Appendix F = 6
- 0-25% homes meet the fire-resistant landscaping guidelines in the Appendix F = 9

WATER SUPPLY

HYDRANTS - Items 19, 20 and 21

- 500 GPM hydrants available on < 660' spacing = 2
- 500 GPM hydrants available = 4
- < 500 GPM hydrants available = 6
- No hydrants = 8

DRAFT SOURCES - Item 22

- Accessible Sources Available Within Hoselay Distance = 2
- Draft Sources Available Within 5 mi. via primary access roads = 4
- Draft Sources Require Development = 6
- Draft Sources Unavailable = 8

HELICOPTER DIP SPOTS - Item 23

- Under 2 min. turnaround (<1 mi.) = 1
- Within 2-5 min. turnaround (1-2 mi.) = 2
- Within 6 min. turnaround (3 mi.) = 3
- Beyond 6 min. turnaround or Unavailable = 4

STRUCTURAL FIRE PROTECTION - Items 24 and 25

- <= 5 min. from fire department = 5; if VFC = 10
- 6-15 min. from fire department = 10; if VFC = 15
- 16-30 min. from fire department = 15; if VFC = 20
- No RFD, FSA, municipal fire district or VFC? = 20

HOMEOWNER CONTACT - Items 26 and 27

- Central contact - formal/well organized group (e.g., a homeowners assoc.) = 5
- Less central contact - an informal/loosely organized group (e.g., a civic club or development office) = 10
- Multiple groups - different contacts representing different parts of the community = 15
- No organized contacts = 20

FIRE OCCURRENCE - Item 28

- .00-.10 Fires/1000 ac./10 yr. = 5
- .11-.20 Fires/1000 ac./10 yr. = 10
- .21-.40 Fires/1000 ac./10 yr. = 15
- .40+ Fires/1000 ac./10 yr. = 20

TOTAL SCORE ______

<= 110	low risk - low priority
111-135	moderate risk - moderate priority
136-150	high risk - high priority
151-170	very high risk - very high priority
>= 171	extreme risk - extreme priority

RISK RATING PLANNED DEVELOPMENT
FORM A - DATA COLLECTION FORM
(Rev. 3/93)

RATING AREA:____________________DATE:__________RATED BY:_______________________

1) **NUMBER OF PRIMARY ACCESS ROADS** ________

2) **NUMBER OF ALTERNATIVE ACCESS ROUTES** ________

3) **WIDTH OF ROAD SURFACE + SHOULDER ON PRIMARY ACCESS ROADS** ________

4) **MAXIMUM ROAD GRADE IN THE AREA (PRIMARY, ALT., SECONDARY)** ________

5) **SECONDARY ROADS END AS:**

Loops or 90'+ Diameter Cul de Sacs ________

70-89' Diameter Cul de Sacs or Hammerhead "T" (40' Min.) ________

< 70' Diameter Cul de Sacs ________

Dead Ends - No Cul de Sac ________

6) **BRIDGES ON PRIMARY ACCESS ROADS ARE:**

> 40 Ton Capacity ________

20-40 Ton Capacity ________

< 20 Ton Capacity ________

No Bridges ________

7) **BRIDGES ON SECONDARY ROADS ARE:**

20-40 Ton Capacity ________

< 20 Ton Capacity ________

No Bridges ________

8) PREDOMINANT SLOPE IN AND AROUND THE INHABITED AREA IS:

0 - 10% ______

11 - 20% ______

21 - 30% ______

> 30% ______

9) PREDOMINANT ASPECT IS:

North (316 degrees through 45 degrees) ______

East (46 degrees through 135 degrees) ______

Level ______

West (226 degrees through 315 degrees) ______

South (136 degrees through 225 degrees) ______

10) DANGEROUS TOPOGRAPHIC FEATURES PRESENT ARE:

None ______

Adjacent Steep Slopes ______

Draws/Ravines ______

Chimneys, Canyons, Saddles ______

11) PREDOMINANT FUEL TYPE IS:

Grass will be the main fuel type in the rating area around more than 90% of planned structures. ______

Low brush fields, or open timber stands will exist in the rating area around more than 10% of planned structures. ______

Dense timber stands or high brush fields will exist in the rating area around more than 10% of planned structures. ______

Slash and\or bugkilled timber stands will exist in rating area and won't be removed by development or dense stands of lodgepole pine trees will remain around more than 10% of planned structures. ______

12) RISKS PRESENT ARE:

Campgrounds/Campsites/Picnic Grounds ______

Children (playgrounds, schools, etc.) ______

Commercial Businesses ______

Debris Burning ______

Domestic Wood Heat ______

Farming/Ranching ______

Mills ______

Mines ______

Powerlines ______

Railroads ______

Recreation Sites (gun clubs, 4x4/motorbike areas, kegger sites, etc.) ______

Travel Routes (highways, etc.) ______

Other(s) - Describe each ______

13) WORST-CASE ELECTRICAL SERVICE IS:

All utilities planned for development or existing in rating area are underground. ______

Rating area utilities will include well maintained above ground powerlines with cleared rights-of-way. Trees or improvements which could blow over into powerlines do not exist or are properly maintained. ______

Rating area utilities include above ground powerlines. Fuel build-up is present in existing/planned rights-of-way, or improvements exist which could blow over onto powerlines. ______

14) HOW MANY HOMES ARE PLANNED FOR THE DEVELOPMENT? ________

15) HOW MANY HOMES WILL HAVE FIRE RESISTANT ROOFING? ________

16) HOMES ARE SPACED:

> 100' Apart ________

60 - 100' Apart ________

< 60' Apart ________

17) HOW MANY HOMES WILL MEET THE FIRE-RESISTANT LANDSCAPING GUIDELINES (See Appendix F) ________

18) DOES THE PROJECT DESIGN INCORPORATE GREENBELTS OR FUEL BREAKS? Yes ________ No ________

19) WILL HYDRANTS BE AVAILABLE? Yes ________ No ________

20) IF YES, AT WHAT SPACING? ________

21) IF YES, ARE THE HYDRANTS PLACED AS DESIRED BY THE FIRE DEPARTMENT? Yes ________ No ________

22) IF HYDRANTS PLANNED, ARE THEY 500(+) GPM? Yes ________ No ________

23) DRAFT SOURCES ARE:

Accessible By Hoselay ________

Within 5 Miles Via Primary Access Roads ________

Available, But Need To Be Developed ________

Distant or Unavailable ________

24) HELICOPTER DIP SPOTS ARE:

Under 2 minute turnaround (< 1 mi.) ________

Within 2-5 Minute turnaround (1-2 mi.) ________

Within 6 minute turnaround (3 mi.) ________

Distant or Unavailable ________

25) IS RATING AREA IN A RURAL FIRE DISTRICT, FIRE SERVICE AREA OR MUNICIPAL FIRE DEPARTMENT? **Yes** ________

No ________

26) Fire Dept. can respond w/in 5 minutes - VFC? ________

Fire Dept. can respond in 6-15 minutes - VFC? ________

Fire dept. can respond in 16-30 minutes - VFC? ________

27) WILL THERE BE A WAY TO CONTACT HOMEOWNERS? **Yes** ________

No ________

28) IF YES, WHAT TYPE OF GROUP(S)?

Formal, Well Organized Group ________

Informal, Loosely Organized Group ________

Multiple Groups ________

29) AVERAGE NUMBER OF FIRES/1000 AC./10 YEARS ________

RISK RATING OF PLANNED DEVELOPMENT
FORM B - RATING FORM
(Rev. 3/93)

RATING AREA:____________________ DATE:__________ RATED BY:________________________

ROADS

ROAD ACCESS - Items 1 and 2

- Multiple primary access roads = 0
- Two primary access roads = 1
- One primary + one alternative access road = 2
- One-way in/out = 3
- No primary access roads = 4

ROAD SURFACE WIDTH, PRIMARY ACCESS ROUTES - Item 3

- > 28’ Road Surface + Shoulder = 1
- 28’ Road Surface + Shoulder = 2
- 16 - < 28’ Road Surface + Shoulder = 3
- < 16’ Road Surface + Shoulder = 4

MAXIMUM ROAD GRADE - Item 4

- 0-5% = 1
- 6-8% = 2
- > 8-10% = 3
- > 10% = 4

SECONDARY ROAD ENDINGS - Item 5

- Loops or 90’+ Diameter Cul de Sacs = 1
- Cul de Sac Diameter 70-89’ or Hammerhead "T" (40’ Min.) = 2
- Cul de Sac Diameter < 70’ = 3
- Dead Ends - No Cul de Sac = 4

BRIDGES - Items 6 and 7

- No Bridges = 0
- 40 Ton (+) limit on access bridges = 1
- 20-39 Ton limit on all access bridges = 2
- < 20 Ton limit any access bridge = 4

TOPOGRAPHY

SLOPE - Item 8

- 0-10%	= 2
- 11-20%	= 4
- 21-30%	= 6
- > 30%	= 8

ASPECT - Item 9

- North (315 degrees through 45 degrees)	= 0
- East (46 degrees through 135 degrees)	= 1
- Level	= 2
- West (226 degrees through 315 degrees)	= 3
- South (136 degrees through 225 degrees)	= 4

MOST DANGEROUS FEATURE - Item 10

- None	= 2
- Adjacent Steep Slopes	= 4
- Draws/Ravines	= 6
- Chimneys, Canyons, Saddles	= 8

FUELS

FUEL TYPE - Item 11

- Grass around >90% of structure	= 5
- Low brush field, or open timber around >10% of structure	= 10
- Dense conifer or brush field exist around >10% of structures	= 15
- Slash, bugkill, dense lodgepole pine exist around >10% of sructures	= 20

RISK SOURCES - Total from Item 12

- 0 - 4 Risk Sources Present	= 0
- 5 - 8 Risk Sources Present	= 5
- 9 -12 Risk Sources Present	= 7
- 13+ Risk Sources Present	= 10

ELECTRICAL UTILITIES - Item 13

- All Underground = 0
- Above Ground/Underground Combination (Well Maintained) = 5
- Above Ground (Poorly Maintained) = 10

STRUCTURES

ROOF MATERIAL - Item 14 AND 15

- 90-100% of homes have metal, composition, tile or other fire resistant roofing = 0
- 80-89% of homes have metal, composition, tile or other fire resistant roofing = 5
- 75-79% of homes have metal, composition, tile or other fire resistant roofing = 7
- < 75% of homes have metal, composition, tile or other fire resistant roofing = 10

DENSITY OF HOMES - Item 16

- (For 0-30% slope)
  - > 100' between homes = 0
  - 60-100' between homes = 4
  - < 60' between homes = 8

- (For 31-50% slope)
  - > 100' between homes = 2
  - 60-100' between homes = 6
  - < 60' between homes = 10

LANDSCAPING - Item 14 and 17

- 76-100% homes meet the fire-resistant landscaping guidelines in Appendix F = 5
- 51-75% homes meet the fire-resistant landscaping guidelines in Appendix F = 10
- 26-50% homes meet the fire-resistant landscaping guidelines in Appendix F = 15
- 0-25% homes meet the fire-resistant landscaping guidelines in Appendix F = 20

GREENBELTS AND FUELBREAKS - Item 18

- Project design incorporates greenbelts and/or fuel breaks = 0
- Project design does not incorporate greenbelts and/or fuel breaks = 10

WATER SUPPLY

HYDRANTS - Items 19, 20, 21 and 22

- 500 GPM hydrants available on < 660' spacing placed as desired by FD = 0
- 500 GPM hydrants available on < 660' spacing = 2
- 500 GPM hydrants available = 4
- < 500 GPM hydrants available = 6
- No hydrants = 8

DRAFT SOURCES - Item 23

- Accessible Sources Available Within Hoselay Distance = 2
- Draft Sources Available Within 5 mi. via primary access roads = 4
- Draft Sources Require Development = 6
- Draft Sources Unavailable = 8

HELICOPTER DIP SPOTS - Item 24

- Under 2 min. turnaround (<1 mi.) = 1
- Within 2-5 min. turnaround (1-2 mi.) = 2
- Within 6 min. turnaround (3 mi.) = 3
- Beyond 6 min. turnaround or unavailable = 4

STRUCTURAL FIRE PROTECTION - Items 25 and 26

- < = 5 min. from fire department = 5; if VFC = 10
- 6-15 min. from fire department = 10; if VFC = 15
- 16-30 min. from fire department = 15; if VFC = 20
- No RFD, FSA, municipal fire district or VFC? = 20

HOMEOWNER CONTACT - Items 27 and 28

-	Central contact - formal/well organized group (e.g., a homeowners assoc.)	= 5
-	Less central contact - an informal/loosely organized group (e.g., a civic club or development office)	= 10
-	Multiple groups - different contacts representing different parts of the community	= 15
-	No organized contacts	= 20

FIRE OCCURRENCE - Item 29

-	.00-.10 Fires/1000 ac./10 yr.	= 5
-	.11-.20 Fires/1000 ac./10 yr.	= 10
-	.21-.40 Fires/1000 ac./10 yr.	= 15
-	.40+ Fires/1000 ac./10 yr.	= 20

TOTAL SCORE ________

< = 101	low risk - low priority
102 - 124	moderate risk - moderate priority
125 - 139	high risk - high priority
140 - 158	very high risk - very high priority
> = 159	extreme risk - extreme priority

VEGETATION REDUCTION STANDARDS
0% TO 10% SLOPE

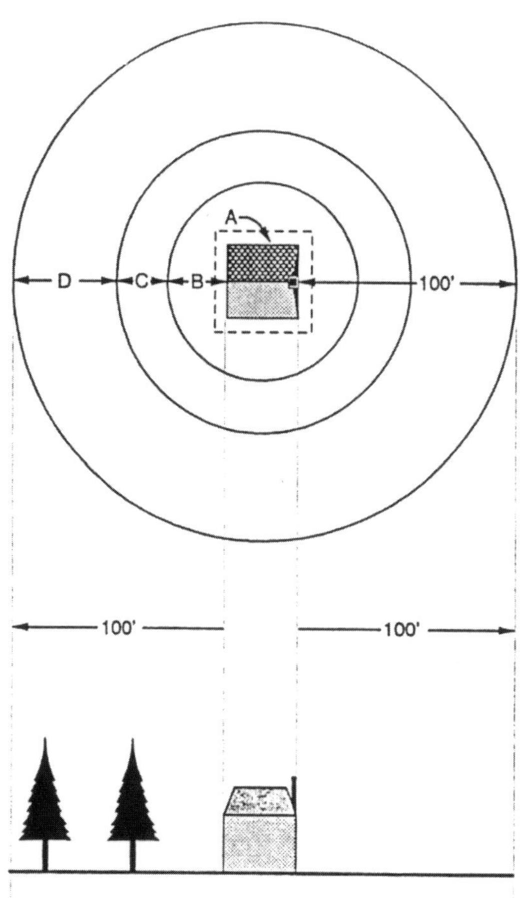

A = THE FIRST 3 FEET OF B
- Maintain an area of non-combustible material - flowers, plants, concrete, gravel, mineral soil etc..

B = 10 FEET
- Remove all trees and downed woody fuels.

C = 20 FEET
- Thin trees out to 10 feet between their crowns.
- Prune the limbs of all remaining trees to 15 feet or one third the total height whichever is less.
- Maintain surface vegetation at 3 inches or less.
- Remove all downed wood fuels.

D = 70 FEET
- Thin trees out to 10 feet between their crowns
- Prune the limbs of all remaining trees to 15 feet or one third the total height whichever is less.
- Maintain surface vegetation at 12 inches or less.
- Remove all downed woody fuels more than 3 inches in diameter.

VEGETATION REDUCTION STANDARDS
10% TO 20% SLOPE

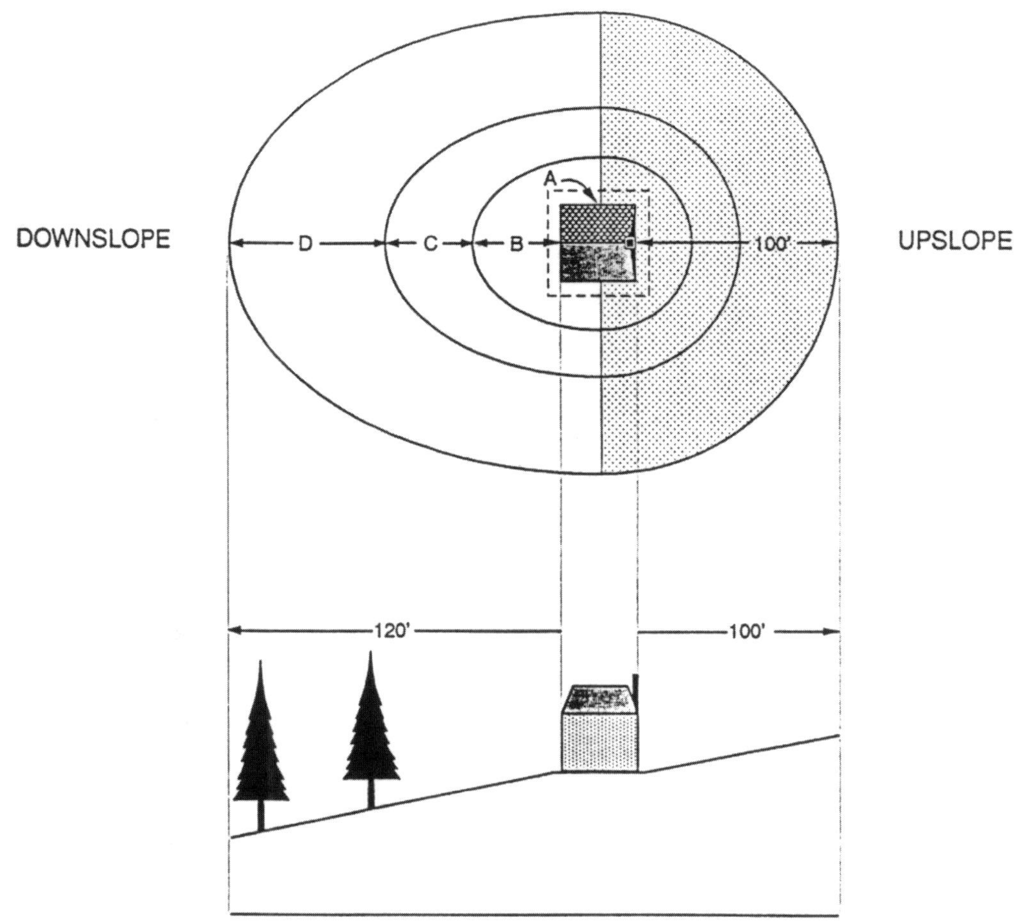

The shaded areas (up-slope) of B, C, D remain a constant distant of 10' 20' and 70' respectively. The shaded area begins from the mid-section of a structure. The unshaded areas (down-slope) of B, C, and D increase with slope as detailed below:

A = THE FIRST 3 FEET OF B

- Maintain an area of non-combustible material - flowers, plants, concrete, gravel, mineral soil etc..

B = 15 FEET

- Remove all trees and downed woody fuels.

C = 25 FEET

- Thin trees out to 10 feet between their crowns.
- Prune the limbs of all remaining trees to 15 feet or one third total height whichever is less.
- Maintain surface vegetation at 3 inches or less.
- Remove all downed wood fuels.

D = 80 FEET

- Thin trees out to 10 feet between their crowns
- Prune the limbs of all remaining trees to 15 feet or one third total height whichever is less.
- Maintain surface vegetation at 12 inches or less.
- Remove all downed woody fuels more than 3 inches in diameter.

Appendix F - Page 2

VEGETATION REDUCTION STANDARDS
20% TO 30% SLOPE

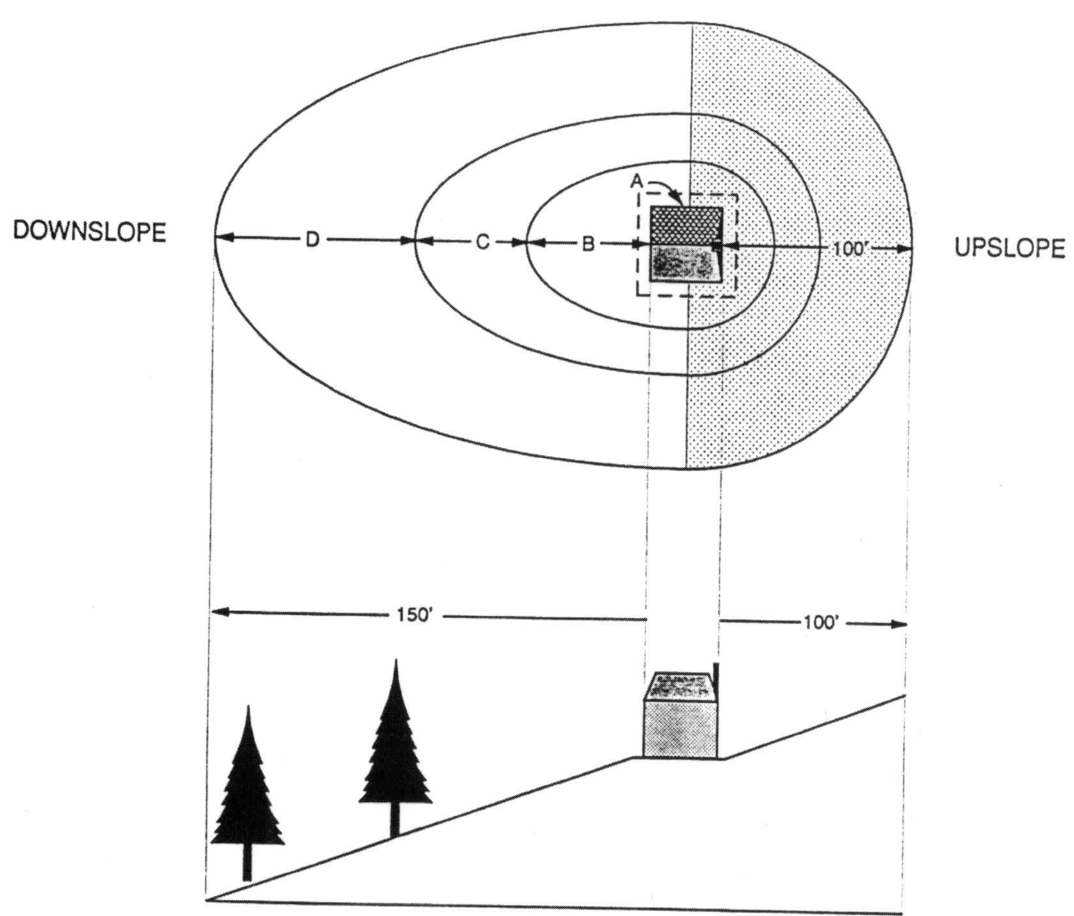

The shaded areas (up-slope) of B, C, D remain a constant distant of 10' 20' and 70' respectively. The shaded area begins from the mid-section of a structure. The unshaded areas (down-slope) of B, C, and D increase with slope as detailed below:

A = THE FIRST 3 FEET OF B

- **Maintain an area of non-combustible material - flowers, plants, concrete, gravel, mineral soil etc..**

B = 20 FEET

- **Remove all trees and downed woody fuels.**

C = 30 FEET

- **Thin trees out to 10 feet between their crowns.**
- **Prune the limbs of all remaining trees to 15 feet or one third total height whatever is less.**
- **Maintain surface vegetation at 3 inches or less.**
- **Remove all downed wood fuels.**

D = 100 FEET

- **Thin trees out to 10 feet between their crowns**
- **Maintain surface vegetation at 12 inches or less.**
- **Remove all downed woody fuels more than 3 inches in diameter.**

VEGETATION REDUCTION STANDARDS
THINNING AND PRUNING GUIDE

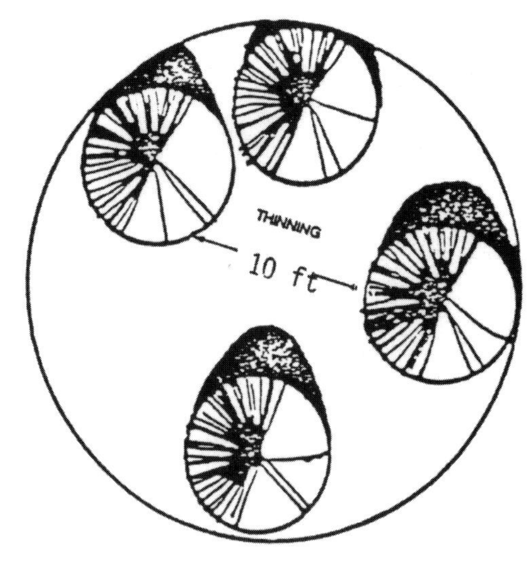

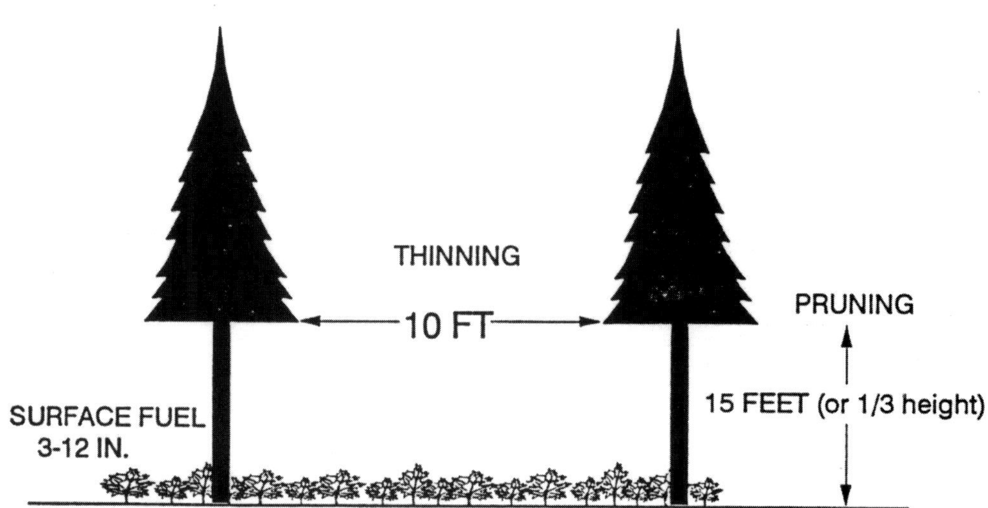

In areas where vegetation modification is prescribed use the following guidelines:

A. **THINNING**
- **Thin trees out to 10 feet between their crowns.**

B. **PRUNING**
- **Prune the limbs of all remaining trees to 15 feet or one third total height whichever is less.**

C. **SURFACE VEGETATION**
- **Maintain surface vegetation at 3" to 12" as detailed.**

Appendix E
Wildland Firefighter 3 Qualifications

NFPA 1051
Standard for Wildland Fire Fighter Professional Qualifications
1995

Chapter 1 Administration

1-4 **Definitions.**
Wildland Fire Fighter III. The person responsible for supervising and directing a single wildland fire suppression resource, such as a hand crew or an engine. The Wildland Fire Fighter III has met the requirements defined in Chapter 5.

Chapter 5 Wildland Fire Fighter III

5-1 **General.** For certification at the Wildland Fire Fighter III level, the Wildland Fire Fighter II shall meet the job performance requirements defined in Sections 5-1 to 5-5 of this standard.

5-1.1 *Prerequisite Knowledge:* Wildland Fire Fighter III's role within the incident management system.

5-2 **Human Resource Management.**

5-2.1 **Description of Duty.** Utilize human resources to accomplish assignments in a safe and efficient manner and supervise personnel during out-of-service periods.

5-2.1.1* *Prerequisite Knowledge:* Basic human resource management, principles of supervision.

5-2.2 Evaluate the physical and mental condition of assigned personnel, given agency personnel performance standards, so that members are capable of performing assigned tasks safely and individuals not meeting the standards are referred to the Wildland Fire Fighter III's supervisor.

5-2.2.1 *Prerequisite Knowledge:* Agency personnel performance standards, crew evaluation.

5-2.3 Verify the qualifications of assigned personnel, given an assignment or task, so that individual fire fighter skills are appropriate to the specific assignment, and deficiencies are identified and reported to the Wildland Fire Fighter III's supervisor.

5-2.3.1 *Prerequisite Knowledge:* Knowledge of the agency's qualifications standards.

5-2.4* Communicate with assigned personnel, given an incident action plan, so that they are informed of pertinent fire assignment and safety information throughout the assignment.

5-2.5 Provide for medical treatment, given an injured or ill crew member and agency policies and procedures, so that the appropriate emergency treatment is provided, evacuation is arranged for if necessary, the Wildland Fire Fighter III's supervisor is notified, and required reports are completed.

5-2.5.1 *Prerequisite Knowledge:* Medical evacuation procedures, agency accident and illness reporting procedures.

5-2.6 Evaluate job performance of assigned personnel, given agency standards, so that the information is provided to the individual being evaluated and all required forms are completed.

5-2.6.1 *Prerequisite Knowledge:* Basic performance evaluation.

5-2.7 Monitor assigned personnel during out-of-service periods, given agency guidelines for out-of-service periods, so that the personnel are physically and mentally prepared for the next assignment.

5-3 **Presuppression.** No job performance requirements at this level.

5-4 **Mobilization.**

5-4.1 **Definition of Duty.** Collection of agency dispatch information, assembly of assigned resources, travel to designated location, and check-in.

5-4.2 Obtain complete information from agency dispatch, given agency standard operating procedures, so that travel route, assignment, time needed, and point of contact are determined.

5-4.2.1 *Prerequisite Knowledge:* Agency dispatch, travel, and accident and equipment breakdown procedures.

5-5 **Suppression.**

5-5.1 **Definition of Duty.** Supervision of a single resource involved in wildland fire suppression.

5-5.2 Size up an incident, given fuels, topography, and weather conditions, so that an incident action plan is developed for fire control with an emphasis on personnel safety.

5-5.3* Develop an initial fire report, given incident information and agency policies and procedures, so that required incident information is communicated to the agency communications center upon arrival.

5-5.4 Deploy resources to suppress a wildland fire, given an assignment, personnel, equipment, and agency policies and procedures, so that appropriate suppression actions are taken, agency policies related to special interest areas are met, and safety of personnel is ensured.

5-5.4.1* *Prerequisite Knowledge:* Fireline construction and location, burning out techniques, air operations, use of heavy equipment, radio communications protocol, and deployment of crew in fireline construction to handle specific problems.

5-5.4.2 *Prerequisite Skills:* Heavy equipment and aircraft use and use of communications equipment.

5-5.5 Evaluate incident conditions, given a wildland fire, so that progress, changes in fuels, topography, weather, fire behavior, and other significant events are identified and communicated to the supervisor, and to assigned and adjoining personnel.

5-5.5.1* *Prerequisite Knowledge:* Intermediate fire behavior.

5-5.5.2 *Prerequisite Skills:* Collect wildland fire weather information.

5-5.6 Communicate with supervisors, crew members, and adjoining personnel, given a wildland fire incident, so that progress, changes in conditions, fire behavior, and other significant events are current.

5-5.7 Deploy resources to mop up a wildland fire, given a controlled fire, personnel, equipment, and agency policies and procedures, so that appropriate mop up actions are taken and agency policies related to special interest areas are met.

5-5.8 Complete wildland fire suppression operations, given a wildland fire that has been controlled and mopped up and agency policies and procedures, so that the fire area is secure and resources are returned to service.

Appendix A Explanatory Material

A-5-2.1.1 See NWCG S-201, *Supervisory Concepts and Techniques* (1994) and NFPA 1021, *Standard for Fire Officer Professional Qualifications,* for additional supervisory information.

A-5-2.4 See NWCG PMS 410-1, *Fireline Handbook,* for additional information on briefing and communicating with crew members.

A-5-5.3 Basic information in a report should include:
(a) fire name,
(b) location,
(c) access,

(d) terrain,
(e) size of fire,
(f) anticipated control problems,
(g) cause (known, suspected),
(h) values threatened,
(i) anticipated time of control,
(j) weather,
(k) resources on the fire,
(l) resources needed, if any,
(m) fire behavior.

A-5-5.4.1 Additional information on the use of heavy equipment and aircraft can be found in NWCG S-213, *Tractor Use/Tractor Boss;* S-214, *Tanker Use/Tanker Boss;* S-230, *Single Resource Boss-Crew;* S-234, *Firing Methods and Procedures;* and S-270, *Basic Air Operations.*

A-5-5.5.1 For additional information regarding intermediate fire behavior, see NWCG S-290, *Intermediate Wildland Fire Behavior* and PMS 427, *Look Up, Look Down, Look Around.*

Appendix B NWCG Publications for Wildland Fire Suppression Personnel

Wildland Fire Fighter III

S-213, Tractor Use/Tractor Boss

S-214, Tanker Use/Tanker Boss

S-230, Single Resource Boss

S-260, Fire Business Management Principles

S-270, Basic Air Operations

S-290, Intermediate Fire Behavior

I-2000, Intermediate Incident Command System

P-151, Wildfire Origin and Cause Determination

Appendix F
Wildland Firefighter 4 Qualifications

NFPA 1051
Standard for Wildland Fire Fighter Professional Qualifications
1995

Chapter 1 Administration

1-4 Definitions.

Wildland Fire Fighter IV. The person responsible for managing all aspects of a wildfire incident, involving relatively few resources, who has met the requirements defined in Chapter 6. Resources vary from a single fire fighter to several single resources possibly of different types or kinds.

Chapter 6 Wildland Fire Fighter IV

6-1 General.

6-1.1 For certification at the Wildland Fire Fighter IV level, the Wildland Fire Fighter III shall meet the job performance requirements defined in Sections 6-1 to 6-5 of this standard.

6-1.1.1* *Prerequisite Knowledge:* Authority and responsibility of cooperating jurisdictional agencies, formal and informal agreements between jurisdictional agencies, the incident management system used by the jurisdiction, and the Wildland Fire Fighter IV's role within that system.

6-1.2* Respond to requests for incident information from the news media, given policies and procedures of the authority having jurisdiction, so that response is accurate, appropriate, and timely.

6-1.2.1 *Prerequisite Knowledge:* Agency procedures for media relations.

6-2 **Human Resource Management.** No additional job performance requirements at this level.

6-3 **Presuppression.** No additional job performance requirements at this level.

6-4 **Mobilization.** No additional job performance requirements at this level.

6-5 **Suppression.**

6-5.1 **Definition of Duty.** Command and overall management of resources in the suppression of a wildland fire.

6-5.2* Formulate an incident action plan, given a wildland fire and available resources, so that incident objectives are set and strategies and tactics are applied according to agency policies and procedures.

6-5.2.1* *Prerequisite Knowledge:* Availability and capability of resources and agency suppression priorities based on values at risk.

6-5.2.2* *Prerequisite Skill:* Identification of values at risk, objective setting, and selection of appropriate wildland fire suppression strategies.

6-5.3* Establish an incident command post (ICP), given authority having jurisdiction policies and procedures, so that the location is appropriate, properly identified, and communicated to personnel.

6-5.4* Maintain incident records, given agency policies and procedures and applicable forms, so that required information is documented.

6-5.4.1 *Prerequisite Knowledge:* Agency incident documentation procedures.

6-5.5* Arrange for the logistical support needs for assigned resources, given a wildland fire, assigned resources, and agency policies and procedures, so that personnel and equipment needs are met in accordance with agency policies.

6-5.6 Analyze incident needs, given assigned resources and incident status, so that excess resources are identified and released in accordance with agency policies and procedures.

6-5.6.1 *Prerequisite Knowledge:* Authority having jurisdiction policies and procedures related to resource release.

6-5.7* Provide incident information to the incoming replacement incident commander, given an extended attack wildland fire, so that the transition of command is completed and the new incident commander has the information necessary to operate.

6-5.7.1 *Prerequisite Knowledge:* Authority having jurisdiction policies and procedures for transition in command.

6-5.8 Complete personnel time and equipment use records, given agency policies, procedures, and related forms, so that the information is accurate and in compliance with standards established by the authority having jurisdiction.

6-5.8.1* *Prerequisite Knowledge:* Basic wildland fire business management.

6-5.8.2 *Prerequisite Skills:* Written communication.

6-5.9* Prepare final incident reports, given an extinguished wildland fire and agency policies and procedures, so that the reports are complete, accurate, and submitted on time.

Appendix A Explanatory Material

A-6-1.1.1 The Wildland Fire Fighter IV is responsible for all initial attack incident activities, including the development and implementation of strategic decisions; approving the ordering and release of resources; and for coordination with other agencies, jurisdictions, and affected entities and organizations.

In instances involving multi-agency response, unified command may be used. Unified command is an element of an incident management system that provides a method for all agencies, or individuals who have jurisdictional or functional responsibility, to jointly manage an incident through a common set of objectives. See NWCG pamphlet and video *Unified Command,* NFES 1466 and 1507.

A-6-1.2 The Wildland Fire Fighter IV is expected to respond to media inquires on a limited basis. It is essential that the Wildland Fire Fighter IV can determine when to refer media requests to the proper authority within the agency. NWCG *Strategic Communications for Wildland Fire Management* training package (NFES 2252, 2253, 2254, 2255, 2265, 2266, 2267) provides information on this subject.

A-6-5.2 The initial incident action plan is based on size up of the situation, including such things as:

(a) analyzing hazards to fire fighters,
(b) estimated rates of spread,
(c) actual and potential threat to values,
(d) incident objectives in priority order,
(e) strategies for protecting values and for suppressing the fire.

The plan outlines the most appropriate method of attack with the resources available, including such things as:

(a) tactical directions to achieve incident objectives,
(b) a coordinated sequence of events,

(c) resource assignments,
(d) immediate support needs.

It emphasizes safety information covering all hazards and relevant safety principles.
ICS Form 201 is the first written documentation for the incident and serves as a briefing document for incoming incident commanders.
Reference NWCG 410-1, *Fireline Handbook,* Chapter 1, Initial Attack.

A-6-5.2.1 Suppression priorities should consider values at risk such as life, property, natural resources, special interest areas, etc. Additional information about suppression resource capabilities is found in PMS 410-1, *Fireline Handbook.*

A-6-5.2.2 See S-200, *Initial Attack Incident Commander;* and S-205, *Fire Operations in the Urban Interface,* for additional information.

A-6-5.3 In many incidents of this size, the ICP could be a vehicle or other easily identified point. The location of the ICP should be determined with due consideration given to safety, access to incoming resources, and communications needs.

A-6-5.4 Incident records should contain essential incident information such as:

(a) incident objectives and strategy,
(b) changes in the situation,
(c) tactical decisions,
(d) resource summary,
(e) organization,
(f) sketch map,
(g) safety problems and hazards,
(h) equipment use, and
(i) other agency information needs.

The ICS Form 201 (incident briefing) is an example of a format to document this information. In many initial actions, control is effected very quickly and the need for written documentation is minimal. However, in longer duration incidents, it is much more important, and, if the incident expands into an extended attack or more complex situation, good documentation is essential to provide for an efficient transition.

A-6-5.5 Logistical needs for initial action incidents are usually minimal, however some items to consider are:

(a) water supply,
(b) fuel for equipment, and
(c) meals and drinking water.

Should the initial action continue for extended periods of time, additional logistical requirements could include:

(a) sanitation facilities,
(b) lighting,
(c) provisions for sleeping, and
(d) relief resources.

A-6-5.7 Early recognition that a wildland fire will not be controlled by the initial attack forces is important. As soon as the Wildland Fire Fighter IV recognizes that additional forces are needed or knows additional forces are enroute, he or she must withdraw from direct fireline suppression and:

(a) Establish an incident command post (ICP)/check-in point to receive, brief, and assign incoming forces.

(b) Document incident status and resource information in writing (for example: ICS Briefing Form 201).

(c) Sketch a map of the fire and identify resource assignments.

(d) Document the fire organization.

(e) Keep track of all resources that are on the scene, enroute, and ordered.

(f) Document strategy, tactics, and current actions.

(g) If available, assign a status/check-in recorder to handle this documentation when:
1. The fire is expanding rapidly.
2. Numerous resources are arriving or are being ordered.
3. Radio contact is constant.

(h) Keep the designated officer, dispatch, the incoming replacement IC, or other higher level officer informed of:
1. Status of the fire.
2. Progress of the suppression forces.
3. Additional resources needed.
4. Weather conditions, especially changes.
5. Special situations such as values threatened.

(i) As additional forces arrive:
1. Divide the fire into areas of responsibility such as right and left flank, or Division A and Division B.
2. Assign individuals responsibility for these areas.

See NWCG PMS 410-1, *Fireline Handbook,* for additional information on the transition of command.

A-6-5.8.1 For additional information on business management, see NWCG S-260, *Fire Business Management Principles.*

A-6-5.9 Final incident reports could include personnel and equipment reports, personnel evaluations, financial documents, fire reports, accident forms, and post incident evaluations.

Appendix B NWCG Publications for Wildland Fire Suppression Personnel

Wildland Fire Fighter IV

S-200, Initial Attack Incident Commander

S-205, Fire Operations in the Urban Interface

S-234, Firing Methods and Procedures

GLOSSARY/INDEX

Glossary

A

Action Plan
See Incident Action Plan.

Adiabatic
Process of thermodynamic change of state in which no heat is added or subtracted from a system; compression always results in warming, expansion in cooling.

Adjustable Fog Nozzle
Nozzle designed to allow the discharge pattern to be adjusted from straight stream to full fan fog; suitable for applying water, wet water, or foam solution. Some adjustable fog nozzles allow the rate of flow to be adjusted as well.

Aeration
Introduction of air into a foam solution to create bubbles that result in finished foam.

Aerial Fuels
Standing and supported live and dead combustibles not in direct contact with the ground and consisting mainly of foliage, twigs, branches, stems, cones, bark, and vines.

Aerial Ignition
Use of an airborne incendiary device to assist in backfiring, burning out, or prescribed fires. Devices are normally carried in or suspended from helicopters.

Air Attack
Using fixed-wing aircraft or helicopters to apply fire retardants or extinguishing agents on a wildland fire. Aircraft can also be used to transport crews, supplies, and equipment or provide medical evacuation and reconnaissance.

Air Drop
Water, short-term fire retardant, or long-term fire retardant cascaded from an air tanker or helicopter.

Air Mass
Extensive body of air, usually 1,000 miles (1 600 km) or more across, having the same properties of temperature and moisture in a horizontal plane.

Air Support Group Supervisor
Individual responsible to the air operations branch director for logistical support and management of helibase and helispot operations and maintenance of a liaison with fixed-wing aircraft bases.

Air Tactical Group Supervisor
Individual responsible to the air operations branch director for the coordination of fixed-wing and/or rotary-wing aircraft operations over an incident.

Air Tanker
Fixed-wing aircraft used to drop retardant or water during an air attack.

Anchor Point
Point from which a fireline is begun; usually a natural or man-made barrier that will prevent fire spread and the possibility of a crew being "flanked" while constructing the fireline. Typical anchor points are roads, lakes, ponds, streams, earlier burns, rock slides, and cliffs.

Angle of Approach
On a vehicle, the smallest angle made between the road surface and a line drawn from the front point of ground contact of the front tire to any projection of the apparatus ahead of the front axle — the front overhang.

Angle of Departure
On a vehicle, the smallest angle made between the road surface and a line drawn from the rear point of ground contact of the rear tire to any projection of the apparatus behind the rear axle — the rear overhang.

Apparatus
Motor-driven vehicle or group of vehicles designed and constructed for the purpose of fighting fires; may be of different types such as engines, water tenders, and ladder trucks.

Area Ignition
Ignition or nearly simultaneous ignition of several individual fires that influence the main fire and each other in a way that produces a hot, fast-moving fire or blowup throughout the area.

Arson
Crime of maliciously and intentionally starting a fire or causing an explosion. Precise legal definitions vary among jurisdictions, wherein it is defined by statutes and judicial decisions.

Ash
Powdery residue left when organic material is burned completely or is oxidized by chemical means.

Aspect
Compass direction toward which a slope faces.

Atmospheric Pressure
Force exerted by the weight of the atmosphere per unit area.

Atmospheric Stability
Degree to which vertical motion in the atmosphere is enhanced or suppressed. Vertical motions and smoke dispersion are enhanced in an unstable atmosphere. Stability suppresses vertical motion and limits smoke dispersion. *Also see* Inversion.

Attack Line
Hoseline connected to a pump discharge of a fire apparatus ready for use in attacking a fire (may or may not be preconnected); contrasted to supply lines connecting a water supply with a pump.

Automatic Aid
Written agreement between two or more agencies to automatically dispatch predetermined resources to any fire or other emergency reported in the geographic area covered by the agreement. These areas are generally where the boundaries between jurisdictions meet or where jurisdictional "islands" exist.

B

Backfire
Fire set along the inner edge of a control line to consume the fuel in the path of a wildland fire and/or change the direction of force of the fire's convection column.

Backfiring
Offensive tactic associated with indirect attack; intentionally setting fire to fuels inside the control line to slow, knock down, or contain a rapidly spreading fire. The intent is for the backfire to meet the advancing fire some distance from the control line. Backfiring provides a wide defense perimeter and may be further employed to change the force of the convective column; makes possible a strategy of locating control lines at places where the fire can be fought on the firefighter's terms.

Backing Fire
Fire spreading (or ignited to spread) into (against) the wind or downslope. A fire spreading on level ground in the absence of wind is a backing fire.

Baffle
Partition placed in vehicular or aircraft water tanks to reduce shifting of the water load when starting, stopping, or turning.

Barrier
Any obstruction of the spread of fire; typically an area or strip devoid of combustible fuel.

Batch-Mixing
Making of foam solution by pouring an appropriate amount of foam concentrate into a water tank.

Belt Weather Kit
Belt-mounted case with pockets fitted for anemometer, compass, sling psychrometer, slide rule, water bottle, pencils, and book of weather report forms.

Berm
Outside or downhill side of a ditch or trench; a mound or wall of earth.

Black
Area already burned by a wildland fire. Also called Burn.

Blacklining
Assuring that there are no unburned fuels adjacent to the control line by burning out such areas; burning out adjacent to a control line to widen and strengthen it.

Blowup
Sudden dangerous increase in fireline intensity typically caused by strong or erratic wind, steep slopes, large open areas, and easily ignited fuels. Blowup is sufficient to preclude direct attack or to change the incident action plan; often accompanied by violent convection and may have other characteristics of a fire storm. *Also see* Flare-Up.

Boneyarding
During mop-up, spreading materials that are no longer burning in an area within the black that has been cleared of all burning or hot fuels.

Booster Line
Noncollapsible rubber-covered, rubber-lined hose usually wound on a reel and mounted somewhere on an engine or water tender. This hose is most commonly found in ½-, ¾-, and 1-inch (13 mm, 19 mm, and 25 mm) diameters and is used for extinguishing low-intensity fires and mop-up. Also called Hard Line.

Breakover
See Slopover.

British Thermal Unit (Btu)
Amount of heat energy required to raise the temperature of 1 pound of water 1 degree Fahrenheit. One Btu = 1.055 kilojoules (kJ).

Brush
Collective term that refers to stands of vegetation dominated by shrubby, woody plants or low-growing trees, usually of a type undesirable for livestock or timber management.

Brush Hook
Heavy cutting tool designed primarily to cut brush at the base of the stem; used in much the same way as an axe; has a wide blade generally curved to protect the blade from being dulled by rocks.

Bulldozer
Any tracked vehicle with a blade for exposing mineral soil. Also called Dozer. *Also see* Dozer Tender and Dozer Transport.

Burn
See Black.

Burning Out
Setting fire inside a control line to consume fuel between the edge of the fire and the control line. Burning out is done on a small scale in order to consume unburned fuel and aid control-line construction. Burning out should not be confused with "backfiring," which is a larger-scale tactic to eliminate large areas of unburned fuels in the path of a fire or to change the direction of force of a convection column.

C

CAFS
Abbreviation for Compressed Air Foam System.

Canopy
Level or area containing the crowns of the tallest vegetation present (living or dead), usually above 20 feet (6 m).

Centrifugal Pump
Pump with one or more impellers that utilizes centrifugal force to move the water.

Char
Carbonaceous material formed mainly on the surface of organic matter, commonly wood, that has not burned completely; the solid remains of burned organic material.

Check-In
Process or location used by assigned resources to report in at an incident.

Chimney
Steep, narrow draws or canyons in which heated air rises rapidly as it would in a flue pipe.

Class A Foam
Foam specially designed for use on Class A combustibles; hydrocarbon-based surfactant, essentially a wetting agent that reduces the surface tension of water and allows it to soak into combustible materials easier than plain water.

Class A Fuels
Ordinary combustible solids such as wood, grass, rubber, cloth, paper, and plastics.

Clear Text
Use of plain English, including certain standard words and phrases, in radio communications transmissions.

Cold Front
Leading edge of a relatively cold air mass that displaces and may cause warmer air to rise. If the lifted air contains enough moisture, cloudiness, precipitation, and even thunderstorms may result. As fronts move through a region, the winds at a given location experience a marked shift in direction.

Cold Trailing
Method of controlling a partly dead fire edge by carefully inspecting and feeling with the hand to detect any fire, digging out every live spot, and trenching any live edge; done to ensure no further advance of a fire.

Combination Nozzle
Nozzle designed to provide either a solid stream or a fixed spray pattern suitable only for mop-up. Not to be confused with an adjustable fog nozzle.

Combustion
Self-sustaining process of rapid oxidation of a fuel, which produces heat and light.

Command
(1) Act of directing and/or controlling resources by virtue of explicit legal, agency, or delegated authority. (2) Term used on the radio to designate the incident commander.

Command Post
See Incident Command Post.

Compressed Air Foam System (CAFS)
Generic term used to describe foam-generation systems consisting of an air compressor (or other air source), a water pump, and foam solution.

Conduction
Transfer of heat energy from one body to another through direct contact or an intervening medium from a region of high temperature to a region of low temperature.

Confine a Fire
To restrict the fire within determined boundaries established either prior to the fire or during the fire. Also called Confinement.

Confinement
See Confine a Fire.

Contain a Fire
To take suppression action that can reasonably be expected to check the fire spread under prevailing and predicted conditions. Also called Containment.

Contained Fire
Fire whose progress has been stopped but for which the control line is not yet finished.

Containment
See Contain a Fire.

Continuous Fuels
Fuels distributed uniformly over an area, thereby providing a continuous path for fire to spread. *Also see* Fuel Continuity.

Control
Point in time when the perimeter spread of a wildland fire has been halted and can reasonably be expected to hold under foreseeable conditions.

Control a Fire
To complete control line around a fire, any spot fire therefrom, and any interior island to be saved; to burn out any unburned area adjacent to the fire side of the control lines; and to cool down all hot spots that are immediate threats to the control line until the lines can reasonably be expected to hold under foreseeable conditions.

Controlled Burning
Fires intentionally set in vegetative fuels for the purpose of burning debris or accumulations of wildland fuels; may be done as part of a fuel-management program to prevent or reduce the rate of spread of wildland fires. *Also see* Prescribed Burning.

Control Line
Inclusive term for all constructed or natural barriers and treated fire edges used to control a fire.

Convection
Transfer of heat by the movement of fluids or gases, usually in an upward direction.

Convection Column
Rising column of heated air or gases above a continuing heat or fire source.

Cooperative Agreement
Written agreement between fire protection agencies agreeing to cooperate in actions or share resources for a common good.

Creeping Fire
Fire burning with a low flame height and spreading slowly.

Crew
Organized group of firefighters under the leadership of a crew leader or other designated supervisor; sometimes referred to as a “company” in municipal fire departments.

Crown Fire
Fire that advances from top to top of trees or shrubs more or less independent of a surface fire. Crown fires are sometimes classed as running or dependent to distinguish the degree of independence from the surface fire.

Crown Out
Fire that rises from ground level into the tree crowns and advances from treetop to treetop.

Cumulonimbus Clouds
Ultimate growth of a cumulus cloud into an anvil-shaped cloud with considerable vertical development, usually with fibrous ice crystal tops, and usually accompanied by lightning, thunder, hail, and strong winds.

Cumulus Clouds
Principal low-cloud type in the form of individual cauliflower-like cells of sharp nonfibrous outline and less vertical development than cumulonimbus clouds.

D

DBH
Abbreviation for Diameter at Breast Height.

Dead Fuels
Fuels with no living tissue in which moisture content is governed almost entirely by atmospheric moisture (relative humidity and precipitation), dry-bulb temperature, and solar radiation.

Defensive Strategy
Defending exposed life, property, and resources in the path of a fire (as opposed to offensive strategies such as flanking, pincer, backfiring, etc.).

Diameter at Breast Height (DBH)
Means by which the relative size of trees is expressed. If a tree trunk measured at breast height is 20 inches (508 mm) or more in diameter, it is considered a large tree. A tree with a DBH of less than that is considered a small tree.

Direct Attack
To attack a wildland fire directly at the burning edge.

Diurnal
Daily; especially pertaining to cyclic actions of the atmosphere that are completed within 24 hours and that recur every 24 hours.

Dozer
See Bulldozer.

Dozer Tender
Any ground vehicle (service unit) with personnel capable of maintenance, minor repairs, and limited fueling of bulldozers.

Dozer Transport
Heavy vehicle carrying bulldozer to incident.

Draft
Drawing water from static sources into a pump that is above the level of the water supply. This is done by removing the air from the pump and allowing atmospheric pressure to push water through a noncollapsible suction hose into the pump.

Drift Smoke
Smoke that has been transported from its point of origin and in which convective, columnar motion no longer dominates.

Drop Zone
Target area for air tankers, helicopters, and cargo dropping.

Dry Adiabatic Lapse Rate
Rate of decrease in temperature with height of a mass of dry air lifted adiabatically through an atmosphere in hydrostatic equilibrium.

Dry Air Mass
Portion of the atmosphere that has a relatively low dew point temperature and where the formation of clouds, fog, or precipitation is unlikely.

Dry Hydrant
Permanently installed pipe that has pumping-engine suction connections installed at water sources to speed drafting operations.

Dry Thunderstorm
Storm, including lightning, during which little or no rain reaches the ground.

Duff
Partly decomposed and matted leaves, twigs, and bark beneath the litter of freshly fallen twigs, needles, and leaves.

Durable Agents
See Gelling Agents.

Dust Devil
See Whirlwind.

E

Eductor
Venturi device that uses water pressure to draw foam concentrate into a water stream for mixing; also enables a pump to draw water from an auxiliary source.

Enhanced Strike Team
Engine strike team to which a water tender is assigned to operate as part of the team.

Environmental Lapse Rate
Rate of temperature change with elevation determined by the vertical distribution of temperature at a given time and place.

Escape Route
Pathway to safety. It can lead to an already burned area, a previously constructed safety area, a meadow that will not burn, or a natural rocky area that is large enough to take refuge without being burned. When escape routes deviate from a defined physical path, they must be clearly marked (flagged).

Expansion
Ratio of the volume of foam in its aerated state to the original volume of the nonaerated foam solution.

Exposure
(1) Property that may be endangered by a fire burning in another structure or by a wildland fire. (2) Direction in which a slope faces. (3) General surroundings of a site with special reference to its openness to winds and sunshine.

Extended-Attack Fire
Wildland fire that has not been contained or controlled by initial-attack forces and for which more fire fighting resources are arriving, en route, or being ordered by the incident commander; situation in which a fire cannot be controlled by initial-attack resources within a reasonable period of time.

Exterior Fire Protection
Protection of structures from the exterior with no interior fire fighting.

F

Fingers
Long, narrow extensions of a fire projecting from the main body.

Fire
Rapid oxidation of combustible materials accompanied by a release of energy in the form of heat and light.

Fire Behavior
Manner in which a fire reacts to the variables of fuel, weather, and topography.

Fire-Behavior Forecast
Prediction of probable fire behavior, usually prepared by a fire-behavior officer in support of fire-suppression or prescribed-burning operations.

Firebreak
Any natural or constructed barrier that is devoid of vegetation and stops or slows the advance of a wildland fire.

Fire Curtain
Aluminized device on a rod designed to be unrolled to reflect radiant heat from operators or crew members on some apparatus, bulldozers, or tractor-plows.

Fire Edge
Boundary of a fire at a given moment.

Fire Front
Part of a fire within which continuous flaming combustion is taking place; assumed to be the leading edge of the fire perimeter. In surface fires, the fire front may be mainly smoldering combustion.

Fire Hazard Severity Rating System
System of adjectives used to describe fire danger to the public. Adjectives range from low to extreme.

Fireline
Part of a control line that is scraped or dug to mineral soil; also, a general term for the area where fire fighting activities are taking place, the wildland equivalent of the term "fireground" as used in structural fire fighting.

Fireline Intensity
Rate of heat energy released per unit time per unit length of fire front. Numerically, it is the product of the heat of combustion, quantity of fuel consumed in the fire front, and the rate of spread of a fire in Btu per second per foot (kilojoules per second per meter) of fire front.

Fire Plow
Heavy-duty plowshare or disc plow pulled by a tractor to construct a fireline.

Fire Prevention
All activities concerned with minimizing the incidence of fires.

Fire Protection
Actions taken to limit the adverse environmental, social, political, economic, and life-threatening effects of fire.

Fire Retardant
Any substance, except plain water, that reduces flammability of fuels or slows their rate of combustion by chemical or physical action.

Fire Season
Period(s) of the year during which fires are likely to occur, spread, and damage wildland values sufficient to warrant organized fire suppression.

Fire Shelter
Aluminized tent carried by firefighters offering personal protection by means of reflecting radiant heat and providing a volume of breathable air in a fire-entrapment situation.

Fire Storm
Violent convection caused by a large continuous area of intense fire; often characterized by destructively violent surface indrafts, near and beyond the perimeter, and sometimes by tornado-like whirls.

Fire Stream
Stream of water or other water-based extinguishing agent after it leaves the nozzle until it reaches the desired point.

Fire Suppressant
Any agent used to extinguish the flaming and smoldering phases of combustion by direct application to the burning fuel.

Fire Suppression
All work and activities connected with fire-extinguishing operations, beginning with discovery and continuing until a fire is completely extinguished.

Fire Swatter
Fire-suppression tool consisting of a flap of belting fabric fastened to a long handle used in direct attack for beating out flames along a fire edge.

Fire Weather
Weather conditions that influence fire ignition, behavior, and suppression.

Fire Weather Forecast
Weather prediction specially prepared for use in wildland fire control.

Firewhirl
Spinning vortex column of ascending hot air and gases rising from a fire and carrying smoke, debris, and flame aloft.

Firing Out
Act of lighting fire with a torch, fusee, etc., to accomplish burning out or backfiring. *Also see* Burning Out, Backfire, and Backfiring.

Flame Depth
Depth of the fire front; horizontal distance between leading and trailing edge of fire front.

Flame Height
Average maximum vertical extension of flames at the leading edge of the fire front. Occasional flashes that rise above the general level of flames are not considered. This distance is less than the flame length if flames are tilted due to wind or slope.

Flame Length
Distance between the flame tip and the midpoint of the flame depth at the base of the flame (generally the ground surface); an indicator of fire intensity.

Flammability
Fuel's susceptibility to ignition.

Flanking Attack
Attacking a fire by working along the flanks either simultaneously or successively from an anchor point.

Flanks of a Fire
Parts of a fire's perimeter that are roughly parallel to the main direction of spread.

Flare-Up
Any sudden acceleration in rate of spread or intensification of a fire. Unlike blowup, a flare-up is of relatively short duration and does not radically change existing control plans. *Also see* Blowup.

Flashy Fuels
Wildland fuels that are easily ignited and that burn rapidly when dry. Some examples are grass, leaves, pine needles, fern, tree moss, and some kinds of slash.

Floating Pump
Small, portable pump that floats on the water source.

Foam
Extinguishing agent formed by mixing a form concentrate with water and aerating the solution for expansion. *Also see* Class A Foam.

Foehn Wind
Type of general wind that occurs when stable, high-pressure air is forced across and then down the lee slopes of a mountain range. The descending air is warmed and dried due to adiabatic compression; locally called by various names such as Santa Ana, Mono, Chinook, etc. Also called Gravity Wind.

Forestry Hose
Unlined, single-jacket, lightweight hose with lightweight couplings.

Frequency-Sharing Agreement
Written agreement between agencies that are licensed to use a communications frequency that allows the other agency to use the frequency under specified conditions.

Friendly Fire
Fully contained and controlled fire started for useful and nondestructive purposes.

Front
In meteorology, the boundary between two air masses of differing densities.

Frontal Winds
Winds generated by the movement of an air mass (front) across the earth's surface.

Fuel
Flammable and combustible substances available for a wildland fire to consume.

Fuel Break
Wide strip or block of land on which the native vegetation has been modified so that fires burning into them can be more readily extinguished. It may or may not have a fireline constructed in it prior to fire occurrence.

Fuel Characteristics
Factors that make up fuels such as compactness, loading, horizontal continuity, vertical arrangement, chemical content, size and shape, and moisture content.

Fuel Continuity
Degree or extent of continuous or uninterrupted distribution of fuel particles in a fuel bed, thus affecting a fire's ability to sustain combustion and spread. This applies to aerial fuels as well as surface fuels.

Fuel Loading
Amount of fuel present expressed quantitatively in terms of weight of fuel per unit area. This may be available fuel (consumable fuel) or total fuel and is usually dry weight.

Fuel Management
Manipulation of fuel prior to an incident to prevent the occurrence or slow the spread of wildland fire. Also called Vegetation Management or Weed Abatement.

Fuel Model
Simulated fuel complex for which all fuel descriptors required for the solution of a mathematical rate-of-spread model have been specified.

Fuel Moisture
Quantity of moisture in fuel expressed as a percentage of the weight when thoroughly dried at 212°F (100°C).

Fuel Volume
Quantity of fuel per unit area; usually expressed in tons per acre (tonnes per hectare).

Fusee
Colored flare designed as a railway warning device used to ignite backfires and other prescribed fires.

G

Gelling Agents
Superabsorbent liquid polymers capable of absorbing hundreds of times their own weight in water. These gels can be used as fire suppressants and fire retardants. Gels function by entrapping water in their structure rather than air, as is the case with fire fighting foams. Also called Durable Agents.

General Winds
Large-scale winds caused by high- and low-pressure systems but generally influenced and modified in the lower atmosphere by terrain.

Gradient Wind
Upper-level winds that flow around high- and low-pressures cells. Gradient winds flow clockwise around high-pressure cells and counterclockwise around low-pressure cells.

Gravity Wind
See Foehn Wind.

Green
Area of unburned fuels, not necessarily green in color, adjacent to but not involved in a wildland fire.

Greenbelt
Landscaped and perhaps irrigated fuel break that is regularly maintained; sometimes put to an additional use (for example, golf course, park, playground, pasture). Greenbelts may also be dedicated but unmaintained open space within or between developments.

Ground Fire
See Surface Fire.

Ground Fuel
See Surface Fuel.

H

Hand Crew
Individuals who have been organized and trained for operational assignments on an incident who primarily use hand tools to clear vegetation.

Handline
(1) Fireline constructed with hand tools. (2) Small hoseline (2½ inch [65 mm] or less) that can be handled and maneuvered without mechanical assistance.

Hard Line
See Booster Line.

Head of a Fire
Most active part of a wildland fire; the forward-advancing part.

Heat Release Rate (HRR)
(1) Total amount of heat produced per unit mass of fuel consumed per unit time. (2) Amount of heat released to the atmosphere from the convective-lift fire phase of a fire per unit time.

Heat Transfer
Flow of heat from a hot substance to a cold substance. This flow may be accomplished by convection, conduction, or radiation.

Heavy Equipment
Ground vehicles such as bulldozers, tractors, and plows used in the suppression of wildland fires and their transport vehicles. Heavy equipment does not include fire apparatus.

Heavy Fuels
Massive natural cover fuels such as logs, snags, and large limbs. Heavy fuels are not easy to ignite; once ignited, they burn slowly and hot.

Heel
Rear portion of a wildland fire. Also called Rear.

Helibase
Main location on an incident for parking, fueling, maintaining, and loading helicopters.

Helispot
Temporary landing spot for helicopters.

Hose Lay
(1) Arrangement of connected lengths of fire hose and accessories on the ground at a wildland fire beginning at the first pumping unit and ending at the point of water delivery. (2) Connected lengths of hose from water source to pumping engine.

Hoseline Tee
Fitting that may be installed between lengths of hose to provide an independently controlled outlet for a branch line.

Hotshot Crew
Highly trained fire fighting crew used primarily in handline construction.

Hot Spot
Particularly active part of a wildland fire.

Hotspotting
Checking the spread of fire at points of more rapid spread or special threat only.

HRR
Abbreviation for Heat Release Rate.

I

IAP
Abbreviation for Incident Action Plan.

IC
Abbreviation for Incident Commander.

ICS
Abbreviation for Incident Command System.

Incident
Occurrence, either human-caused or natural phenomenon, that requires action by emergency services personnel to prevent or minimize loss of life or damage to property and/or natural resources.

Incident Action Plan (IAP)
Contains objectives reflecting the overall incident strategy and specific tactical actions for the next operational period; may be oral or written. When written, the plan may have a number of forms as attachments.

Incident Base
Location at the incident where the primary logistics functions are coordinated and administered (formerly called "fire camp"). Incident name or other designator is added to the term "base." The incident command post may be colocated with the base. There is only one base per incident.

Incident Commander (IC)
Person in charge of and responsible for the management of all incident operations.

Incident Command Post
Location at which the incident commander and command staff direct, order, and control resources at an incident; may be colocated with the incident base.

Incident Command System (ICS)
System by which facilities, equipment, personnel, procedures, and communications are organized to operate within a common organizational structure designed to aid in the management of resources at emergency incidents.

Incident Management System
System described in NFPA 1561, *Standard on Fire Department Incident Management System*, that defines the roles, responsibilities, and standard operating procedures used to manage emergency operations. Such systems may also be referred to as Incident Command Systems (ICS).

Indirect Attack
Controlling the fire by locating the control line along natural firebreaks some distance from the approaching fire and burning out the intervening fuels.

Infrared Radiation
Radiation with a wavelength outside the visible spectrum at the red end of the spectrum. Thermal radiation from free-burning fires is an example of infrared radiation.

Initial Action
See Initial Attack.

Initial Attack
Control efforts taken by the resources that are the first to arrive at an incident. Also called Initial Action.

Intensity
See Fireline Intensity.

Inversion
Increase of temperature with height in the atmosphere. Vertical motion in the atmosphere is inhibited allowing for smoke buildup. A "normal" atmosphere has temperature decreasing with height. *Also see* Atmospheric Stability.

Island
Unburned area within a fire perimeter.

Knock Down
To reduce the flame or heat on the more vigorously burning parts of a fire edge.

L

Ladder Fuels
Fuels that provide vertical continuity between strata, thereby allowing fire to carry from surface fuels into the crowns of trees or shrubs with relative ease. They help initiate and assure the continuation of crowning.

Lapse Rate
Change of an atmospheric variable (temperature unless specified otherwise) with height.

Life Safety
Refers to the joint consideration of the life and physical well-being of individuals, both civilians and firefighters.

Light Fuels
Fast-drying fuels, with a comparatively high surface-area-to-volume ratio, that are generally less than ¼ inch (6.35 mm) in diameter and have a time lag of 1 hour or less. These fuels readily ignite and are rapidly consumed by fire when dry.

Litter
Top layer of forest floor composed of loose debris of dead sticks, branches, twigs, and recently fallen leaves or needles; little altered in structure by decomposition.

Live Fuels
Living plants, such as trees, grasses, and shrubs, in which the seasonal moisture content cycle is controlled largely by internal physiological mechanisms rather than by external weather influences.

Local Winds
Winds that are generated over a comparatively small area by local terrain and weather. They differ from those that would be appropriate to the general pressure pattern or that possess some other peculiarity.

Lookout
(1) Person designated to detect and report fires from a vantage point. (2) Location from which fires can be detected and reported. (3) Fire crew member assigned to observe the fire and warn the crew when there is danger of becoming trapped.

M

Medium Fuels
Material available to burn in a geographic area that is in the midrange of size such as various brush species; generally excludes short grasses and large trees.

Mineral Soil
Soil containing little or no combustible material.

Mobile Attack
Suppressing fire along a fire edge by driving mobile apparatus along the perimeter and simultaneously applying fire streams to knock down the fire. Also called Pump and Roll.

Moist Adiabatic Lapse Rate
Rate of decrease in temperature with increasing height of an air mass.

Mop-Up
Act of making a fire safe after it is controlled such as extinguishing or removing burning material along or near the control line, felling dead trees (snags), and trenching logs to prevent rolling.

Mutual Aid
Direct assistance from one fire agency to another during an emergency based upon a prearrangement between agencies involved and generally made upon the request of the receiving agency.

N

National Fire Danger Rating System (NFDRS)
Multiple index matrix designed to provide fire-control and land-management personnel with a systematic means of assessing various aspects of fire danger on a day-to-day basis.

National Wildland Fire Coordinating Group (NWCG)
Currently made up of the Department of Agriculture Forest Service (FS); four Department of the Interior agencies (Bureau of Land Management [BLM], National Park Service [NPS], Bureau of Indian Affairs [BIA], and the Fish and Wildlife Service [FWS]); the United States Fire Administration (USFA); and state forestry agencies through the National Association of State Foresters (NASF). The group's purpose is to coordinate programs of

the participating wildfire management agencies to avoid duplication and to provide a means of constructively working together. Thus it facilitates the coordination and effectiveness of wildland fire activities and provides a forum to discuss, recommend action, or resolve issues and problems of substantive nature. NWCG is the certifying body for all courses in the National Fire Curriculum.

Natural Barrier
Any area where lack of flammable material obstructs the spread of wildland fires.

NFDRS
Abbreviation for National Fire Danger Rating System.

NWCG
Abbreviation for National Wildland Fire Coordinating Group.

O

Offensive Strategy
Generally refers to direct attack on the fire perimeter by crews, engines, aircraft, or an aggressive indirect attack such as backfiring.

Operations Section
Section responsible for all tactical operations at the incident; includes branches, divisions and/or groups, task forces, strike teams, single resources, and staging areas.

Operations Section Chief
Person responsible to the incident commander for managing all tactical operations directly applicable to accomplishing the incident objectives.

Origin
See Point of Origin.

Overhaul
Searching for and extinguishing any hidden or remaining fire once the main body of fire has been extinguished.

P

Paracargo
Anything intentionally dropped or intended for dropping from any aircraft by parachute, other retarding devices, or free fall.

Parallel Attack
Constructing a fireline parallel to a wildland fire's edge. After the line is constructed, the fuel inside the line is burned out.

Patrol
(1) To travel over a given route to prevent, detect, and suppress fires. (2) To go back and forth vigilantly over a length of control line during and/or after construction to prevent slopovers, suppress spot fires, and extinguish overlooked hot spots.

Penetrant
Water with added chemicals called "wetting agents" that increase water's spreading and penetrating properties due to a reduction in surface tension. Also called Wet Water.

Perimeter
Entire outer edge or boundary of a fire.

Personal Protective Equipment (PPE)
Basic protective equipment for wildland fire suppression includes a helmet, protective footwear, gloves, flame-resistant clothing, and fire shelter as defined in NFPA 1977, *Standard on Protective Clothing and Equipment for Wildland Fire Fighting.*

Pincer Attack
Direct attack around a fire in opposite directions by two or more attack units with the ultimate intent of pinching off (stopping) the head of the fire.

Point of Origin
Point of original ignition of a fire.

Portable Fire Pump
Small, gasoline-driven pump that can be carried to a water source by one or two firefighters or other conveyance over difficult terrain.

Positive-Displacement Pump
Gear pump or piston pump that moves a specified quantity of water through the pump chamber with each stroke or cycle; capable of pumping air and therefore is self-priming, but must have pressure relief provisions if plumbing or hoses have shutoff nozzles or valves.

PPE
Abbreviation for Personal Protective Equipment.

Precipitation
Any or all forms of water particles, liquid or solid, that fall from the atmosphere.

Preconnect
Hard suction hose or discharge hose carried connected to a pump, eliminating delay when hose and nozzles must be connected and attached at a fire.

Prefire Planning
See Pre-Incident Planning.

Pre-Incident Planning
Act of preparing to handle an incident at a particular location or a particular type of incident before an incident occurs. Sometimes also referred to as Prefire Planning.

Prescribed Burning
Controlled application of fire to wildland fuels in either their natural or modified state under specified environmental conditions that allows the fire to be confined to a predetermined area and to produce the fire behavior and fire characteristics required to attain planned fire treatment and resource-management objectives. A written plan that describes specifically planned results and specific conditions as part of a vegetation-management program.

Presuppression
Activities in advance of fire occurrence to ensure effective suppression action.

Pretreat
Use of water, foam, or retardant along a control line or on anything in order to try to save it in advance of a fire; often used where ground cover or terrain is considered best for control action.

Priming Device (Primer)
Any device, usually a positive-displacement pump, used to exhaust the air from inside a centrifugal pump and the attached hard suction to create a partial vacuum within to allow atmospheric pressure to force water from a static source through the suction hose into the centrifugal pump.

Progressive Hose Lay
Laying hose from a fire pump to a fire's edge, extinguishing fire as far as the hose will reach, connecting another section, advancing while extinguishing fire as far as the hose will reach, connecting another section, etc. Used when mobile attack is not possible.

Progressive Line Construction
System of organizing workers to build a fireline in which they advance without changing relative positions in line.

Pump and Roll
See Mobile Attack.

Pyrolysis
Thermal or chemical decomposition of fuel because of heat; the preignition combustion phase of burning during which heat energy is absorbed by the fuel, which in turn gives off flammable tars, pitches, and gases.

R

Radiation
Transfer of heat through intervening space by infrared thermal waves.

Rappelling
Technique of landing firefighters from helicopters in hover, which involves sliding down ropes with the aid of descent-control devices.

Rate of Spread (ROS)
Relative activity of a fire in extending its horizontal dimensions. Expressed as rate of increase of the total perimeter of a fire, as rate of forward spread of the fire front, or as rate of increase in area, depending on the intended use of the information. Usually expressed in chains or acres (hectares) per hour for a specific period in the fire's history.

Rear
See Heel.

Reburn
Burning of an area that has been previously burned but that contains flammable fuel that ignites when burning conditions are more favorable; area that has reburned.

Reconnaissance
To examine an area to obtain information about current and probable fire behavior and other related fire-suppression information.

Rehabilitation
(1) Activities necessary to repair environmental damage or disturbance caused by wildland fire or the fire-suppression activity. (2) Allowing firefighters to rest, rehydrate, and recover during an incident.

Relative Humidity
Percentage of moisture in the air compared to the maximum amount of moisture that air will hold at that temperature.

Resources
All personnel and major items of equipment that are available, or potentially available, for assignment to incidents.

Retardant
See Fire Retardant.

Retardant Drop
Fire retardant cascaded from an air tanker or helicopter.

ROS
Abbreviation for Rate of Spread.

Rotor Blast
Air turbulence occurring under and around the rotors of an operating helicopter. Also called Rotor Downwash from the main rotor.

Rotor Downwash
See Rotor Blast.

Running Fire
Behavior of a fire spreading rapidly with a well-defined head.

S

Saddle
Depression or pass in a ridgeline; low area on a ridgeline between two higher points.

Safe Refuge
See Safety Zone.

Safety Island
See Safety Zone.

Safety Officer
Member of the command staff responsible to the incident commander for monitoring and assessing hazardous and unsafe conditions and developing measures for assessing personnel safety on the incident.

Safety Zone
Recently burned area or one cleared of vegetation used for escape in the event a line is outflanked or a spot fire outside a control line renders the line unsafe. In firing operations, crews progress so as to maintain a safety zone close at hand, allowing the fuels inside the control line to be consumed before going ahead. Sometimes called Safety Island or Safe Refuge.

Salvo Drop
Air tanker dropping its entire load of fire retardant at one time.

Scratch Line
Unfinished preliminary control line hastily established or constructed as an emergency measure to check the spread of fire.

Secondary Line
Any fireline that is constructed at a distance from the fire perimeter concurrently with or after a line has already been constructed on or near the perimeter of the fire; generally constructed as an insurance measure in case a fire escapes control by the primary line.

Set
(1) Individual incendiary fire. (2) Point or points of origin of an incendiary fire. (3) Material left to ignite an incendiary fire at a later time. (4) Individual lightning or railroad fires, especially when several are started within a short time. (5) Burning material at the points deliberately ignited for backfiring, slash burning, prescribed burning, and other purposes.

Shelter in Place
Remaining in a structure or vehicle when a fire moves through rather than attempting to use roads that may be blocked or untenable because of fire; opposite of evacuation.

Size-Up
Ongoing process of observation and evaluation of existing factors that are used to develop objectives, strategy, and tactics for fire suppression.

Slash
Debris left after logging, pruning, thinning, or brush cutting; includes logs, chunks, bark, branches, stumps, and broken understory trees or brush.

Sling Psychrometer
Meteorological instrument used to determine relative humidity.

Slope
Natural or artificial topographic incline; degree of deviation from horizontal.

Slope Winds
Small-scale convective winds that occur due to local heating and cooling of a natural incline of the ground.

Slopover
Fire edge that crosses a control line. Also called Breakover.

Slurry
Thick mixture formed when a fire-retardant chemical is mixed with water and a viscosity agent.

Smoke Jumper
Firefighter who travels to remote wildland fires and other emergencies by aircraft and parachutes to the scene.

Smoldering
Fire burning without flame and barely spreading.

Snag
Standing dead tree or part of a dead tree from which at least the leaves and smaller branches have fallen.

Span of Control
Maximum number of subordinates that can be effectively supervised; ranges from three to seven individuals or functions, with five generally established as optimum.

Split Drop
Two retardant drops made from one compartment at a time from an air tanker with a multicompartment tank.

Spot Fire
Fires starting outside the perimeter of a main fire typically caused by flying sparks or embers. *Also see* Spotting.

Spotting
Behavior of a fire producing sparks or embers that are carried by the wind to start new fires beyond the main fire. *Also see* Spot Fire.

Stable Atmosphere
Condition of the atmosphere in which the temperature decrease with increasing altitude is less than the dry adiabatic lapse rate. In this condition, the atmosphere tends to suppress large-scale vertical motion.

Staging
Process by which available resources are held in reserve at a location away from the incident while awaiting assignment.

Staging Area
Temporary incident location from which resources must be able to respond within three minutes of being assigned. Staging area managers report to the incident commander or operations section chief if established.

Static Water Supply
Supply of water at rest that does not provide a pressure head for fire suppression but may be employed as a suction source for fire pumps (for example, water in a reservoir, pond, or cistern).

Strategy
Overall objectives for an incident established by the incident commander.

Strike Team
Specified combinations of the same kind and type of resources with common communications and a leader. Strike teams are most often composed of either engines, hand crews, or bulldozers, but they may be composed of any resource of the same kind and type. Exception — *see* Enhanced Strike Team.

Structural Triage
Process of inspecting and classifying structures according to their *defensibility/indefensibility* based on their situation, their construction, and the immediately adjacent fuels.

Structure Protection Group/Sector Supervisor
Individual responsible for supervising assigned strike teams, firefighters, or single resources in the defense of structures from wildland fire.

Subsurface Fire
Fire that consumes the organic material beneath the ground such as a peat fire or roots burning.

Subsurface Fuel
All combustible materials below the surface litter, such as tree or shrub roots, peat, and sawdust, that normally support smoldering combustion without flame.

Suppressant
See Fire Suppressant.

Suppression
See Fire Suppression.

Surface Fire
Wildland fire that burns loose debris of the surface; includes dead branches, fallen leaves, needles, duff, stubble, grass, and low vegetation. Also called Ground Fire.

Surface Fuel
Fuel that contacts the surface of the ground; consists of duff, leaf and needle litter, dead branch material, downed logs, bark, tree cones, and low-stature living plants. These are the materials normally scraped away to construct a fireline. Sometimes called Ground Fuel.

Sustained Attack
Continuing fire-suppression action until fire is under control.

T

Tactics
Deploying and directing resources on an incident to accomplish the objectives determined by the selected strategy.

Tandem
Two or more units of any kind working one in front of the other to accomplish a specific fire-suppression job. Term can be applied to combinations of hand crews, engines, bulldozers, or aircraft.

Target Hazard
Facilities in which there is a great potential likelihood of life or property loss.

Task Force
Any combination of single resources, within a reasonable span of control, assembled for a particular tactical need with common communications and a leader.

Thermal Belt
Elevation on a mountainous slope that typically experiences the least variation in diurnal temperatures and has the highest average temperatures and, thus, the lowest relative humidity. Its presence is most evident during clear weather with light wind.

Topography
Land surface configuration.

Torching
Vertical phenomenon in which a surface fire ignites the foliage of a tree or bush that becomes entirely involved in fire very quickly. A torching fire may or may not initiate a crown fire.

Tractor-Plow
Any tractor with a plow for exposing mineral soil.

Trail Drop
Dropping fire suppressant sequentially from tanks in aircraft so equipped; generally used in light fuels.

Trenching
Digging a trench across a slope to catch any burning material that could roll downhill and cross the control line.

Triage
See Structural Triage.

Turbulence
Irregular motion of the atmosphere usually produced when air flows over a comparatively uneven surface such as the surface of the earth; when two currents of air flow past or over each other in different directions or at different speeds.

Typical Tool Order
Order in which hand-crew members are assigned tools for varying types of wildland fuels. The types of tools necessary will be different for each fuel type.

U

Undercut Line
Fireline below a fire burning on a slope; should be trenched to catch rolling material. Also called Underslung Line.

Underslung Line
See Undercut Line.

Unfriendly Fire
Uncontained and uncontrolled fire of intentional or accidental origin that may cause injury or damage.

Unified Command
Unified team effort in the Incident Command System that allows all agencies with responsibility for the incident, either geographical or functional, to manage the incident by establishing a common set of incident objectives and strategies. This is accomplished without losing or abdicating authority, responsibility, or accountability. In unified command there is a single incident command post and a single operations chief at any given time.

V

Vegetation Management
See Fuel Management.

Venturi Principle
When a fluid is forced under pressure through a restricted orifice, there is a decrease in the pressure exerted against the sides of the constriction and a corresponding increase in the velocity of the fluid. This creates a slight vacuum into which any surrounding fluid will be drawn.

Virga
Precipitation that evaporates before reaching the ground.

W

Warm Front
Leading edge of a relatively warm air mass that moves in such a way that warm air displaces colder air. Winds associated with warm-frontal activity are usually light, and mixing is limited. The atmosphere is relatively stable when compared to cold-frontal activity.

Water Tender
Any ground vehicle capable of transporting large quantities of water; still called "tankers" in some regions.

Weed Abatement
See Fuel Management.

Wet Line
Line of water or water and chemical retardant sprayed along the ground that serves as a temporary fire-stop or containment line from which to ignite or stop a low-intensity fire.

Wetting Agent
Additive that reduces the surface tension of water (producing wet water), causing it to spread and penetrate more effectively, and that may produce foam through mechanical means. *Also see* Penetrant.

Wet Water
See Penetrant.

Whirlwind
Small rotating windstorm of limited extent containing sand or dust. Also called Dust Devil.

Wildland Fire
Unplanned and unwanted fire requiring suppression action; an uncontrolled fire, usually spreading through vegetative fuels. These fires can threaten structures or other improvements.

Wildland Firefighter I (NFPA)
Person, at the first level of progression, who has demonstrated the knowledge and skills necessary to function safely as a member of a wildland fire-suppression crew. The Wildland Firefighter I works under direct supervision.

Wildland Firefighter II (NFPA)
Person, at the second level of progression, who has demonstrated the skills and depth of knowledge necessary to function under general supervision. This person shall function safely and effectively as a member of a wildland fire-suppression crew of equally or less experienced firefighters to accomplish a series of tasks. The Wildland Firefighter II can be called upon to provide leadership and temporary supervision of a small crew. The Wildland Firefighter II maintains direct communications with a supervisor.

Wildland Firefighter III (NFPA)
Person responsible for supervising and directing a single wildland fire-suppression resource such as a hand crew or an engine.

Wildland Firefighter IV (NFPA)
Person responsible for managing all aspects of a wildland fire incident, involving relatively few resources. Resources vary from a single firefighter to several single resources possibly of different types or kinds.

Wildland/Urban Interface
Line, area, or zone where structures and other human development meet or intermingle with undeveloped wildland or vegetative fuels.

Wind
Horizontal movement of air relative to the surface of the earth.

Index

Indexed by Kari Bero

T

U

V